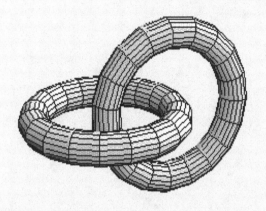

Mathematical Methods for
Partial Differential Equations

by J.H. Heinbockel
Emeritus Professor of Mathematics
Old Dominion University

Trafford rev. 11/12/2018

 www.trafford.com

North America & international
toll-free: 1 888 232 4444 (USA & Canada)
fax: 812 355 4082

PREFACE

This is an introductory text on mathematical methods used in the study of partial differential equations and boundary value problems. The text contains material suitable for a two semester upper division undergraduate or beginning graduate course in fundamental concepts associated with partial differential equations. It is assumed that the reader has background knowledge from the subject areas of calculus, differential equations, and Laplace transforms. The material emphasizes mathematical methods and qualitative theory rather than numerical methods. The material is suitable for mathematicians, engineers, physicist and scientist who desire an understanding of the basics associated with partial differential equations and boundary value problems.

Chapter one reviews concepts from the study of linear ordinary differential equations (ODE's) and introduces some elementary ideas associated with first and second order partial differential equations (PDE's). Chapter two introduces orthogonal functions and Sturm-Liouville systems. Chapter three uses orthogonal functions to develop Fourier series. Knowledge of the Fourier series leads to the Fourier integral representation of functions. Chapter four introduces the heat equation and applications for both finite and infinite domains. The method of separation of variables is introduced along with applications of Fourier series and Fourier integrals. Chapter five introduces the wave equation with selected applications. Chapter six introduces the Laplace equation, Poisson's equation and the Helmholtz equation along with selected applications. The chapters four, five and six emphasizes separation of variables, Fourier series and Fourier integrals. Chapter seven introduces the Heaviside unit step function and Dirac delta function along with techniques associated with Laplace, Fourier exponential, Fourier sine, Fourier cosine and Hankel transforms together with selected finite transforms associated with Sturm-Liouville systems. Chapter eight develops basic concepts associated with Green's functions for ordinary differential equations. Chapter nine develops Green's functions for partial differential equations.

At the end of each chapter there is a variety of exercises where solution techniques can be applied. In addition, there are many exercise problems where new concepts are introduced and so students should be encouraged to read all the exercises. The example problems are concluded using the blackbox notation represented by the symbol ■.

The Appendix A contains units of measurements from the Système International d'Unités along with some selected physical constants. The Appendix B contains solutions to selected exercises. All footnotes associated with influential scientists whose theories have contributed to the mathematics in this text have been collected into the Appendix C. The Appendix D contains a short table of definite integrals.

The author wishes to give special thanks to Drs. John Tweed, John Swetits, John Adam and Gordon Melrose for reviewing portions of the text. Their comments and suggestions for improving the material has been invaluable.

J.H. Heinbockel, 2003

Mathematical Methods for Partial Differential Equations

Chapter 1
Introduction

In this introduction we review some topics from ordinary differential equations (ODE's) because many partial differential equations (PDE's) can be reduced to a study of ordinary differential equations. We then introduce some important definitions and terminology associated with first and second order partial differential equations. This is followed by an introduction to some solution techniques associated with easy to solve partial differential equations.

Ordinary differential equations.

We begin with a review of some solution techniques for ordinary differential equations. The differential equations considered are linear differential equations of the form

$$L(y) = a_0(x)\frac{d^n y}{dx^n} + a_1(x)\frac{d^{n-1}y}{dx^{n-1}} + \cdots + a_{n-1}(x)\frac{dy}{dx} + a_n(x)y = F(x), \qquad (1.1)$$

where the coefficients $a_0(x), a_1(x), \ldots, a_n(x)$ are continuous functions over some interval $a \le x \le b$ and $a_0(x)$ is not identically zero over this interval. The general procedure for solving such equations is to first solve the n*th*-order homogeneous equation $L(y) = 0$ and obtain a fundamental set of solutions $\{y_1(x), y_2(x), \ldots, y_n(x)\}$. The general solution of the homogeneous equation is a linear combination of these n-independent solutions. It is called a complementary solution and written as

$$y_c = c_1 y_1(x) + c_2 y_2(x) + \cdots + c_n y_n(x) \qquad (1.2)$$

where $c_1, c_2, \ldots, c_n$ are arbitrary constants. Next find any particular solution y_p of the nonhomogeneous differential equation $L(y) = F(x)$. One usually tries to use the method of undetermined coefficients or the method of variation of parameters to construct a particular solution. The general solution of equation (1.1) is then written $y = y_c + y_p$.

In applied problems the solutions to differential equations of the form given by equation (1.1) are required to satisfy certain auxiliary conditions. The number of these auxiliary conditions is in most applications equal to the order of the differential equation. For example, when the second order differential equation

$$a_0(x)\frac{d^2 y}{dx^2} + a_1(x)\frac{dy}{dx} + a_2(x)y = 0 \qquad a \le x \le b \qquad (1.3)$$

is subjected to the auxiliary conditions at a single point $x = a$ of the form

$$y(a) = \alpha \qquad y'(a) = \beta$$

where α and β are given constants, the differential equation plus auxiliary conditions is referred to as an initial value problem (IVP). Initial value problems usually have unique solutions. Whenever the differential equation (1.3) is subjected to auxiliary conditions at two different points, say $x = a$ and $x = b$, having the form

$$\begin{aligned} c_{11}y(a) + c_{12}y'(a) &= \alpha & c_{11}^2 + c_{12}^2 \neq 0 \\ c_{21}y(b) + c_{22}y'(b) &= \beta & c_{21}^2 + c_{22}^2 \neq 0 \end{aligned} \tag{1.4}$$

where $c_{11}, c_{12}, c_{21}, c_{22}, \alpha, \beta$ are given constants, then the auxiliary conditions are called boundary conditions (BC). In this case the differential equation (1.3) together with the boundary conditions (1.4) is called a boundary value problem (BVP). Solutions to boundary value problems depend upon the boundary conditions. If not properly formulated the boundary value problem might not have a solution. In other formulations the boundary value problem might have an infinite number of solutions. In still other circumstances there will exist a unique solution and so boundary value problems must be carefully analyzed to see (i) if a solution exists and (ii) how many solutions exist. Boundary conditions of the form

$$\begin{aligned} c_{11}y(a) + c_{12}y'(a) + c_{13}y(b) + c_{14}y'(b) &= \alpha \\ c_{21}y(b) + c_{22}y'(b) + c_{23}y(a) + c_{24}y'(a) &= \beta, \end{aligned}$$

where the c_{ij}, $i = 1, 2$, $j = 1, 2, 3, 4$, and α, β are constants, are called mixed boundary conditions, while the boundary conditions given by equations (1.4) are called unmixed boundary conditions. Mixed boundary value problems are usually much harder to solve.

First Order Equations.

Consider the first order linear equation

$$L(y) = \frac{dy}{dx} + p(x)y = q(x). \tag{1.5}$$

To solve this differential equation we first solve the homogeneous equation

$$\frac{dy}{dx} + p(x)y = 0 \tag{1.6}$$

to obtain the complementary solution y_c. One can separate the variables in equation (1.6) to obtain

$$\frac{dy}{y} = -p(x)\,dx. \tag{1.7}$$

Define the function

$$P(x) = \int_0^x p(\xi)\,d\xi \qquad \text{with} \qquad \frac{dP}{dx} = p(x), \tag{1.8}$$

then the equation (1.7) has the integral $\ln y = -P(x) + C$ where C is a constant of integration. We solve for y

$$e^{\ln y} = e^{-P(x)+C} = e^{-P(x)}e^C \qquad \text{or} \qquad y = C_1 e^{-P(x)},$$

where $C_1 = e^C$ is some new constant. This produces the complementary solution

$$y_c = C_1 e^{-P(x)}.$$

We use the method of variation of parameters and assume a particular solution of the form $y_p = u(x)e^{-P(x)}$ where C_1 in the complementary solution has been replaced by an unknown function $u(x)$. This assumed solution has the derivative

$$\frac{dy_p}{dx} = u(x)e^{-P(x)}(-p(x)) + \frac{du}{dx}e^{-P(x)}.$$

Substitute y_p and $\frac{dy_p}{dx}$ into the nonhomogeneous differential equation (1.5) and simplify to obtain the differential equation which defines the function $u = u(x)$. This differential equation is

$$\frac{du}{dx}e^{-P(x)} = q(x).$$

Separate the variables in this equation and solve for u to obtain

$$u = u(x) = \int_0^x q(\xi)e^{P(\xi)}\,d\xi.$$

This produces the particular solution

$$y_p = e^{-P(x)}\int_0^x q(\xi)e^{P(\xi)}\,d\xi.$$

The general solution to the equation (1.5) is then

$$y = y_c + y_p = e^{-P(x)}\left[C_1 + \int_0^x q(\xi)e^{P(\xi)}\,d\xi\right]. \tag{1.9}$$

An alternate method for solving the first order linear equation (1.5) is to multiply the equation by the integrating factor $e^{P(x)}$, where $P(x) = \int_0^x p(\xi)\,d\xi$, to obtain the exact differential equation

$$\frac{d}{dx}\left(e^{P(x)}y\right) = e^{P(x)}q(x).$$

This equation is easily integrated to obtain the solution given by equation (1.9).

Second Order Equations

Consider a second order linear homogeneous differential equation having the form of equation (1.3). We assume $\{y_1(x), y_2(x)\}$ is a fundamental set of solutions to the homogeneous second order differential equation

$$L(y) = a_0(x)\frac{d^2y}{dx^2} + a_1(x)\frac{dy}{dx} + a_2(x)y = 0.$$

This produces the complementary solution

$$y_c = c_1 y_1(x) + c_2 y_2(x) \tag{1.10}$$

where c_1, c_2 are arbitrary constants. The method of variation of parameters requires one to assume a particular solution to the nonhomogeneous differential equation

$$L(y) = a_0(x)\frac{d^2y}{dx^2} + a_1(x)\frac{dy}{dx} + a_2(x)y = F(x) \tag{1.11}$$

of the form

$$y_p = u(x)y_1(x) + v(x)y_2(x) \tag{1.12}$$

where $u(x)$ and $v(x)$ are functions replacing the constants c_1, c_2 in equation (1.10). The functions u, v are selected to satisfy the equations

$$
\begin{aligned}
u'(x)y_1(x) + v'(x)y_2(x) &= 0 \\
u'(x)y_1'(x) + v'(x)y_2'(x) &= \frac{F(x)}{a_0(x)}, \qquad a_0(x) \neq 0
\end{aligned}
\tag{1.13}
$$

from which $u'(x) = \frac{du}{dx}$ and $v'(x) = \frac{dv}{dx}$ can be determined. Using Cramer's[†] rule we solve for u' and v' and find

$$u'(x) = \frac{du}{dx} = \frac{\begin{vmatrix} 0 & y_2(x) \\ F(x)/a_0(x) & y_2'(x) \end{vmatrix}}{\begin{vmatrix} y_1(x) & y_2(x) \\ y_1'(x) & y_2'(x) \end{vmatrix}}, \qquad v'(x) = \frac{dv}{dx} = \frac{\begin{vmatrix} y_1(x) & 0 \\ y_1'(x) & F(x)/a_0(x) \end{vmatrix}}{\begin{vmatrix} y_1(x) & y_2(x) \\ y_1'(x) & y_2'(x) \end{vmatrix}}$$

which simplifies to

$$\frac{du}{dx} = \frac{-y_2(x)F(x)}{a_0(x)W(x)} \quad \text{and} \quad \frac{dv}{dx} = \frac{y_1(x)F(x)}{a_0(x)W(x)} \tag{1.14}$$

where $W = W(x) = y_1(x)y_2'(x) - y_2(x)y_1'(x) \neq 0$ is the nonzero Wronskian[†] associated with the fundamental set of solutions. These equations are then integrated

---

[†] See Appendix C

to obtain the desired functions $u = u(x)$ and $v = v(x)$. A simple integration gives the solution to the equations (1.14) as

$$u = u(x) = -\int_\alpha^x \frac{y_1(\xi)F(\xi)}{a_0(\xi)W(\xi)}\,d\xi, \qquad v = v(x) = \int_\alpha^x \frac{y_1(\xi)F(\xi)}{a_0(\xi)W(\xi)}\,d\xi, \qquad (1.15)$$

where α is some constant initial value which is usually selected away from any singularities associated with the integrals. The resulting solution of the nonhomogeneous differential equation is the particular solution

$$y_p = y_p(x) = y_1(x)\int_\alpha^x \frac{-y_2(\xi)F(\xi)}{p(\xi)W(\xi)}\,d\xi + y_2(x)\int_\alpha^x \frac{y_1(\xi)F(\xi)}{p(\xi)W(\xi)}\,d\xi$$

which can also be written in the form

$$y_p = y_p(x) = \int_\alpha^x \frac{[y_2(x)y_1(\xi) - y_1(x)y_2(\xi)]F(\xi)}{p(\xi)W(\xi)}\,d\xi. \qquad (1.16)$$

The general solution of equation (1.11) is then written as $y = y_c + y_p$ or

$$y = c_1 y_1(x) + c_2 y_2(x) + \int_\alpha^x \frac{[y_2(x)y_1(\xi) - y_1(x)y_2(\xi)]F(\xi)}{p(\xi)W(\xi)}\,d\xi. \qquad (1.17)$$

Harmonic Motion

One of the many easy to solve second order differential equations is the equation

$$\frac{d^2F}{dx^2} + \lambda F = 0 \qquad \text{with } \lambda \text{ constant.} \qquad (1.18)$$

This equation will arise in the study of many partial differential equations in Cartesian coordinates. The differential equation contains a parameter λ and so we consider the following three cases of negative, zero and positive values which can be assigned to the constant λ.

Case 1: $\lambda = -\omega^2$ $(\omega > 0)$. The differential equation $\frac{d^2F}{dx^2} - \omega^2 F = 0$ is a differential equation with constant coefficients and so one can assume an exponential solution $F = e^{mx}$. This gives the characteristic equation $m^2 - \omega^2 = 0$ with the characteristic roots $m = \omega$ and $m = -\omega$. A fundamental set of solutions is given by $\{e^{\omega x}, e^{-\omega x}\}$. Once we know a fundamental set of solutions we can then form linear combinations of these functions and generate an infinite set of other solutions. Recall a linear combination is nothing more than multiplying the given functions by arbitrary constants (which may or may not be complex constants) and adding the results. From the above fundamental set we can form the linear combinations

$$\sinh \omega x = \frac{e^{\omega x} - e^{-\omega x}}{2}, \qquad \cosh \omega x = \frac{e^{\omega x} + e^{-\omega x}}{2}.$$

Many other linear combinations can be formed. One more example is,

$$\sinh \omega(x - x_0) = \frac{e^{-\omega x_0} e^{\omega x} - e^{\omega x_0} e^{-\omega x}}{2}, \qquad \cosh \omega(x - x_0) = \frac{e^{-\omega x_0} e^{\omega x} + e^{\omega x_0} e^{-\omega x}}{2}$$

with x_0 constant. These are additional forms for solutions. We summarize these results with the following.

Solutions to the differential equation $\dfrac{d^2 F}{dx^2} - \omega^2 F = 0$ can be written in many forms. Four possible forms are

$$F = F(x) = c_1 e^{\omega x} + c_2 e^{-\omega x}$$

$$F = F(x) = c_1 e^{\omega(x - x_0)} + c_2 e^{-\omega(x - x_0)}$$

$$F = F(x) = c_1 \sinh \omega x + c_2 \cosh \omega x$$

$$F = F(x) = c_1 \sinh \omega(x - x_0) + c_2 \cosh \omega(x - x_0)$$

where x_0, c_1 and c_2 are arbitrary constants.

The particular final form selected for the general solution depends upon initial or boundary conditions assigned to the problem. The final form selected is usually a form which simplifies any additional algebra associated with boundary or initial conditions.

Case 2: $\lambda = 0$. The differential equation $\dfrac{d^2 F}{dx^2} = 0$ is easily solved by integrating twice to obtain $F = F(x) = c_1 + c_2 x$. Another form for this solution is the shifted form $F = F(x) = K_1 + K_2(x - x_0)$ where x_0 is constant. One can also assume an exponential solution $F = e^{mx}$ and obtain the characteristic equation $m^2 = 0$ with characteristic roots $\{0, 0\}$. This gives the fundamental set $\{1, x\}$ and so the general solution is $y = c_1(1) + c_2 x$. The particular final form selected is a matter of choice. When the shifted form is expanded we have $c_2 = K_2$ and the combination $K_1 - K_2 x_0$ is treated as a new constant and labeled as c_1.

Case 3: $\lambda = \omega^2$ $(\omega > 0)$. The differential equation $\dfrac{d^2 F}{dx^2} + \omega^2 F = 0$ is a differential equation with constant coefficients. We assume an exponential solution

$F = e^{mx}$ which gives the characteristic equation $m^2 + \omega^2 = 0$ with characteristic roots $m = i\omega, m = -i\omega$, where $i^2 = -1$. These roots produce the fundamental set $\{e^{i\omega x}, e^{-i\omega x}\}$. Linear combinations of functions from the fundamental set will generate additional solution forms. For example, using the Euler identity

$$e^{i\theta} = \cos\theta + i\sin\theta \qquad (1.19)$$

one can generate the solutions

$$\sin\omega x = \frac{e^{i\omega x} - e^{-i\omega x}}{2i}, \qquad \cos\omega x = \frac{e^{i\omega x} + e^{-i\omega x}}{2}. \qquad (1.20)$$

The particular linear combinations

$$\sin\omega(x - x_0) = \frac{e^{-i\omega x_0}e^{i\omega x} - e^{i\omega x_0}e^{-i\omega x}}{2i}$$
$$\cos\omega(x - x_0) = \frac{e^{-i\omega x_0}e^{i\omega x} + e^{i\omega x_0}e^{-i\omega x}}{2}, \qquad (1.21)$$

with x_0 constant, gives the shifted solutions for the sine and cosine solutions. These results are summarized for later use.

Solutions to the differential equation

$$\frac{d^2F}{dx^2} + \omega^2 F = 0$$

can be written in many forms. Four possible forms are

$$F = F(x) = c_1 e^{i\omega x} + c_2 e^{-i\omega x}$$
$$F = F(x) = c_1 e^{i\omega(x - x_0)} + c_2 e^{-i\omega(x - x_0)}$$
$$F = F(x) = c_1 \sin\omega x + c_2 \cos\omega x$$
$$F = F(x) = c_1 \sin\omega(x - x_0) + c_2 \cos\omega(x - x_0)$$

where x_0, c_1 and c_2 are arbitrary constants.

Cauchy-Euler equation

The Cauchy[†]-Euler[†] equation

$$a_0 x^2 \frac{d^2y}{dx^2} + a_1 x \frac{dy}{dx} + a_2 y = 0 \qquad (1.22)$$

[†] See Appendix C

where a_0, a_1, a_2 are constants, is related to a differential equation with constant coefficients by way of the transformation $t = \ln x$. Using the derivatives

$$\frac{dy}{dx} = \frac{dy}{dt}\frac{dt}{dx} = \frac{dy}{dt}\left(\frac{1}{x}\right) \quad \text{and} \quad \frac{d^2y}{dx^2} = \frac{dy}{dt}\left(\frac{-1}{x^2}\right) + \frac{d^2y}{dt^2}\left(\frac{1}{x^2}\right)$$

the equation (1.22) becomes

$$a_0\frac{d^2y}{dt^2} + (a_1 - a_0)\frac{dy}{dt} + a_2y = 0. \tag{1.23}$$

Assuming an exponential solution of $y = e^{mt}$ to equation (1.23) gives the characteristic equation

$$a_0m^2 + (a_1 - a_0)m + a_2 = 0. \tag{1.24}$$

This characteristic equation is derivable directly from equation (1.22) by assuming a solution $y = x^m$ since $y = e^{mt} = e^{m\ln x} = x^m$. The roots of the characteristic equation determine the types of solutions that can exist for the Cauchy-Euler equation. We examine the following cases.

Case 1: If the characteristic roots are real and distinct $m = \alpha, m = \beta$, then the differential equation (1.22) has the fundamental set of solutions $\{x^\alpha, x^\beta\}$ and so the general solution can be written as the linear combination $y = c_1x^\alpha + c_2x^\beta$ where c_1, c_2 are arbitrary constants.

Case 2: If the characteristic roots are repeated roots $m = \alpha, m = \alpha$ then a fundamental set of solutions for the transformed equation (1.23) is given by $\{e^{\alpha t}, te^{\alpha t}\}$. The transformation $t = \ln x$ gives the fundamental set of solutions $\{x^\alpha, x^\alpha \ln x\}$ to the original Cauchy-Euler equation (1.22). The general solution of equation (1.22) can then be written as $y = c_1x^\alpha + c_2x^\alpha \ln x$ where c_1, c_2 are arbitrary constants.

Case 3: If the characteristic roots of equation (1.24) are imaginary roots of the form $m = \alpha + i\beta$, $m = \alpha - i\beta$, with $i^2 = -1$, a fundamental set of solutions to the transformed differential equation (1.23) is given by $\{e^{\alpha t}\cos \beta t, e^{\alpha t}\sin \beta t\}$ which by way of the transformation $t = \ln x$, gives the fundamental set of solutions to the Cauchy-Euler equation as $\{x^\alpha \cos(\beta \ln x), x^\alpha \sin(\beta \ln x)\}$. The general solution to the Cauchy-Euler differential equation (1.22) can then be written as the linear combination $y = c_1x^\alpha \cos(\beta \ln x) + c_2x^\alpha \sin(\beta \ln x)$ where c_1, c_2 are arbitrary constants.

Bessel functions   The differential equation

$$t^2 \frac{d^2 z}{dt^2} + t \frac{dz}{dt} + (t^2 - \nu^2)z = 0 \tag{1.25}$$

is known as Bessel's[†] differential equation.   This equation arises in many applied problems involving cylindrical coordinates and we shall have to deal with it in our study of solutions to certain partial differential equations. The quantity ν in equation (1.25) is called a parameter and can be any real number. The Bessel equation (1.25) has a regular singular point at $t = 0$ and so one can assume a Frobenius[†] type solution $z = \sum_{n=0}^{\infty} c_n t^{n+r}$. In this way one can calculate solutions $J_\nu(t)$ and $J_{-\nu}(t)$, for ν not an integer, where

$$J_\nu(t) = \sum_{m=0}^{\infty} \frac{(-1)^m}{m! \, \Gamma(\nu + m + 1)} \left(\frac{t}{2}\right)^{2m+\nu} \tag{1.26}$$

is defined as a Bessel function of the first kind of order ν. In equation (1.26) the function Γ is called a Gamma function. It is defined

$$\Gamma(\alpha) = \int_0^\infty e^{-t} t^{\alpha-1}\, dt, \qquad \alpha > 0 \tag{1.27}$$

with the properties

$$\Gamma(\alpha + 1) = \alpha \Gamma(\alpha) \quad \text{and} \quad \Gamma(n + m + 1) = (n + m)! \tag{1.28}$$

for m and n integers. For the special values $n = 0, -1, -2, \ldots$ the Gamma function $\Gamma(n)$ is undefined. For these special values the function $1/\Gamma(n)$ is defined to be zero. A sketch of the Gamma function is illustrated in the figure 1-1.

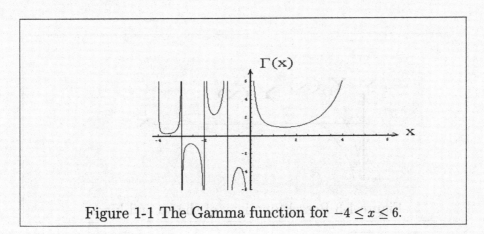

Figure 1-1 The Gamma function for $-4 \le x \le 6$.

[†]   See Appendix C

10

For $\nu = m$ an integer, one finds the functions $J_m(x)$ and $J_{-m}(x)$ satisfy the relation $J_{-m}(x) = (-1)^m J_m(x)$ and so these functions are no longer linearly independent functions. For any value of ν the Weber[†] and Schlafi[†] ratio

$$Y_\nu(t) = \frac{J_\nu(t)\cos\nu\pi - J_{-\nu}(t)}{\sin\nu\pi} \tag{1.29}$$

is a linear combination of $J_\nu(t)$ and $J_{-\nu}(t)$ which can be used to define Bessel functions of the second kind of order ν. The limiting process $Y_n(t) = \lim\limits_{\nu \to n} Y_\nu(t)$ is used to define Bessel functions of the second kind with integer values n. Some texts refer to Bessel functions of the second kind of order ν as Neumann[†] functions of order ν and are denoted using the notation $N_\nu(x)$. Graphs of selected Bessel functions are given in the figure 1-2 and figure 1-3. Note the zeros of the Bessel functions are not equally spaced. If we denote by ξ_{kn} the nth zero of the kth order Bessel function, then one can write $J_k(\xi_{kn}) = 0$, for $n = 1, 2, 3, \ldots$.

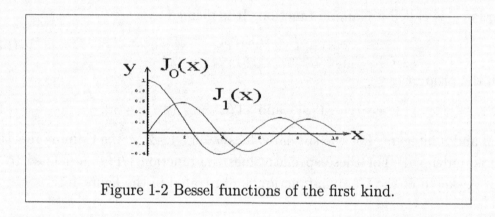

Figure 1-2 Bessel functions of the first kind.

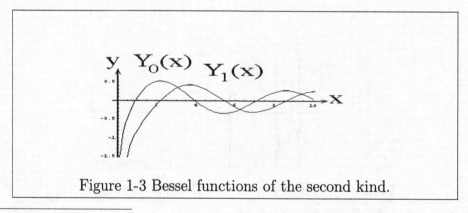

Figure 1-3 Bessel functions of the second kind.

[†]   See Appendix C

The zeros of the Bessel functions can be found in the many handbooks.[1] Listed for later reference are the first five zeros for the J_0, J_1, Y_0, Y_1 Bessel functions.

First five zero's of the Bessel functions

$$\mathbf{J_0(x), J_1(x), Y_0(x), Y_1(x)}$$

$J_0(x) = 0$	for x=	$2.4048, 5.5201, 8.6537, 11.7915, 14.9309, \ldots$
$J_1(x) = 0$	for x=	$3.8317, 7.0156, 10.1735, 13.3237, 16.4706 \ldots$
$Y_0(x) = 0$	for x=	$0.8936, 3.9577, 7.0861, 10.2223, 13.3611, \ldots$
$Y_1(x) = 0$	for x=	$2.1971, 5.4297, 8.5960, 11.7492, 14.8974, \ldots$

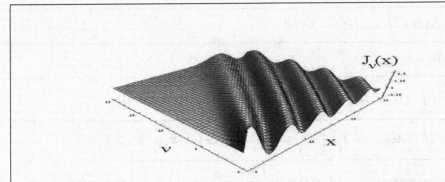

Figure 1-4 Bessel function of the first kind as a function of ν and x.

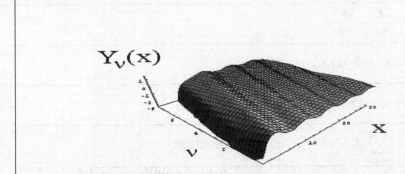

Figure 1-5 Bessel function of the second kind as a function of ν and x.

---

[1] See for example: M. Abramowitz and I.A. Stegun, Handbook of Mathematical Functions, Dover Publications, 1972.

Table 1-1 Bessel Function Properties

1.	$\dfrac{d}{dx}\left[x^{-\nu}J_\nu(x)\right] = -x^{-\nu}J_{\nu+1}(x)$ **or** $J_\nu'(x) = \dfrac{\nu}{x}J_\nu(x) - J_{\nu+1}(x)$
2.	$\dfrac{d}{dx}[x^\nu J_\nu(x)] = x^\nu J_{\nu-1}(x)$ **or** $J_\nu'(x) = J_{\nu-1}(x) - \dfrac{\nu}{x}J_\nu(x)$
3.	$J_{\nu-1}(x) + J_{\nu+1}(x) = \dfrac{2\nu}{x}J_\nu(x), \qquad Y_{\nu-1}(x) + Y_{\nu+1}(x) = \dfrac{2\nu}{x}Y_\nu(x)$
4.	$J_{\nu-1}(x) - J_{\nu+1}(x) = 2J_\nu'(x)$
5.	$\displaystyle\int x^\nu J_{\nu-1}(x)\,dx = x^\nu J_\nu(x) + C$
6.	$\displaystyle\int x^{-\nu}J_{\nu+1}(x)\,dx = -x^{-\nu}J_\nu(x) + C$
7.	$\displaystyle\int J_{\nu+1}(x)\,dx = \int J_{\nu-1}(x)\,dx - 2J_\nu(x) + C$
8.	$\dfrac{d}{dx}J_0(x) = -J_1(x), \qquad\qquad \dfrac{d}{dx}[J_0(\lambda x)] = -\lambda J_1(\lambda x)$
9.	$(\beta^2 - \alpha^2)\displaystyle\int_a^b xJ_n(\alpha x)J_n(\beta x)\,dx = \left[x\left(\alpha J_n(\beta x)J_n'(\alpha x) - \beta J_n(\alpha x)J_n'(\beta x)\right)\right]_a^b$
10.	$J_{\frac{1}{2}}(x) = \sqrt{\dfrac{2}{\pi x}}\sin x, \qquad J_{\frac{3}{2}}(x) = \sqrt{\dfrac{2}{\pi x}}\left(\dfrac{\sin x}{x} - \cos x\right), \qquad Y_{\frac{1}{2}}(x) = -J_{-\frac{1}{2}}(x)$
11.	$J_{-\frac{1}{2}}(x) = \sqrt{\dfrac{2}{\pi x}}\cos x, \qquad J_{-\frac{3}{2}}(x) = -\sqrt{\dfrac{2}{\pi x}}\left(\dfrac{\cos x}{x} + \sin x\right), \qquad Y_{\frac{3}{2}}(x) = J_{-\frac{3}{2}}(x)$
12.	$\displaystyle\int_a^b xJ_k^2(\lambda x)\,dx = \dfrac{b^2}{2}\left\{[J_k'(\lambda b)]^2 + J_k^2(\lambda b) - \dfrac{k^2}{\lambda^2 b^2}J_k^2(\lambda b)\right\} - \dfrac{a^2}{2}\left\{[J_k'(\lambda a)]^2 + J_k^2(\lambda a) - \dfrac{k^2}{\lambda^2 a^2}J_k^2(\lambda a)\right\}$
13.	$J_0(x) = 1 - \dfrac{x^2}{2^2} + \dfrac{x^4}{2^4 2!} - \dfrac{x^6}{2^6 3!} + \cdots$
14.	$x\dfrac{d}{dx}J_\nu(\lambda_n x) = \nu J_\nu(\lambda_n x) - \lambda_n x J_{\nu+1}(\lambda_n x)$
15.	$2\displaystyle\int xJ_\nu^2(x)\,dx = x^2[J_\nu'(x)]^2 + (x^2 - \nu^2)[J_\nu(x)]^2 + c$
16.	$\exp\left(\dfrac{x}{2}\left[t - \dfrac{1}{t}\right]\right) = \displaystyle\sum_{n=-\infty}^{\infty} J_n(x)t^n$ Generating function
17.	$\displaystyle\int_0^b xJ_k(\lambda_n x)J_k(\lambda_m x)\,dx = \dfrac{b^2}{2}J_{k+1}^2(\lambda_n b)\delta_{mn}$ for m, n integers and λ_n satisfying $J_k(\lambda_n b) = 0$ for $n = 1, 2, 3, \ldots$

The Bessel functions can be of integer order, fractional order or noninteger nonfractional orders. The figure 1-4 illustrates the behavior of the Bessel function $J_\nu(x)$ when plotted to form the surface $z = J_\nu(x)$ as a function of ν and x. Note the behavior of the surface and observe the function is bounded for large positive arguments x and it approaches zero for large values of the order ν. The figure 1-5 illustrates the surface $z = Y_\nu(x)$. The surface plot illustrates that the function $Y_\nu(x)$ becomes unbounded as x approaches zero.

The table 1-1 lists several important properties of the Bessel functions. In table 1-1 observe that in all of the differentiation and integration properties the variable x is a dummy variable and can be replace by some other symbol. For example in entry number 4, if we replace x by βx we can write

$$J_\nu'(\beta x) = \frac{1}{2\beta}\left[J_{\nu-1}(\beta x) - J_{\nu+1}(\beta x)\right].$$

Here the prime notation always means differentiation with respect to the argument of the function. For example, $J_\nu'(\xi) = \frac{d}{d\xi}J_\nu(\xi)$. If we let $\xi = \beta x$, then by chain rule differentiation we have

$$\frac{d}{dx}\left[J_\nu(\beta x)\right] = \frac{d}{d\xi}J_\nu(\xi)\frac{d\xi}{dx} = J_\nu'(\beta x)\beta.$$

Note in particular the implied differential relations from entry 8 in this table.

$$J_0'(x) = -J_1(x) \quad \text{and} \quad \frac{d}{dx}\left[J_0(\lambda x)\right] = -\lambda J_1(\lambda x). \tag{1.30}$$

Both Bessel functions $J_n(x)$ and $Y_n(x)$ satisfy the same differential equation, and so they possess many of the same properties. Hence in many of the properties listed in table 1-1 one can replace $J_n(x)$ by $Y_n(x)$.

A generating function for a set of functions $\{\phi_n(x)\}$ is a function $g(x,t)$ which has some kind of a power series expansion in the variable t such that the coefficient of the nth term of the power series is $\phi_n(x)$. One possible form for a generating function is

$$g(x,t) = \sum_{n=0}^{\infty} \phi_n(x)t^n = \phi_0(x) + \phi_1(x)t + \phi_2(x)t^2 + \cdots.$$

The table 1-1 contains a generating function for the Bessel functions of the first kind $J_n(x)$. Note that the summation index for the given generating function varies over all values of the index n.

14

Example 1-1. (Bessel function properties.)

Show $\qquad 2\int xJ_\nu^2(x)\,dx = x^2[J_\nu'(x)]^2 + (x^2-\nu^2)[J_\nu(x)]^2 + c$

where c is a constant of integration.

Solution: Multiply Bessel's equation by $2y'$ and obtain

$$2x^2y''y' + 2x(y')^2 + 2x^2yy' - 2\nu^2yy' = 0. \qquad (1.31)$$

Use of the identity

$$\frac{d}{dx}\left[x^2(y')^2\right] = 2x^2y'y'' + 2x(y')^2$$

enables us to express equation (1.31) in the form

$$d[x^2(y')^2] = -2x^2yy'\,dx + 2\nu^2yy'\,dx.$$

This form is easily integrated to produce

$$x^2(y')^2 = -\int 2x^2yy'\,dx + \nu^2y^2 + c, \qquad (1.32)$$

where c is a constant of integration. Integrate the remaining integral in equation (1.32) by parts to obtain the equation

$$2\int xy^2\,dx = x^2(y')^2 + (x^2-\nu^2)y^2 + c$$

which for $y = J_\nu(x)$ reduces to

$$2\int xJ_\nu^2(x)\,dx = x^2\left[J_\nu'(x)\right]^2 + (x^2-\nu^2)\left[J_\nu(x)\right]^2 + c.$$

which is entry number 15 in the table 1-1.

∎

For future reference we summarize solution properties of Bessel's equation.

For ν different from an integer, the Bessel differential equation

$$t^2\frac{d^2z}{dt^2} + t\frac{dz}{dt} + (t^2-\nu^2)z = 0 \qquad (1.33)$$

has the general solution $\quad z = c_1J_\nu(t) + c_2J_{-\nu}(t) \qquad (1.34)$

For all values of ν the general solution can be written

$$z = c_1J_\nu(t) + c_2Y_\nu(t) \qquad (1.35)$$

where c_1 and c_2 denote arbitrary constants.

The substitutions $t = \lambda x$ and $y = z$ transforms the Bessel differential equation (1.33) into the form

$$x^2 \frac{d^2 y}{dx^2} + x \frac{dy}{dx} + (\lambda^2 x^2 - \nu^2) y = 0 \tag{1.36}$$

which for all values of ν has the general solution

$$y = c_1 J_\nu(\lambda x) + c_2 Y_\nu(\lambda x) \tag{1.37}$$

where c_1 and c_2 denote arbitrary constants.

The substitutions $t = \beta x^\gamma$ and $y = x^\alpha z$ transforms the equations (1.33) and (1.35). The new differential equation is

$$x^2 \frac{d^2 y}{dx^2} + (1 - 2\alpha) x \frac{dy}{dx} + [(\beta \gamma x^\gamma)^2 + \alpha^2 - \nu^2 \gamma^2] y = 0 \tag{1.38}$$

with general solution

$$y = x^\alpha [c_1 J_\nu(\beta x^\gamma) + c_2 Y_\nu(\beta x^\gamma)] \tag{1.39}$$

where c_1 and c_2 denote arbitrary constants.

A more general form of Bessel's equation results using the substitutions $z = t^{(1-a)/2} e^{-(b/r)t^r} y$ and $t = \left(\frac{sx}{\sqrt{d}} \right)^{1/s}$. It can be shown with some difficulty that the equation

$$x^2 \frac{d^2 y}{dx^2} + x(a + 2bx^r) \frac{dy}{dx} + [c + dx^{2s} + b(a + r - 1)x^r + b^2 x^{2r}] y = 0$$

has the solution

$$y = x^\alpha e^{-\beta x^r} [c_1 J_\nu(\lambda x^s) + c_2 Y_\nu(\lambda x^s)]$$

where $\alpha = \dfrac{1-a}{2}$, $\beta = \dfrac{b}{r}$, $\lambda = \dfrac{\sqrt{d}}{s}$, $\nu = \dfrac{\sqrt{(1-a)^2 - 4c}}{2s}$ with the restrictions that $(1-a)^2 - 4c \geq 0$ $\quad d > 0 \quad r \neq 0 \quad s \neq 0$.

Modified Bessel's Equation

In Bessel's equation (1.25) replace t by it, where $i^2 = -1$, to obtain the modified Bessel equation

$$t^2 \frac{d^2 z}{dt^2} + t \frac{dz}{dt} - (t^2 + \nu^2) z = 0. \tag{1.40}$$

16

The functions

$$I_\nu(x) = i^{-\nu} J_\nu(ix) \quad \text{and} \quad I_{-\nu}(x) = i^\nu J_{-\nu}(ix) \tag{1.41}$$

are solutions of the modified Bessel equation and are called modified Bessel functions of the first kind of order ν. The functions

$$K_\nu(x) = \frac{\pi}{2\sin(\nu\pi)} \left[I_{-\nu}(x) - I_\nu(x) \right] \quad n \neq 0,1,2,3,\ldots$$

$$K_n(x) = \lim_{\nu \to n} K_\nu(x) \quad n = 0,1,2,3,\ldots \tag{1.42}$$

are also solutions of the modified Bessel equation and are called modified Bessel functions of the second kind of order ν. The figures 1-6 and 1-7 illustrate these modified Bessel functions for selected values of ν.

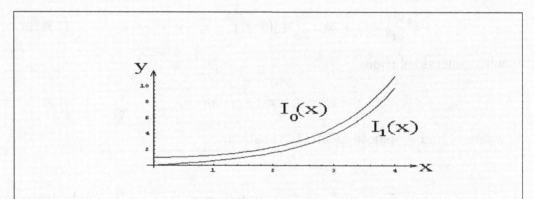

Figure 1-6 Modified Bessel functions of the first kind of orders 0 and 1.

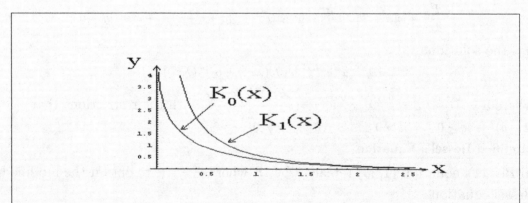

Figure 1-7 Modified Bessel functions of the second kind of orders 0 and 1.

For all values of ν the general solution to the modified Bessel differential equation

$$x^2 \frac{d^2y}{dx^2} + x\frac{dy}{dx} - (x^2 + \nu^2)y = 0 \qquad (1.43)$$

can be written

$$y = c_1 I_\nu(x) + c_2 K_\nu(x) \qquad (1.44)$$

where c_1 and c_2 are arbitrary constants.

A change of variables can be made in the differential equation (1.43) to obtain the following more general result involving a parameter λ.

For all values of ν the modified Bessel differential equation

$$x^2 \frac{d^2y}{dx^2} + x\frac{dy}{dx} - (\lambda^2 x^2 + \nu^2)y = 0 \qquad (1.45)$$

has the general solution

$$y = c_1 I_\nu(\lambda x) + c_2 K_\nu(\lambda x) \qquad (1.46)$$

where c_1 and c_2 are arbitrary constants.

Legendre functions

The Legendre[†] differential equation is written

$$(1 - x^2)\frac{d^2y}{dx^2} - 2x\frac{dy}{dx} + n(n+1)y = 0 \qquad -1 \leq x \leq 1 \qquad (1.47)$$

where n is a constant. By assuming a power series solution of the form

$$y = \sum_{m=0}^{\infty} c_m x^m$$

one can obtain the general solution

$$y = y(x) = c_0 U_n(x) + C_1 V_n(x) \qquad (1.48)$$

[†] See Appendix C

where $U_n(x)$ and $V_n(x)$ are series for the region $-1 \le x \le 1$ given by

$$U_n(x) = 1 - n(n+1)\frac{x^2}{2!} + n(n-2)(n+1)(n+3)\frac{x^4}{4!}$$
$$- n(n-2)(n-4)(n+1)(n+3)(n+5)\frac{x^6}{6!} + \cdots$$
$$V_n(x) = x - (n-1)(n+2)\frac{x^3}{3!} + (n-1)(n-3)(n+2)(n+4)\frac{x^5}{5!}$$
$$- (n-1)(n-3)(n-5)(n+2)(n+4)(n+6)\frac{x^7}{7!} + \cdots . \tag{1.49}$$

Observe that if $U_n(x)$ is a solution of the Legendre equation, then $CU_n(x)$ is also a solution for any constant C. Similarly, if $V_n(x)$ is a solution, then $CV_n(x)$ is also a solution for any constant C. The series solutions given by equations (1.49) are such that when n equals an even integer (say $n = 2m$), then $U_{2m}(x)$ is a polynomial of degree $2m$, and when n equals an odd integer (say $n = 2m+1$), then $V_{2m+1}(x)$ is a polynomial of degree $2m+1$. To represent all polynomial solutions $P_n(x)$, for $n = 0, 1, 2, 3, \ldots$ we select constants C such that all polynomial solutions satisfy the end condition $P_n(1) = 1$. The resulting polynomial solutions are called Legendre polynomials of order n or Legendre functions of the first kind or zonal harmonics. The first few polynomial solutions are

$$P_0(x) = 1 \qquad\qquad P_3(x) = \frac{1}{2}(5x^3 - 3x)$$
$$P_1(x) = x \qquad\qquad P_4(x) = \frac{1}{8}(35x^4 - 30x^2 + 3) \tag{1.50}$$
$$P_2(x) = \frac{1}{2}(3x^2 - 1) \qquad P_5(x) = \frac{1}{8}(63x^5 - 70x^3 + 15x)$$

and are illustrated in the figure 1-8. A recursive formula for generating higher order Legendre polynomials is given by

$$nP_n(x) = (2n-1)xP_{n-1}(x) - (n-1)P_{n-2}(x). \tag{1.51}$$

These Legendre polynomial solutions can be represented by the series

$$P_n(x) = \sum_{m=0}^{[\frac{n}{2}]} \frac{(-1)^m(2n-2m)!}{2^n m!(n-2m)!(n-m)!}x^{n-2m} \tag{1.52}$$

where $[\frac{n}{2}]$ is $\frac{n}{2}$ or $\frac{n-1}{2}$ whichever is an integer. Another form for these polynomial solutions is

$$P_n(x) = \frac{1}{2^n n!}\frac{d^n}{dx^n}\left[(x^2 - 1)^n\right] \tag{1.53}$$

which is known as Rodrigues'[†] formula for the Legendre polynomials. The function $g(x,t) = (1 - 2xt + t^2)^{-1/2}$ is known as a generating function because when expanded in a power series in the variable t it has the form

$$g(x,t) = (1 - 2xt + t^2)^{-1/2} = \sum_{n=0}^{\infty} P_n(x) t^n \tag{1.54}$$

valid for $|x| \leq 1$ and $|t| < 1$, where the Legendre polynomials are the coefficients of the powers of t in the series expansion of the generating function.

For the conditions of n being either an even or odd integer one of the solutions given by equations (1.49) is a polynomial while the other solution is an infinite series which converges in the interval $-1 < x < 1$. Under these conditions Legendre functions of the second kind $Q_n(x)$, for $n = 0, 1, 2, \ldots$, are constructed by taking constants times these infinite series and are defined by

$$Q_n(x) = \begin{cases} U_n(1) V_n(x), & \text{for } n \text{ even} \\ -V_n(1) U_n(x), & \text{for } n \text{ odd} \end{cases} \tag{1.55}$$

The first few Legendre functions of the second kind have the form

$$\begin{aligned}
Q_0(x) &= \frac{1}{2} \ln \frac{1+x}{1-x} & Q_3(x) &= P_3(x) Q_0(x) - \frac{5}{2} x^2 + \frac{2}{3} \\
Q_1(x) &= x Q_0(x) - 1 & Q_4(x) &= P_4(x) Q_0(x - \frac{35}{8} x^3 + \frac{55}{24} x \\
Q_2(x) &= P_2(x) Q_0(x) - \frac{3}{2} x & Q_5(x) &= P_5(x) Q_0(x) - \frac{63}{8} x^4 + \frac{49}{8} x^2 - \frac{8}{15}
\end{aligned} \tag{1.56}$$

and are illustrated in the figure 1-9. These functions are not defined at the end points $x = -1$ and $x = 1$. A recursive formula for generating higher order Legendre functions of the second kind is given by

$$n Q_n(x) = (2n - 1) x Q_{n-1}(x) - (n - 1) Q_{n-2}(x). \tag{1.57}$$

[†] See Appendix C

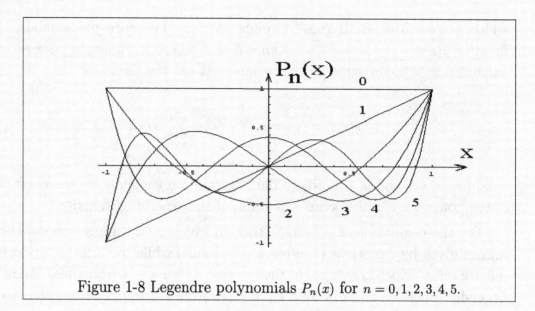

Figure 1-8 Legendre polynomials $P_n(x)$ for $n = 0, 1, 2, 3, 4, 5$.

To summarize we can write the following.

For positive integers n, the general solution to the Legendre differential equation

$$(1 - x^2)\frac{d^2y}{dx^2} - 2x\frac{dy}{dx} + n(n+1)y = 0 \qquad -1 \le x \le 1 \qquad (1.58)$$

can be written as

$$y = c_1 P_n(x) + c_2 Q_n(x) \qquad (1.59)$$

where c_1 and c_2 are arbitrary constants.

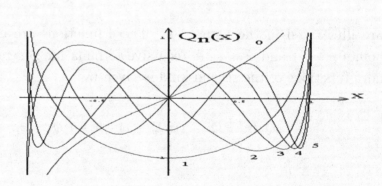

Figure 1-9 Legendre functions of second kind $Q_n(x)$ for $n = 0, 1, 2, 3, 4, 5$.

The change of variable $x = \cos\theta$ transforms the Legendre differential equation to the form

$$\frac{d^2y}{d\theta^2} + \cot\theta\frac{dy}{d\theta} + n(n+1)y = 0 \tag{1.60}$$

which can also be represented in the form

$$\frac{1}{\sin\theta}\frac{d}{d\theta}\left(\sin\theta\frac{dy}{d\theta}\right) + n(n+1)y = 0$$

and the general solution can be written

$$y = c_1 P_n(\cos\theta) + c_2 Q_n(\cos\theta) \tag{1.61}$$

where c_1 and c_2 are arbitrary constants. This form of Legendre's differential equation arises in many problems involving spherical coordinates.

Listed in the table 1-2 are some properties of the Legendre functions.

Table 1-2. Legendre Function Properties

1.	$g(x,t) = (1 - 2xt + t^2)^{-1/2} = \sum_{n=0}^{\infty} P_n(x)t^n$ Generating function
2.	$P_n(-x) = (-1)^n P_n(x)$ $P_n(-1) = (-1)^n$
3.	$P_{2m+1}(0) = 0$ $P'_{2m}(0) = 0$ $P_{2n}(0) = \dfrac{(-1)^n(2n)!}{2^{2n}(n!)^2}$
4.	$(2n+1)xP_n(x) = (n+1)P_{n+1}(x) + nP_{n-1}(x)$
5.	$xP'_n(x) = P'_{n+1}(x) - (n+1)P_n(x)$
6.	$\displaystyle\int_{-1}^{1} P_n^2(x)\,dx = \dfrac{2}{2n+1},\quad n = 0,1,2,\ldots$
7.	$\int_{-1}^{1} P_n(x)P_m(x)\,dx = \frac{2}{2n+1}\delta_{mn}$ m,n integers

Example 1-2.   (**Recurrence formula for Legendre Polynomials.**)

One way to derive the recurrence formula given by entry number 4 in the above table is as follows. Use the operator $D = \frac{d}{dx}$ and let $u = x^2 - 1$ and verify the following differentiation properties.

$$D[u^n] = 2xnu^{n-1}$$

$$D[u^{n+1}] = 2x(n+1)u^n$$

$$D^2[u^{n+1}] = 2(n+1)D[xu^n] = 2(n+1)[xDu^n + u^n] \tag{1.62}$$

$$D^2[u^{n+1}] = 2(n+1)[u^n + 2nx^2u^{n-1} - 2nu^{n-1} + 2nu^{n-1}]$$

$$D^2[u^{n+1}] = 2(n+1)[(2n+1)u^n + 2nu^{n-1}]$$

and

$$D[xu^n] = xDu^n + u^n$$

$$D^2[xu^n] = D[xDu^n + u^n] = xD^2u^n + 2Du^n$$

$$D^3[xu^n] = D[xD^2u^n + 2Du^n] = xD^3u^n + 3D^2u^n \tag{1.63}$$

$$\cdots$$

$$D^n[xu^n] = D[xD^{n-1}u^n + (n-1)D^{n-2}u^n] = xD^nu^n + nD^{n-1}u^n.$$

We can now write the Rodrigues formula (1.53), with n replaced by $n+1$ in the operator form

$$2^{n+1}(n+1)!P_{n+1}(x) = D^{n+1}u^{n+1} = D^{n-1}\left(D^2u^{n+1}\right)$$

$$= D^{n-1}\left[2(n+1)\left((2n+1)u^n + 2nu^{n-1}\right)\right]$$

$$= 2(n+1)\left[(2n+1)D^{n-1}u^n + 2nD^{n-1}u^{n-1}\right]$$

which simplifies to

$$2^n n! P_{n+1}(x) = (2n+1)D^{n-1}u^n + 2n\left(2^{n-1}(n-1)!P_{n-1}(x)\right). \tag{1.64}$$

One can also write the Rodrigues formula in the form

$$2^{n+1}(n+1)!P_{n+1}(x) = D^{n+1}u^{n+1} = D^n\left[Du^{n+1}\right] = D^n\left[2x(n+1)u^n\right]$$

which simplifies to

$$2^n n! P_{n+1}(x) = D^n[xu^n] = xD^nu^n + nD^{n-1}u^n$$

$$2^n n! P_{n+1}(x) = x\left(2^n n! P_n(x)\right) + nD^{n-1}u^n. \tag{1.65}$$

Now eliminate the term $D^{n-1}u^n$ using equations (1.65) and (1.64) to obtain the recurrence relation

$$(n+1)P_{n+1}(x) + nP_{n-1}(x) = (2n+1)xP_n(x) \qquad n = 1, 2, 3, \ldots \tag{1.66}$$

The adjoint equation.

The majority of ordinary differential equations we will consider in this text can be written in the general second order linear form

$$L(y) = p(x)\frac{d^2y}{dx^2} + q(x)\frac{dy}{dx} + r(x)y = 0 \tag{1.67}$$

where $p(x), q(x), r(x)$ and the solution $y(x)$ are well defined and continuous over some interval $a \le x \le b$. The differential operator

$$L(y) = p(x)\frac{d^2y}{dx^2} + q(x)\frac{dy}{dx} + r(x)y \tag{1.68}$$

is a linear differential operator having the property

$$L(c_1 y_1 + c_2 y_2) = c_1 L(y_1) + c_2 L(y_2)$$

for all functions y_1, y_2 and constants c_1, c_2. The adjoint operator $L^*(y)$ associated with the operator $L(y)$ is defined

$$L^*(y) = \frac{d^2}{dx^2}\left(p(x)y\right) - \frac{d}{dx}\left(q(x)y\right) + r(x)y. \tag{1.69}$$

If $L^*(y) = L(y)$ then L is called a self-adjoint operator. The various derivative terms can be expressed

$$\frac{d}{dx}(p(x)y) = p(x)\frac{dy}{dx} + p'(x)y$$
$$\frac{d^2}{dx^2}(p(x)y) = p(x)\frac{d^2y}{dx^2} + 2p'(x)\frac{dy}{dx} + p''(x)y$$
$$\frac{d}{dx}(q(x)y) = q(x)\frac{dy}{dx} + q'(x)y$$

and so the adjoint operator can be written in the alternative form

$$L^*(y) = p(x)\frac{d^2y}{dx^2} + (2p'(x) - q(x))\frac{dy}{dx} + (p''(x) - q'(x) + r(x))y.$$

In order for the operator L to be self-adjoint we equate the adjoint and original operator and require $L^*(y) = L(y)$. If $L(y) = L^*(y)$, then the coefficients in the differential operator L must satisfy

$$2p'(x) - q(x) = q(x) \quad \text{and} \quad p''(x) - q'(x) + r(x) = r(x)$$

or

$$p'(x) = q(x) \quad \text{and} \quad p''(x) = q'(x).$$

This implies a self-adjoint operator can be written in the form

$$L(y) = \frac{d}{dx}\left(p(x)\frac{dy}{dx}\right) + r(x)y. \tag{1.70}$$

or

$$L(y) = p(x)\frac{d^2y}{dx^2} + p'(x)\frac{dy}{dx} + r(x)y. \tag{1.71}$$

Observe that every linear second order differential equation

$$L_1(y) = a_0(x)\frac{d^2y}{dx^2} + a_1(x)\frac{dy}{dx} + a_2(x)y = 0 \tag{1.72}$$

can be written in self-adjoint form. The form of the differential equation (1.72) can be changed by multiplying the equation (1.72) by some nonzero function $\mu(x)$ to obtain

$$L(y) = \mu(x)a_0(x)\frac{d^2y}{dx^2} + \mu(x)a_1(x)\frac{dy}{dx} + \mu(x)a_2(x)y = 0 \tag{1.73}$$

and selecting the function $\mu(x)$ such that

$$\mu(x)a_0(x) = p(x) \tag{1.74}$$

$$\mu(x)a_1(x) = p'(x) \tag{1.75}$$

$$\mu(x)a_2(x) = r(x). \tag{1.76}$$

The resulting form of the equation (1.73) will then be a self-adjoint form. Differentiating equation (1.74) must give equation (1.75) and so μ must satisfy

$$p'(x) = \mu(x)a_0'(x) + \mu'(x)a_0(x) = \mu(x)a_1(x)$$

$$\text{or} \qquad \mu'(x)a_0(x) = (a_1(x) - a_0'(x))\mu(x)$$

$$\text{which implies} \quad \frac{\mu'(x)}{\mu(x)} = \frac{a_1(x)}{a_0(x)} - \frac{a_0'(x)}{a_0(x)}.$$

This can be integrated to produce $\ln\mu = \int \frac{a_1(x)}{a_0(x)} - \ln a_0(x)$ where constants of integration have been omitted since they will not affect the results. Solving for μ we find

$$\mu(x) = \frac{1}{a_0(x)}\exp\left(\int \frac{a_1(x)}{a_0(x)}\,dx\right), \qquad a_0(x) \neq 0. \tag{1.77}$$

This shows a general linear second order differential equation can always be written in self-adjoint form by selecting an appropriate multiplication factor and then relabeling the coefficients.

Example 1-3. (Bessel equation.)

Write the Bessel differential equation

$$x^2\frac{d^2y}{dx^2} + x\frac{dy}{dx} + (x^2 - \nu^2)y = 0, \qquad \nu \quad \text{constant},$$

in self-adjoint form.

Solution: Here $a_0(x) = x^2$, $a_1(x) = x$ and $a_2(x) = x^2 - \nu^2$ so the selected multiplication factor is $\mu = \frac{1}{x^2}\exp\left(\int \frac{x}{x^2}\,dx\right) = \frac{1}{x}$. Multiplying the given equation by μ produces the self-adjoint form

$$x\frac{d^2y}{dx^2} + \frac{dy}{dx} + (x - \frac{\nu^2}{x})y = 0$$

or

$$\frac{d}{dx}\left(x\frac{dy}{dx}\right) + (x - \frac{\nu^2}{x})y = 0.$$

∎

The self-adjoint operator L, given by equation (1.70), has the following property. For any two continuous and differential functions $u = u(x)$ and $v = v(x)$ there exists the Lagrange[†] identity

$$uL(v) - vL(u) = \frac{d}{dx}\left[p(x)\left(u\frac{dv}{dx} - v\frac{du}{dx}\right)\right] \tag{1.78}$$

which can be readily verified. An integration of the Lagrange identity produces the Green's[†] identity

$$\begin{aligned}\int_a^b [uL(v) - vL(u)]\,dx &= p(x)\left[u(x)v'(x) - v(x)u'(x)\right]_a^b \\ &= p(b)\left[u(b)v'(b) - v(b)u'(b)\right] \\ &\quad - p(a)\left[u(a)v'(a) - v(a)u'(a)\right].\end{aligned} \tag{1.79}$$

which has many applications.

Example 1-4. (Bessel function properties.) Assume $\beta \neq \alpha$, and show

$$(\beta^2 - \alpha^2)\int xJ_n(\alpha x)J_n(\beta x)\,dx = x\left[\alpha J_n'(\alpha x)J_n(\beta x) - \beta J_n'(\beta x)J_n(\alpha x)\right] + c$$

where c is a constant of integration.

[†] See Appendix C

Solution: The function $u = u(x) = J_n(\alpha x)$ is a solution of the Bessel equation

$$x^2 u'' + x u' + (\alpha^2 x^2 - n^2) u = 0 \qquad (1.80)$$

and the function $v = v(x) = J_n(\beta x)$ is a solution of the Bessel equation

$$x^2 v'' + x v' + (\beta^2 x^2 - n^2) v = 0. \qquad (1.81)$$

The equations (1.80) and (1.81) can be written in the self-adjoint forms as

$$L(u) = \frac{d}{dx}[xu'] - \frac{n^2}{x}u = -\alpha^2 x u \qquad \text{and}$$

$$L(v) = \frac{d}{dx}[xv'] - \frac{n^2}{x}v = -\beta^2 x v.$$

The Lagrange identity for the above self-adjoint operator L becomes

$$v L(u) - u L(v) = \frac{d}{dx}[x(vu' - uv')]. \qquad (1.82)$$

We calculate

$$v L(u) - u L(v) = -\alpha^2 x u v + \beta^2 x u v = (\beta^2 - \alpha^2) x u v, \quad \text{with}$$

$$u' = \frac{d}{dx}[J_n(\alpha x)] = \alpha J_n'(\alpha x) \quad \text{and}$$

$$v' = \frac{d}{dx}[J_n(\beta x)] = \beta J_n'(\beta x),$$

and find that equation (1.82) evaluates to

$$(\beta^2 - \alpha^2) x J_n(\alpha x) J_n(\beta x) = \frac{d}{dx}\{x[\alpha J_n(\beta x) J_n'(\alpha x) - \beta J_n(\alpha x) J_n'(\beta x)]\}. \qquad (1.83)$$

When equation (1.83) is integrated, the desired result is obtained, which is also given as entry 9 in the Table 1-1. ∎

Partial differential equations

A general m-th order partial differential equation in n-independent variables $x_1, x_2, \ldots, x_n$ and one dependent variable u can be written in the functional form

$$F\left(x_1, x_2, \ldots, x_n, u, \frac{\partial u}{\partial x_1}, \ldots, \frac{\partial u}{\partial x_n}, \frac{\partial^2 u}{\partial x_1^2}, \frac{\partial^2 u}{\partial x_1 \partial x_2}, \ldots, \frac{\partial^m u}{\partial x_n^m}\right) = 0. \qquad (1.84)$$

Here F represents some equation relating an unknown function $u = u(x_1, x_2, \ldots, x_n)$ and its derivatives. The partial differential equation is called an m-th ordered partial differential equation based upon the highest ordered derivative occurring in the equation. In equation (1.84) the variable x_n is sometimes replaced by a time variable t. Some examples of partial differential equations are given below.

$$L(u) = \frac{\partial u}{\partial x} + u = 0 \qquad \text{first order linear equation}$$
$$L(u) = \frac{\partial^2 u}{\partial t^2} - \frac{\partial^2 u}{\partial x^2} = 0 \qquad \text{second order linear equation}$$
$$L(u) = u\frac{\partial^2 u}{\partial x^2} + u^3 = 0 \qquad \text{second order nonlinear equation}$$
$$L(u) = \frac{\partial^2 u}{\partial x^2} + \frac{\partial^2 u}{\partial y^2} + \frac{\partial^2 u}{\partial z^2} = 0 \qquad \text{second order linear equation}$$
$$L(u) = x\frac{\partial^3 u}{\partial x^3} + \frac{\partial^2 u}{\partial y^2} + u = f(x,y) \qquad \text{third order linear equation}$$
$$L(u) = u\frac{\partial u}{\partial x} + \frac{\partial u}{\partial y} + u = x^2 \qquad \text{first order nonlinear equation}$$

Here $L(u)$ denotes a partial differential operator. If the operator L satisfies

$$L(c_1 u + c_2 v) = c_1 L(u) + c_2 L(v) \qquad (1.85)$$

for all constants c_1, c_2 and smooth‡ differentiable functions u, v, then the partial differential equation is called linear. If the partial differential equation is not linear it is called a nonlinear equation.

A linear first order partial differential equation in two independent variables x, y and dependent variable z has the general form

$$L(z) = A(x,y)\frac{\partial z}{\partial x} + B(x,y)\frac{\partial z}{\partial y} + C(x,y)z = G(x,y) \qquad x, y \in R \qquad (1.86)$$

where A, B, C, G are real functions of x, y which have continuous first derivatives over some region R of the x,y-plane. We assume the functions A and B cannot both be zero on the region R. Sometimes in applied problems one of the variables x or y is replaced by time t.

Example 1-5. (Special first order equation.)
Solve the partial differential equation $L(z) = \dfrac{\partial z}{\partial x} = 0$ for $z = z(x,y)$.
Solution: The partial derivative sign indicates the presence of more than one independent variable and so an integration gives $z = z(x,y) = f(y)$ where $f(y)$ is an arbitrary function of y. Here $f(y)$ is considered as a constant when there is a differentiation with respect to the x-variable. This simple example indicates that general solutions to some partial differential equations can differ from general solutions to ordinary differential equations by the fact that arbitrary functions are used instead of arbitrary constants.

■

---

‡ A function $f(x)$ that has a continuous derivative over an interval $[a, b]$ is called smooth over the interval.

The operator $L(\ )$ in equation (1.86) is a linear operator because it satisfies the equation (1.85) for all differentiable functions u, v and constants c_1, c_2. The equation $L(z) = 0$ is called a homogeneous equation while the equation $L(z) = G$ is called a nonhomogeneous equation. A function $z = z(x, y)$ with continuous partial derivatives is called a solution of the equation (1.86) if it satisfies the partial differential equation for all x, y in the region R. The general procedure for solving equation (1.86) is to first solve the homogeneous equation $L(z) = 0$ and call the solution z_H. If this homogeneous solution z_H contains an arbitrary function such that $L(z_H) = 0$ for all arbitrary functions, then the solution is called a general solution to the homogeneous equation. Next find any particular solution z_P to the nonhomogeneous equation $L(z) = G$, then $z = z_H + z_P$ is a general solution of the equation (1.86).

Example 1-6.   (First order equation)
   Solve the partial differential equation

$$L(u) = \frac{\partial z}{\partial x} + z = x^2 \qquad \text{for } z = z(x, y)$$

Solution: We first solve the homogeneous equation $\dfrac{\partial z}{\partial x} + z = 0$. Here we hold y-constant and treat the partial differential equation like an ordinary differential equation. There is an integrating factor e^x so the equation can be written in the form of an exact differential

$$e^x \frac{\partial z}{\partial x} + e^x z = \frac{\partial}{\partial x} (e^x z) = 0.$$

This equation is easily integrated to obtain the solution

$$e^x z = f(y) \quad \text{or} \quad z = z_H = e^{-x} f(y)$$

where $f(y)$ is an arbitrary function of y. By the method of undetermined coefficients we assume a particular solution of the form $z_P = ax^2 + bx + c$ where a, b, c are undetermined coefficients. Substituting z_P into the nonhomogeneous equation gives

$$\frac{\partial z}{\partial x} + z = 2ax + b + ax^2 + bx + c = x^2.$$

Equating like powers of x gives the system of equations

$$a = 1 \qquad 2a + b = 0 \qquad b + c = 0$$

with solution $a = 1, b = -2, c = 2$. This gives the particular solution $z_P = x^2 - 2x + 2$ and the general solution

$$z = z_H + z_P = e^{-x}f(y) + x^2 - 2x + 2$$

where $f(y)$ is treated as an arbitrary smooth function. ∎

Characteristic curves

First order partial differential equations of the form

$$A(x, y, z)\frac{\partial z}{\partial x} + B(x, y, z)\frac{\partial z}{\partial y} = C(x, y, z) \tag{1.87}$$

are called quasilinear equations of which the linear equation (1.86) is a special case. Associated with this quasilinear partial differential equation is the vector field

$$\vec{V} = \vec{V}(x, y, z) = A(x, y, z)\hat{i} + B(x, y, z)\hat{j} + C(x, y, z)\hat{k}. \tag{1.88}$$

where $\hat{i}$, $\hat{j}$, $\hat{k}$ are unit vectors in the directions of the x, y and z axes respectively. Think of the solution $z = z(x, y)$ to equation (1.87) as a smooth surface[‡] with normal vector

$$\vec{N} = \frac{\partial z}{\partial x}\hat{i} + \frac{\partial z}{\partial y}\hat{j} - \hat{k} \tag{1.89}$$

at some point (x, y, z) on the surface. One can then view the partial differential equation (1.87) as representing the vector dot product $\vec{N} \cdot \vec{V} = 0$. This dot product gives the physical interpretation that the vector $\vec{V}$ lies in the tangent plane to the point on the surface where the normal vector is constructed. If there exists a curve C on the surface $z = z(x, y)$ having the property that for each point P on the curve C, the tangent to C at point P has the same direction as the vector field $\vec{V}$ at the point P, then the curve is called a characteristic curve. That is, a characteristic curve C is a field line associated with the vector field $\vec{V} = \vec{V}(x, y, z)$. For $\vec{r} = \vec{r}(t) = x\hat{i} + y\hat{j} + z\hat{k}$ the position vector to a point P on a characteristic curve C, the tangent direction is given by $\frac{d\vec{r}}{dt} = \frac{dx}{dt}\hat{i} + \frac{dy}{dt}\hat{j} + \frac{dz}{dt}\hat{k}$ and must have the same direction as the vector field $\vec{V}$ at the point P. Therefore, $\frac{d\vec{r}}{dt}$ must be proportional to $\vec{V}$ and we can write

$$\frac{d\vec{r}}{dt} = K\vec{V} \quad \text{or} \quad \frac{dx}{dt}\hat{i} + \frac{dy}{dt}\hat{j} + \frac{dz}{dt}\hat{k} = K\left[A(x, y, z)\hat{i} + B(x, y, z)\hat{j} + C(x, y, z)\hat{k}\right]$$

where K is a proportionality constant. Equating like components gives

---

[‡] A smooth surface has at each point a well defined normal which varies continuously over points of the surface.

$$\frac{dx}{dt} = KA(x,y,z), \qquad \frac{dy}{dt} = KB(x,y,z), \qquad \frac{dz}{dt} = KC(x,y,z) \qquad (1.90)$$

or the ratios

$$\frac{dx}{A(x,y,z)} = \frac{dy}{B(x,y,z)} = \frac{dz}{C(x,y,z)} = K\,dt \qquad (1.91)$$

represent differential equations satisfied by the characteristic curves or field lines. The equations (1.91) are called subsidiary equations associated with the partial differential equation (1.87) and define the characteristic curves associated with the partial differential equation. If $x = x(t)$ and $y = y(t)$ are solutions of the subsidiary equations (1.90), then on the surface $z = z(x,y)$ we have

$$\frac{dz}{dt} = \frac{\partial z}{\partial x}\frac{dx}{dt} + \frac{\partial z}{\partial y}\frac{dy}{dt} = K\left(A(x,y,z)\frac{\partial z}{\partial x} + B(x,y,z)\frac{\partial z}{\partial x}\right) = KC(x,y,z).$$

This shows the characteristic curves defined by the first two equations of (1.90) must satisfy the third equation. The characteristic curves lie upon the solution surface and satisfy the equation describing the surface.

Using the equations (1.91) we try to obtain two linearly independent solutions of the subsidiary equations

$$u_1(x,y,z) = c_1 \qquad u_2(x,y,z) = c_2$$

which represent two families of surfaces. One way to determine if the solutions are linearly independent is to check the cross product $\nabla u_1 \times \nabla u_2$ at common points of intersection of the above surfaces. The gradients $\nabla u_1 = \operatorname{grad} u_1$ and $\nabla u_2 = \operatorname{grad} u_2$ represent normals to the surfaces $u_1 = c_1$ and $u_2 = c_2$. For linear independence of the solutions we don't want both normals in the same direction at a common point of intersection of the surfaces, hence we require $\nabla u_1 \times \nabla u_2 \neq 0$. Observe when the two independent families of surfaces intersect there is created a bundle of characteristic curves or field lines. These characteristic curves result because the values for c_1 and c_2 vary. That is, for every combination of c_1, c_2 there results a characteristic curve. The figure 1-10(a) is an attempt to get you to visualize a bundle of these characteristic curves cut by a plane. Each dot represents a characteristic curve. These curves fill up the space. The solution $z = z(x,y)$ of the partial differential equation (1.87) is thus built up from some combinations of these characteristic curves. One can say the characteristic curves are the building blocks which construct the surface. The general solution can be represented in any of the forms

$$F(u_1, u_2) = 0 \qquad u_1 = f(u_2) \qquad u_2 = g(u_1)$$

where F, f and g represent arbitrary functions.

Example 1-7. **(First order equation)**

Solve the partial differential equation $x\dfrac{\partial z}{\partial x} - y\dfrac{\partial z}{\partial y} = x^2 y$ for $z = z(x,y)$.

Solution: The subsidiary equations associated with the given partial differential equation are

$$\frac{dx}{x} = \frac{dy}{-y} = \frac{dz}{x^2 y}.$$

From the first two terms we have the differential equation

$$\frac{dx}{x} = \frac{dy}{-y} \quad \text{or} \quad \ln x + \ln y = \ln c_1.$$

This gives the solution $u_1 = u_1(x,y,z) = xy = c_1$ as a family of cylinders. From the ratios

$$\frac{dx}{x} = \frac{dz}{x^2 y}$$

we find, using the previous result $xy = c_1$, the differential form $c_1 dx = dz$. This equation is easily integrated to produce the family of solution surfaces given by $u_2 = u_2(x,y,z) = z - c_1 x = z - x^2 y = c_2$. The solution can now be represented in several different ways. One form is

$$F(xy, z - x^2 y) = 0$$

while other forms are

$$z = f(xy) + x^2 y \quad \text{or} \quad xy = g(z - x^2 y)$$

where F, f and g represent arbitrary functions.

Let us check the solution $F(xy, z - x^2 y) = 0$ and see if it satisfies the given equation. Let $u = xy$ and $v = z - x^2 y$ and treat z as a function of x and y. We differentiate the solution function F with respect to x and y to obtain the derivatives

$$\frac{\partial F}{\partial u}\left(\frac{\partial u}{\partial x} + \frac{\partial u}{\partial z}\frac{\partial z}{\partial x}\right) + \frac{\partial F}{\partial v}\left(\frac{\partial v}{\partial x} + \frac{\partial v}{\partial z}\frac{\partial z}{\partial x}\right) = 0$$

$$\frac{\partial F}{\partial u}\left(\frac{\partial u}{\partial y} + \frac{\partial u}{\partial z}\frac{\partial z}{\partial y}\right) + \frac{\partial F}{\partial v}\left(\frac{\partial v}{\partial y} + \frac{\partial v}{\partial z}\frac{\partial z}{\partial y}\right) = 0.$$

(1.92)

Now eliminate $\frac{\partial F}{\partial u}$ and $\frac{\partial F}{\partial v}$ from these equations to obtain

$$\frac{\partial(u,v)}{\partial(z,y)}\frac{\partial z}{\partial x} + \frac{\partial(u,v)}{\partial(x,z)}\frac{\partial z}{\partial y} = \frac{\partial(u,v)}{\partial(y,x)}$$

(1.93)

where

$$\frac{\partial(u,v)}{\partial(\xi,\eta)} = \begin{vmatrix} \frac{\partial u}{\partial \xi} & \frac{\partial u}{\partial \eta} \\ \frac{\partial v}{\partial \xi} & \frac{\partial v}{\partial \eta} \end{vmatrix} = \frac{\partial u}{\partial \xi}\frac{\partial v}{\partial \eta} - \frac{\partial u}{\partial \eta}\frac{\partial v}{\partial \xi} \qquad (1.94)$$

denotes the Jacobian of u, v with respect to ξ, η. It is readily verified that the equation (1.93) simplifies to the given partial differential equation.

To verify the solution $z = x^2 y + f(xy)$, for arbitrary f, is a solution of the given partial differential equation we form the derivatives

$$\frac{\partial z}{\partial x} = 2xy + f'(xy)y \qquad\qquad \frac{\partial z}{\partial y} = x^2 + f'(xy)x$$

where prime denotes differentiation with respect to the argument of the function. We substitute z and the derivatives into the partial differential equation to show the equation is satisfied.

∎

The problem of obtaining two independent integrals to the system of differential equations (1.91) can at times be difficult. One can always use the theory of proportions to construct other differential equations. Recall the theory of proportions,

$$\text{if} \quad \frac{dx}{A} = \frac{dy}{B} = \frac{dz}{C}, \quad \text{then} \quad \frac{\lambda dx + \mu dy + \nu dz}{\lambda A + \mu B + \nu C} = \frac{dx}{A} = \frac{dy}{B} = \frac{dz}{C}$$

for arbitrary constants λ, μ, ν. In this way other differential ratios can be formed that may or may not be easier to integrate.

Example 1-8. (First order equation)
Solve the quasilinear equation $z\dfrac{\partial z}{\partial x} + y\dfrac{\partial z}{\partial y} = x$.
Solution: Here the subsidiary equations are

$$\frac{dx}{z} = \frac{dy}{y} = \frac{dz}{x} = K\,dt.$$

Using the theory of proportions we obtain

$$\frac{dx + dz}{x + z} = \frac{dy}{y}$$

which can be integrated to produce the solution family of surfaces $\dfrac{x+z}{y} = c_1$.
From the ratios

$$\frac{dx}{z} = \frac{dz}{x} \quad \text{we write} \quad x\,dx = z\,dz$$

which can be easily integrated to obtain a second independent family of surfaces $x^2 - z^2 = c_2$. The intersection of these two families produces the field lines or characteristic curves from which specialized solutions can be constructed. The general solution can then be written in either of the forms

$$F(\frac{x+z}{y}, x^2 - z^2) = 0 \qquad \text{or} \qquad \frac{x+z}{y} = f(x^2 - z^2)$$

where F and f represent arbitrary differentiable functions.

■

Example 1-9. (First order equation.)

Solve the first order linear equation $\qquad \alpha \frac{\partial z}{\partial x} + \beta \frac{\partial z}{\partial y} + \gamma z = 0$

where α, β, γ are constants.

Solution: The subsidiary equations are $\dfrac{dx}{\alpha} = \dfrac{dy}{\beta} = \dfrac{dz}{-\gamma z}$

Case 1: $(\alpha \neq 0)$ Two independent solutions of the subsidiary equations are

$$\beta x - \alpha y = c_1 \qquad \text{and} \qquad z e^{\frac{\gamma}{\alpha} x} = c_2.$$

This gives the general solution $c_2 = f(c_1)$ or

$$z = e^{-\frac{\gamma}{\alpha} x} f(\beta x - \alpha y).$$

Alternatively, on can select the solutions

$$\beta x - \alpha y = c_1 \qquad \text{and} \qquad z e^{\frac{\gamma}{\beta} y} = c_3$$

which gives the general solution $c_3 = f(c_1)$ or

$$z = e^{-\frac{\gamma}{\beta} y} f(\beta x - \alpha y) \quad \beta \neq 0 \tag{1.95}$$

where f represents an arbitrary function.

Case 2: $(\alpha = 0)$ The subsidiary equations

$$\frac{dx}{0} = \frac{dy}{\beta} = \frac{dz}{-\gamma z}$$

produce the solutions $x = c_1$ and $z e^{\frac{\gamma}{\beta} y} = c_2$ which gives the general solution $c_2 = g(c_1)$ or

$$z = e^{-\frac{\gamma}{\beta} y} g(x), \quad \beta \neq 0 \tag{1.96}$$

where g represents an arbitrary function. The equation (1.96) is a special case of equation (1.95) when $\alpha = 0$, where $f(\beta x) = g(x)$ is an arbitrary function. These solutions can be verified by substitution into the given equations to show they are satisfied.

■

Example 1-10. **(First order equation)**

Solve the first order partial differential equation $(y - z)\dfrac{\partial z}{\partial x} + (z - x)\dfrac{\partial z}{\partial y} = x - y$

Solution: The subsidiary equations are

$$\frac{dx}{y - z} = \frac{dy}{z - x} = \frac{dz}{x - y} = K\, dt.$$

Note that

$$x\, dx + y\, dy + z\, dz = 0 \quad \text{and} \quad dx + dy + dz = 0.$$

This gives the independent solutions

$$x^2 + y^2 + z^2 = c_1 \quad \text{and} \quad x + y + z = c_2.$$

The general solution can therefore be expressed in either of the forms

$$x + y + z = f(x^2 + y^2 + z^2) \quad \text{or} \quad F(x + y + z, x^2 + y^2 + z^2) = 0$$

where f, F represent arbitrary functions.

■

Example 1-11. **(First order equation)**

We can turn the previous example around. Given the general function $x + y + z = f(x^2 + y^2 + z^2)$, where f is an arbitrary smooth function, then find the partial differential equation satisfied by this function.

Solution: We treat z as a function of x, y and use implicit differentiation to obtain the derivatives

$$
\begin{aligned}
x + y + z &= f(x^2 + y^2 + z^2) \\
1 + \frac{\partial z}{\partial x} &= f'(x^2 + y^2 + z^2)\left(2x + 2z\frac{\partial z}{\partial x}\right) \\
1 + \frac{\partial z}{\partial y} &= f'(x^2 + y^2 + z^2)\left(2y + 2z\frac{\partial z}{\partial y}\right).
\end{aligned}
\tag{1.97}
$$

Now use algebra to eliminate the arbitrary functions f, f' from the equations (1.97) to obtain the differential equation given in the previous example.

If the general function is given in the form $F(x+y+z, x^2 + y^2 + z^2) = 0$, where F is arbitrary, we still treat z as a function of x, y and use implicit differentiation.

Let $u = x + y + z$ and $v = x^2 + y^2 + z^2$, then differentiate the given general solution with respect to x and y giving

$$\frac{\partial F}{\partial u}\left(1 + \frac{\partial z}{\partial x}\right) + \frac{\partial F}{\partial v}\left(2x + 2z\frac{\partial z}{\partial x}\right) = 0$$
$$\frac{\partial F}{\partial u}\left(1 + \frac{\partial z}{\partial y}\right) + \frac{\partial F}{\partial v}\left(2y + 2z\frac{\partial z}{\partial y}\right) = 0.$$
(1.98)

Now use algebra to eliminate the partial derivatives $\frac{\partial F}{\partial u}$ and $\frac{\partial F}{\partial v}$ to obtain the partial differential equation.

∎

Cauchy problem.

The Cauchy problem associated with quasilinear first order partial differential equations in two independent variables is to find a surface $z = z(x,y)$ such that the function z satisfies the given partial differential equation and also passes through a given space curve Γ. The space curve Γ is assumed to be a noncharacteristic curve, for if it were a characteristic curve the Cauchy problem might be unsolvable or an infinite number of solutions might result. The prescribed space curve Γ is usually converted to a parametric form

$$x = x(\tau), \qquad y = y(\tau), \qquad z = z(\tau).$$
(1.99)

A Cauchy problem is characterized by two parts: Part(a) we are required to obtain a solution to a first order quasilinear equation and then part(b) construct a particular solution which passes through a prescribed curve Γ. To accomplish this we first solve the subsidiary equations and construct two independent families of solution surfaces

$$u_1 = u_1(x,y,z) = c_1, \qquad u_2 = u_2(x,y,z) = c_2.$$
(1.100)

This produces the general solution

$$F(c_1, c_2) = F(u_1(x,y,z), u_2(x,y,z)) = 0$$
(1.101)

where F is an arbitrary function. If the solution surface is to contain the prescribed curve Γ, then when the equations (1.99) are substituted into the equation (1.101) we require that the function F be selected such that

$$F(u_1(x(\tau),y(\tau),z(\tau)), u_2(x(\tau),y(\tau),z(\tau))) = 0.$$

This is equivalent to substituting the equations (1.99) into the equations (1.100) to obtain

$$u_1 = u_1(x(\tau), y(\tau), z(\tau)) = c_1, \qquad u_2 = u_2(x(\tau), y(\tau), z(\tau)) = c_2. \qquad (1.102)$$

and then eliminating the parameter τ between these equations to obtain a specific function F, such that $F(c_1, c_2) = 0$, between the constants c_1 and c_2. The desired solution surface is then given by equation (1.101). If Γ is a characteristic curve, then it is possible that one will not be able to eliminate the parameter τ from the equations (1.102).

Example 1-12. (Cauchy problem)
Find the solution surface of the partial differential equation

$$x\frac{\partial z}{\partial x} + y\frac{\partial z}{\partial y} = z$$

which passes through the curve $z(x, 1) = \sin x$.
Solution: The subsidiary equations are

$$\frac{dx}{x} = \frac{dy}{y} = \frac{dz}{z}$$

with independent solutions

$$u_1 = \frac{x}{y} = c_1, \qquad u_2 = \frac{x}{z} = c_2. \qquad (1.103)$$

One parametric representation of the prescribed space curve Γ is

$$x = t, \qquad y = 1, \qquad z = \sin t. \qquad (1.104)$$

Substituting these values into the equations (1.103) we obtain

$$t = c_1, \qquad t = c_2 \sin t.$$

We now eliminate t from these equations and find a relation between c_1 and c_2. We find

$$c_1 = c_2 \sin c_1.$$

This gives the solution surface which contains the given curve Γ. Substituting for c_1 and c_2 from the equations (1.103) gives the solution surface

$$\frac{x}{y} = \frac{x}{z}\sin\left(\frac{x}{y}\right) \quad \text{or} \quad z = y\sin\left(\frac{x}{y}\right).$$

Suppose the integrals

$$\frac{x}{y} = c_1 \qquad \text{and} \qquad \frac{y}{z} = c_3$$

were selected as independent integrals of the subsidiary equations. In this case when we substitute in the values of x, y, z from the equations (1.104) we obtain

$$t = c_1, \qquad 1 = c_3 \sin t.$$

Eliminating t from these equations produces the same solution surface as above.
Similarly, if the surfaces

$$\frac{y}{z} = c_3, \qquad \frac{x}{z} = c_2$$

are selected as two independent integrals of the subsidiary equations, then substituting the values for x, y, z from equations (1.104) and eliminating the parameter t from the resulting equations we again obtain the same solution surface as above.

One can alternatively write the general solution in the form

$$\frac{x}{z} = f(\frac{x}{y}) \qquad (1.105)$$

where f is arbitrary. If we substitute the parametric values for x, y, z into this equation we find the function f must have the form

$$f(t) = \frac{t}{\sin t}.$$

By using this value of f in the equation (1.105) we obtain after simplification the solution surface $z = y \sin \left(\frac{x}{y} \right)$. ∎

Example 1-13. (Cauchy problem)
Find the solution surface of the partial differential equation

$$x \frac{\partial z}{\partial x} + y \frac{\partial z}{\partial y} = z$$

which passes through the curve $x = t$, $y = t$, $z = t$.
Solution: This is the same problem as the previous example with a different curve Γ. We select the first set of independent solutions from the above example.

These are $\dfrac{x}{y} = c_1,$ $\dfrac{x}{z} = c_2.$ For this example the given curve Γ is a characteristic curve formed by the intersection of the planes $\dfrac{x}{y} = c_1,$ $\dfrac{x}{z} = c_2$ when $c_1 = 1$ and $c_2 = 1.$ Hence, when we substitute the parametric values into these equations we obtain $1 = c_1$ $1 = c_2$ and so we cannot eliminate t from these equations. This was why we replaced the restriction on the Cauchy problem that the given curve Γ be noncharacteristic. Also note from the theory of proportions we can form from the subsidiary equations

$$\frac{\alpha dx + \beta dy}{\alpha x + \beta y} = \frac{dz}{z}$$

for arbitrary constants α and $\beta.$ An integration produces the solution surface $z = \alpha_0 x + \beta_0 y$ where α_0, β_0 are new arbitrary constants. If $x = t, y = t, z = t$ is to be a characteristic curve then the constants α_0 and β_0 must satisfy $\alpha_0 + \beta_0 = 1.$ This gives an infinite number of solution surfaces. Another reason for requiring the given curve Γ be noncharacteristic for Cauchy problems.

∎

Example 1-14. (Cauchy problem and characteristic curves.)
Find the solution surface $z = z(x, y)$ which satisfies

$$x\frac{\partial z}{\partial x} + y\frac{\partial z}{\partial y} = 1$$

which passes through the curve $x = 1,$ $y = \tau,$ $z = \sin \tau.$
Solution: The characteristic curves which make up the solution surface satisfy the subsidiary equations (1.90) which for this problem are given by

$$\frac{dx}{dt} = x, \quad \frac{dy}{dt} = y, \quad \frac{dz}{dt} = 1 \tag{1.106}$$

where we have selected $K = 1.$ To see how the solution surface is constructed from characteristic curves we will draw some characteristic curves coming off of the given initial curve. We select the initial condition for the above differential equations such that at $t = 0$ we will have $x = 1,$ $y = \tau,$ $z = \sin \tau.$ Solving the equations (1.106) subject to these initial conditions produces the solutions

$$x = e^t, \quad y = \tau e^t, \quad z = t + \sin \tau. \tag{1.107}$$

The surface $\vec{r} = \vec{r}(t, \tau) = e^t \hat{i} + \tau e^t \hat{j} + (t + \sin \tau)\hat{k}$ for $0 \le t \le 1$ and $0 \le \tau \le 2\pi$ is illustrated in the figure 1-10 where the heavy curve denotes the given initial sine curve in the plane $x = 1$ and the curves emanating from this curve are the characteristics curves associated with the partial differential equation. The figure 1-10 shows how the solution surface is constructed from these characteristic curves. By eliminating t and τ from the equations (1.107) we obtain the surface $z = \ln x + \sin \dfrac{y}{x}$ which satisfies the Cauchy problem.

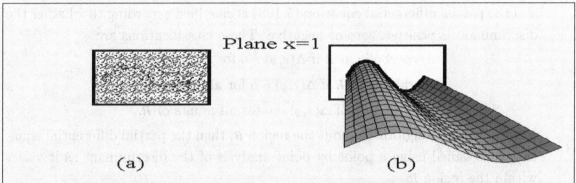

Plane x=1

(a) (b)

Figure 1-10 (a) Section of the plane $x = 1$ intersecting characteristics curves.
(b) Surface constructed using characteristic curves through sine curve.

Linear second order equations

Linear second order partial differential equations in two independent variables x, y and dependent variable u have the general form

$$L(u) = A(x,y)\frac{\partial^2 u}{\partial x^2} + 2B(x,y)\frac{\partial^2 u}{\partial x \partial y} + C(x,y)\frac{\partial^2 u}{\partial y^2}$$
$$+ D(x,y)\frac{\partial u}{\partial x} + E(x,y)\frac{\partial u}{\partial y} + F(x,y)u = G(x,y) \qquad x,y \in R \qquad (1.108)$$

for $x, y \in R$ where A, B, C, D, E, F, G are real valued functions which possess second derivatives which are continuous on the region R where the solution is desired. Special cases of the above equation are the heat equation, the wave equation and Laplace's[†] equation which have the forms

$$L(u) = \frac{\partial u}{\partial t} - K\frac{\partial^2 u}{\partial x^2} = G(x,t) \qquad \text{Heat equation,} \quad K \text{ constant.}$$

$$L(u) = \frac{\partial^2 u}{\partial t^2} - \alpha^2\frac{\partial^2 u}{\partial x^2} = G(x,t) \qquad \text{Wave equation,} \quad \alpha^2 \text{ constant.}$$

$$L(u) = \frac{\partial^2 u}{\partial x^2} + \frac{\partial^2 u}{\partial y^2} = 0 \qquad \text{Laplace's equation}$$

---

† See Appendix C

Observe in the heat equation and wave equation the time variable t has replaced y in equation (1.108).

The partial differential equation (1.108) is classified in a manner similar to how conic sections are classified. The discriminant of the operator L is defined

$$\Delta = \Delta(x,y) = B^2(x,y) - A(x,y)C(x,y) \qquad (1.109)$$

and the partial differential equation (1.108) is classified according to whether the discriminant is positive, zero or negative. These classifications are

Hyperbolic in R if $\Delta(x,y) > 0$ for all points of R.

Parabolic in R if $\Delta(x,y) = 0$ for all points of R.

Elliptic in R if $\Delta(x,y) < 0$ for all points of R.

If $\Delta(x,y)$ changes sign throughout the region R, then the partial differential equation is classified using a point by point analysis of the discriminant as it varies within the region R.

The above classifications of the partial differential equation (1.108) remain invariant under a single-valued continuous transformation

$$\xi = \xi(x,y), \qquad \eta = \eta(x,y) \qquad (1.110)$$

with Jacobian

$$J = \frac{\partial(\xi,\eta)}{\partial(x,y)} = \begin{vmatrix} \frac{\partial \xi}{\partial x} & \frac{\partial \xi}{\partial y} \\ \frac{\partial \eta}{\partial x} & \frac{\partial \eta}{\partial y} \end{vmatrix}$$

different from zero. This assures the existence of an inverse transformation

$$x = x(\xi,\eta), \qquad y = y(\xi,\eta).$$

The invariance of the classifications of linear second order partial differential equation is demonstrated by representing the equation (1.108) in terms of the new variables ξ and η. We treat $u = u(\xi,\eta)$ and use the subscript notation

$$u_x = \frac{\partial u}{\partial x}, \quad u_y = \frac{\partial u}{\partial y}, \quad u_{xy} = \frac{\partial^2 u}{\partial x \partial y}, \quad \xi_x = \frac{\partial \xi}{\partial x}, \quad \xi_y = \frac{\partial \xi}{\partial y}, \quad \xi_{xx} = \frac{\partial^2 \xi}{\partial x^2}, \quad \xi_{yy} = \frac{\partial^2 \xi}{\partial y^2}, \quad \text{etc.}$$

to denote partial derivatives. We then substitute the derivatives

$$u_x = u_\xi \xi_x + u_\eta \eta_x$$

$$u_{xx} = u_\xi \xi_{xx} + (u_{\xi\xi}\xi_x + u_{\xi\eta}\eta_x)\xi_x + u_\eta \eta_{xx} + (u_{\eta\xi}\xi_x + u_{\eta\eta}\eta_x)\eta_x$$

$$u_y = u_\xi \xi_y + u_\eta \eta_y$$

$$u_{yy} = u_\xi \xi_{yy} + (u_{\xi\xi}\xi_y + u_{\xi\eta}\eta_y)\xi_y + u_\eta \eta_{yy} + (u_{\eta\xi}\xi_y + u_{\eta\eta}\eta_y)\eta_y$$

$$u_{xy} = u_\xi \xi_{xy} + (u_{\xi\xi}\xi_y + u_{\xi\eta}\eta_y)\xi_x + u_\eta \eta_{xy} + (u_{\eta\xi}\xi_y + u_{\eta\eta}\eta_y)\eta_x$$

into equation (1.108) to express its form in terms of the new variables ξ and η. We find this new form is

$$L(u) = \bar{A}\frac{\partial^2 u}{\partial \xi^2} + 2\bar{B}\frac{\partial^2 u}{\partial \xi \partial \eta} + \bar{C}\frac{\partial^2 u}{\partial \eta^2} + \bar{D}\frac{\partial u}{\partial \xi} + \bar{E}\frac{\partial u}{\partial \eta} + \bar{F}u = \bar{G} \qquad (1.111)$$

where

$$\begin{aligned}
\bar{A} &= A\xi_x^2 + 2B\xi_x\xi_y + C\xi_y^2 \\
\bar{B} &= A\xi_x\eta_x + B(\xi_x\eta_y + \xi_y\eta_x) + C\xi_y\eta_y \\
\bar{C} &= A\eta_x^2 + 2B\eta_x\eta_y + C\eta_y^2 \\
\bar{D} &= D\xi_x + A\xi_{xx} + 2B\xi_{xy} + E\xi_y + C\xi_{yy} \\
\bar{E} &= D\eta_x + A\eta_{xx} + 2B\eta_{xy} + E\eta_y + C\eta_{yy} \\
\bar{F} &= F \\
\bar{G} &= G
\end{aligned}$$

where A, B, C, D, E, F, G are also converted to the ξ, η coordinates. It is now an algebraic exercise to show

$$(\bar{B}^2 - \bar{A}\bar{C}) = \left[\frac{\partial(\xi, \eta)}{\partial(x, y)}\right]^2 (B^2 - AC) \qquad \text{or} \qquad \bar{\Delta} = \left[\frac{\partial(\xi, \eta)}{\partial(x, y)}\right]^2 \Delta. \qquad (1.112)$$

The equation (1.112) shows that if the Jacobian $\frac{\partial(\xi, \eta)}{\partial(x, y)} \neq 0$, then the discriminant keeps the same sign in the ξ, η coordinate system. The heat equation, wave equation and Laplace's equation are special forms associated with the parabolic, hyperbolic and elliptic classifications respectively. These special forms are called canonical forms.

Canonical forms

Define

$$f(\xi, \eta) = A\xi_x\eta_x + B(\xi_x\eta_y + \xi_y\eta_x) + C\xi_y\eta_y \qquad (1.113)$$

then the transformed partial differential equation (1.111) can be written in the form

$$\begin{aligned}
L(u) = f(\xi, \xi)\frac{\partial^2 u}{\partial \xi^2} + f(\xi, \eta)\frac{\partial^2 u}{\partial \xi \partial \eta} + f(\eta, \eta)\frac{\partial^2 u}{\partial \eta^2} \\
+ \bar{D}\frac{\partial u}{\partial \xi} + \bar{E}\frac{\partial u}{\partial \eta} + \bar{F}u = \bar{G}.
\end{aligned} \qquad (1.114)$$

The equation

$$f(\psi, \psi) = A\psi_x^2 + 2B\psi_x\psi_y + C\psi_y^2 = 0 \qquad (1.115)$$

is called the characteristic equation associated with the operator L of equation (1.114) and the sum of the first three terms in equation (1.114)

$$f(\xi,\xi)\frac{\partial^2 u}{\partial \xi^2} + f(\xi,\eta)\frac{\partial^2 u}{\partial \xi \partial \eta} + f(\eta,\eta)\frac{\partial^2 u}{\partial \eta^2} \tag{1.116}$$

is called the principal part of the operator L. By analyzing the characteristic equation (1.115) we find it is possible to control the form given by the principal part of the operator L. The principal part can be made to take on certain standardized forms called canonical forms. We consider the following cases:

Case 1: $(A \neq 0,\ A > 0)$ In the special case $A \neq 0$ we can factor the characteristic equation and write it in the form

$$f(\psi,\psi) = A(\psi_x - \alpha_1(x,y)\psi_y)(\psi_x - \alpha_2(x,y)\psi_y) = 0 \tag{1.117}$$

where $\alpha_1(x,y)$ and $\alpha_2(x,y)$ are the roots of the equation $A\alpha^2 + 2B\alpha + C = 0$ given by

$$\alpha_1(x,y) = \frac{-B(x,y) + \sqrt{\Delta(x,y)}}{A(x,y)}, \qquad \alpha_2(x,y) = \frac{-B(x,y) - \sqrt{\Delta(x,y)}}{A(x,y)}$$

which have the sum and product relations

$$\alpha_1 + \alpha_2 = \frac{-2B}{A} \qquad \text{and} \qquad \alpha_1 \alpha_2 = \frac{C}{A}.$$

The equation (1.117) implies that if we select the function ξ or η to satisfy the characteristic equation (1.115) then the coefficient terms $\bar{A} = f(\xi,\xi) = 0$ and $\bar{C} = f(\eta,\eta) = 0$. Observe that the equations

$$\bar{A} = f(\xi,\xi) = A\xi_x^2 + 2B\xi_x\xi_y + C\xi_y^2 = 0$$
$$\bar{C} = f(\eta,\eta) = A\eta_x^2 + 2B\eta_x\eta_y + C\eta_y^2 = 0$$

can be written in the alternate form

$$A\left(\xi_x/\xi_y\right)^2 + 2B\left(\xi_x/\xi_y\right) + C = 0$$
$$A\left(\eta_x/\eta_y\right)^2 + 2B\left(\eta_x/\eta_y\right) + C = 0$$

from which one can obtain the roots

$$\xi_x/\xi_y = \alpha_1 = \frac{-B + \sqrt{B^2 - AC}}{A}$$
$$\eta_x/\eta_y = \alpha_2 = \frac{-B - \sqrt{B^2 - AC}}{A}$$

from which the functions ξ and η can be determined.

In order for ξ or η to satisfy the characteristic equation they must be selected to satisfy one of the factors associated with equation (1.117). This requires ξ and η be selected as solutions of the first order partial differential equations

$$\xi_x - \alpha_1(x,y)\xi_y = 0, \qquad \eta_x - \alpha_2(x,y)\eta_y = 0 \qquad (1.118)$$

The partial differential equation for ξ has the subsidiary equations

$$\frac{dx}{1} = \frac{dy}{-\alpha_1(x,y)} = \frac{d\xi}{0}$$

which define the characteristic curves. Denote by $u(x,y) = c_1$ the solution of the differential equation $\frac{dy}{dx} = -\alpha_1(x,y)$. The ratio $\frac{d\xi}{0}$ implies $\xi = c_2 = $ constant, is a second independent solution. These functions produce the solution $\xi = g(u(x,y))$, where g represents an arbitrary function.

Similarly the first order partial differential equation for η has the subsidiary equations

$$\frac{dx}{1} = \frac{dy}{-\alpha_2(x,y)} = \frac{d\eta}{0}.$$

Let $v(x,y) = c_1$ denote the solution of the differential equation $\frac{dy}{dx} = -\alpha_2(x,y)$ combined with $\eta = c_2$ enables one to construct the general solution $\eta = g(v(x,y))$, where g represents an arbitrary function. The simplest solutions to the partial differential equations for ξ and η are obtained by selecting $g(x) = x$. This gives the special transformation equations

$$\xi = u(x,y), \qquad \eta = v(x,y) \qquad (1.119)$$

which represents a two parameter family of characteristics for hyperbolic equations. Note that the transformation equations which convert second order partial differential equations to canonical forms are not unique. The transformation equations (1.119) will then reduce the equation (1.108) to the canonical form

$$2f(\xi,\eta)\frac{\partial^2 u}{\partial\xi\partial\eta} + \bar{D}\frac{\partial u}{\partial\xi} + \bar{E}\frac{\partial u}{\partial\eta} + \bar{F}u = \bar{G}$$

or

$$\frac{\partial^2 u}{\partial\xi\partial\eta} + F(\xi,\eta,u,\frac{\partial u}{\partial\xi},\frac{\partial u}{\partial\eta}) = 0. \qquad (1.120)$$

This is a canonical form associated with hyperbolic equations. Our original assumption that $\Delta > 0$, implies the coefficient $f(\xi,\eta) \neq 0$ since the transformed discriminant must also be positive with $\bar{\Delta} = \bar{B}^2 > 0$.

An alternate canonical form for hyperbolic equations is obtained using the additional transformations

$$\bar{y} = \frac{1}{2}(\xi - \eta), \qquad \bar{x} = \frac{1}{2}(\xi + \eta)$$

which reduces the equation (1.120) to the alternate canonical form

$$\frac{\partial^2 u}{\partial \bar{x}^2} - \frac{\partial^2 u}{\partial \bar{y}^2} + 4G(\bar{y}, \bar{x}, \frac{\partial u}{\partial \bar{y}}, \frac{\partial u}{\partial \bar{x}}) = 0.$$

Case 2: $(A \neq 0, \Delta = 0)$ In the case where the discriminant is zero, we find parabolic equations have equal roots associated with the characteristic equation and

$$\alpha_1 = \alpha_2 = \alpha(x, y) = \frac{-B(x, y)}{A(x, y)}$$

so that equation (1.117) reduces to

$$f(\psi, \psi) = A(\psi_x - \alpha(x, y)\psi_y)^2 = 0.$$

We can proceed exactly as we did in case 1 above. Let $u = u(x, y) = c_1$ denote the solution of $\frac{dy}{dx} = -\alpha(x, y)$, then the transformation $\xi = u(x, y)$ represents a one parameter family of characteristics which will make the term $\bar{A} = f(\xi, \xi)$ equal to zero in equation (1.114). Note also the hypothesis $\Delta = 0$ requires that

$$\bar{\Delta} = \bar{B}^2 - \bar{A}\bar{C} = f(\xi, \eta)^2 - f(\xi, \xi)f(\eta, \eta) = f(\xi, \eta)^2 = 0$$

since we have selected ξ such that $\bar{A} = f(\xi, \xi) = 0$. This reduces the equation (1.108) to the canonical form

$$f(\eta, \eta)\frac{\partial^2 u}{\partial \eta^2} + \bar{D}\frac{\partial u}{\partial \xi} + \bar{E}\frac{\partial u}{\partial \eta} + \bar{F}u = \bar{G}$$

or

$$\frac{\partial^2 u}{\partial \eta^2} + F(\xi, \eta, u, \frac{\partial u}{\partial \xi}, \frac{\partial u}{\partial \eta}) = 0.$$

Here $\eta = \eta(x, y)$ can be selected as any function which satisfies

$$\frac{\partial(\xi, \eta)}{\partial(x, y)} = \begin{vmatrix} \xi_x & \xi_y \\ \eta_x & \eta_y \end{vmatrix} \neq 0.$$

Under these circumstances the term $\bar{C} = f(\eta, \eta)$ will be different from zero.

Case 3: $(A \neq 0, \ \Delta < 0)$ In the case where the discriminant is negative, we find elliptic equations have complex roots associated with the characteristic equation and

$$\alpha_1(x,y) = \frac{-B(x,y) + i\sqrt{-\Delta(x,y)}}{A(x,y)}, \qquad \alpha_2(x,y) = \frac{-B(x,y) - i\sqrt{-\Delta(x,y)}}{A(x,y)}$$

where i is an imaginary unit satisfying $i^2 = -1$. Following the example given in case 1, we find the differential equations associated with the subsidiary equations are also complex-valued. Observe that for complex-valued functions the characteristic equation becomes

$$f(\xi + i\eta, \xi + i\eta) = A(\xi_x + i\eta_x)^2 + 2B(\xi_x + i\eta_x)(\xi_y + i\eta_y) + C(\xi_y + i\eta_y)^2 = 0$$

which simplifies to

$$f(\xi + i\eta, \xi + i\eta) = f(\xi, \xi) + 2i\, f(\xi, \eta) - f(\eta, \eta) = 0.$$

Hence if $f(\xi + i\eta, \xi + i\eta) = 0$ we require both the real part and imaginary part of this complex-valued function equal zero. This produces $f(\xi, \xi) = f(\eta, \eta)$ and $f(\xi, \eta) = 0$. Therefore, if the solution of the complex-valued differential equation $\frac{dy}{dx} = -\alpha_1(x,y)$ is given by $u(x,y) + iv(x,y) = c_1$, then the functions $\xi = u(x,y)$ and $\eta = v(x,y)$ will transform the partial differential equation (1.108) to the canonical form

$$f(\xi, \xi)\left[\frac{\partial^2 u}{\partial \xi^2} + \frac{\partial^2 u}{\partial \eta^2}\right] + \bar{D}\frac{\partial u}{\partial \xi} + \bar{E}\frac{\partial u}{\partial \eta} + \bar{F}u = \bar{G}$$

or

$$\left[\frac{\partial^2 u}{\partial \xi^2} + \frac{\partial^2 u}{\partial \eta^2}\right] + G(\xi, \eta, u, \frac{\partial u}{\partial \xi}, \frac{\partial u}{\partial \eta}) = 0.$$

Here $f(\xi, \xi) \neq 0$ because the discriminant, by the hypothesis $\Delta < 0$, must satisfy $\bar{\Delta} = \bar{B}^2 - \bar{A}\bar{C} = f(\xi, \eta)^2 - f(\xi, \xi)f(\eta, \eta) = -f(\xi, \xi)f(\eta, \eta) = -f(\xi, \xi)^2 < 0$. Equations which are elliptic have no real characteristic curves.

Constant coefficients

Some linear second order partial differential equations with constant coefficients are easily solved. Define the operators

$$D_x = \frac{\partial}{\partial x}, \quad D_y = \frac{\partial}{\partial y}, \quad D_x^2 = \frac{\partial^2}{\partial x^2}, \quad D_y^2 = \frac{\partial^2}{\partial y^2}, \quad D_x D_y = \frac{\partial^2}{\partial x \partial y}$$

and write the partial differential equation (1.108) in the operator form

$$L(u) = (AD_x^2 + 2BD_xD_y + CD_y^2 + DD_x + ED_y + F)u = G.$$

The principle of superposition for a general linear nth order PDE states that if $u_1, u_2, u_3, \ldots, u_n$ are each solutions of the homogeneous partial differential equation $L(u) = 0$, then any linear combination

$$u_H = c_1 u_1 + c_2 u_2 + c_3 u_3 + \cdots + c_n u_n$$

is also a solution for arbitrary constants $c_1, c_2, \ldots, c_n$. If we denote by u_P any particular solution of the nonhomogeneous equation $L(u) = G$, then $u = u_H + u_P$ is also a solution of $L(u) = G$. If under certain special conditions the second order homogeneous linear partial differential equation $L(u) = 0$ has a solution involving two arbitrary functions, then the solution is called a general solution. Here it is implied that the arbitrary functions are at least twice differentiable in order to satisfy the given second order partial differential equation.

Example 1-15.   (Factorable operators.)
The second order linear partial differential equation with constant coefficients

$$L(u) = \frac{\partial^2 u}{\partial x^2} - \frac{\partial^2 u}{\partial y^2} = 12x^2$$

can be written in the operator form

$$(D_x^2 - D_y^2)u = 12x^2$$

where the operators can be factored as

$$(D_x + D_y)(D_x - D_y)u = 12x^2 \quad \text{or} \quad (D_x - D_y)(D_x + D_y)u = 12x^2.$$

The homogeneous equation

$$(D_x + D_y)(D_x - D_y)u = 0 \quad \text{or} \quad (D_x - D_y)(D_x + D_y)u = 0$$

can be treated in terms of first order partial differential equations. For example, if $(D_x + D_y)u = 0$ the subsidiary equations are

$$\frac{dx}{1} = \frac{dy}{1} = \frac{du}{0}$$

which gives the solution families $x - y = c_1$ and $u = c_2$. The general solution is $c_2 = f(c_1)$ or $u = f(x-y)$ where f is an arbitrary function. Similarly, if $(D_x - D_y)u = 0$ we obtain the subsidiary equations

$$\frac{dx}{1} = \frac{dy}{-1} = \frac{du}{0}$$

with independent solutions $x + y = c_1$ and $u = x_2$ which produces the general solution $c_2 = g(c_1)$ or $u = g(x+y)$ where g represents an arbitrary function. Here f, g are arbitrary twice differentiable functions. The general solution of the homogeneous second order partial differential equation can then be written as

$$u = f(x - y) + g(x + y).$$

Holding y-constant we can integrate the given nonhomogeneous equation to obtain

$$\frac{\partial^2 u}{\partial x^2} = 12x^2, \qquad \frac{\partial u}{\partial x} = 4x^3, \qquad u = x^4$$

as a particular solution. Here we want any particular solution and so we neglect all constants of integration and obtain the simplest. The general solution of the original partial differential equation can be expressed as

$$u = u_H + u_P = f(x - y) + g(x + y) + x^4. \qquad f, g \text{ arbitrary functions.}$$

∎

Example 1-16. (Constant coefficients.)
Solve the partial differential equation

$$L(u) = L_1 L_2 u = (\alpha_1 D_x + \beta_1 D_y + \gamma_1)(\alpha_2 D_x + \beta_2 D_y + \gamma_2)u = 0$$

where the coefficients $\alpha_1, \alpha_2, \beta_1, \beta_2, \gamma_1, \gamma_2$ are all constants.
Solution: For constant coefficients the above operators are commutative so if $L_1 = (\alpha_1 D_x + \beta_1 D_y + \gamma_1)$ and $L_2 = (\alpha_2 D_x + \beta_3 D_y + \gamma_2)$, then $L_1 L_2 = L_2 L_1$. We consider solutions associated with various values assigned to the constant coefficients.

  Case 1: $(\alpha_1 \neq 0, \ \alpha_2 \neq 0)$ Using the results from the example 1-9 we find the solution to $L_1 u = 0$ is obtained from the subsidiary equations

$$\frac{dx}{\alpha_1} = \frac{dy}{\beta_1} = \frac{du}{-\gamma_1 u} \tag{1.121}$$

while the solution of $L_2 u = 0$ is obtained from the subsidiary equations

$$\frac{dx}{\alpha_2} = \frac{dy}{\beta_2} = \frac{du}{-\gamma_2 u} \tag{1.122}$$

The general solution to the equation $L_1 u = 0$ can be written as

$$u = e^{-\frac{\gamma_1}{\alpha_1} x} f(\beta_1 x - \alpha_1 y), \qquad \alpha_1 \neq 0$$

and the solution to $L_2 u = 0$ can be written

$$u = e^{-\frac{\gamma_2}{\alpha_2}x} g(\beta_2 x - \alpha_2 y), \qquad \alpha_2 \neq 0$$

where f, g are arbitrary differentiable functions. The general solution is therefore

$$u_H = e^{-\frac{\gamma_1}{\alpha_1}x} f(\beta_1 x - \alpha_1 y) + e^{-\frac{\gamma_2}{\alpha_2}x} g(\beta_2 x - \alpha_2 y). \qquad (1.123)$$

This shows the given equation can be split into two first order differential equations in either of the forms:

Form 1: If $\quad L_2 L_1 u = 0,$ $\qquad\qquad$ Form 2: If $\quad L_1 L_2 u = 0,$

$\qquad$ then let $\quad L_1 u = v \qquad$ or $\qquad$ then let $\quad L_2 u = w$

$\qquad\qquad\qquad L_2 v = 0$ $\qquad\qquad\qquad\qquad\qquad L_1 w = 0.$

The solution of either system produces the result given by equation (1.123).

Case 2: $(\alpha_1 \neq 0, \; L_1 = L_2)$ This situation is similar to ordinary differential equations with constant coefficients where for repeated roots you multiply a fundamental solution by the independent variable to obtain a second linearly independent solution. If we know the solution to $L_1 u = 0$ is given by

$$u = e^{-\frac{\gamma_1}{\alpha_1}x} f(\beta_1 x - \alpha_1 y),$$

then for a repeated factor another solution to $L_2 L_1 u = 0$ is either

$$u = x e^{-\frac{\gamma_1}{\alpha_1}x} g(\beta_1 x - \alpha_1 y) \quad \text{or} \quad u = y e^{-\frac{\gamma_1}{\alpha_1}x} g(\beta_1 x - \alpha_1 y) \qquad (1.124)$$

where g represents an arbitrary function. These results are readily verified by differentiation and substitution of the above solutions into the given differential equation. One form for the general solution is

$$u_H = e^{-\frac{\gamma_1}{\alpha_1}x} f(\beta_1 x - \alpha_1 y) + x e^{-\frac{\gamma_1}{\alpha_1}x} g(\beta_1 x - \alpha_1 y) \qquad (1.125)$$

where f, g are arbitrary functions.

Case 3: $(\alpha_1 = 0, \; \alpha_2 \neq 0)$ In this case the partial differential equation $L_1 u = 0$ reduces to $\beta_1 \dfrac{\partial u}{\partial u} + \gamma_1 u = 0$ with solution $u = e^{-\frac{\gamma_1}{\beta_1}y} f(\beta_1 x)$. The solution to $L_2 L_1 u = 0$ is written

$$u_H = e^{-\frac{\gamma_1}{\beta_1}y} f(\beta_1 x) + e^{-\frac{\gamma_2}{\alpha_2}x} g(\beta_2 x - \alpha_2 y) \qquad (1.126)$$

where f, g are arbitrary functions.

Case 4: ($\alpha_1 = \alpha_2 = 0$) The two first order partial differential equations

$$L_2 L_1 u = (\beta_1 D_y + \gamma_1)(\beta_2 D_y + \gamma_2)u = 0$$

has the general solution

$$u_H = e^{-\frac{\gamma_1}{\beta_1}y} f(\beta_1 x) + e^{-\frac{\gamma_2}{\beta_2}y} g(\beta_2 x)$$

where f, g are arbitrary functions.   ∎

The problem of finding a solution to a partial differential equation involving time t, which is to be constructed over some spatial domain R, and such that the solution is required to satisfy a given condition at time $t = 0$, is referred to as an initial value problem. If in addition one requires values of the solution to satisfy prescribed conditions on the boundary of the region R, then the problem is called a boundary and initial value problem. If the partial differential equation is time independent, and its solution is required to satisfy prescribed conditions on the boundary of R, then the partial differential equation plus boundary conditions is called a boundary value problem.

Define the boundary operator

$$B(u) = \alpha \frac{\partial u}{\partial n} + \beta u \tag{1.127}$$

where α, β are constants. Boundary conditions can then be written in the form

$$B(u) = f(x, y)\Big|_{u, f} \text{ evaluated for } x, y \text{ on the boundary.}$$

where f is a specified condition to be satisfied on the boundary. In the boundary operator the derivative $\frac{\partial u}{\partial n} = \operatorname{grad} u \cdot \hat{n}$ is called a normal derivative, where $\hat{n}$ represents a unit exterior normal to the boundary at the point where the normal derivative is to be calculated.

Boundary conditions of the type

$$B(u) = u = f\Big|_{(x,y) \text{ on boundary of } R}$$

where $\alpha = 0$, $\beta = 1$ and f is a specified function, are called Dirichlet[†] boundary conditions or boundary value problems of the first kind. Dirichlet boundary

---

[†] See Appendix C

conditions requires the solution take on prescribed values on the boundary of the region R.

Boundary conditions of the type

$$B(u) = \frac{\partial u}{\partial n} = f\Big|_{(x,y) \text{ on boundary of } R}$$

where $\alpha = 1, \beta = 0$ and f is a specified function, are called Neumann conditions or boundary value problems of the second kind. Neumann conditions require the normal flux across the boundary be specified.

Boundary conditions of the type

$$B(u) = \frac{\partial u}{\partial n} + hu = f\Big|_{(x,y) \text{ on boundary of } R}$$

where $\alpha = 1, \beta = h$ and f is a specified function, are called Robin[†] conditions or boundary value problems of the third kind. These type of conditions represent a linear combination of the Dirichlet and Neumann conditions and represent heat loss or evaporation at a boundary being specified.

Mixed boundary conditions arise when one portion of the surface is assigned one type of boundary condition while the remainder of the surface satisfies some other type of boundary condition. An example of a mixed boundary condition for the heat equation would be the temperature is specified over one portion of the surface S_1 while heat loss is specified over some other portion of the surface S_2, where $S_1 \cap S_2 = \emptyset$.

Dirichlet, Neumann, Robin and Mixed boundary conditions can be written using the boundary operator

$$B(u) = \alpha \frac{\partial u}{\partial n} + \beta u, \qquad \alpha, \beta \text{ constants.} \tag{1.128}$$

Boundary conditions of the form

$$B(u) = f, \qquad \begin{array}{c} u, f \text{ evaluated for} \\ x, y \text{ on the boundary} \end{array} \quad f \neq 0 \tag{1.129}$$

are called nonhomogeneous boundary conditions, while boundary conditions of the form

$$B(u) = 0, \qquad u \text{ evaluated on the boundary} \tag{1.130}$$

are called homogeneous boundary conditions.

We will use the principal of superposition for both discrete and continuous parameter systems. In particular if $u_1, u_2, \ldots$ are solutions of a linear nth order homogeneous partial differential equation $L(u) = 0$, then the sum

$$u = \sum_{n=1}^{\infty} c_n u_n \qquad (1.131)$$

with $c_1, c_2, \ldots$ constants, is also a solution. A more generalized superposition principle, involving functions containing a parameter, uses integration. For example, if $u = u(x, y, z; \alpha)$ is a solution of $L(u) = 0$ where $\alpha > 0$ is a parameter, then the continuous sum

$$u(x, y, z) = \int_0^{\infty} C(\alpha) u(x, y, z; \alpha) \, d\alpha \qquad (1.132)$$

is also a solution, where $C(\alpha)$ is an arbitrary function of the parameter α.

Arbitrary functions

Not all partial differential equations have solutions which involve arbitrary functions. Let two functions $\xi = \xi(x, y)$ and $\eta = \eta(x, y)$ satisfy the conditions

(i) Both ξ, η are real-valued functions with continuous first and second derivatives.

(ii) The function η cannot be identically zero.

(iii) The function ξ cannot be constant for all x, y values.

Definition: If the function $u = u(x, y) = \eta(x, y) f(\xi(x, y))$ is a solution of the linear second order partial differential equation

$$L(u) = A \frac{\partial^2 u}{\partial x^2} + 2B \frac{\partial^2 u}{\partial x \partial y} + C \frac{\partial u}{\partial y} + D \frac{\partial u}{\partial x} + E \frac{\partial u}{\partial y} + Fu = 0$$

for an arbitrary twice differentiable function f, then the functions ξ, η are called a functionally invariant pair.

Example 1-17. (Wave equation.)

Does the wave equation $\dfrac{\partial^2 u}{\partial x^2} = \dfrac{1}{c^2} \dfrac{\partial^2 u}{\partial t^2}$, with c constant, have a functionally invariant pair ξ, η?

Solution: Assume a solution to the wave equation of the form $u = \eta(x, t) f(\xi(x, t))$

52

where f is arbitrary. This assumed solution has the derivatives

$$
\begin{aligned}
u_x &= \eta f'(\xi)\xi_x + \eta_x f(\xi) \\
u_{xx} &= \eta f'(\xi)\xi_{xx} + \eta f''(\xi)\xi_x^2 + \eta_x f'(\xi)\xi_x + \eta_x f'(\xi)\xi_x + \eta_{xx} f(\xi) \\
u_t &= \eta f'(\xi)\xi_t + \eta_t f(\xi) \\
u_{tt} &= \eta f'(\xi)\xi_{tt} + \eta f''(\xi)\xi_t^2 + \eta_t f'(\xi)\xi_t + \eta_t f'(\xi)\xi_t + \eta_{tt} f(\xi)
\end{aligned}
\tag{1.133}
$$

We substitute these derivatives into the given partial differential equation and rearrange terms to find that if the assumed solution is to satisfy the equation then the following equation must be identically zero for arbitrary f values.

$$
\eta f''(\xi)[c^2\xi_x^2 - \xi_t^2] + f'(\xi)[c^2\eta\xi_{xx} - \eta\xi_{tt} + 2c^2\eta_x\xi_x - 2\eta_t\xi_t] + f(\xi)[c^2\eta_{xx} - \eta_{tt}] = 0 \quad (1.134)
$$

Therefore, for arbitrary f values, the terms inside the brackets must be zero or

$$
c^2\xi_x^2 - \xi_t^2 = 0 \qquad \text{Characteristic equation}
$$

$$
\eta(c^2\xi_{xx} - \xi_{tt}) + 2(c^2\eta_x\xi_x - \eta_t\xi_t) = 0
$$

$$
c^2\eta_{xx} - \eta_{tt} = 0.
$$

The functions $\xi = k_1 x + k_2 t$, $\eta = 1$ satisfy these equations if k_1, k_2 are constant and are selected such that

$$
c^2 k_1^2 - k_2^2 = 0 \qquad \text{or} \qquad (ck_1 - k_2)(ck_1 + k_2) = 0.
$$

Thus, if we select $k_2 = ck_1$ or $k_2 = -ck_1$ we find $u = f(k_1(x+ct))$ and $u = f(k_1(x-ct))$ are solutions for arbitrary functions f, which implies $u = f(x+ct)$ and $u = f(x-ct)$ are solutions of the wave equation for arbitrary functions f. Here $\{k_1 x + k_2 t, 1\}$ are a functionally invariant pair for the correct selection of the constants k_1, k_2.

∎

Example 1-18. (Heat equation.)

Does the heat equation $\dfrac{\partial^2 u}{\partial x^2} = \dfrac{1}{K}\dfrac{\partial u}{\partial t}$ have a functionally invariant pair (ξ, η)?

Solution: Assume a solution of the form $u = \eta(x,t)f(\xi(x,t))$ and substitute into the given equation. (Here we can use the derivatives from the previous example.) We find that if $u = \eta f(\xi)$ is to be a solution for arbitrary f then the following equation must be identically satisfied.

$$
\eta f''(\xi)[\xi_x^2] + f'(\xi)[\eta\xi_{xx} + 2\eta_x\xi_x - \frac{1}{K}\eta\xi_t] + f(\xi)[\eta_{xx} - \frac{1}{K}\xi_t] = 0.
$$

For arbitrary values of f the terms inside the brackets must be identically zero or

$$\xi_x^2 = 0 \quad \text{Characteristic equation}$$

$$\eta(\xi_{xx} - \frac{1}{K}\xi_t) + 2\eta_x\xi_x = 0$$

$$\eta_{xx} - \frac{1}{K}\xi_t = 0.$$

The first equation gives $\xi_x = 0$ which integrates to give $\xi = \xi(t)$. This reduces the second equation to $-\frac{\eta}{K}\xi_t = 0$, which in turn implies $\xi =$ constant. Here the heat equation will not have solutions in terms of arbitrary functions as a functionally invariant pair does not exist.

■

Example 1-19. (Laplace equation.)
Does the Laplace equation $\dfrac{\partial^2 u}{\partial x^2} + \dfrac{\partial^2 u}{\partial y^2} = 0$ have a functionally invariant pair (ξ, η)?
Solution: Assume a solution $u = \eta(x, y)f(\xi(x, y))$ and substitute into the Laplace equation. We find that if $u = \eta f(\xi)$ is a solution to the given partial differential equation, then the following equation must be identically satisfied.

$$\eta f''(\xi)[\xi_x^2 + \xi_y^2] + f'(\xi)[\eta(\xi_{xx} + \xi_{yy}) + 2(\eta_x\xi_x + \eta_y\xi_y)] + f(\xi)[\eta_{xx} + \eta_{yy}] = 0$$

This equation is satisfied for arbitrary f values if the terms inside the brackets are identically zero or

$$\xi_x^2 + \xi_y^2 = 0 \quad \text{Characteristic equation}$$

$$\eta(\xi_{xx} + \xi_{yy}) + 2(\eta_x\xi_x + \eta_y\xi_y) = 0 \tag{1.135}$$

$$\eta_{xx} + \eta_{yy} = 0$$

Here the characteristic equation $\xi_x^2 + \xi_y^2 = 0$ has imaginary solutions. The only real solutions for this equation requires $\xi =$ constant. Hence, the Laplace equation does not have real solutions in terms of arbitrary functions as a functionally invariant pair does not exist. If we admit complex-valued functions, then note that $\{\xi = k_1 x + k_2 y, \eta = 1\}$ satisfies the equations (1.135) whenever $k_2 = \pm i k_1$. In this case the function $u = f(x + iy) + g(x - iy)$ for arbitrary f, g is a solution of the Laplace equation, where $i^2 = -1$. In this special case we find real solutions must be constructed from certain combinations of complex valued functions.

■

Exercises 1

▶ **1.**

(a) Find the general solution to the differential equation $\dfrac{d^2u}{dx^2} = 0$.

(b) Solve the differential equation $\dfrac{d^2u}{dx^2} = 0$, $a \le x \le b$ subject to the boundary conditions $u(a) = T_a$ and $u(b) = T_b$ where T_a, T_b are constants.

▶ **2.**

(a) Show the general solution $u = u(x, y)$ of the partial differential equation $\dfrac{\partial^2 u}{\partial x^2} = 0$ is given by $u = u(x, y) = C_1(y)x + C_2(y)$ where C_1 and C_2 represent arbitrary functions of y. Compare this solution with your answer to the problem number 1. Comment upon the differences between the solutions to an ordinary and a partial differential equation.

(b) Solve the partial differential equation $\dfrac{\partial^2 u}{\partial y^2} = 0$ for $u = u(x, y)$.

(c) Solve the partial differential equation $\dfrac{\partial^2 u}{\partial x \partial y} = 0$ for $u = u(x, y)$.

(d) Solve the partial differential equation $\dfrac{\partial^2 u}{\partial x^2} = 0$ for $u = u(x, t)$ for $0 < x < L$, $t > 0$ subject to the boundary conditions $u(0, t) = A(t)$ and $u(L, t) = B(t)$.

▶ **3.**

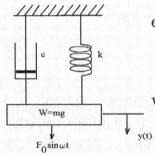

The spring-mass viscous damper illustrated has the equation of motion

$$m\frac{d^2y}{dt^2} + c\frac{dy}{dt} + ky = F_0 \sin(\omega t)$$

where m, c, k, F_0, ω are constants.

(a) Find the solution in the cases:

    (i) $c^2 > 4km$ (ii) $c^2 = 4km$ (iii) $c^2 < 4km$

(b) Show the steady state solution to the spring-mass system can be represented in the form $y = A\sin(\omega t - \phi)$ where

$$A = \frac{F_0}{\sqrt{(k - m\omega^2)^2 + (c\omega)^2}} \qquad \tan\phi = \frac{c\omega}{k - m\omega^2}$$

(c) Define $\omega_c^2 = k/m$, $A_0 = F_0/k$, and define the dimensionless variables

$$\eta = \frac{c\omega_c}{2k} \qquad \lambda = \frac{\omega}{\omega_c} \qquad Y = \frac{A}{A_0}$$

where λ is a frequency ratio, Y is an amplitude ratio and η is a damping ratio. Show that $Y = \dfrac{1}{\sqrt{(1-\lambda^2)^2 + (2\eta\lambda)^2}}$. Verify the following graph and write out your interpretation as to what the graph means.

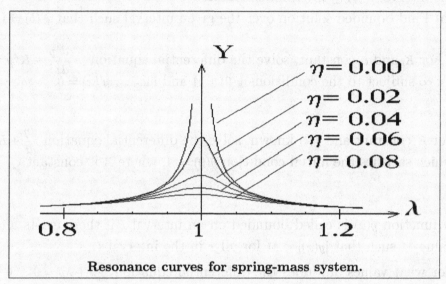

Resonance curves for spring-mass system.

▶ **4.**   Solve the differential equation $\dfrac{d}{dx}\left[f(x)\dfrac{dy}{dx}\right] - \dfrac{\lambda}{f(x)}y = 0$ for the cases
(i) $\lambda = -\omega^2$, (ii) $\lambda = 0$, (iii) $\lambda = \omega^2$.
Hint: Make the substitution $z = \displaystyle\int_a^x \dfrac{d\xi}{f(\xi)}$ and use chain rule differentiation.

▶ **5.**

Solve the differential equation $r^2\dfrac{d^2F}{dr^2} + 2r\dfrac{dF}{dr} - n(n+1)F = 0$ where n is an integer.

▶ **6.**
(a) Solve the differential equation $\dfrac{d}{dx}\left(x\dfrac{du}{dx}\right) = 0$ over the interval $1 < x < 5$ subject to the boundary conditions $u(1) = A$ and $u(5) = B$ where A, B are constants.

(b) Show the differential equation $\dfrac{d}{dx}\left(\kappa(x)\dfrac{du}{dx}\right) = 0$ over the interval $a < x < b$ has the general solution $u = u(x) = c_2 + c_1\displaystyle\int_a^x \dfrac{d\xi}{\kappa(\xi)}$ where $\kappa(x) > 0$ is some general function of x which is nonzero and well behaved over the interval $a < x < b$. Find the constants c_1 and c_2 if the given equation is subject to the boundary conditions $u(a) = A$ and $\frac{du}{dx}\big|_{x=b} = B$ where A and B are constants.

▶ **7.**

(a) Find all solutions to the differential equation $r\dfrac{d^2 F}{dr^2} + \dfrac{dF}{dr} + n^2 r F = 0$, over the interval $0 \le r \le b$, with n constant.

(b) Find bounded solution over the given interval such that $F(b) = 1$.

▶ **8.** For K and α constant, solve the differential equation $\dfrac{d^2 y}{dx^2} - K^2 y = \alpha$ $0 < x < \infty$ subject to the conditions $y(0) = A$ and $\lim_{x \to \infty} y'(x) = 0$.

▶ **9.**

For K constant and $f(t)$ known, solve the differential equation $\dfrac{dy}{dt} + Ky = f(t)$ which satisfies the initial condition $y(0) = A$, where A is constant.

▶ **10.**

A function $y(x)$ is called bounded on an interval I if there exists a constant value M such that $|y(x)| < M$ for all x in the interval I.

(a) For what values of λ does the differential equation $\frac{d^2 y}{dx^2} + \lambda y = 0$ $0 \le x \le L$, possess bounded solutions? What are the bounded solutions?

(b) Find all bounded solutions to the differential equation
$$r^2 \dfrac{d^2 F}{dr^2} + r\dfrac{dF}{dr} + (\lambda^2 r^2 - \nu^2)F = 0, \qquad 0 \le r \le b$$

(c) Find all bounded solutions to the differential equation
$$\dfrac{1}{\sin \theta}\dfrac{d}{d\theta}\left(\sin \theta \dfrac{dy}{d\theta}\right) + n(n+1)y = 0, \qquad 0 \le \theta \le \pi$$

(d) Find all bounded solutions to the differential equation
$$\rho^2 \dfrac{d^2 y}{d\rho^2} + \rho\dfrac{dy}{d\rho} - 4y = 0, \qquad 0 \le \rho \le 6$$

▶ **11.**

(a) Find the first five values of $\lambda > 0$ such that $J_0(6\lambda) = 0$.

(b) For $b > 0$ constant, find the first five values of $\lambda > 0$ such that $J_0(\lambda b) = 0$.

(c) If $J_0(\xi_{0j}) = 0$ for $j = 1, 2, 3, \ldots$, then for $b > 0$ constant find values of $\lambda > 0$ such that $J_0(\lambda b) = 0$. How many values of λ are there?

▶ **12.** **(Stability of a vertical column.)**

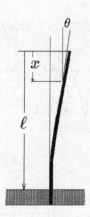

Let ℓ denote the length of a column of wire with uniform cross section and let m denote its uniform mass per unit length. If θ denotes the angular deflection from the vertical at a distance x measured from the top of the column, the differential equation for small angular displacements is given by $EI\dfrac{d^2\theta}{dx^2} = -gm\,x\,\theta$ where EI is the flexural rigidity, g is the acceleration of gravity.

(a) For m, g, E, I constants, let $\sigma^2 = \frac{gm}{EI}$ and solve the differential equation for θ as a function of x.

(b) Apply the boundary conditions that $\theta'(0) = 0$ (bending moment is zero) and $\theta(\ell) = 0$ (zero deflection at $x = \ell$).

(c) Show the column will not bend from the vertical until $\frac{2}{3}\sigma\ell^{3/2}$ equals or exceeds the first zero of $J_{-1/3}(\xi)$.

Hint: Expand $J_{1/3}, J_{-1/3}$ and use $J_{-1/3}(x) = 0$ for $x \approx 1.8664$.

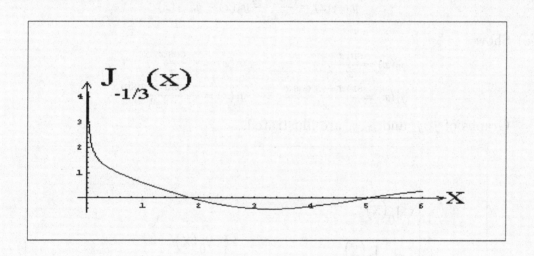

▶ **13.** **(Spherical Bessel functions.)**

(a) Show the general solution of the differential equation

$$r^2\frac{d^2v}{dr^2} + r\frac{dv}{dr} + [k^2r^2 - (\ell + \frac{1}{2})^2]v = 0, \qquad (13a)$$

where $k \neq 0$ is constant and ℓ is an integer, is given by

$$v = v(r) = c_1 J_{\ell+\frac{1}{2}}(kr) + c_2 Y_{\ell+\frac{1}{2}}(kr). \qquad (13b)$$

58

(b) Make the change of variable $v = \sqrt{r}u$ in equation (13a) to obtain

$$r^2 \frac{d^2u}{dr^2} + 2r\frac{du}{dr} + [k^2r^2 - \ell(\ell+1)]u = 0 \tag{13c}$$

(c) Define the spherical Bessel functions of the first kind by $\quad j_\ell(x) = \sqrt{\frac{\pi}{2x}} J_{\ell+\frac{1}{2}}(x)$

and spherical Bessel functions of the second kind by $\quad y_\ell(x) = \sqrt{\frac{\pi}{2x}} Y_{\ell+\frac{1}{2}}(x)$

for $\ell = 0, \pm 1, \pm 2, \pm 3, \ldots$ Show equation (13c) has the general solution

$$u = C_1 j_\ell(kr) + C_2 y_\ell(kr) \tag{13d}$$

where C_1 and C_2 are arbitrary constants.

(d) Using entries from table 1-1 show the spherical Bessel functions satisfy the recurrence formulas

$$j_{n+1}(x) = \frac{(2n+1)}{x} j_n(x) - j_{n-1}(x)$$
$$y_{n+1}(x) = \frac{(2n+1)}{x} y_n(x) - y_{n-1}(x) \tag{13e}$$

(e) Show

$$j_0(x) = \frac{\sin x}{x} \qquad\qquad y_0(x) = -\frac{\cos x}{x}$$
$$j_1(x) = \frac{\sin x - x\cos x}{x^2} \qquad\qquad y_1(x) = \frac{-\cos x - x\sin x}{x^2}$$

Graphs of j_0, j_1 and y_0, y_1 are illustrated.

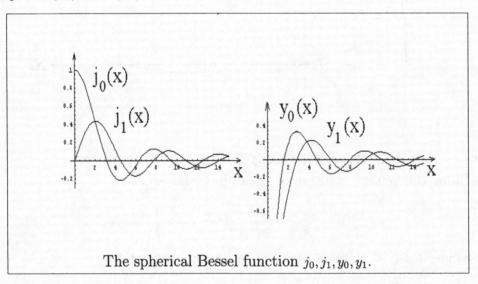

The spherical Bessel function j_0, j_1, y_0, y_1.

► **14.** (Modified spherical Bessel functions.)

(a) Show the general solution of the differential equation

$$r^2\frac{d^2v}{dr^2} + r\frac{dv}{dr} - [\lambda^2 r^2 + (\ell+\frac{1}{2})^2]v = 0, \tag{14a}$$

where $\lambda \neq 0$ is constant and ℓ is an integer, is given by

$$v = v(r) = c_1 I_{\ell+\frac{1}{2}}(\lambda r) + c_2 K_{\ell+\frac{1}{2}}(\lambda r). \tag{14b}$$

(b) Make the change of variable $v = \sqrt{r}u$ in equation (14a) and show it transforms to the differential equation

$$r^2\frac{d^2u}{dr^2} + 2r\frac{du}{dr} - [\lambda^2 r^2 + \ell(\ell+1)]u = 0 \tag{14c}$$

(c) Define the modified spherical Bessel functions of the first kind by

$$i_\ell(x) = \sqrt{\frac{\pi}{2x}}I_{\ell+\frac{1}{2}}(x)$$

and modified spherical Bessel functions of the second kind by

$$k_\ell(x) = \sqrt{\frac{\pi}{2x}}K_{\ell+\frac{1}{2}}(x)$$

for $\ell = 0, \pm 1, \pm 2, \pm 3, \dots$ Show equation (14c) has the general solution

$$u = C_1 i_\ell(\lambda r) + C_2 k_\ell(\lambda r) \tag{14d}$$

where C_1 and C_2 are arbitrary constants.

(d) Show the modified spherical Bessel functions satisfy the recurrence formulas

$$i_{n+1}(x) = -\frac{(2n+1)}{x}i_n(x) + i_{n-1}(x)$$

$$k_{n+1}(x) = -\frac{(2n+1)}{x}k_n(x) + k_{n-1}(x) \tag{14d}$$

(e) Show

$$i_0(x) = \frac{\sinh x}{x} \qquad\qquad k_0(x) = \frac{\pi}{2x}e^{-x}$$

$$i_1(x) = \frac{-\sinh x + x\cosh x}{x^2} \qquad k_1(x) = \frac{\pi}{2x}e^{-x}\left(1 + \frac{1}{x}\right)$$

Graphs of i_0, i_1 and k_0, k_1 are illustrated.

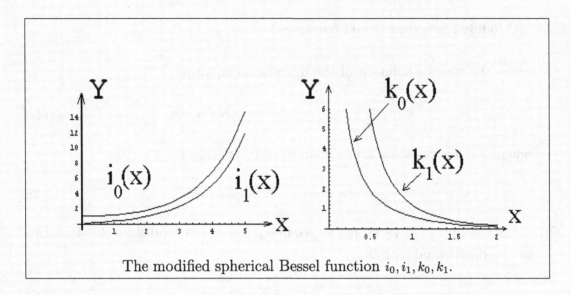

The modified spherical Bessel function i_0, i_1, k_0, k_1.

▶ **15.**

(a) Find the general solution to the differential equation
$$\frac{1}{r}\frac{d}{dr}\left(r\frac{du}{dr}\right) = 0, \qquad 0 < r < b.$$

(b) Solve the differential equation $\dfrac{1}{r}\dfrac{d}{dr}\left(r\dfrac{du}{dr}\right) = 0, \qquad a < r < b$
subject to the boundary conditions $u(a) = T_a$ and $u(b) = T_b$

▶ **16.**

(a) Find the general solution to the differential equation
$$\frac{1}{\rho^2}\frac{d}{d\rho}\left(\rho^2\frac{du}{d\rho}\right) = 0, \qquad 0 < \rho < b.$$

(b) Solve the differential equation $\dfrac{1}{\rho^2}\dfrac{d}{d\rho}\left(\rho^2\dfrac{du}{d\rho}\right) = 0, \qquad a < \rho < b$ subject to the
boundary conditions $u(a) = T_a$ and $u(b) = T_b$.

▶ **17.**

Show that solutions of the form $y_m(z) = AJ_m(\omega z) + BY_m(\omega z)$, where A, B and
ω are constants, can be written in terms of the Hankel functions $H_m^{(1)}, H_m^{(2)}$ of
the first and second kind of order m which are defined

$$H_m^{(1)}(z) = J_m(z) + \imath Y_m(z), \qquad H_m^{(2)}(z) = J_m(z) - \imath Y_m(z)$$

where $\imath^2 = -1$. These types of solutions are used when boundedness conditions
near the origin are not required and one has some knowledge of complex
variable theory.

▶ **18.**

Use the method of variation of parameters to solve the differential equation

$$\frac{d^2F}{dt^2} + \omega^2 F = h(t) \qquad\qquad F(0) = 0 \quad \text{and} \quad F'(0) = 0.$$

where $h(t)$ is some known function of t and ω is a given constant.

▶ **19.**

Assume the differential equation $\frac{dy}{dx} - iy = 0$, with $i^2 = -1$ and subject to the initial condition $y(0) = 1$, has a unique solution.

(a) Show that $y = e^{ix}$ is a solution satisfying the differential equation and initial condition.

(b) Show $y = \cos x + i \sin x$ is a solution satisfying the differential equation and initial conditions.

(c) If the solution is unique, what conclusion can you draw?

▶ **20.** Classify the given partial differential equations

$$\text{(a)} \quad \frac{\partial^2 u}{\partial x^2} - \frac{\partial^2 u}{\partial y^2} = f(x,y) \qquad\qquad \text{(c)} \quad \frac{\partial^2 u}{\partial x^2} + \frac{\partial u}{\partial x} + \frac{\partial u}{\partial y} = 0$$

$$\text{(b)} \quad \frac{\partial^2 u}{\partial x^2} + \frac{\partial^2 u}{\partial y^2} = 0 \qquad\qquad \text{(d)} \quad \frac{\partial^2 u}{\partial x^2} + \frac{\partial^2 u}{\partial x \partial y} + \frac{\partial^2 u}{\partial y^2} = 0$$

▶ **21.** Solve the given partial differential equations for $z = z(x,y)$.

$$\text{(a)} \quad \frac{\partial z}{\partial y} = x \qquad\qquad \text{(c)} \quad \frac{\partial z}{\partial x} = y$$

$$\text{(b)} \quad \frac{\partial^2 z}{\partial y^2} = 1 \qquad\qquad \text{(d)} \quad \frac{\partial^2 z}{\partial x^2} = 1$$

▶ **22.** Find the general solution $z = z(x,y)$.

$$\text{(a)} \quad \frac{\partial z}{\partial x} + 3z = x \qquad\qquad \text{(c)} \quad \frac{\partial z}{\partial x} + \frac{\partial z}{\partial y} = z$$

$$\text{(b)} \quad \frac{\partial z}{\partial x} + \frac{\partial z}{\partial y} + 2z = x \qquad\qquad \text{(d)} \quad 3\frac{\partial z}{\partial x} - 4\frac{\partial z}{\partial y} = x$$

▶ **23.**

Find the general solution to the given first order partial differential equations.

$$\text{(a)} \quad x\frac{\partial z}{\partial x} + y\frac{\partial z}{\partial y} = z \qquad\qquad \text{(c)} \quad (x+y)\frac{\partial z}{\partial x} + (x+y)\frac{\partial z}{\partial y} = z$$

$$\text{(b)} \quad y\frac{\partial z}{\partial x} - x\frac{\partial z}{\partial y} = yz \qquad\qquad \text{(d)} \quad y\frac{\partial z}{\partial x} + x\frac{\partial z}{\partial y} = 1$$

▶ **24.** Given the partial differential equation $x^2\dfrac{\partial z}{\partial x} + y^2\dfrac{\partial z}{\partial y} - (x+y)z = 0$.

Verify the given functions represent solutions to this equation.

(a) $F(\dfrac{xy}{z}, \dfrac{x-y}{z})$ where F is an arbitrary smooth function.

(b) $z = xy\, f(\dfrac{x-y}{z})$ where f is an arbitrary smooth function.

(c) $z = xy\, f(\dfrac{1}{y} - \dfrac{1}{x})$ where f is an arbitrary smooth function.

▶ **25.** Find solutions to the given partial differential equations

(a) $(D_x^2 - D_y^2)u = x$

(b) $(D_x + D_y + 1)(D_x + D_y - 1)u = 0$

(c) $(D_x - D_y)^2 u = 0$

(d) $(D_x^2 + 2D_xD_y + D_y^2 - D_x - D_y - 2)u = 0$

(e) $(2D_x + D_y - 1)(D_x + 2D_y + 1)u = 0$

(f) $(D_x + D_y)^2 u = 0$

▶ **26.** **(Exponential solutions)**

Consider the case where the linear homogeneous partial differential equation

$$L(u) = (AD_x^2 + 2BD_xD_y + CD_y^2 + DD_x + ED_y + F)u = 0 \qquad (26a)$$

has constant coefficients. Assume a solution of the form $u = e^{\alpha x + \beta y}$, where α and β are constants.

(a) Show u is a solution provided α, β satisfy

$$A\alpha^2 + 2B\alpha\beta + C\beta^2 + D\alpha + E\beta + F = 0 \qquad (26b)$$

(b) Assume the equation (26b) defines β as a function of α which can be written in the form $\beta = f(\alpha)$. Show $u = g(\alpha)e^{\alpha x + f(\alpha)y}$ is also a solution of equation (26a) for arbitrary functions g.

(c) Show by superposition $u = \displaystyle\int g(\alpha)e^{\alpha x + f(\alpha)y}\,d\alpha$ is also a solution of equation (26a). Hint: Interchange the order of the operations of differentiation and integration. $\dfrac{\partial}{\partial x}\displaystyle\int (\;)\,d\alpha = \int \dfrac{\partial}{\partial x}(\;)\,d\alpha$

▶ **27.** Assume a solution $u = e^{\alpha x + \beta t}$ to the heat equation $\dfrac{\partial^2 u}{\partial x^2} = \dfrac{1}{K}\dfrac{\partial u}{\partial t}$.

(a) Show $u = u(x,t;\alpha) = e^{\alpha x + \alpha^2 Kt}$ is a solution.

(b) Let $\alpha = in$, $\alpha^2 = -n^2$, $\beta = -Kn^2$, where $i^2 = -1$, and show

$$u = u(x,t) = \sum_{n=0}^{\infty} (A_n \cos nx + B_n \sin nx)\, e^{-Kn^2 t}$$

is a solution of the heat equation for arbitrary constants A_n, B_n.

▶ **28.** Given a function $u = u(x,t)$ which satisfies the heat equation $\dfrac{\partial^2 u}{\partial x^2} = \dfrac{1}{K}\dfrac{\partial u}{\partial t}$. Determine if the given functions are also solutions of the heat equation. Justify your answers.

(a) $v = v(x,t) = u(x - \xi, t)$, for constant ξ.

(b) $v = v(x,t) = \dfrac{\partial u}{\partial x}$

(c) $w = w(x,t) = \dfrac{\partial u}{\partial t}$

(d) $v = v(x,t) = \displaystyle\int_{-\infty}^{\infty} f(\xi)u(x - \xi, t)\, d\xi$ where f is arbitrary.

(e) $v = v(x,t) = u(\sqrt{\alpha}\,x, \alpha t)$, where $\alpha > 0$ is constant.

▶ **29.** Solve the given Cauchy problems

(a) $\dfrac{\partial u}{\partial x} = \cos y, \quad u(1,y) = y$

(b) $\dfrac{\partial^2 u}{\partial x^2} = y^2 \quad u(0,y) = y, \quad u(1,y) = \cos y$

▶ **30.** Find the general solution for the partial differential equation

$$\frac{\partial^2 u}{\partial x^2} + 4\frac{\partial^2 u}{\partial x \partial y} + 3\frac{\partial^2 u}{\partial y^2} = 48e^{3x+y}$$

Hint: Assume a particular solution of the form $u_p = Ae^{3x+y}$ and solve for A.

▶ **31.** Consider the partial differential equation $\dfrac{\partial z}{\partial x} = \dfrac{\partial z}{\partial y}$ subject to the boundary condition $z(0,y) = 2e^{3y}$. Assume a solution of the form $z = z(x,y) = F(x)G(y)$ where the variables are separated.

(a) Show that if x, y are independent variables, then F and G must be solutions of the ordinary differential equations

$$\frac{F'(x)}{F(x)} = \frac{G'(y)}{G(y)} = \lambda$$

where λ is called a separation constant.

(b) Solve the above differential equations and show $z = z(x,y) = Ce^{\lambda(x+y)}$ is a solution where C is an arbitrary constant.

(c) Is the function $z = z(x,y;\lambda) = C(\lambda)e^{\lambda(x+y)}$, where $C = C(\lambda)$ is a function of λ, a solution of the partial differential equation?

(d) Find a solution satisfying the given boundary condition.

▶ **32.** Partial differential equations can arise by eliminating arbitrary constants from a function in a manner analogous to the formation of ordinary differential equations associated with functions having arbitrary constants. For example, consider the function $z = ax^2 + by^2$ where a, b are arbitrary constants.

(a) If we treat z as a function of x and y, show when $z = ax^2 + by^2$ is differentiated with respect to x and y there results

$$\frac{\partial z}{\partial x} = 2ax \qquad \frac{\partial z}{\partial y} = 2by$$

Show when a, b are eliminated from $z = ax^2 + by^2$ by solving for a and b from the above derivatives, there results the partial differential equation

$$x\frac{\partial z}{\partial x} + y\frac{\partial z}{\partial y} = 2z.$$

(b) If we treat x as a function of y and z, show when $z = ax^2 + by^2$ is differentiated with respect to y and z there results

$$0 = 2ax\frac{\partial x}{\partial y} + 2by \qquad 1 = 2ax\frac{\partial x}{\partial z}.$$

Show when a, b are eliminated from $z = ax^2 + by^2$ by solving for a and b from the above derivatives, there results the partial differential equation

$$y\frac{\partial x}{\partial y} + 2z\frac{\partial x}{\partial z} = x.$$

(c) If we treat y as a function of x and z, show when $z = ax^2 + by^2$ is differentiated with respect to x and z there results

$$0 = 2ax + 2by\frac{\partial y}{\partial x} \qquad 1 = 2by\frac{\partial y}{\partial z}.$$

Show when a, b are eliminated from $z = ax^2 + by^2$ by solving for a and b from the above derivatives, there results the partial differential equation

$$x\frac{\partial y}{\partial x} + 2z\frac{\partial y}{\partial z} = y.$$

(d) Show $z = ax^2 + by^2$ is a solution of each of the partial differential equations given in parts (a)(b) and (c) above. The derived partial differential equations are called equivalent equations. They differ only in what is being assumed for the dependent and independent variables.

▶ **33.** The partial differential operator

$$L(u) = A(x,y)\frac{\partial^2 u}{\partial x^2} + 2B(x,y)\frac{\partial^2 u}{\partial x \partial y} + C(x,y)\frac{\partial^2 u}{\partial y^2}$$

$$+D(x,y)\frac{\partial u}{\partial x} + E(x,y)\frac{\partial u}{\partial y} + F(x,y)u \tag{33a}$$

has the adjoint operator

$$L^*(u) = \frac{\partial^2}{\partial x^2}(A(x,y)u) + \frac{\partial^2}{\partial x \partial y}(2B(x,y)u) + \frac{\partial^2}{\partial y^2}(C(x,y)u)$$

$$-\frac{\partial}{\partial x}(D(x,y)u) - \frac{\partial}{\partial y}(E(x,y)u) + F(x,y)u \tag{33b}$$

(a) Calculate $L^*(u)$ and show

$$L^*(u) = A\frac{\partial^2 u}{\partial x^2} + 2B\frac{\partial^2 u}{\partial x \partial y} + C\frac{\partial^2 u}{\partial y^2}$$

$$+\left(2\frac{\partial A}{\partial x} + 2\frac{\partial B}{\partial y} - D\right)\frac{\partial u}{\partial x} + \left(2\frac{\partial B}{\partial x} + 2\frac{\partial C}{\partial y} - E\right)\frac{\partial u}{\partial y} \tag{33c}$$

$$+\left(\frac{\partial^2 A}{\partial x^2} + 2\frac{\partial^2 B}{\partial x \partial y} + \frac{\partial^2 C}{\partial y^2} - \frac{\partial D}{\partial x} - \frac{\partial E}{\partial y} + F\right)u$$

(b) Show $L = L^*$ is a self-adjoint operator if the following conditions are satisfied.

$$\frac{\partial A}{\partial x} + \frac{\partial B}{\partial y} = D \qquad\qquad \frac{\partial B}{\partial x} + \frac{\partial C}{\partial y} = E$$

(c) Show that if L is self-adjoint it can be written in the form

$$L(u) = \frac{\partial}{\partial x}\left(A\frac{\partial u}{\partial x} + B\frac{\partial u}{\partial y}\right) + \frac{\partial}{\partial y}\left(B\frac{\partial u}{\partial x} + C\frac{\partial u}{\partial y}\right) + Fu \tag{33d}$$

(d) Verify that for arbitrary differentiable functions u, v the Lagrange identity is given by

$$vL(u) - uL^*(v) = \frac{\partial Q}{\partial x} - \frac{\partial P}{\partial y} \tag{33e}$$

where

$$Q = A\left(v\frac{\partial u}{\partial x} - u\frac{\partial v}{\partial x}\right) + B\left(v\frac{\partial u}{\partial y} - u\frac{\partial v}{\partial y}\right) + \left(D - \frac{\partial A}{\partial x} - \frac{\partial B}{\partial y}\right)uv$$

$$P = B\left(u\frac{\partial v}{\partial x} - v\frac{\partial u}{\partial x}\right) + C\left(u\frac{\partial v}{\partial y} - v\frac{\partial u}{\partial y}\right) + \left(\frac{\partial B}{\partial x} + \frac{\partial C}{\partial y} - E\right)uv$$

▶ **34.** Given $L(u) = \dfrac{\partial^2 u}{\partial x \partial y} + \alpha(x,y)\dfrac{\partial u}{\partial x} + \beta(x,y)\dfrac{\partial u}{\partial y} + \gamma(x,y)u$ for (x,y) belonging to the region R illustrated.

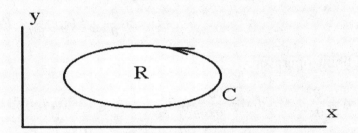

(a) Find $L^*(u)$

(b) Show for arbitrary differentiable functions u,v that

$$vL(u) - uL^*(v) = \frac{1}{2}\frac{\partial}{\partial x}\left(v\frac{\partial u}{\partial y} - u\frac{\partial v}{\partial y} + 2\alpha uv\right) + \frac{1}{2}\frac{\partial}{\partial y}\left(v\frac{\partial u}{\partial x} - u\frac{\partial v}{\partial x} + 2\beta uv\right)$$

(c) Show for C a simple-closed curve[‡] in a region R

$$\iint_R [vL(u) - uL^*(v)]\, dx\, dy = \frac{1}{2}\oint_C \left(u\frac{\partial v}{\partial x} - v\frac{\partial u}{\partial x} - 2\beta uv\right) dx + \frac{1}{2}\oint_C \left(v\frac{\partial u}{\partial y} - u\frac{\partial v}{\partial y} + 2\alpha uv\right) dy$$

where the line integrals are taken in the positive sense.

Hint: Use Green's theorem in the plane

$$\iint_R \left(\frac{\partial N}{\partial x} - \frac{\partial M}{\partial y}\right) dx\, dy = \oint_C M(x,y)\, dx + N(x,y)\, dy = \oint_C (N(x,y)\hat{i} - M(x,y)\hat{j})\cdot\hat{n}\, ds$$

where ds is an element of arc length along the curve and $\hat{n}$ is a unit exterior normal to the boundary curve that can be determined from the unit tangent $\hat{t}$ to the curve, since $\hat{n} = \hat{t}\times\hat{k} = \dfrac{d\vec{r}}{ds}\times\hat{k} = \dfrac{dy}{ds}\hat{i} - \dfrac{dx}{ds}\hat{j}$ with $\hat{n}\cdot\hat{i} = \dfrac{dy}{ds}$ and $\hat{n}\cdot\hat{j} = -\dfrac{dx}{ds}$. The boundary curve C is understood to be a simple closed path enclosing the region R and the line integral around C is in the positive sense. Note Green's theorem in the plane is a special case of Stokes' theorem

$$\iint_S (\nabla\times\vec{F})\cdot\hat{n}\, d\sigma = \oint_C \vec{F}\cdot d\vec{r}.$$

[‡] A parametric curve $\{x(t), y(t)\}$, $a \le t \le b$ has end points $(x(a), y(a))$ and $(x(b), y(b))$. A curve is called closed if its end points coincide. If (x_0, y_0) is a point on the curve, which is not an end point, such that there exists more than one value of the parameter t such that $(x(t), y(t)) = (x_0, y_0)$, then the point (x_0, y_0) is called a multiple point or a point where the curve crosses itself. A curve is called a simple closed curve if it has no multiple points. Simple closed curves are defined by one-to-one mappings.

▶ **35.** Show that by an integration of Lagrange's identity over the region R illustrated in the previous problem, there is obtained the Green's formula

$$\iint_R (vL(u) - uL^*(v))\, dxdy = \iint_R \left(\frac{\partial Q}{\partial x} - \frac{\partial P}{\partial y} \right)\, dxdy$$

$$= \oint_C P(x,y)\, dx + Q(x,y)\, dy = \oint_C (Q(x,y)\hat{i} - P(x,y)\hat{j}) \cdot \hat{n}\, ds$$

where the line integral around C is taken in the positive sense.

▶ **36.**

(a) Show the Lagrange identity associated with the wave operator

$$L(u) = \frac{\partial^2 u}{\partial t^2} - c^2 \frac{\partial^2 u}{\partial x^2},$$

where c is a constant, is given by

$$vL(u) - uL^*(v) = \frac{\partial Q}{\partial x} - \frac{\partial P}{\partial t}$$

where $Q = -c^2 \left(v\dfrac{\partial u}{\partial x} - u\dfrac{\partial v}{\partial x} \right)$ and $P = u\dfrac{\partial v}{\partial t} - v\dfrac{\partial u}{\partial t}$.

(b) Show the Green's formula associated with the operator $L(u)$ is given by

$$\iint_R [vL(u) - uL^*(v)]\, dxdt = \oint_C \left(u\frac{\partial v}{\partial t} - v\frac{\partial u}{\partial t} \right)\, dx - c^2 \left(v\frac{\partial u}{\partial x} - u\frac{\partial v}{\partial x} \right)\, dt$$

$$= \oint_C \left\{ -c^2 \left(v\frac{\partial u}{\partial x} - u\frac{\partial v}{\partial x} \right) \hat{i} - \left(u\frac{\partial v}{\partial t} - v\frac{\partial u}{\partial t} \right) \hat{j} \right\} \cdot \hat{n}\, ds.$$

▶ **37.**

(a) Verify the heat operator $L(u) = \dfrac{\partial u}{\partial t} - c^2 \dfrac{\partial^2 u}{\partial x^2}$, where c is a constant, has the Lagrange identity $vL(u) - uL^*(v) = \dfrac{\partial Q}{\partial x} - \dfrac{\partial P}{\partial t}$ where $Q = -c^2 \left(v\dfrac{\partial u}{\partial x} - u\dfrac{\partial v}{\partial x} \right)$ and $P = -uv$.

(b) Show the Green's formula associated with the operator $L(u)$ is given by

$$\iint_R [vL(u) - uL^*(v)]\, dxdt = \oint_C -uv\, dx - c^2 \left(v\frac{\partial u}{\partial x} - u\frac{\partial v}{\partial x} \right)\, dt$$

$$= \oint_C \left\{ -c^2 \left(v\frac{\partial u}{\partial x} - u\frac{\partial v}{\partial x} \right) \hat{i} + uv\hat{j} \right\} \cdot \hat{n}\, ds$$

▶ **38.**

(a) Show the Lagrange identity associated with the Laplacian operator

$$L(u) = \frac{\partial^2 u}{\partial x^2} + \frac{\partial^2 u}{\partial y^2}$$

is given by

$$vL(u) - uL^*(v) = \frac{\partial Q}{\partial x} - \frac{\partial P}{\partial y}$$

where $Q = v\dfrac{\partial u}{\partial x} - u\dfrac{\partial v}{\partial x}$ and $P = u\dfrac{\partial v}{\partial y} - v\dfrac{\partial u}{\partial y}$.

(b) Show the Green's formula associated with the operator $L(u)$ is given by

$$\iint_R (vL(u) - uL^*(v))\, dxdy = \oint_C \left(u\frac{\partial v}{\partial y} - v\frac{\partial u}{\partial y} \right) dx + \left(v\frac{\partial u}{\partial x} - u\frac{\partial v}{\partial x} \right) dy$$

$$= \oint_C v\left(\frac{\partial u}{\partial x}\hat{i} + \frac{\partial u}{\partial y}\hat{j} \right) \cdot \hat{n}\, ds - u\left(\frac{\partial v}{\partial x}\hat{i} + \frac{\partial v}{\partial y}\hat{j} \right) \cdot \hat{n}\, ds$$

$$= \oint_C (v\nabla u \cdot \hat{n} - u\nabla v \cdot \hat{n})\, ds$$

$$= \oint_C \left(v\frac{\partial u}{\partial n} - u\frac{\partial v}{\partial n} \right) ds$$

where

$$\frac{\partial u}{\partial n} = \nabla u \cdot \hat{n} \qquad \frac{\partial v}{\partial n} = \nabla v \cdot \hat{n}$$

are normal derivatives or directional derivatives on the boundary curve C in the direction of the outward normal to the region R and ds is an element of arc length.

▶ **39.**

(a) Show that $\dfrac{d}{dx}\left[x^\nu J_\nu(x) \right] = x^\nu J_{\nu-1}(x)$

(b) Show that $\dfrac{d}{dx}\left[x^{-\nu} J_\nu(x) \right] = -x^{-\nu} J_{\nu+1}(x)$

(c) Show parts (a) and (b) imply

$$\frac{dJ_\nu(x)}{dx} = J_{\nu-1}(x) - \frac{\nu}{x}J_\nu(x) \qquad \text{and} \qquad \frac{dJ_\nu(x)}{dx} = -J_{\nu+1}(x) + \frac{\nu}{x}J_\nu(x)$$

(d) Add and subtract the results in part (c) to show

$$\frac{dJ_\nu(x)}{dx} = \frac{1}{2}\left[J_{\nu-1}(x) - J_{\nu+1}(x) \right] \quad \text{and} \quad J_{\nu+1}(x) = \frac{2\nu}{x}J_\nu(x) - J_{\nu-1}(x)$$

(e) Show that $\dfrac{d}{dx}\left[x^\nu Y_\nu(x) \right] = x^\nu Y_{\nu-1}(x)$ Hint: Use the Weber-Schlafi ratio.

Chapter 2
Orthogonal Functions

The term dyadic is used in referring to something that has two parts. For example, one can imagine the mathematical operation of a dot product of two vectors

$$\vec{A} = A_1\hat{i} + A_2\hat{j} + A_3\hat{k} \qquad \text{and} \qquad \vec{B} = B_1\hat{i} + B_2\hat{j} + B_3\hat{k}$$

as being represented by a dyadic operator box which has two inputs and an output. Such an operator box is illustrated in the figure 2-1.

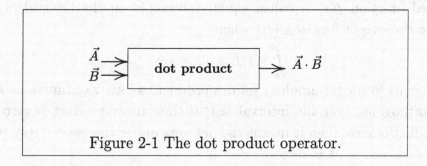

Figure 2-1 The dot product operator.

The figure 2-1 illustrates two vector inputs into an operator box. The operation inside the box is the dot product. The output from the operator box is a scalar representing the dot product of the two vector inputs. Recall that when the output from this operator box is zero, then the vectors are said to be orthogonal. Other types of dyadic operators can be defined and illustrated as an operator box. Consider the dyadic operator illustrated in the figure 2-2.

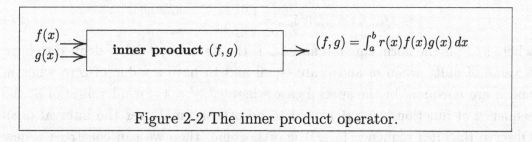

Figure 2-2 The inner product operator.

In figure 2-2 the inputs are real functions $f(x)$ and $g(x)$ of a single real variable and the output is a weighted integral over the domain $a \leq x \leq b$ of the input functions. The function $r(x)$ is called the weight function of the integral. The weight function is never negative. This type of operation is called an inner product and denoted by the notation

$$(f, g) = \int_a^b r(x) f(x) g(x) \, dx. \tag{2.1}$$

The inner product of a function with itself is called a norm squared and written

$$(f, f) = \|f\|^2 = \int_a^b r(x) f^2(x) \, dx. \tag{2.2}$$

We assume the input functions $f(x)$ and $g(x)$ are nonzero, bounded and such that the resulting inner product integral exists. We also assume that the norm squared is nonzero unless $f(x)$ is identically zero for all x over the interval. A real-valued function $f(x)$ is called square-integrable on the interval (a, b) with respect to the weight function $r(x)$ when

$$\int_a^b r(x) f^2(x) \, dx < +\infty. \tag{2.3}$$

Analogous to the terminology for dot products we say two functions $f(x)$ and $g(x)$ are orthogonal over the interval (a, b) if their inner product is zero. If the inner product is zero, then it means the net area under the $y = r(x) f(x) g(x)$ curve between $x = a$ and $x = b$ is zero.

A set or sequence of functions $\{f_1(x), f_2(x), \ldots, f_n(x), \ldots, f_m(x), \ldots\}$ is said to be orthogonal over an interval (a, b) with respect to a weight function $r(x) > 0$ if for all integer values of n and m, with $n \neq m$, the inner product of f_m with f_n satisfies

$$(f_m, f_n) = \int_a^b r(x) f_m(x) f_n(x) \, dx = 0 \qquad m \neq n. \tag{2.4}$$

Here the inner product is zero for all combinations of m and n values with $m \neq n$. If the sequence of functions $\{f_n(x)\}$, $n = 0, 1, 2, \ldots$ is an orthogonal sequence we can write for integers m and n that the inner product satisfies the relation

$$(f_m, f_n) = \|f_n\|^2 \delta_{mn} = \begin{cases} 0 & m \neq n \\ \|f_n\|^2 & m = n \end{cases} \tag{2.5}$$

where $\|f_n\|^2$ is the norm squared and δ_{mn} is the Kronecker† delta defined to have a value of unity when m and n are equal and to have a value of zero when m and n are unequal. In the special case where $\|f_n\|^2 = 1$, for all values of n, the sequence of functions $\{f_n(x)\}$ is said to be orthonormal over the interval (a, b). Observe that if a sequence $\{g_n(x)\}$ is orthogonal, then we can construct a new orthonormal sequence $\{f_n(x)\}$, defined by $f_n(x) = \dfrac{g_n(x)}{\|g_n(x)\|}$. The proof is left as an exercise.

---

† See Appendix C

Example 2-1. (Orthogonal set)

The set of functions $\{f_n = \sin \frac{n\pi x}{L}\}$ for $n = 1, 2, 3, \ldots$ is a set of orthogonal functions over the interval $(0, L)$ with respect to the weight function $r(x) = 1$. This is shown by calculating the inner products

$$(f_m, f_n) = (\sin \frac{m\pi x}{L}, \sin \frac{n\pi x}{L}) = \int_0^L \sin \frac{m\pi x}{L} \sin \frac{n\pi x}{L} \, dx$$

$$= \frac{L}{\pi} \left[\frac{\sin(m-n)\frac{\pi x}{L}}{2(m-n)} - \frac{\sin(m+n)\frac{\pi x}{L}}{2(m+n)} \right]_0^L = 0, \qquad m \neq n$$

The norm squared for each function is given by

$$(f_m, f_m) = ||f_m||^2 = \int_0^L \sin^2 \frac{m\pi x}{L} \, dx = \int_0^L \frac{1}{2} \left(1 - \cos \frac{2m\pi x}{L} \right) dx$$

$$= \left[\frac{x}{2} - \frac{L}{4m\pi} \sin \frac{2m\pi x}{L} \right]_0^L = \frac{L}{2}$$

for $m = 1, 2, 3, \ldots$. These results are written in inner product notation as

$$(\sin \frac{m\pi x}{L}, \sin \frac{n\pi x}{L}) = \frac{L}{2} \delta_{mn} = \begin{cases} 0 & m \neq n \\ \frac{L}{2} & m = n \end{cases} \tag{2.6}$$

Note also the sequence $\{\sqrt{\frac{2}{L}} \sin \frac{n\pi x}{L}\}$ is an orthonormal sequence.

∎

Example 2-2. (Orthogonal set)

For $n = 1, 2, 3, \ldots$ the set of functions $\{1, \cos \frac{n\pi x}{L}\}$, is a set of orthogonal functions over the interval $(0, L)$ with respect to the weight function $r(x) = 1$. This is shown by calculating the inner products

$$(1, \cos \frac{n\pi x}{L}) = \int_0^L \cos \frac{n\pi x}{L} \, dx = \frac{L}{n\pi} \sin \frac{n\pi x}{L} \Big]_0^L = 0, \quad n = 1, 2, 3, \ldots$$

$$(\cos \frac{m\pi x}{L}, \cos \frac{n\pi x}{L}) = \int_0^L \cos \frac{m\pi x}{L} \cos \frac{n\pi x}{L} \, dx$$

$$= \frac{L}{\pi} \left[\frac{\sin(m-n)\frac{\pi x}{L}}{2(m-n)} + \frac{\sin(m+n)\frac{\pi x}{L}}{2(m+n)} \right]_0^L = 0, \qquad m \neq n.$$

The norm squared is $\qquad ||1||^2 = \int_0^L dx = L.$

and for $n = 1, 2, 3 \ldots$, $\qquad ||\cos \frac{n\pi x}{L}||^2 = \int_0^L \cos^2 \frac{n\pi x}{L} \, dx$

$$= \int_0^L \frac{1}{2} \left(1 + \cos \frac{2n\pi x}{L} \right) dx$$

$$= \left[\frac{x}{2} + \frac{L}{4n\pi} \sin \frac{2n\pi x}{L} \right]_0^L = \frac{L}{2}.$$

These results can be summarized using the inner product notation

$$(\cos \frac{m\pi x}{L}, \cos \frac{n\pi x}{L}) = \begin{cases} 0, & m \neq n \\ \frac{L}{2}, & m = n \neq 0 \\ L, & m = n = 0 \end{cases} . \tag{2.7}$$

The set of orthogonal functions was written as $\{1, \cos \frac{n\pi x}{L}\}$ for $n = 1, 2, 3, \ldots$ rather than in the form $\{\cos \frac{n\pi x}{L}\}$ for $n = 0, 1, 2, 3, \ldots$ because we wanted to emphasize that the case $n = 0$ had a different norm squared value.

∎

The inner product associated with real continuous functions $f(x)$, $g(x)$ and sets of real functions $\{f_k(x)\}$ and $\{g_k(x)\}$ satisfies the following properties:

$$(f, g) = (g, f)$$

$$(cf, g) = c(f, g) \quad \text{and} \quad (f, cg) = c(f, g) \quad \text{where } c \text{ is a constant.}$$

$$(f_1 + f_2, g) = (f_1, g) + (f_2, g) \quad \text{and} \quad (f, g_1 + g_2) = (f, g_1) + (f, g_2) \tag{2.8}$$

$$(\sum_{k=1}^{n} f_k, g) = \sum_{k=1}^{n} (f_k, g) \quad \text{and} \quad (f, \sum_{k=1}^{n} g_k) = \sum_{k=1}^{n} (f, g_k)$$

$$\|f\| = 0 \quad \text{if and only if} \quad f = 0.$$

Gram-Schmidt Orthogonalization Process

A linear combination of a set of functions $\{f_1(x), f_2(x), \ldots, f_n(x)\}$ is written

$$c_1 f_1(x) + c_2 f_2(x) + \cdots + c_n f_n(x)$$

where $c_1, c_2, \ldots, c_n$ are arbitrary constants. If nonzero constants can be found such that some linear combination of the functions $\{f_m(x)\}$ gives zero, then the set of functions is said to be linearly dependent. If the only linear combination that produces zero, for all values of x, requires that $c_1 = c_2 = \ldots = c_n = 0$, then the set of function $\{f_m(x)\}$, $m = 1, \ldots, n$ is said to be linearly independent. An infinite set is said to be linearly independent if every finite subset is linearly independent. Observe that if the set $\{f_m(x)\}$ is an orthogonal set, then it is a linearly independent set. To show this we assume for some finite value n we have a linear combination which equals zero. This gives

$$c_1 f_1(x) + c_2 f_2(x) + \cdots + c_n f_n(x) = 0.$$

Now multiply both sides of this equation by the function $r(x)f_k(x)$ where k is fixed and has some value between 1 and n, and then integrate over the interval where the functions are orthogonal. This produces the sum of inner products which must equal zero

$$c_1(f_1, f_k) + c_2(f_2, f_k) + \cdots + c_k(f_k, f_k) + \cdots c_n(f_n, f_k) = 0. \tag{2.9}$$

Examine each inner product in equation (2.9) and note the only nonzero term on the left-hand side of this equation is the kth term with nonzero inner product $c_k(f_k, f_k) = c_k\|f_k\|^2$. The equation (2.9) reduces to $c_k\|f_k\|^2 = 0$ which implies, for $\|f_k\| \neq 0$, that $c_k = 0$. Hence, for $k = 1, \ldots, n$, we can show all c_k are zero which implies the set is linearly independent.

An orthogonal set of functions $\{g_n(x)\}$ can be constructed from a set of nonorthogonal linearly independent functions $\{f_n(x)\}$. This is accomplished by defining the functions

$$g_0(x) = f_0(x) \quad \text{and} \quad g_1(x) = f_1(x) - c_{01}g_0(x)$$

where the constant c_{01} is select to make the inner product (g_0, g_1) equal to zero. The next function is constructed

$$g_2(x) = f_2(x) - c_{02}g_0(x) - c_{12}g_1(x)$$

where the constants c_{02}, c_{12} are selected to make the inner products (g_0, g_2) and (g_1, g_2) both zero. The next function is constructed

$$g_3(x) = f_3(x) - c_{03}g_0(x) - c_{13}g_1(x) - c_{23}g_2(x)$$

where the constants c_{03}, c_{13}, c_{23} are selected to make the three inner products $(g_0, g_3),(g_1, g_3)$ and (g_2, g_3) all zero. Continuing in this manner the nth constructed function has the form

$$g_n(x) = f_n(x) - \sum_{k=0}^{n-1} c_{kn}g_k(x)$$

where each of the constants c_{kn}, $k = 0, \ldots, n-1$ are selected to make the inner products $(g_m, g_n) = 0$ for $m = 0, 1, 2, \ldots, n-1$. This process is known as the Gram-Schmidt[†] orthogonalization process. By examining the general term we can require the general inner product equal zero or

$$(g_m, g_n) = (g_m, f_n - \sum_{k=0}^{n-1} c_{kn}g_k(x)) = 0$$

$$(g_m, g_n) = (g_m, f_n) - \sum_{k=0}^{n-1} c_{kn}(g_m, g_k) = 0$$

---

† See Appendix C

for m having some fixed value which is less than or equal to $n-1$. Now examine the summation term and note the only nonzero term in this sum occurs when $k = m$ and consequently

$$(g_m, g_n) = (g_m, f_n) - c_{mn}\|g_m\|^2 = 0$$

which implies $\quad c_{kn} = \dfrac{(g_k, f_n)}{\|g_k\|^2} \quad$ for $k = 0, 1, 2, \ldots, n-1$. We thus find that each of the unknown coefficients in the Gram-Schmidt orthogonalization process is given by an inner product divided by a norm squared.

Example 2-3. (Orthogonal set)

From the set of linearly independent functions $f_n(x) = x^n$ for $n = 0, 1, 2, \ldots$, construct a set of orthogonal functions over the interval $(-1, 1)$ with respect to the weight function $r(x) = 1$. We begin the Gram-Schmidt orthogonalization process by setting $g_0(x) = f_0(x) = 1$ and calculating $\|1\|^2 = 2$, then set

$$g_1(x) = f_1(x) - c_{01}g_0(x) \quad \text{with } c_{01} \text{ selected such that } (g_1, g_0) = 0$$

we find $\quad g_1(x) = x - c_{01}(1) \quad$ with $\quad c_{01} = \dfrac{(x, 1)}{\|1\|^2} = 0$

or $\quad g_1(x) = x \quad$ with $\quad \|x\|^2 = (x, x) = \displaystyle\int_{-1}^{1} x^2\, dx = \dfrac{2}{3}$.

Observe that the constant c_{01} is given by an inner product divided by a norm squared. The next term in the orthogonal set is

$$g_2(x) = f_2(x) - c_{02}g_0(x) - c_{12}g_1(x)$$
$$g_2(x) = x^2 - c_{02}(1) - c_{12}x$$

where the coefficients are selected to make both the inner products (g_0, g_2) and (g_1, g_2) equal to zero. We find these coefficients are given by an inner product divided by a norm squared or

$$c_{02} = \frac{(x^2, 1)}{\|1\|^2} = \frac{1}{3}, \qquad c_{12} = \frac{(x^2, x)}{\|x\|^2} = 0.$$

This gives

$$g_2(x) = x^2 - \frac{1}{3} \qquad \text{with} \qquad \|g_2\|^2 = \frac{8}{45}.$$

The next term is given by

$$g_3(x) = f_3(x) - c_{03}g_0(x) - c_{13}g_1(x) - c_{23}g_2(x)$$
$$g_3(x) = x^3 - c_{03}(1) - c_{13}x - c_{23}\left(x^2 - \frac{1}{3}\right)$$

where again the coefficients are given by inner products divided by a norm squared or

$$c_{03} = \frac{(x^3, 1)}{||1||^2} = 0, \qquad c_{13} = \frac{(x^3, x)}{||x||^2} = \frac{3}{5}, \qquad c_{23} = \frac{(x^3, x^2 - \frac{1}{3})}{||x^2 - \frac{1}{3}||^2} = 0$$

which gives

$$g_3(x) = x^3 - \frac{3}{5}x \qquad \text{and} \qquad ||g_3||^2 = \frac{8}{175}.$$

Continuing in this manner one can generate the additional terms

$$g_4(x) = x^4 - \frac{6}{7}x^2 + \frac{3}{35} \quad \text{with} \quad ||g_4||^2 = \frac{128}{11025}$$

$$g_5(x) = x^5 - \frac{10}{9}x^3 + \frac{5}{21}x \quad \text{with} \quad ||g_5||^2 = \frac{128}{43659}$$

$$\cdots$$

∎

Orthogonal sets of functions are extremely important and can be used to help construct solutions to partial differential equations. Their occurrence is not accidental as we shall show in the next section.

Sturm-Liouville Systems

A regular Sturm-Liouville[†] system consists of the linear homogeneous differential equation

$$L(y) = \frac{d}{dx}\left(p(x)\frac{dy}{dx}\right) + q(x)y = -\lambda r(x)y \qquad (2.10)$$

over an interval $a \le x \le b$, containing a parameter λ, and subject to boundary conditions at each end point of the form

$$\beta_1 y(a) + \beta_2 \frac{dy(a)}{dx} = 0 \qquad \text{and} \qquad \beta_3 y(b) + \beta_4 \frac{dy(b)}{dx} = 0 \qquad (2.11)$$

where $\beta_1, \beta_2, \beta_3, \beta_4$ are real constants independent of λ. It is to be understood that for a regular Sturm-Liouville problem the constants β_1, β_2, as well as β_3, β_4, cannot both be zero simultaneously. This can be expressed $\beta_1^2 + \beta_2^2 \ne 0$ and $\beta_3^2 + \beta_4^2 \ne 0$. The differential equation (2.10) has a parameter λ and the operator $L(y)$ is a self-adjoint differential operator. The coefficients in the differential equation (2.10) must be such that $p(x), p'(x), q(x), r(x)$ are real and continuous with the requirement $p(x) > 0$ and $r(x) > 0$ over the solution interval $a \le x \le b$.

[†] See Appendix C

Whenever the boundary conditions given by equation (2.11) are replaced by periodic boundary conditions of the form

$$y(a) = y(b), \qquad y'(a) = y'(b), \qquad (2.12)$$

where the prime $'$ denotes differentiation with respect to the argument of the function, then the set of equations (2.10) and (2.12) is called a periodic Sturm-Liouville system.

Note the parameter λ occurring in equation (2.10) can be written in different forms. For example one can replace λ by $\sqrt{\lambda}, \lambda^2, -\lambda^2, -2\lambda$ or some other representation of a constant. The form selected is usually made in order that the differential equation take on an easier form to solve or the choice of a particular form makes some resulting algebra easier.

Associated with the Sturm-Liouville equation (2.10) we introduce the following concepts, notations and properties. Some of the properties are given without proofs as these proofs can be found in other textbooks.

1. We desire nonzero continuous solutions to the Sturm-Liouville system. We find there can be no solutions, a unique solution or an infinite number of solutions to the Sturm-Liouville system given by the equation (2.10). The number and type solutions depends upon the value selected for λ. The parameter λ is to be selected to obtain nonzero solutions.

2. Values of λ for which nonzero solutions exist are called eigenvalues. The set of eigenvalues associated with a Sturm-Liouville problem is called the spectrum of the problem. If all eigenvalues are real and there exists an infinite number of them, they can be labeled $\lambda_1, \lambda_2, \ldots, \lambda_n, \ldots$ where λ_1 is the smallest eigenvalue and $\lambda_n \to \infty$ as $n \to \infty$. The corresponding nonzero solution functions are called eigenfunctions. The German word 'eigen' means peculiar or specific. (The German word for eigenvalue is 'eigenwert'.) To each eigenvalue λ_n there corresponds an eigenfunction which is denoted using a subscript notation as $y_n(x) = y(x; \lambda_n)$. Observe that if y_n is an eigenfunction, then cy_n is also an eigenfunction for nonzero constants c. Sometimes it will be convenient to label the lowest eigenvalue as λ_0 and start the indexing at zero rather than one.

3. The eigenfunctions $y_n(x)$, $n = 1, 2, 3, \ldots$ have exactly $n - 1$ zeros on the interval $a \leq x \leq b$.

4. The set of eigenfunctions are orthogonal over the interval (a, b) with respect to the weight function $r(x)$. That is, the inner product associated with two different eigenfunctions satisfies

$$(y_n, y_m) = \int_a^b r(x)y_n(x)y_m(x)\,dx = 0 \quad \text{for } m \neq n \tag{2.13}$$

and the norm squared is nonzero for each value of n

$$(y_n, y_n) = ||y_n||^2 = \int_a^b r(x)y_n^2(x)\,dx \neq 0. \tag{2.14}$$

5. The eigenfunctions form a complete set. We assume that $f(x)$ is a piecewise smooth function. That is, a function which is piecewise continuous‡ over an interval and possess derivatives $f'(x)$ which are also piecewise continuous. It can be shown that under certain conditions, the function $f(x)$ can be represented by a series involving the eigenfunctions $y_n(x)$ having the form

$$f(x) = \sum_{n=n_0}^{\infty} c_n y_n(x). \tag{2.15}$$

In particular, if $f(x)$ is square-integrable with respect to the weight function $r(x)$, then we say the eigenfunctions form a complete set if

$$\lim_{k \to \infty} \int_a^b r(x) \left(f(x) - \sum_{n=n_0}^{k} c_n y_n(x) \right)^2 dx = 0 \tag{2.16}$$

where n_0 is the starting index having a value of 0 or 1 depending upon the indexing assigned to the eigenvalues. The resulting series (2.15) is called a generalized Fourier† series and the constants c_n are called Fourier coefficients. In addition it can be shown that the Fourier coefficients are given by an inner product divided by a norm squared. Thus, for $n = 1, 2, 3, \ldots$ one can write

$$c_n = \frac{(f, y_n)}{||y_n||^2} = \frac{\int_a^b r(x)f(x)y_n(x)\,dx}{\int_a^b r(x)y_n^2(x)\,dx} \tag{2.17}$$

6. The Sturm-Liouville system is called a regular system if the coefficients in the differential equation satisfy $p(x) > 0, q(x), r(x) > 0$, are real and continuous

‡ A function $f(x)$ is piecewise continuous over an interval if the interval can be subdivided into a finite number of subintervals and inside each subinterval $f(x)$ is continuous with finite limits at the end points of each subinterval.

† See Appendix C

everywhere, and boundary conditions of the type given by equations (2.11) exist.

7. If any one of the regularity conditions given in property 6 is not satisfied, then the Sturm-Liouville problem is called singular. For example, if $p(x)$ vanishes at an end point or if one of the functions $p(x), q(x), r(x)$ becomes infinite at an end point, or all the boundary condition constants are zero, or one or both of the end points a,b becomes infinite, then the Sturm-Liouville system is said to be singular.

To prove the orthogonality property we start with the hypothesis that for different eigenvalues λ_n and λ_m there are two different nonzero solutions $y_n(x)$ and $y_m(x)$ of the differential equation $L(y) = -\lambda r(x)y$, where $L(y)$ is the self-adjoint differential operator defined by equation (2.10). Therefore we can write

$$\text{for } \lambda = \lambda_m, \qquad L(y_m(x)) = -\lambda_m r(x)y_m(x)$$
$$\text{and for } \lambda = \lambda_n, \qquad L(y_n(x)) = -\lambda_n r(x)y_n(x). \tag{2.18}$$

One can employ the Green's identity given by equation (1.79) and make the substitutions $u = y_n(x)$ and $v = y_m(x)$, then the equations (2.18) simplify the resulting Green's identity and one obtains the relations

$$\int_a^b [y_n L(y_m) - y_m L(y_n)]\,dx = p(x)\left[y_n(x)y_m'(x) - y_m(x)y_n'(x)\right]_a^b$$

$$\int_a^b [y_n(-\lambda_m r(x)y_m) - y_m(-\lambda_n r(x)y_n)]\,dx = p(x)\left[y_n(x)y_m'(x) - y_m(x)y_n'(x)\right]_a^b$$

$$(\lambda_n - \lambda_m)\int_a^b r(x)y_n(x)y_m(x)\,dx = p(b)\left[y_n(b)y_m'(b) - y_m(b)y_n'(b)\right]$$
$$- p(a)\left[y_n(a)y_m'(a) - y_m(a)y_n'(a)\right]. \tag{2.19}$$

Observe that the left hand side of the equation (2.19) has an integral which represents the inner product of the two eigenfunctions. To show this inner product is zero, so that the eigenfunctions are orthogonal for $n \neq m$, we want the right-hand side of equation (2.19) to be zero. Toward this end we consider the following cases:

Case 1. If $p(a) = 0$ and $p(b) = 0$, then the right-hand side is zero and consequently the inner product satisfies $(y_n, y_m) = 0$ and so the eigenfunctions must be orthogonal. Note that this requires a singular Sturm-Liouville problem and further, no boundary conditions are required.

Case 2. If $p(a) = 0$ but $p(b) \neq 0$ then we must require both solutions y_n and y_m to satisfy the boundary conditions

$$\beta_3 y_n(b) + \beta_4 y_n'(b) = 0 \qquad \text{and} \qquad \beta_3 y_m(b) + \beta_4 y_m'(b) = 0. \qquad (2.20)$$

This is again a singular Sturm-Liouville problem. In order to have nonzero boundary conditions it is necessary that the determinant of the coefficients in equations (2.20) be equal to zero. This implies for β_3, β_4 different from zero we must have

$$y_n(b) y_m'(b) - y_n'(b) y_m(b) = 0.$$

This condition together with the condition $p(a) = 0$ makes the right-hand side of equation (2.19) zero, which shows the inner product of the eigenfunctions is zero and hence the functions $y_n(x)$ and $y_m(x)$ are orthogonal.

Case 3. If $p(b) = 0$ but $p(a) \neq 0$ then we require both y_n and y_m satisfy boundary conditions of the type

$$\beta_1 y_n(a) + \beta_2 y_n'(a) = 0 \qquad \text{and} \qquad \beta_1 y_m(a) + \beta_2 y_m'(a) = 0. \qquad (2.21)$$

Again, for β_1, β_2 nonzero the determinant of the coefficients requires

$$y_n(a) y_m'(a) - y_m(a) y_n'(a) = 0$$

and hence the right-hand side of equation (2.19) is zero and the eigenfunctions are orthogonal. Note these conditions require a singular Sturm-Liouville problem.

Case 4. Here we assume both $p(a) \neq 0$ and $p(b) \neq 0$. In this case we require both y_n and y_m satisfy boundary conditions at the end points such that

$$\begin{aligned}
\beta_1 y_n(a) + \beta_2 y_n'(a) &= 0, & \beta_3 y_n(b) + \beta_4 y_n'(b) &= 0 \\
\beta_1 y_m(a) + \beta_2 y_m'(a) &= 0, & \beta_3 y_m(b) + \beta_4 y_m'(b) &= 0.
\end{aligned} \qquad (2.22)$$

These conditions imply both

$$y_n(a) y_m'(a) - y_m(a) y_n'(a) = 0 \qquad \text{and} \qquad y_n(b) y_m'(b) - y_m(b) y_n'(b) = 0$$

which in turn make the right-hand side of equation (2.19) zero and consequently the eigenfunctions are orthogonal. Note this requires a regular Sturm-Liouville problem.

The previously listed property of Sturm-Liouville systems given by equation (2.17) follows directly from the property 4 and equation (2.15). Observe that if a piecewise smooth function $f(x)$ is represented as a series of eigenfunctions, then we are representing $f(x)$ in the form of a series

$$f(x) = \sum_{n=1}^{\infty} c_n y_n(x) = c_1 y_1(x) + c_2 y_2(x) + c_3 y_3(x) + \cdots \tag{2.23}$$

with constants c_n to be determined. (Here it is assumed the indexing of the summation begins with n=1.) Now multiply both sides of equation (2.23) by $r(x) y_m(x)\, dx$ and integrate both sides of the resulting equation from a to b. There results the series of inner products

$$(f, y_m) = c_1(y_1, y_m) + c_2(y_2, y_m) + \cdots + c_m(y_m, y_m) + \cdots . \tag{2.24}$$

The set of functions $\{y_n\}$ are orthogonal over the interval (a, b) so the only nonzero term on the right-hand side of equation (2.24) is the term with the index m. Hence the equation (2.24) simplifies to

$$(f, y_m) = c_m(y_m, y_m) = c_m \|y_m\|^2 \tag{2.25}$$

which shows the coefficients of the series expansion are given by an inner product divided by a norm squared or

$$c_m = \frac{(f, y_m)}{\|y_m\|^2} = \frac{\int_a^b r(x) f(x) y_m(x)\, dx}{\int_a^b r(x) y_m^2(x)\, dx} \qquad \text{for} \quad m = 1, 2, 3, \ldots . \tag{2.26}$$

We will have more to say about these types of expansions in the next chapter.

Example 2-4. **(Orthogonal Trigonometric Functions)**
Solve the Sturm-Liouville problem

$$y'' + \lambda y = 0, \qquad 0 \leq x \leq L, \qquad y(0) = 0 \quad \text{and} \quad y(L) = 0$$

Solution: This is a regular Sturm-Liouville problem with coefficients $p(x) = 1$, $q(x) = 0$ and $r(x) = 1$. The differential equation is already in self-adjoint form. We desire to find values of λ for which there exists nonzero solutions. We examine the cases $\lambda = -\omega^2$ of negative eigenvalues, $\lambda = 0$ of a zero eigenvalue, and $\lambda = \omega^2$ of positive eigenvalues. Here ω is always assumed to be a positive constant.

Case 1 $\lambda = -\omega^2$. The solution of the differential equation $y'' - \omega^2 y = 0$ is represented in the form $y = C_1 \sinh(\omega x) + C_2 \cosh(\omega x)$, with C_1, C_2 constants, as the algebra will be easier for this form of the solution. We have

$$y(0) = 0 \quad \text{requires} \quad C_1 \sinh(0) + C_2 \cosh(0) = 0 \quad \text{or} \quad C_2 = 0$$

$$y(L) = 0 \quad \text{requires} \quad C_1 \sinh(L) = 0 \quad \text{or} \quad C_1 = 0.$$

This shows that only the trivial solution exists. We do not want the trivial solution and so we go on to the next case.

Case 2 $\lambda = 0$. To solve the differential equation $y'' = 0$ we integrate twice to obtain $y = y(x) = C_1 x + C_2$, with C_1, C_2 constants. The boundary conditions require

$$y(0) = 0 \quad \text{or} \quad y(0) = C_2 = 0. \quad \text{At } x = L \text{ we want}$$

$$y(L) = 0 \quad \text{or} \quad y(L) = C_1 L = 0 \quad \text{which requires } C_1 = 0.$$

This gives $y = 0$ as the solution. This is the trivial solution and so we go on to the next case.

Case 3 $\lambda = \omega^2$. The solutions to the differential equation $y'' + \omega^2 y = 0$ are written in the form $y = y(x) = C_1 \sin \omega x + C_2 \cos \omega x$ where C_1, C_2 are arbitrary constants. The boundary conditions require

$$y(0) = 0 \quad \text{or} \quad y(0) = C_2 = 0$$

$$y(L) = 0 \quad \text{or} \quad y(L) = C_1 \sin \omega L = 0.$$

If $C_1 = 0$ we get the trivial solution. We don't want this so let $C_1 = 1$ for convenience. We are then left with the requirement $\sin \omega L = 0$. The figure 2-3 illustrates a graph of the sine curve which oscillates between $+1$ and -1.

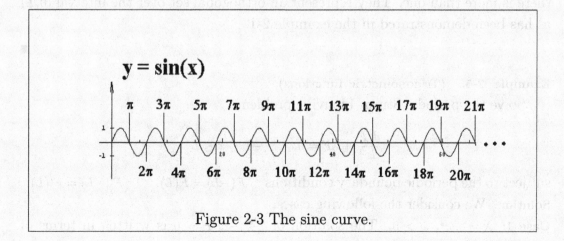

Figure 2-3 The sine curve.

Observe the sine curve has an infinite number of zeros that are equally spaced at multiples of π. Hence the equation $\sin \omega L = 0$ has an infinite number of solutions. The solutions of $\sin \omega L = 0$ require that

$$\omega L = n\pi \quad \text{for} \quad n = 1, 2, 3, \ldots, \tag{2.27}$$

and so we write ω using an index notation

$$\omega = \omega_n = \frac{n\pi}{L} \quad \text{for} \quad n = 1, 2, 3, \ldots$$

These values were selected to satisfy the required boundary conditions. The corresponding eigenvalues are also written using an index notation

$$\lambda = \lambda_n = \omega_n^2 = \left(\frac{n\pi}{L}\right)^2 \quad \text{for} \quad n = 1, 2, 3, \ldots$$

Remark 1. Here we have changed notations for ω and λ by placing a subscript n on them to emphasize that there is more than one value which gives a nonzero solution.

Remark 2. We did not include $n = 0$ as this value gives $\omega = 0$ which in turn gives $y = 0$ which is the trivial solution. Recall that we want nonzero solutions to the Sturm-Liouville problem. Also the case for $\lambda = 0$ was previously discussed.

Associated with each eigenvalue there is an eigenfunction which we write as

$$y_n(x) = y(x; \lambda_n) = \sin \omega_n x = \sin \frac{n\pi x}{L} \quad \text{for} \quad n = 1, 2, 3, \ldots \tag{2.28}$$

These eigenfunctions are also written using an index notation to emphasize that there is more than one. They represent an orthogonal set over the interval $(0, L)$ as has been demonstrated in the example 2-1.

∎

Example 2-5.   (Trigonometric functions)

Solve the periodic Sturm-Liouville problem

$$\frac{d^2 F}{dx^2} + \lambda F = 0, \qquad -L \le x \le L$$

subject to the periodic boundary conditions $F(-L) = F(L), \qquad F'(-L) = F'(L)$.

Solution: We consider the following cases.

Case 1: $\lambda = -\omega^2$, $\omega > 0$. The solution of $\frac{d^2 F}{dx^2} - \omega^2 F = 0$ is written in terms of hyperbolic functions $F = F(x) = c_1 \sinh \omega x + c_2 \cosh \omega x$. We select this form for the solution because the subsequent algebra will be easier. Applying the boundary conditions we find

$$F(-L) = c_1 \sinh(-\omega L) + c_2 \cosh(-\omega L) = F(L) = c_1 \sinh \omega L + c_2 \cosh \omega L$$

$$F'(-L) = c_1 \omega \cosh(-\omega L) + c_2 \omega \sinh(-\omega L) = F'(L) = c_1 \omega \cosh \omega L + c_2 \omega \sinh \omega L.$$

The solution of this system gives $c_1 = c_2 = 0$ and so F turns out to produce the trivial solution. We discard this case.

Case 2: $\lambda = 0$. The solution to $\frac{d^2 F}{dx^2} = 0$ is $F = F(x) = c_1 x + c_2$. The periodic boundary conditions require

$$F'(-L) = c_1 = F'(L) = c_1 \qquad \text{and}$$

$$F(-L) = c_1(-L) + c_2 = F(L) = c_1 L + c_2.$$

This requires $c_1 = 0$, and the constant c_2 be arbitrary. We thus have $\lambda = 0$ is an eigenvalue with corresponding eigenfunction $F(x) = c_2$, with c_2 arbitrary and different from zero. Using the property 2, that any constant times an eigenfunction is also an eigenfunction, we can set $c_2 = 1$ and label the eigenfunction $F_0(x) = 1$, because if $F_0(x)$ is an eigenfunction then any constant times $F_0(x)$ is still an eigenfunction.

Case 3: $\lambda = \omega^2$, $\omega > 0$. The solution to the equation $\frac{d^2 F}{dx^2} + \omega^2 F = 0$ is given by

$$F = F(x) = c_1 \cos \omega x + c_2 \sin \omega x$$

$$\text{with} \qquad F' = F'(x) = -c_1 \omega \sin \omega x + c_2 \omega \cos \omega x.$$

The periodic boundary conditions require

$$F(-L) = c_1 \cos(-\omega L) + c_2 \sin(-\omega L) = F(L) = c_1 \cos \omega L + c_2 \sin \omega L$$

which requires $2c_2 \sin \omega L = 0$ with c_1 arbitrary. The derivative condition requires $F'(-L) = -c_1 \omega \sin(-\omega L) + c_2 \omega \cos(-\omega L) = F(L) = -c_1 \omega \sin \omega L + c_2 \omega \cos \omega L$ which implies $2c_1 \sin \omega L = 0$, with c_2 arbitrary. Thus, c_1 and c_2 can both be arbitrary if we require $\sin \omega L = 0$. This equation has an infinite number of solutions. We write

$$\omega L = n\pi \quad \text{for} \quad n = 1, 2, 3, \ldots \quad \text{or} \quad \omega = \omega_n = \frac{n\pi}{L} \quad \text{for} \quad n = 1, 2, 3, \ldots$$

Here we have relabeled the eigenvalue solutions to reflect the fact that there are an infinite number of them. The eigenvalues are therefore given by

$$\lambda = \lambda_n = \omega_n^2 = \frac{n^2 \pi^2}{L^2} \quad \text{for} \quad n = 1, 2, 3, \ldots$$

The corresponding eigenfunctions are $\{\sin \frac{n\pi x}{L}, \cos \frac{n\pi x}{L}\}$ for $n = 1, 2, 3, \ldots$. Note the function $F(x) = c_1 \cos \frac{n\pi x}{L} + c_2 \sin \frac{n\pi x}{L}$ is a solution for arbitrary values of c_1 and c_2. We let $c_1 = 1$ and $c_2 = 0$ giving only cosine terms and we let $c_1 = 0$ and $c_2 = 1$ to obtain only sine terms. In summary we have shown

84

The periodic Sturm-Liouville problem

$$\frac{d^2F}{dx^2} + \lambda F = 0, \quad -L \le x \le L \tag{2.29}$$

with periodic boundary conditions $F(-L) = F(L)$ and $F'(-L) = F(L)$ has the eigenvalues $\lambda_0 = 0$, $\lambda_n = \frac{n^2\pi^2}{L^2}$ for $n = 1, 2, 3, \ldots$ with corresponding eigenfunctions $\{1, \sin\frac{n\pi x}{L}, \cos\frac{n\pi x}{L}\}$. These eigenfunctions satisfy the orthogonality conditions

$$(1, \sin\frac{n\pi x}{L}) = \int_{-L}^{L} \sin\frac{n\pi x}{L}\, dx = 0 \quad n = 1, 2, 3, \ldots$$

$$(1, \cos\frac{n\pi x}{L}) = \int_{-L}^{L} \cos\frac{n\pi x}{L}\, dx = 0$$

$$(\sin\frac{n\pi x}{L}, \sin\frac{m\pi x}{L}) = \int_{-L}^{L} \sin\frac{n\pi x}{L}\sin\frac{m\pi x}{L}\, dx = 0 \quad n \ne m \tag{2.29)(a)}$$

$$(\cos\frac{n\pi x}{L}, \cos\frac{m\pi x}{L}) = \int_{-L}^{L} \cos\frac{n\pi x}{L}\cos\frac{m\pi x}{L}\, dx = 0 \quad n \ne m$$

$$(\cos\frac{n\pi x}{L}, \sin\frac{m\pi x}{L}) = \int_{-L}^{L} \cos\frac{n\pi x}{L}\sin\frac{m\pi x}{L}\, dx = 0 \quad \text{for all } n, m \text{ values.}$$

with norm squared given by

$$(1, 1) = \|1\|^2 = \int_{-L}^{L} dx = 2L$$

$$(\sin\frac{n\pi x}{L}, \sin\frac{n\pi x}{L}) = \|\sin\frac{n\pi x}{L}\|^2 = \int_{-L}^{L} \sin^2\frac{n\pi x}{L}\, dx = L \tag{2.29)(b)}$$

$$(\cos\frac{n\pi x}{L}, \cos\frac{n\pi x}{L}) = \|\cos\frac{n\pi x}{L}\|^2 = \int_{-L}^{L} \cos^2\frac{n\pi x}{L}\, dx = L$$

Example 2-6. (Orthogonal complex functions)
A set of complex functions $\phi_n(x)$ is called orthogonal in the hermitian sense if

$$(\phi_n(x), \phi_m(x)) = \int_a^b r(x)\phi_n(x)\overline{\phi_m(x)}\, dx = 0, \quad m \ne n, \tag{2.30}$$

where $\overline{\phi_m(x)}$ is the complex conjugate of $\phi_m(x)$. For example, the set of functions $\phi_n(x) = e^{in\pi x/L}$ are orthogonal in the hermitian sense over the interval $(-L, L)$ with respect to the weight function $r(x) = 1$. It is left as an exercise to show

$$(\phi_n, \phi_m) = \int_{-L}^{L} e^{in\pi x/L} e^{-im\pi x/L}\, dx = \begin{cases} 0, & n \neq m \\ 2L, & n = m \end{cases}.$$

∎

Example 2-7. (**Bessel function boundary condition y(b) = 0**)
Consider the Sturm-Liouville problem involving Bessel's differential equation

$$L_1(y) = x^2 \frac{d^2y}{dx^2} + x\frac{dy}{dx} + (\lambda^2 x^2 - k^2)y = 0, \quad 0 \leq x \leq b \tag{2.31}$$

where k and λ are constants. This equation is subject to the boundary condition $y(b) = 0$. The given equation is Bessel's equation which is not in self-adjoint form. Here $a_0(x) = x^2$, $a_1(x) = x$ and $a_2(x) = \lambda^2 x^2 - k^2$. Using equation (1.77) we find that if we multiply the given equation by $\mu = \frac{1}{x}$ we obtain the self-adjoint form

$$L(y) = x\frac{d^2y}{dx^2} + \frac{dy}{dx} + \left(\lambda^2 x - \frac{k^2}{x}\right)y = 0$$

or

$$L(y) = \frac{d}{dx}\left(x\frac{dy}{dx}\right) - \frac{k^2}{x}y = -\lambda^2 xy, \quad 0 < x < b. \tag{2.32}$$

This is a singular Sturm-Liouville problem with λ^2 replacing λ and weight function $r(x) = x$. Note that this equation is singular because $p(0) = 0$. In this case only one boundary condition is needed in order to insure orthogonality of the solution set. The coefficients are given by $p(x) = x$, $q(x) = -\frac{k^2}{x}$ and weight function $r(x) = x$. The only bounded solution of this equation on the interval $0 \leq x \leq b$ is given by some constant times $y = y(x;\lambda) = J_k(\lambda x)$. Let ξ_{kn}, for $n = 1,2,3,\ldots$, denote the nth zero of the kth order Bessel function. That is we define ξ_{kn} as those values which satisfy $J_k(\xi_{kn}) = 0$ for $n = 1,2,3,\ldots$. The first couple of zero's $\xi_{k1}, \xi_{k2}, \ldots$ are illustrated in the figure 2-4.

If we require $y(b) = J_k(\lambda b) = 0$, then we must select $\lambda b = \xi_{kn}$ or $\lambda = \lambda_n = \frac{\xi_{kn}}{b}$ for $n = 1,2,3,\ldots$ These are the unequally spaced eigenvalues for this Sturm-Liouville problem. The corresponding eigenfunctions are

$$y_n(x) = y(x;\lambda_n) = J_k(\lambda_n x) = J_k(\frac{\xi_{kn}}{b}x), \quad n = 1,2,3,\ldots$$

These are orthogonal functions over the interval $(0,b)$ with respect to the weight function x and have the inner product

$$(y_n, y_m) = \int_0^b xy_n(x)y_m(x)\, dx = \begin{cases} 0 & m \neq n \\ \|y_n\|^2 & m = n \end{cases} \tag{2.33}$$

and norm squared

$$(y_n, y_n) = \|y_n\|^2 = \int_0^b x J_k^2(\lambda_n x)\, dx = \frac{b^2}{2} J_{k+1}^2(\lambda_n b). \tag{2.34}$$

Note the weight function $r(x) = x$ is required for representing the inner product integral.

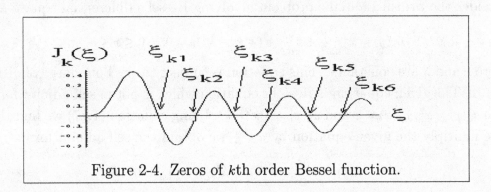

Figure 2-4. Zeros of kth order Bessel function.

The norm squared result is calculated as follows. Examine the entry 15 from table 1 of chapter 1 and replace x by $\lambda_n x$ and ν by k to obtain

$$2\lambda_n^2 \int_0^b x J_k^2(\lambda_n x)\, dx = \left[\lambda_n^2 x^2 \left[J_k'(\lambda_n x)\right]^2 + (\lambda_n^2 x^2 - k^2)\left[J_k(\lambda_n x)\right]^2\right]_0^b. \tag{2.35}$$

The boundary condition requires $J_k(\lambda_n b) = 0$, so the integral (2.35) simplifies to

$$2 \int_0^b x J_k^2(\lambda_n x)\, dx = b^2 \left[J_k'(\lambda_n b)\right]^2. \tag{2.36}$$

Now use entry number 14 of table 1, chapter 1, and remember that the prime notation means differentiation with respect to the argument of the function. We find

$$x\frac{d}{dx} J_k(\lambda_n x) = x J_k'(\lambda_n x)\lambda_n = k J_k(\lambda_n x) - \lambda_n x J_{k+1}(\lambda_n x). \tag{2.37}$$

Now evaluate the equation (2.37) at $x = b$, using the given boundary condition to obtain

$$J_k'(\lambda_n b) = -J_{k+1}(\lambda_n b). \tag{2.38}$$

Substitute this result into the equation (2.36) to obtain the norm squared given by equation (2.34). The series expansion

$$f(x) = \sum_{n=1}^{\infty} c_n y_n(x) = \sum_{n=1}^{\infty} c_n J_k(\lambda_n x) \tag{2.39}$$

is called a Fourier-Bessel series with coefficients given by an inner product divided by a norm squared

$$c_n = \frac{(f, y_n)}{\|y_n\|^2} = \frac{\int_0^b x f(x) J_k(\lambda_n x)\, dx}{\int_0^b x J_k^2(\lambda_n x)\, dx}.$$

∎

Example 2-8. (Bessel function with different boundary condition)
We solve the singular Sturm-Liouville problem

$$\frac{d}{dx}\left(x\frac{dy}{dx}\right) - \frac{k^2}{x}y = -\lambda^2 xy, \qquad 0 \le x \le b \tag{2.40}$$

subject to the boundary condition

$$by'(b) + \beta y(b) = 0, \tag{2.41}$$

where β is a constant. Boundary conditions having the form $c_1 y'(b) + c_2 y(b) = 0$, with c_1, c_2 constants, can be converted to the form of equation (2.41) if one multiplies the boundary condition by b/c_1 and defines $bc_2/c_1 = \beta$ as a new constant. The only bounded solutions of the Bessel equation over the interval $0 \le x \le b$ is given by some constant times $y = y(x) = J_k(\lambda x)$. The boundary condition given by equation (2.41) requires λ be chosen such that the equation

$$(\lambda b)J_k'(\lambda b) + \beta J_k(\lambda b) = 0. \tag{2.42}$$

Let the roots of this equation be denoted by the eigenvalues

$$\lambda = \lambda_n = \frac{\alpha_{kn}}{b} \quad \text{for} \quad n = 1, 2, 3, \ldots \tag{2.43}$$

where α_{kn} are the roots satisfying

$$\alpha_{kn} J_k'(\alpha_{kn}) + \beta J_k(\alpha_{kn}) = 0 \quad \text{for} \quad n = 1, 2, 3, \ldots. \tag{2.44}$$

The Table 1-1 aids in writing the equation (2.44) in the alternative form

$$(k + \beta)J_k(\alpha_{kn}) - \alpha_{kn}J_{k+1}(\alpha_{kn}) = 0. \tag{2.45}$$

The corresponding eigenfunctions are then $y_n(x) = J_k(\lambda_n x)$. These functions satisfy the orthogonality condition

$$(y_n, y_m) = \int_0^b x J_k(\lambda_n x) J_k(\lambda_m x)\, dx = \begin{cases} 0, & m \ne n \\ \|y_n\|^2, & m = n \end{cases} \tag{2.46}$$

where the norm squared is given by

$$(y_n, y_n) = ||y_n||^2 = \int_0^b x J_k^2(\lambda_n x)\, dx = \left[\frac{\beta^2 + \lambda_n^2 b^2 - k^2}{2\lambda_n^2}\right] J_k^2(\lambda_n b). \tag{2.47}$$

To obtain this result for the norm squared we used entry number 12, from the table 1 of chapter 1, with $a = 0$ to obtain

$$\int_0^b x J_k^2(\lambda_n x)\, dx = \frac{b^2}{2}\left\{[J_k'(\lambda b)]^2 + J_k^2(\lambda b) - \frac{k^2}{\lambda^2 b^2} J_k^2(\lambda b)\right\}. \tag{2.48}$$

We can now use the boundary condition equation (2.42) to eliminate the derivative term in equation (2.48)to obtain

$$\int_0^b x J_k^2(\lambda_n x)\, dx = \frac{b^2}{2}\left\{[-\frac{\beta J_k(\lambda_n b)}{\lambda_n b}]^2 + J_k^2(\lambda b) - \frac{k^2}{\lambda^2 b^2} J_k^2(\lambda b)\right\}$$
$$= \left\{\frac{\beta^2 + \lambda_n^2 b^2 - k^2}{2\lambda_n^2}\right\} J_k^2(\lambda_n b). \tag{2.49}$$

Remark: Here we have assumed that $k \neq 0$. If $k = 0$ and $\beta = 0$, then we find $\lambda_0 = 0$ is an eigenvalue corresponding to the eigenfunction $y_0(x) = J_0(0) = 1$. In this case the eigenvalues are the roots of the equation $J_0'(\lambda b) = 0$ or $J_1(\lambda b) = 0$ and the norm squared is

$$||y_0||^2 = ||1||^2 = \int_0^b x\, dx = \frac{b^2}{2}. \tag{2.50}$$

This is the only case where $\lambda = 0$ can be an eigenvalue of the given Sturm-Liouville system. A series expansion of the form

$$f(x) = \sum_{n=1}^\infty c_n y_n(x) = \sum_{n=1}^\infty c_n J_k(\lambda_n x) \tag{2.51}$$

is similar in form the equation (2.39) , however, the eigenvalues are different.

■

Example 2-9.   (Bessel function on interval $a \leq x \leq b$)
Solve the Sturm-Liouville system

$$\frac{d}{dx}\left(x\frac{dy}{dx}\right) - \frac{k^2}{x}y = -\lambda^2 xy, \qquad 0 < a < x < b \tag{2.52}$$

subject to the boundary conditions $y(a) = 0$ and $y(b) = 0$. The domain for the solution of the Bessel equation is away from the origin so that the general solution can be written $y(x) = c_1 J_k(\lambda x) + c_2 Y_k(\lambda x)$ where c_1 and c_2 are arbitrary constants. We select the constants c_1 and c_2 such that the boundary condition $y(b) = 0$ is automatically satisfied. We select $c_1 = Y_k(\lambda b)$ and $c_2 = -J_k(\lambda b)$ and obtain the solution

$$y(x) = y(x; \lambda) = U_k(\lambda_n x) = Y_k(\lambda b) J_k(\lambda x) - J_k(\lambda b) Y_k(\lambda x). \tag{2.53}$$

Now the eigenvalues $\lambda = \lambda_n$ are chosen as the roots of the equation

$$y(a) = U_k(\lambda_n a) = Y_k(\lambda_n b) J_k(\lambda_n a) - J_k(\lambda_n b) Y_k(\lambda_n a) = 0. \tag{2.54}$$

The corresponding eigenfunctions are $y_n(x) = U_k(\lambda_n x)$ which satisfy the orthogonality condition

$$(y_n, y_m) = \int_a^b x U_k(\lambda_n x) U_k(\lambda_m x)\, dx = \begin{cases} 0, & m \neq n \\ \|y_n\|^2, & m = n \end{cases} \tag{2.55}$$

where the norm squared is

$$(y_n, y_n) = \|y_n\|^2 = \int_a^b x U_k^2(\lambda_n x)\, dx = \frac{1}{2}\left\{ b^2 [U_k'(\lambda_n b)]^2 - a^2 [U_k'(\lambda_n a)]^2. \right\} \tag{2.56}$$

To show how this last integral originates we consider the Bessel equation

$$\frac{d}{dx}\left(x \frac{dy}{dx} \right) + (\lambda^2 x - \frac{m^2}{x}) y = 0.$$

Multiply this equation by $2xy'$ to obtain

$$2xy' \frac{d}{dx}(xy') + (2\lambda^2 x^2 - 2m^2) yy' = 0.$$

Now integrate from a to b giving

$$\int_a^b \left[\frac{d}{dx}[x^2(y')^2] + (\lambda^2 x^2 - m^2)\frac{d}{dx}(y^2) \right] dx = 0.$$

Now integrate all the terms. The middle term is to be integrated by parts. The result is

$$x^2(y')^2 \Big|_a^b + \lambda^2 \left[x^2 y^2 \Big|_a^b - 2\int_a^b xy^2\, dx \right] - m^2 y^2 \Big|_a^b = 0$$

which can be expressed in the form

$$\begin{aligned} \int_a^b xy^2\, dx &= \frac{1}{2\lambda^2}\left[x^2(y')^2 + \lambda^2 x^2 y^2 - m^2 y^2 \right]_a^b \\ &= \frac{1}{2\lambda^2}\left[b^2 (y'(b))^2 + \lambda^2 b^2 y^2(b) - m^2 y^2(b) \right] \\ &\quad - \frac{1}{2\lambda^2}\left[a^2 (y'(a))^2 + \lambda^2 a^2 y^2(a) - m^2 y^2(a) \right]. \end{aligned} \tag{2.57}$$

Here $y(x)$ is any solution of Bessel's equation. Our solution satisfies $y(a) = 0$ and $y(b) = 0$ with

$$y'(b) = \lambda_n U_k'(\lambda_n b) \quad \text{and} \quad y'(a) = \lambda_n U_k'(\lambda_n a) \tag{2.58}$$

which simplifies the equation (2.57) to the norm squared result cited earlier. This result can also be written in the alternate form

$$\|U_k(\lambda_n x)\|^2 = \frac{2\left(J_k^2(\lambda_n a) - J_k^2(\lambda_n b)\right)}{\pi^2 \lambda_n^2 J_k^2(\lambda_n a)}. \tag{2.59}$$

Series expansions of the form

$$f(x) = \sum_{n=1}^{\infty} c_n y_n(x) = \sum_{n=1}^{\infty} c_n U_k(\lambda_n x) \tag{2.60}$$

generate another type of Fourier-Bessel expansion. ∎

Example 2-10. (Legendre equation)

Find the eigenvalues and eigenfunctions associated with the singular Sturm-Liouville problem

$$(1 - x^2)\frac{d^2 y}{dx^2} - 2x\frac{dy}{dx} + \lambda y = 0 \qquad -1 \le x \le 1$$

where there are no boundary conditions. Note this equation is already in self-adjoint form and can be written

$$\frac{d}{dx}\left[(1 - x^2)\frac{dy}{dx}\right] + \lambda y = 0$$

which is now in the form of equation (2.10). Here $p(x) = 1 - x^2$ is zero at both the boundary end points, $q(x) = 0$ and the weight function is $r(x) = 1$. For $\lambda = \lambda_n = n(n+1)$ we obtain the Legendre equation. The only bounded solutions on the interval $-1 \le x \le 1$ are the Legendre polynomials $P_n(x)$ for $n = 0, 1, 2, \ldots$. These functions are orthogonal over the interval $(-1, 1)$ with respect to the weight function $r(x) = 1$. To calculate the norm squared we can use the generating function $g(x, t)$ given in table 2 from chapter 1. We form the quantity

$$g(x, \xi)g(x, \eta) = \sum_{n=0}^{\infty} \sum_{m=0}^{\infty} P_n(x)P_m(x)\xi^n \eta^m \tag{2.61}$$

and then integrate both sides with respect to x from -1 to 1. There results

$$I = \int_{-1}^{1} \frac{1}{\sqrt{1 - 2x\xi + \xi^2}} \frac{1}{\sqrt{1 - 2x\eta + \eta^2}}\, dx = \sum_{n=0}^{\infty} \sum_{m=0}^{\infty} (P_n, P_m)\xi^n \eta^m \tag{2.62}$$

where $(P_n, P_m) = \int_{-1}^{1} P_n(x)P_m(x)\,dx$ denotes the resulting inner product terms. From a table of integrals we find the left-hand side of equation (2.62) can be evaluated to obtain

$$I = \frac{1}{\sqrt{\xi\eta}} \ln \left[\frac{\sqrt{a-1} - \sqrt{b-1}}{\sqrt{a+1} - \sqrt{b+1}} \right] = \sum_{n=0}^{\infty} \sum_{m=0}^{\infty} (P_n, P_m)\xi^n \eta^m. \qquad (2.63)$$

where $a = \frac{1+\xi^2}{2}$ and $b = \frac{1+\eta^2}{2}$. This equation further simplifies to the form

$$I = \frac{1}{\sqrt{\xi\eta}} \ln \left[\frac{1 + \sqrt{\xi\eta}}{1 - \sqrt{\xi\eta}} \right] = \sum_{n=0}^{\infty} \sum_{m=0}^{\infty} (P_n, P_m)\xi^n \eta^m. \qquad (2.64)$$

Now let $x = \xi\eta$ and expand the left side of equation (2.64) in a Mclaurin[†] series to obtain

$$I = \frac{1}{\sqrt{x}} \ln \left[\frac{1 + \sqrt{x}}{1 - \sqrt{x}} \right] = 2 + \frac{2}{3}x + \frac{2}{5}x^2 + \frac{2}{7}x^3 + \cdots + \frac{2}{2k+1}x^k + \cdots. \qquad (2.65)$$

Also expand the right-hand side of equation (2.64) by summing on m to obtain

$$I = \sum_{n=0}^{\infty} (P_n, P_0)\xi^n + (P_n, P_1)\xi^n \eta + (P_n, P_2)\xi^n \eta^2 + \cdots. \qquad (2.66)$$

The Legendre polynomials are orthogonal over the interval $(-1, 1)$ and so the inner products (P_n, P_m) with $n \neq m$ are zero. Therefore, the only nonzero term in the expansion given by equation (2.66) is the term $(P_n, P_n)\xi^n \eta^n$. The equation (2.66) then becomes

$$I = \|P_0\|^2 + \|P_1\|^2 x + \|P_2\|^2 x^2 + \cdots + \|P_k\|^2 x^k + \cdots \qquad (2.67)$$

where $x = \xi\eta$. Equating like terms from equations (2.67) and (2.65) we find

$$\|P_k\|^2 = (P_k, P_k) = \frac{2}{2k+1}. \qquad (2.68)$$

∎

The singular Sturm-Liouville problem

$$\frac{d}{dx}[(1 - x^2)\frac{dy}{dx}] + \lambda y = 0, \qquad -1 \leq x \leq 1 \qquad (2.69)$$

[†] See Appendix C

has the eigenvalues $\lambda_n = n(n+1)$ and eigenfunctions $y_n = P_n(x)$. The eigenfunctions are orthogonal over the interval $-1 \leq x \leq 1$ and satisfy

$$(P_n, P_m) = \int_{-1}^{1} P_n(x)P_m(x)\, dx = ||P_n||^2 \delta_{nm} \tag{2.70}$$

where $\qquad ||P_n||^2 = \dfrac{2}{2n+1}, \quad n = 0,1,2,3,\ldots \tag{2.71}$

The series expansion $f(x) = \displaystyle\sum_{n=0}^{\infty} c_n P_n(x)$ with $c_n = \dfrac{(f, P_n)}{||P_n||^2}$ is called a Fourier-Legendre series.

In later chapters we will find the Legendre equation occurring in certain problems dealing with spherical coordinates and so we will need the following form of the previous result.

The transformation $x = \cos\theta$ transforms the previous Sturm-Liouville problem into the singular Sturm-Liouville problem

$$\frac{d}{d\theta}\left(\sin\theta \frac{dy}{d\theta}\right) + \lambda \sin\theta\, y = 0, \qquad 0 \leq \theta \leq \pi \tag{2.72}$$

which has the eigenvalues $\lambda_n = n(n+1)$ and corresponding eigenfunctions $y_n = P_n(\cos\theta)$ which are orthogonal over the interval $(0, \pi)$ with respect to the weight function $\sin\theta$ so that the inner products satisfy

$$(P_n, P_m) = \int_{0}^{\pi} P_n(\cos\theta)P_m(\cos\theta)\sin\theta\, d\theta = \frac{2}{2n+1}\delta_{nm} \tag{2.73}$$

Special Orthogonal Functions

The following are some special orthogonal functions $\{\phi_n(x)\}$ which arise in the study of certain partial differential equations occurring in physics, engineering and the physical sciences. Listed for reference purposes is the Sturm-Liouville differential equation, the orthogonality property, the norm squared, selected special values, a recurrence relation to generate additional special values and a graph of some selected functions. A recurrence relation is any equation which relates two or more members from the set of orthogonal functions.

1. **Associated Legendre functions** $P_n^m(x)$, $Q_n^m(x)$.

   Legendre's associated differential equation

   $$(1-x^2)\frac{d^2y}{dx^2} - 2x\frac{dy}{dx} + \left[n(n+1) - \frac{m^2}{1-x^2}\right]y = 0, \quad -1 \leq x \leq 1 \tag{2.74}$$

General solution

$$y = c_1 P_n^m(x) + c_2 Q_n^m(x), \qquad c_1, c_2 \text{ constants}$$

where $P_n^m(x)$ and $Q_n^m(x)$ are defined[‡]

$$P_n^m(x) = (1 - x^2)^{m/2} \frac{d^m}{dx^m} P_n(x)$$

$$Q_n^m(x) = (1 - x^2)^{m/2} \frac{d^m}{dx^m} Q_n(x) \tag{2.75}$$

are functions associated with the mth derivatives of the Legendre functions of the first and second kind. The only bounded solutions on the interval $-1 \le x \le 1$ are given by $y = c_1 P_n^m(x)$ where c_1 is a constant.

Orthogonality

$$(P_n^k, P_m^k) = \int_{-1}^{1} P_n^k(x) P_m^k(x) \, dx = \|P_n^k\|^2 \delta_{nm} \tag{2.76}$$

Norm squared

$$\|P_n^k\|^2 = (P_n^k, P_n^k) = \frac{2}{2n+1} \frac{(n+k)!}{(n-k)!} \tag{2.77}$$

Selected functions

$$P_1^1(x) = \sqrt{1 - x^2}$$

$$P_2^1(x) = 3x\sqrt{1 - x^2}, \qquad P_2^2(x) = 3(1 - x^2),$$

$$P_3^1(x) = \frac{3}{2}(5x^2 - 1)\sqrt{1 - x^2}, \qquad P_3^2(x) = 15x(1 - x^2), \qquad P_3^3(x) = 15(1 - x^2)^{3/2}$$

$$P_4^1(x) = \frac{5}{2}(7x^3 - 3x)\sqrt{1 - x^2}, \qquad P_4^2(x) = \frac{15}{2}(7x^2 - 1)(1 - x^2), \qquad P_4^3(x) = 105x(1 - x^2)^{3/2}$$

Recurrence formula

$$P_n^{m+2}(x) - \frac{2(m+1)x}{\sqrt{1 - x^2}} P_n^{m+1}(x) + [n(n+1) - m(m+1)]P_n^m(x) = 0$$

The figure 2-5 illustrates graphs of the associated Legendre functions for selected values of n and m.

[‡] Associated Legendre polynomials can be defined in other ways.

94

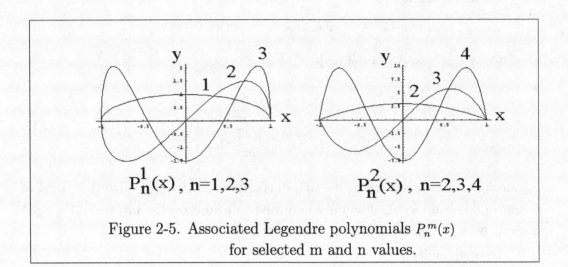

Figure 2-5. Associated Legendre polynomials $P_n^m(x)$
for selected m and n values.

The transformation $x = \cos\theta$ transforms the equation (2.74) to the form

$$\frac{1}{\sin\theta}\frac{d}{d\theta}\left(\sin\theta\frac{dy}{d\theta}\right) + [n(n+1) - \frac{m^2}{\sin^2\theta}]y = 0 \qquad (2.78)$$

with bounded solutions $P_n^m(\cos\theta)$ which are orthogonal over the interval $(0, \pi)$ with respect to the weight function $\sin\theta$. It is left as an exercise to show these functions satisfy the orthogonality property

$$(P_n^m(\cos\theta), P_\ell^m(\cos\theta)) = \int_0^\pi P_n^m(\cos\theta)P_\ell^m(\cos\theta)\sin\theta\,d\theta = \frac{2}{2n+1}\frac{(n+m)!}{(n-m)!}\delta_{n\ell} \qquad (2.79)$$

2. **Hermite Polynomials $H_n(x)$.**

   Hermite[†] differential equation

   $$\frac{d^2y}{dx^2} - 2x\frac{dy}{dx} + 2ny = 0, \quad -\infty < x < \infty$$

   Polynomial solutions[‡] $\quad y = H_n(x) = (-1)^n e^{x^2}\frac{d^n}{dx^n}\left(e^{-x^2}\right)$
   Orthogonality

   $$(H_n, H_m) = \int_{-\infty}^{\infty} e^{-x^2} H_n(x)H_m(x)\,dx = ||H_n||^2\delta_{nm}$$

[†] See Appendix C

[‡] Hermite polynomials can be defined in other ways.

Norm squared $\quad\|H_n\|^2 = \sqrt{\pi}\,2^n\,n!$

Selected functions

$$
\begin{aligned}
H_0(x) &=1, & H_3(x) &=8x^3 - 12x \\
H_1(x) &=2x, & H_4(x) &=16x^4 - 48x^2 + 12 \\
H_2(x) &=4x^2 - 2, & H_5(x) &=32x^5 - 160x^3 + 120x
\end{aligned}
$$

Recurrence formula $\quad H_{n+1}(x) = 2xH_n(x) - 2nH_{n-1}(x) \quad n = 1,2,3,\ldots$

The figure 2-6 illustrates some selected graphs for the scaled Hermite functions $e^{-x^2}H_n(x)$.

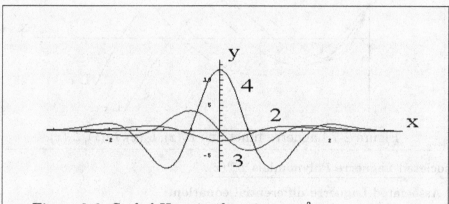

Figure 2-6. Scaled Hermite functions $e^{-x^2}H_n(x)$, $n = 2,3,4$

3. **Laguerre Polynomials $L_n(x)$.**

   Laguerre[†] differential equation

   $$
   x\frac{d^2y}{dx^2} + (1-x)\frac{dy}{dx} + ny = 0, \quad 0 < x < \infty
   $$

   Polynomial solutions[‡] $\quad y = L_n(x) = \dfrac{1}{n!}e^x\dfrac{d^n}{dx^n}\left(x^n e^{-x}\right)$

   Orthogonality $\quad (L_n, L_m) = \displaystyle\int_0^\infty e^{-x}L_n(x)L_m(x)\,dx = \|L_n\|^2\delta_{nm}$

   Norm squared $\quad \|L_n\|^2 = 1$

   Selected functions

   $$
   \begin{aligned}
   L_0(x) &=1, & L_3(x) &=\frac{1}{6}(-x^3 + 9x^2 - 18x + 6) \\
   L_1(x) &=1 - x, & L_4(x) &=\frac{1}{24}(x^4 - 16x^3 + 72x^2 - 96x + 24) \\
   L_2(x) &=\frac{1}{2}(x^2 - 4x + 2), & L_5(x) &=\frac{1}{120}(-x^5 + 25x^4 - 200x^3 + 600x^2 - 600x + 120)
   \end{aligned}
   $$

   [†] See Appendix C

   [‡] Laguerre polynomials can be defined in other ways.

Recurrence formula $\qquad (n+1)L_{n+1}(x) - (2n+1-x)L_n(x) + nL_{n-1}(x) = 0$

The figure 2-7 illustrates some selected graphs of Laguerre functions.

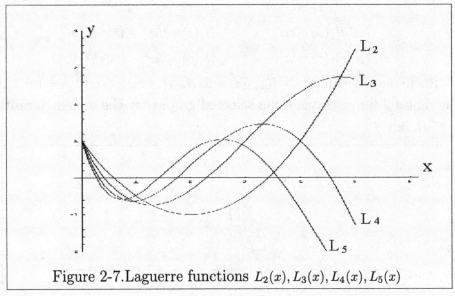

Figure 2-7.Laguerre functions $L_2(x), L_3(x), L_4(x), L_5(x)$

4. **Associated Laguerre Polynomials** $L_n^m(x)$.

Associated Laguerre differential equation.

$$x\frac{d^2y}{dx^2} + (m+1-x)\frac{dy}{dx} + ny = 0, \qquad 0 < x < \infty$$

Polynomial solutions[‡] $\qquad y = L_n^m(x) = \frac{1}{n!}x^{-m}e^x\frac{d^n}{dx^n}\left(e^{-x}x^{n+m}\right)$

Orthogonality

$$(L_n^k, L_m^k) = \int_0^\infty x^k e^{-x} L_n^k(x) L_m^k(x)\,dx = ||L_n^k||^2 \delta_{nm}$$

Norm squared $\qquad ||L_n^m||^2 = (L_n^m, L_n^m) = \frac{\Gamma(1+m+n)}{n!}, \qquad m > -1$

Selected functions

$$L_2^1(x) = \frac{1}{2}(6 - 6x + x^2),$$

$$L_2^2(x) = \frac{1}{2}(12 - 8x + x^2),$$

$$L_3^1(x) = \frac{1}{6}(24 - 36x + 12x^2 - x^3)$$

$$L_3^2(x) = \frac{1}{6}(60 - 6x + 15x^2 - x^3)$$

$$L_3^3(x) = \frac{1}{6}(120 - 90x + 18x^2 - x^3)$$

[‡] Associated Laguerre polynomials can be defined in other ways.

Recurrence formula

$$(n+1)L_{n+1}^m(x) + (x - m - 2n - 1)L_n^m(x) + (n+m)L_{n-1}^m(x) = 0$$

The figure 2-8 illustrates some graphs for selected scaled associated Laguerre functions $x^m e^{-x} L_n^m(x)$.

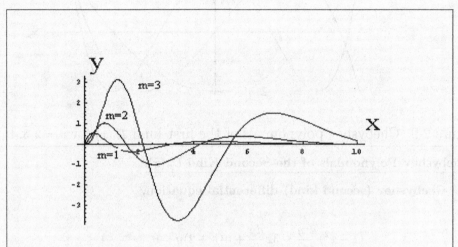

Figure 2-8. Scaled Laguerre functions $x^m e^{-x} L_3^m(x)$ for $m = 1, 2, 3$.

5. Chebyshev Polynomials of the First Kind $T_n(x)$

Chebyshev[†] (first kind) differential equation

$$(1 - x^2)\frac{d^2y}{dx^2} - x\frac{dy}{dx} + n^2 y = 0, \qquad -1 \le x \le 1$$

Polynomial solutions $\qquad y = T_n(x) = \cos(n \arccos x)$

Orthogonality

$$(T_n, T_m) = \int_{-1}^1 \frac{T_n(x)T_m(x)}{\sqrt{1 - x^2}}\, dx = ||T_n||^2 \delta_{nm}$$

Norm squared $\quad ||T_n||^2 = (T_n, T_n) = \begin{cases} \pi & \text{if } n = 0 \\ \frac{\pi}{2} & \text{if } n = 1, 2, 3, \ldots \end{cases}$

Selected functions

$$T_0(x) = 1, \qquad T_2(x) = 2x^2 - 1$$
$$T_1(x) = x, \qquad T_3(x) = 4x^3 - 3x$$

[†] See Appendix C

Recurrence formula $\qquad T_{n+1}(x) - 2xT_n(x) + T_{n-1}(x) = 0$

The figure 2-9 illustrates some function plots for selected Chebyshev polynomials of the first kind $T_n(x)$.

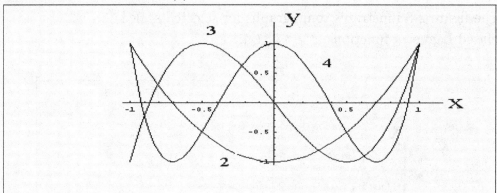

Figure 2-9. Chebyshev polynomials of the first kind $T_n(x)$ for $n = 2, 3, 4$.

6. **Chebyshev Polynomials of the Second Kind $U_n(x)$.**

Chebyshev (second kind) differential equation.

$$(1 - x^2)\frac{d^2y}{dx^2} - 3x\frac{dy}{dx} + n(n+2)y = 0, \qquad -1 \le x \le 1$$

Polynomial solutions $\qquad y = U_n(x) = \dfrac{\sin[(n+1)\arccos x]}{\sin(\arccos x)}$

Orthogonality

$$(U_n, U_m) = \int_{-1}^{1} \sqrt{1 - x^2}\, U_n(x)U_m(x)\, dx = ||U_n||^2 \delta_{nm}$$

Norm squared $\qquad ||U_n||^2 = (U_n, U_n) = \dfrac{\pi}{2}$

Selected functions

$$U_0(x) = 1, \qquad U_2(x) = 4x^2 - 1$$
$$U_1(x) = 2x, \qquad U_3(x) = 8x^3 - 4x$$

Recurrence formula $\qquad U_{n+1}(x) - 2xU_n(x) + U_{n-1}(x) = 0$

The figure 2-10 illustrates some graphs for selected Chebyshev polynomials of the second kind $U_n(x)$.

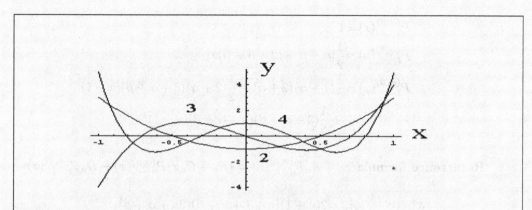

Figure 2-10. Chebyshev polynomials of the second kind $U_n(x)$ for $n = 2, 3, 4$.

7. **Jacobi Polynomials $P_n^{(\alpha,\beta)}(x)$.**

Jacobi[†] differential equation

$$(1 - x^2)\frac{d^2y}{dx^2} + [\beta - \alpha - (\alpha + \beta + 2)x]\frac{dy}{dx} + n(n + \alpha + \beta + 1)y = 0, \qquad -1 < x < 1$$

Polynomial solution[‡]

$$y = P_n^{(\alpha,\beta)}(x) = \frac{(1 + \alpha)_n}{n!}F(-n, \alpha + \beta + n + 1, 1 + \alpha, \frac{1 - x}{2})$$

where $\quad (1 + \alpha)_0 = 1, \qquad (1 + \alpha)_n = (1 + \alpha)(2 + \alpha)\cdots(n + \alpha)$

and for $\gamma > 0$ the function $F(\alpha, \beta, \gamma, x)$ is the Hypergeometric series

$$F(\alpha, \beta, \gamma, x) = 1 + \frac{\alpha\beta}{\gamma}x + \frac{\alpha(\alpha + 1)\beta(\beta + 1)}{\gamma(\gamma + 1)}\frac{x^2}{2!} + \frac{\alpha(\alpha + 1)(\alpha + 2)\beta(\beta + 1)(\beta + 2)}{\gamma(\gamma + 1)(\gamma + 2)}\frac{x^3}{3!} + \cdots$$

Orthogonality

$$(P_n^{(\alpha,\beta)}, P_m^{(\alpha,\beta)}) = \int_{-1}^{1}(1 - x)^\alpha(1 + x)^\beta P_n^{(\alpha,\beta)}(x)P_m^{(\alpha,\beta)}(x)\,dx = ||P_n^{(\alpha,\beta)}||^2\delta_{nm}$$

Norm squared

$$||P_n^{(\alpha,\beta)}||^2 = \frac{2^{\alpha+\beta+1}\Gamma(\alpha + n + 1)\Gamma(\beta + n + 1)}{(\alpha + \beta + 2n + 1)n!\Gamma(\alpha + \beta + n + 1)}$$

[†]   See Appendix C

[‡]   Jacobi polynomials can be defined in other ways.

Selected functions

$$P_0^{(\alpha,\beta)}(x) = 1$$

$$P_1^{(\alpha,\beta)}(x) = \frac{1}{2}[\alpha - \beta + (2 + \alpha + \beta)x]$$

$$P_2^{(\alpha,\beta)}(x) = \frac{1}{2}(1 + \alpha)(2 + \alpha) + \frac{1}{2}(2 + \alpha)(3 + \alpha + \beta)(x - 1)$$

$$+ \frac{1}{8}(3 + \alpha + \beta)(4 + \alpha + \beta)(x - 1)^2$$

Recurrence formula $A_n P_{n+1}^{(\alpha,\beta)}(x) = (B_n + C_n x)P_n^{(\alpha,\beta)}(x) - D_n P_{n-1}^{(\alpha,\beta)}(x)$

where $A_n = 2(n + 1)(n + 1 + \alpha + \beta)(2n + \alpha + \beta)$

$B_n = (2n + 1 + \alpha + \beta)(\alpha^2 - \beta^2)$

$C_n = (2n + \alpha + \beta)(2n + 1 + \alpha + \beta)(2n + 2 + \alpha + \beta)$

$D_n = 2(n + \alpha)(n + \beta)(2n + 2 + \alpha + \beta)$

The figure 2-11 illustrates some selected Jacobi polynomials $P_n^{(\alpha,\beta)}(x)$.

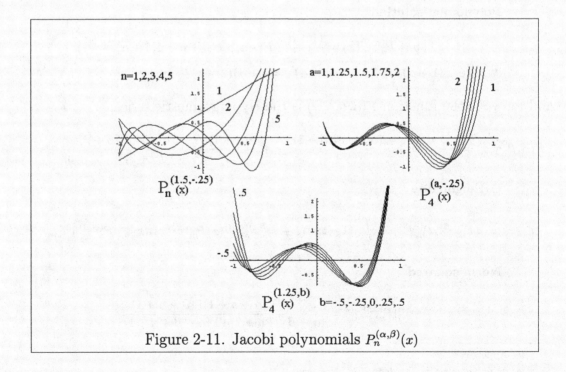

Figure 2-11. Jacobi polynomials $P_n^{(\alpha,\beta)}(x)$

8. Gegenbauer Polynomials $C_n^\nu(x)$.

Gegenbauer[†] differential equation

$$(1 - x^2)\frac{d^2y}{dx^2} - (2\nu + 1)x\frac{dy}{dx} + n(n+2)y = 0, \qquad -1 < x < 1$$

Polynomial solutions

$$y = C_n^\nu(x) = \frac{(2\nu)_n P_n^{(\nu-1/2,\nu-1/2)}(x)}{(\nu + 1/2)_n}(x)$$

where $(\alpha)_0 = 1$ and $(\alpha)_n = \alpha(\alpha + 1)(\alpha + 2)\cdots(\alpha + n)$ and $P_n^{(\alpha,\beta)}$ are Jacobi polynomials.

Orthogonality

$$(C_n^\nu, C_m^\nu) = \int_{-1}^{1}(1 - x^2)^{\nu-1/2}C_n^\nu(x)C_m^\nu(x)\,dx = \|C_n^\nu\|^2\delta_{nm}$$

Norm squared

$$\|C_n^\nu\|^2 = \frac{\pi 2^{1-2\nu}\Gamma(n + 2\nu)}{(n+\nu)n!\left[\Gamma(\nu)\right]^2}, \qquad \nu \neq 0$$

Selected functions

$$C_0^\nu(x) = 1, \qquad C_1^\nu(x) = 2\nu x, \qquad C_2^\nu(x) = \nu[2(1+\nu)x^2 - 1]$$

Recurrence formula

$$(n+1)C_{n+1}^\nu(x) = 2(n+\nu)xC_n^\nu(x) - (n + 2\nu - 1)C_{n-1}^\nu(x)$$

The figure 2-12 illustrates a sketch of some selected Gegenbauer polynomials $C_4^\nu(x)$ over the domain $-1 \leq x \leq 1$.

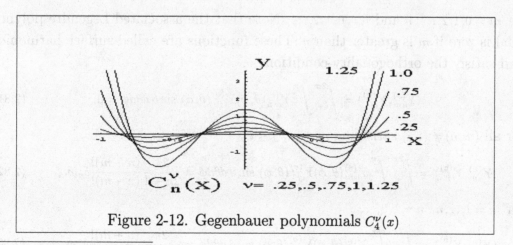

Figure 2-12. Gegenbauer polynomials $C_4^\nu(x)$

[†] See Appendix C

Note that because of scaling and normalization, many of the specialized orthogonal polynomials can have different definitions then what is defined in these notes.

Orthogonality over a multidimensional region

By changing symbols one can create products of orthogonal functions which produce orthogonal functions in two or more variables which are orthogonal over two-dimensional or higher dimensional regions. For example, the set of functions

$$\phi_{nm}(x,y) = \sin\frac{n\pi x}{a} \sin\frac{m\pi y}{b}, \quad 0 \le x \le a, \quad 0 \le y \le b$$

for $m = 1, 2, 3, \ldots$ and $n = 1, 2, 3, \ldots$ is a set of orthogonal functions over the rectangular region $0 \le x \le a$, $0 \le y \le b$ and satisfies the inner product relation

$$(\phi_{nm}, \phi_{ij}) = \frac{ab}{4}\delta_{ni}\delta_{mj}.$$

To show this we let $F_n(x) = \sin\frac{n\pi x}{a}$ and $G_m(y) = \sin\frac{m\pi y}{b}$ and write

$$(\phi_{nm}, \phi_{ij}) = \int_0^a \int_0^b \phi_{nm}(x,y)\phi_{ij}(x,y)\,dxdy = (F_n, F_i)(G_m, G_j)$$

$$= \int_0^a F_n(x)F_i(x)\,dx \int_0^b G_m(y)G_j(y)\,dy = \| F_n \| \delta_{ni} \| G_m \| \delta_{mj} = \frac{ab}{4}\delta_{ni}\delta_{mj}.$$

As another example consider changing variables and using the product of the associated Legendre polynomials with the orthogonal set $\{1, \cos m\phi, \sin m\phi\}$ to create the functions

$$Y_{nm}^{(e)}(\theta, \phi) = P_n^m(\cos\theta)\cos m\phi$$
$$Y_{nm}^{(o)}(\theta, \phi) = P_n^m(\cos\theta)\sin m\phi \tag{2.80}$$

for $m = 0, 1, 2, \ldots, n$ and $n = 0, 1, 2, \ldots$. Note that the associated Legendre polynomial is zero if m is greater than n. These functions are called surface harmonics and satisfy the orthogonality conditions

$$(Y_{nm}^{(e)}, Y_{ij}^{(o)}) = \int_0^{2\pi} \int_0^{\pi} Y_{nm}^{(e)}(\theta, \phi)Y_{ij}^{(o)}(\theta, \phi)\,\sin\theta\,d\theta d\phi = 0 \tag{2.81}$$

for all $(m, n) \ne (i, j)$ satisfying $m \le n$, $j \le i$.

$$(Y_{nm}^{(e)}, Y_{ij}^{(e)}) = \int_0^{2\pi} \int_0^{\pi} Y_{nm}^{(e)}(\theta, \phi)Y_{ij}^{(e)}(\theta, \phi)\,\sin\theta\,d\theta d\phi = \frac{2\pi}{2n+1}\frac{(n+m)!}{(n-m)!}\delta_{ni}\delta_{mj} \tag{2.82}$$

for $m = 1, \ldots n$, $n = 1, 2, \ldots$

$$(Y_{nm}^{(o)}, Y_{ij}^{(o)}) = \int_0^{2\pi} \int_0^{\pi} Y_{nm}^{(o)}(\theta, \phi)Y_{ij}^{(o)}(\theta, \phi)\,\sin\theta\,d\theta d\phi = \frac{2\pi}{2n+1}\frac{(n+m)!}{(n-m)!}\delta_{ni}\delta_{mj} \tag{2.83}$$

for $m = 1, \ldots n$, $n = 1, 2, \ldots$. In the special case $m = 0$ we have the inner product

$$(Y_{n0}^{(e)}, Y_{j0}^{(e)}) = \int_0^{2\pi} \int_0^\pi P_n(\cos\theta) P_j(\cos\theta) \sin\theta d\theta d\phi = \frac{4\pi}{2n+1} \delta_{nj}. \qquad (2.84)$$

Note that $d\sigma = \sin\theta \, d\theta d\phi$ is an element of surface area on a unit sphere and so the above functions are said to be orthogonal over the surface area of a unit sphere. Also note that the above inner products are products of one dimensional inner products. For example,

$$(Y_{nm}^{(e)}, Y_{ij}^{(e)}) = (P_n^m(\cos\theta), P_i^j(\cos\theta))(\cos m\phi, \cos j\phi)$$

with similar results for the other inner products.

We use the Euler identity $e^{im\phi} = \cos m\phi + i\sin m\phi$ where the quantity i is used to denote an imaginary unit with $i^2 = -1$. Recall that in cases where eigenvalues and eigenfunctions are complex quantities we have shown the set of functions $\{e^{in\phi}\}$ for $n = 0, \pm 1, \pm 2, \ldots$ are orthogonal in the hermitian sense over the interval $(0, 2\pi)$ and satisfies the orthogonality property

$$(e^{in\phi}, e^{im\phi}) = \int_0^{2\pi} e^{in\phi} e^{-im\phi} \, d\phi = \begin{cases} 0, & n \neq m \\ 2\pi, & n = m \end{cases} \qquad (2.85)$$

We now make use of the above properties to express the above surface harmonics in a slightly different way. Note that the associated Legendre polynomials, with $\cos\theta$ as an argument, are orthogonal over the interval $0 \leq \theta \leq \pi$ with respect to the weight function $\sin\theta$. The trigonometric functions $\{1, \cos m\phi, \sin m\phi\}$ are orthogonal over the interval $(0, 2\pi)$. Therefore the product functions multiplied by a constant will be orthogonal with respect to the weight function $\sin\theta$ over the region defined by $0 \leq \theta \leq \pi$, and $0 \leq \phi \leq 2\pi$, or orthogonal over the unit sphere. We select a normalization factor and define the functions

$$Y_n^m(\theta, \phi) = \sqrt{\frac{2n+1}{4\pi} \frac{(n-m)!}{(n+m)!}} P_n^m(\cos\theta) e^{im\phi}, \quad 0 \leq \theta \leq \pi, \quad 0 \leq \phi \leq 2\pi \qquad (2.86)$$

for $m = 0, 1, 2, \ldots, n$ and $n = 0, 1, 2, \ldots$. The set of functions $Y_n^m(\theta, \phi)$ are called spherical harmonics and are an important set of orthogonal functions arising in many applications from chemistry and physics. They are complex valued functions and so we use the inner product in the hermitian sense as defined by equation (2.30) and write the inner products in the form

$$\left(Y_n^m, Y_i^j\right) = \int_0^{2\pi} \int_0^\pi Y_n^m(\theta, \phi) \overline{Y_i^j(\theta, \phi)} \sin\theta \, d\theta d\phi = \delta_{mi} \delta_{nj} \qquad (2.87)$$

which is sometimes expressed in the form

$$\left(Y_n^m, Y_i^j\right) = \int_0^{2\pi} \int_{-1}^1 Y_n^m(\mu, \phi)\overline{Y_i^j(\mu, \phi)}\, d\mu d\phi = \delta_{mi}\delta_{nj} \tag{2.88}$$

where $\mu = \cos\theta$, $d\mu = -\sin\theta\, d\theta$ and the over line $\overline{\quad}$ denotes the complex conjugate.

The nonnormalized real $\Re$ and imaginary $\Im$ parts of equation (2.86) are sometimes referred to as even and odd functions of the spherical coordinates θ, ϕ or surface harmonics and written

$$\begin{aligned}
Y_{nm}^{(e)}(\theta, \phi) &= \Re\{P_n^m(\cos\theta)e^{im\phi}\} = P_n^m(\cos\theta)\cos m\phi \\
Y_{nm}^{(o)}(\theta, \phi) &= \Im\{P_n^m(\cos\theta)e^{im\phi}\} = P_n^m(\cos\theta)\sin m\phi
\end{aligned} \tag{2.89}$$

for $m = 0, 1, 2, \ldots, n$ and $n = 0, 1, 2, 3, \ldots$ where the symbols $\Re$ and $\Im$ are used to denote the real part and imaginary parts.

Example 2-11. (Orthogonal functions of two variables)

A set of functions $\phi_{nm}(x, y)$ defined over a region R of the x, y-plane are said to be orthogonal over the region R with respect to a weight function $\omega = \omega(x, y) > 0$ if the inner product satisfies

$$(\phi_{nj}, \phi_{mk}) = \iint_R \omega(x, y)\phi_{nj}(x, y)\phi_{mk}(x, y)\, dxdy = \begin{cases} \|\phi_{nj}\|^2, & (n, j) = (m, k) \\ 0, & (n, j) \neq (m, k) \end{cases}$$

where $\|\phi_{nj}\|^2 = (\phi_{nj}, \phi_{nj})$. For example the set of functions

$$C_{nm}(r, \theta) = J_n\left(\frac{\xi_{mn}r}{a}\right)\cos n\theta, \quad \text{and} \quad S_{nm}(r, \theta) = J_n\left(\frac{\xi_{mn}r}{a}\right)\sin n\theta,$$

for $n = 0, 1, 2, 3, \ldots$ and $m = 1, 2, 3, \ldots$ over the region $R = \{r, \theta \mid 0 \leq r \leq a, 0 \leq \theta \leq 2\pi\}$, with weight function $\omega = r$, where ξ_{nm} represents the mth zero of $J_n(\xi)$. That is, $J_n(\xi_{nm}) = 0$, $m = 1, 2, 3, \ldots$. These functions have the norm squared values

$$\|C_{nm}\|^2 = \|S_{nm}\|^2 = \frac{a^2}{2}\left[J_{n+1}(\xi_{nm})\right]^2 \pi, \quad n = 1, 2, 3, \ldots$$

$$\|C_{0m}\|^2 = \frac{a^2}{2}\left[J_1(\xi_{0m})\right]^2 2\pi \quad m = 1, 2, 3, \ldots$$

∎

Exercises 2

▶ 1.

Show that if the set of functions $\{g_n(x)\}$, $n = 1, 2, 3, \ldots$ is an orthogonal set of functions over an interval (a, b) with respect a weight function $r(x) > 0$, then the set of functions $f_n(x) = \dfrac{g_n(x)}{\|g_n\|}$ is an orthonormal set.

▶ 2.

Show the set of functions $\{\sin nx\}$ $n = 1, 2, 3, \ldots$ is an orthogonal set over the interval $-\pi < x < \pi$. Find the norm squared for each function. Construct the associated orthonormal set of functions.

▶ 3.

Show the set of functions $\{1, \cos nx\}$, for $n = 1, 2, 3, \ldots$ is an orthogonal set over the interval $-\pi < x < \pi$. Find the norm squared for each function. Construct the associated orthonormal set.

▶ 4.

Determine if the set of functions $1, \cos 2x, \cos 4x, \cos 6x, \ldots, \cos 2mx, \ldots$ is orthogonal over the interval $0 < x < \pi$. If the set is an orthogonal set, then find the norm squared for each eigenfunction and construct the associated orthonormal set.

▶ 5.

Use the Gram-Schmidt orthogonalization process to construct an orthogonal set $\{\phi_n(x)\}$, $n = 0, 1, 2, 3, \ldots$ from the functions $\{x^n\}$ for $n = 0, 1, 2, 3, \ldots$ on the interval $(-\infty, \infty)$ with respect to the weight function $r(x) = e^{-x^2}$. Calculate only the functions $\{\phi_0(x), \phi_1(x), \phi_2(x), \phi_3(x)\}$ and show

$$\phi_0(x) = 1, \quad \phi_1(x) = x, \quad \phi_2(x) = x^2 - 1/2, \quad \phi_3(x) = x^3 - \frac{3}{2}x$$

The set of functions $H_n(x) = 2^n \phi_n(x)$ are called Hermite polynomials.
Hint: For $I_m = \int_{-\infty}^{\infty} x^m e^{-x^2}\, dx$ show $I_m = \frac{1}{2}(m-1)I_{m-2}$.
Show also $I_0 = \sqrt{\pi}$ and $I_1 = 0$.

▶ **6.**

Use the Gram-Schmidt orthogonalization process to construct an orthogonal set $\{\phi_n(x)\}$, $n = 0, 1, 2, 3, \ldots$ from the functions $\{x^n\}$ for $n = 0, 1, 2, \ldots$ on the interval $(0, 1)$ with respect to the weight function $r(x) = 1$. Calculate only the functions $\{\phi_0(x), \phi_1(x), \phi_2(x), \phi_3(x)\}$.

▶ **7.**

Use the Gram-Schmidt orthogonalization process to construct an orthogonal set $\{\phi_n(x)\}$, $n = 0, 1, 2, 3, \ldots$ from the functions $\{x^n\}$ for $n = 0, 1, 2, 3, \ldots$ on the interval $(0, \infty)$ with respect to the weight function $r(x) = e^{-x}$. Calculate only the functions $\{\phi_0(x), \phi_1(x), \phi_2(x), \phi_3(x)\}$ and show

$$\phi_0(x) = 1, \quad \phi_1(x) = x - 1, \quad \phi_2(x) = x^2 - 4x + 2, \quad \phi_3(x) = x^3 - 9x^2 + 18x - 6$$

The functions $L_n(x) = \dfrac{(-1)^n}{n!}\phi_n(x)$ are called Laguerre polynomials.

▶ **8.**

Consider the Sturm-Liouville problem to solve $\dfrac{d^2y}{dx^2} + \lambda y = 0, \quad 0 < x < L$ subject to the boundary conditions $y'(0) = 0$ and $y(L) = 0$. Find the eigenvalues and eigenfunctions associated with this problem. Write out the orthogonality relationship satisfied by the eigenfunctions.
Hint: Consider the cases $\lambda = -\omega^2, \lambda = 0$ and $\lambda = \omega^2$.

▶ **9.**

Find the eigenvalues and eigenfunctions associated with the Sturm-Liouville problem

$$\frac{d^2y}{dx^2} + \lambda y = 0, \qquad 0 < x < L$$

subject to the boundary conditions $y(0) = 0$ and $y'(L) = 0$. Write out the orthogonality relationship satisfied by the eigenfunctions.
Hint: Consider the cases $\lambda = -\omega^2, \lambda = 0$ and $\lambda = \omega^2$.

▶ **10.**

Find the eigenvalues and eigenfunctions associated with the Sturm-Liouville problem

$$\frac{d^2y}{dx^2} + \lambda y = 0, \quad 0 < x < 1$$

subject to the boundary conditions $y'(0) = 0$ and $y'(1) = 0$. Write out the orthogonality relationship satisfied by the eigenfunctions.

▶ 11.

Find the eigenvalues and eigenfunctions for the Sturm-Liouville system $\frac{d^2y}{dx^2} + \lambda y = 0$ subject to the boundary conditions $y'(0) = 0$ and $y'(L) = 0$.

▶ 12.

Find the eigenvalues and eigenfunctions associated with the Sturm-Liouville problem

$$\frac{d}{dx}\left(x\frac{dy}{dx}\right) + \frac{\lambda}{x}y = 0, \qquad 1 < x < b$$

subject to the boundary conditions $y'(1) = 0$ and $y(b) = 0$. Write out the orthogonality condition satisfied by the eigenfunctions.

▶ 13.

Find the eigenvalues and eigenfunctions associated with the Sturm-Liouville problem

$$x^2\frac{d^2y}{dx^2} + x\frac{dy}{dx} + \lambda y = 0, \qquad 1 < x < b$$

subject to the boundary conditions $y(1) = 0$ and $y(b) = 0$. Write out the orthogonality relationship satisfied by the eigenfunctions.

Hint: Consider the cases $\lambda = -\omega^2, \lambda = 0$ and $\lambda = \omega^2$.

▶ 14.

Find the eigenvalues and eigenfunctions associated with the Sturm-Liouville problem

$$\frac{d}{dx}\left(x^2\frac{dy}{dx}\right) + \lambda y = 0, \qquad 1 < x < b$$

subject to the boundary conditions $y(1) = 0$ and $y(b) = 0$. Write out the orthogonality relationship satisfied by the eigenfunctions.

Hint: Consider the cases $1 - 4\lambda = -\omega^2, 1 - 4\lambda = 0$ and $1 - 4\lambda = \omega^2$.

▶ 15.

Find the eigenvalues and eigenfunctions associated with the Sturm-Liouville problem

$$\frac{d^2y}{d\theta^2} + \cot\theta\frac{dy}{d\theta} + \lambda y = 0, \qquad 0 \leq \theta \leq \pi$$

subject to boundedness conditions over the given interval. Write out the orthogonality relationship satisfied by the eigenfunctions.

▶ **16.**

Find the eigenvalues and eigenfunctions associated with the Sturm-Liouville problem

$$\frac{d}{dx}\left(x^3\frac{dy}{dx}\right) + \lambda x\, y = 0, \qquad 1 < x < e$$

subject to the boundary conditions $y(1) = 0$ and $y(e) = 0$. Write out the orthogonality relationship satisfied by the eigenfunctions.

▶ **17.**

Review the definitions of the spherical Bessel functions $j_m(x)$ and $y_m(x)$ given in problem 13 from the exercises of Chapter 1.

(a) Write out the differential equation (13c) from the exercise 13, Chapter 1, in self-adjoint form.

(b) Formulate a Sturm-Liouville problem over the interval $0 < r < b$ subject to the boundary condition $y(b) = 0$. Write out the orthogonality condition for the problem you formulate.

▶ **18.**

Find the eigenvalues and eigenfunctions associated with the Sturm-Liouville problem

$$\frac{d^2y}{dx^2} + \lambda y = 0, \qquad 0 < x < 1$$

subject to the boundary conditions $y(0) = 0$ and $y'(1) + hy(1) = 0$ where $h > 0$ is constant. Write out the orthogonality relationship satisfied by the eigenfunctions. Hint: Consider the cases $\lambda = -\omega^2, \lambda = 0$ and $\lambda = \omega^2$. Label the intersection of the graphs $y_2 = \tan\omega$ and $y_1 = -\omega/h$ by $\omega_1, \omega_2, \ldots$. (See figure 2-13 with $L = 1$.)

▶ **19.**

Find the eigenvalues and eigenfunctions associated with the Sturm-Liouville problem

$$\frac{d^2y}{dx^2} + \lambda y = 0, \qquad 0 < x < L$$

subject to the boundary conditions $y(0) = 0$ and $y'(L) + hy(L) = 0$ where $h > 0$ is a constant. Write out the orthogonality relationship satisfied by the eigenfunctions. Hint: Study the graphs of $y_1 = -\omega L/Lh$ and $y_2 = \tan\omega L$ vs. ωL. Then show all points of intersection of these graphs.

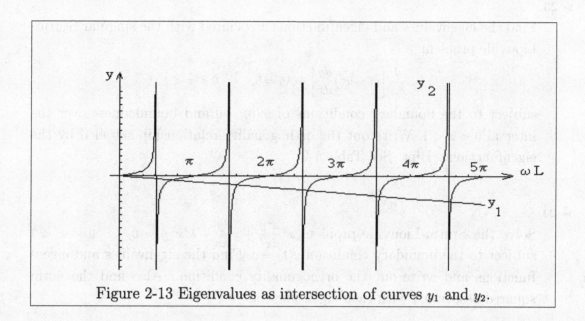

Figure 2-13 Eigenvalues as intersection of curves y_1 and y_2.

▶ **20.** Find the eigenvalues and eigenfunctions associated with the Sturm-Liouville system

$$x^2 \frac{d^2y}{dx^2} + 3x\frac{dy}{dx} + (1+\lambda^2)y = 0, \quad 1 < x < 100$$

subject to the boundary conditions $y(1) = 0$ and $y(100) = 0$.

▶ **21.**

Solve the Sturm-Liouville problem

$$\frac{d^2y}{dx^2} + \frac{dy}{dx} + \lambda y = 0, \quad 0 < x < L$$

subject to the boundary conditions $y(0) = 0$ and $y(L) = 0$.

▶ **22.**

Find the eigenvalues and eigenfunctions associated with the singular Sturm-Liouville problem

$$\frac{d}{dx}\left[(1-x^2)\frac{dy}{dx}\right] + \lambda y = 0, \quad 0 < x < 1$$

subject to the boundary conditions of $y(0) = 0$ and boundedness over the interval $0 < x < 1$. Write out the orthogonality relationship satisfied by the eigenfunctions. Hint: See Table 1-2.

▶ **23.**

Find the eigenvalues and eigenfunctions associated with the singular Sturm-Liouville problem

$$\frac{d}{dx}\left[(1-x^2)\frac{dy}{dx}\right] + \lambda y = 0, \qquad 0 < x < 1$$

subject to the boundary conditions of $y'(0) = 0$ and boundedness over the interval $0 < x < 1$. Write out the orthogonality relationship satisfied by the eigenfunctions. Hint: See Table 1-2.

▶ **24.**

Solve the Sturm-Liouville problem $x^2\dfrac{d^2y}{dx^2} + x\dfrac{dy}{dx} + \lambda^2 x^2 y = 0, \qquad 0 < x < 1$ subject to the boundary condition $y(1) = 0$. Find the eigenvalues and eigenfunctions and write out the orthogonality condition. Also find the norm squared associated with each eigenfunction.

▶ **25.**

Solve the Sturm-Liouville problem $\dfrac{d^2y}{dx^2} + \lambda y = 0, \quad 0 < x < \pi$ subject to the boundary conditions $y(0) = 0$ and $y'(\pi) + y(\pi) = 0$. Find all eigenvalues and eigenfunctions. Write out the orthogonality condition satisfied by the eigenfunctions. Estimate the value of the first eigenvalue.

▶ **26.**

Solve the Sturm-Liouville problem

$$x^2\frac{d^2y}{dx^2} + x\frac{dy}{dx} + (\lambda^2 x^2 - 9)y = 0, \quad 0 < x < c$$

subject to the boundary condition $y(c) = 0$.
(a) Determine the eigenvalues.
(b) Determine the eigenfunctions.
(c) Write out the orthogonality relationship satisfied by the eigenfunctions.
(d) Calculate the norm squared of the eigenfunctions.

▶ **27.**   Find all the eigenvalues and eigenfunctions associated with the Sturm-Liouville system

$$x^2\frac{d^2y}{dx^2} - 5x\frac{dy}{dx} + (9 + \lambda^2)y = 0, \qquad 1 < x < e$$

subject to the boundary conditions $y(1) = 0$ and $y'(e) = 0$.

▶ **28.**

Let $\{f_n(x)\}$, $n = 1,2,3,\ldots$ denote a set of functions which are orthogonal functions over the interval $a \leq x \leq b$ with respect to the weight function $\omega(x) > 0$. Show that if a function $F(x)$ is represented in the form of a series of orthogonal functions,

$$F(x) = c_1 f_1(x) + c_2 f_2(x) + \cdots + c_m f_m(x) + \cdots$$

where $c_1, c_2, \ldots$ are constants, then the constants can be determined from an inner product divided by a norm squared and written in the form

$$c_m = \frac{(F, f_m)}{\| f_m \|^2} = \frac{\int_a^b F(x) f_m(x) \omega(x)\, dx}{\int_a^b f_m^2(x)\omega(x)\, dx}$$

for $m = 1,2,3,\ldots$. Hint: Multiply both sides of the series by $f_1(x)\omega(x)$ and then integrate from a to b. This will enable you to solve for the constant c_1. In a similar fashion solve for c_2, c_3, c_4, etc.

▶ **29.**

(a) Solve the Sturm-Liouville system

$$\frac{d^2 y}{dx^2} + \lambda y = 0, \quad a \leq x \leq b$$
$$y(a) = 0 \quad \text{and} \quad y'(b) + hy(b) = 0, \quad h > 0 \text{ is constant.}$$

Show the eigenfunctions are $y_n(x) = \sin \omega_n(x - a)$ where ω_n are the roots of the equation $h \tan \omega(b - a) = -\omega$.

(b) Show that the eigenfunctions have the norm squared

$$\| y_n \|^2 = \frac{b - a}{2} + \frac{\cos^2 \omega_n(b - a)}{2h}.$$

(c) Sketch the curves $Y_1 = \tan \omega(b - a)$, $Y_2 = -\omega/h$ vs $\omega(b - a)$ and show where they intersect. Explain what the intersections represent.

(d) Write a computer program to find the eigenvalues $\lambda_n = \omega_n^2$, $n = 1,2,\ldots,10$ in the special case where $a = 0$, $b = 1$ and $h = 8$.

(e) Write out the orthogonality relationship satisfied by the above eigenfunctions.

▶ **30.** Find all the eigenvalues and eigenfunctions associated with the Sturm-Liouville system

$$\frac{d^2y}{dx^2} + 2\alpha\frac{dy}{dx} + (\alpha^2 + \lambda)y = 0, \quad 0 < x < L, \quad \alpha \text{ constant}$$

subject to the boundary conditions $y(0) = 0$ and $y(L) = 0$.

▶ **31.** Explain why the given Sturm-Liouville equations are singular.

$$\text{Bessel equation} \quad \frac{d}{dx}\left(x\frac{dy}{dx}\right) - \frac{\nu^2}{x}y = -\lambda^2 xy, \quad 0 < x < b$$

$$\text{Legendre equation} \quad \frac{d}{dx}\left[(1 - x^2)\frac{dy}{dx}\right] = -\lambda y, \quad -1 < x < 1$$

$$\text{Hermite equation} \quad \frac{d}{dx}\left(e^{-x^2}\frac{dy}{dx}\right) = -\lambda e^{-x^2}y, \quad -\infty < x < \infty$$

$$\text{Laguerre equation} \quad \frac{d}{dx}\left(xe^{-x}\frac{dy}{dx}\right) = -\lambda e^{-x}y, \quad 0 < x < \infty$$

$$\text{Chebyshev equation} \quad \frac{d}{dx}\left(\sqrt{1 - x^2}\frac{dy}{dx}\right) = -\frac{\lambda^2}{\sqrt{1 - x^2}}y, \quad -1 < x < 1$$

▶ **32.**

Solve the Sturm-Liouville problem

$$\frac{d}{dr}\left(r\frac{dF}{dr}\right) + \lambda r F = 0, \quad 0 < r < a$$

subject to the boundary conditions $F(a) = 0$ and $|F(0)| < \infty$. Show the orthogonality condition satisfied by the eigenfunctions.

▶ **33.**

(a) Solve the boundary value problem

$$\frac{d^2y}{dx^2} - \frac{dy}{dx} + \lambda e^{2x}y = 0, \quad 0 < x < 1$$

subject to the boundary conditions $y(0) = 0$ and $y(1) = 0$ to determine the eigenvalues and eigenfunctions.

(b) Determine the weight function $r(x)$.

(c) Write out the orthogonality condition satisfied by the eigenfunctions.

Hint: Observe that if $z = e^x - 1$, then $e^{-x}\frac{d}{dx} = \frac{d}{dz}$.

Chapter 3
Fourier Series and Integrals

We have shown the set of functions $\{1, \sin\frac{n\pi x}{L}, \cos\frac{n\pi x}{L}\}$ are eigenfunctions which are orthogonal on the interval $(-L, L)$. A function $f(x)$ is called piecewise smooth on an interval if the interval can be broken up into sections such that on each section the function is continuous and its derivative $f'(x)$ is also continuous. The above set of orthogonal functions can be used in a series to represent piecewise smooth functions $f(x)$ over the interval $(-L, L)$. The series has the form

$$f(x) = a_0 + \sum_{n=1}^{\infty}\left(a_n\cos\frac{n\pi x}{L} + b_n\sin\frac{n\pi x}{L}\right) \tag{3.1}$$

and is called a Fourier trigonometric series representation of the function $f(x)$ on the interval $(-L, L)$. The constants a_0, a_n, b_n are called the Fourier coefficients of the series. Using the orthogonality properties of the eigenfunctions, equations (2.29)(a) and (2.29)(b), we calculate these Fourier coefficients as follows. Integrate all terms in equation (3.1) over the interval $(-L, L)$. The orthogonality properties of the above set gives the a_0 coefficient as an inner product divided by a norm squared

$$a_0 = \frac{(f, 1)}{\|1\|^2} = \frac{1}{2L}\int_{-L}^{L} f(x)\, dx. \tag{3.2}$$

The coefficient a_0 represents the average value of the function $f(x)$ over the interval $(-L, L)$. Now multiply both sides of equation (3.1) by $\cos\frac{m\pi x}{L}$ and integrate over the interval $(-L, L)$. We find the coefficients a_n are given by an inner product divided by a norm squared

$$a_n = \frac{(f, \cos\frac{n\pi x}{L})}{\|\cos\frac{n\pi x}{L}\|^2} = \frac{1}{L}\int_{-L}^{L} f(x)\cos\frac{n\pi x}{L}\, dx \quad n = 1, 2, 3, \ldots \tag{3.3}$$

Here each a_n represents the average of the function $2f(x)\cos\frac{n\pi x}{L}$ over the interval $(-L, L)$. Finally, if we multiply both sides of equation (3.1) by $\sin\frac{m\pi x}{L}$ and integrate over the interval $(-L, L)$. We find the Fourier coefficients b_n are also given by an inner product divided by a norm squared

$$b_n = \frac{(f, \sin\frac{n\pi x}{L})}{\|\sin\frac{n\pi x}{L}\|^2} = \frac{1}{L}\int_{-L}^{L} f(x)\sin\frac{n\pi x}{L}\, dx \quad n = 1, 2, 3, \ldots \tag{3.4}$$

Here each b_n represents the average of the function $2f(x)\sin\frac{n\pi x}{L}$ over the interval $(-L, L)$.

114

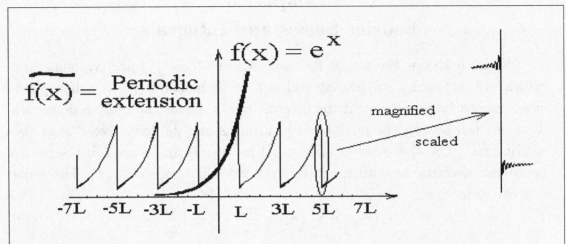

Figure 3-1. Fourier trigonometric representation of the function e^x,
using 35 terms of series, together with the function e^x.

Example 3-1.   (Fourier Series.)
   If one represents the exponential function as a Fourier series
$$e^x = a_0 + \sum_{n=1}^{\infty} \left(a_n \cos \frac{n\pi x}{L} + b_n \sin \frac{n\pi x}{L} \right) \text{ over the interval } (-L, L), \text{ then find the coef-}$$
ficients $a_0, a_n, b_n,\ n = 1, 2, 3, \ldots$
Solution: The Fourier coefficients are calculated from the relations

$$a_0 = \frac{(e^x, 1)}{\|1\|^2} = \frac{1}{2L} \int_{-L}^{L} e^x\, dx = \frac{1}{L} \sinh L$$

$$a_n = \frac{(e^x, \cos \frac{n\pi x}{L})}{\|\cos \frac{n\pi x}{L}\|^2} = \frac{1}{L} \int_{-L}^{L} e^x \cos \frac{n\pi x}{L}\, dx = \frac{2L(-1)^n \sinh L}{L^2 + n^2\pi^2}$$

$$b_n = \frac{(e^x, \sin \frac{n\pi x}{L})}{\|\sin \frac{n\pi x}{L}\|^2} = \frac{1}{L} \int_{-L}^{L} e^x \sin \frac{n\pi x}{L}\, dx = \frac{-2n\pi(-1)^n \sinh L}{L^2 + n^2\pi^2}$$

which gives the Fourier trigonometric series representation of e^x as

$$e^x = \frac{\sinh L}{L} + \sum_{n=1}^{\infty} \left(\frac{2L(-1)^n \sinh L}{L^2 + n^2\pi^2} \cos \frac{n\pi x}{L} - \frac{2n\pi(-1)^n \sinh L}{L^2 + n^2\pi^2} \sin \frac{n\pi x}{L} \right) \tag{3.5}$$

The figure 3-1 illustrates a graphical representation of two curves. The first curve
plotted illustrates the given function $f(x) = e^x$ for all values of x while the second
curve plotted illustrates the Fourier trigonometric series representation of this
function over the interval $(-L, L)$ plus its periodic extension. Note that because
the set of functions $\{1, \sin \frac{n\pi x}{L}, \cos \frac{n\pi x}{L}\}$ are periodic of period $2L$ the Fourier series

given by equation (3.5) only represents e^x on the interval $(-L, L)$. The Fourier series does not represent e^x for all values of x. The interval $(-L, L)$ is called the full Fourier interval. Outside the full Fourier interval the Fourier series gives the periodic extension of the values of $f(x)$ inside the full Fourier interval.

■

Properties of the Fourier trigonometric series

1. The Dirichlet[†] conditions for existence of a Fourier series are: (i) $f(x)$ must be single-valued and periodic with period $2L$. (ii) The function $f(x)$ is bounded with a finite number of maxima and minima and a finite number of discontinuities over the interval $(-L, L)$.

2. The Fourier series, when it exists, represents $f(x)$ on the interval $(-L, L)$ which is called the full Fourier interval.

3. The Fourier series evaluated at points x outside the full Fourier interval gives the periodic extension of $f(x)$ defined over the full Fourier interval.

4. The equal sign in equation (3.1) is sometimes replaced by the symbol $\sim$, representing "corresponds to", because the series on the right-hand side does not always converge to $f(x)$ for all values of x. The Fourier series only represents $f(x)$ on the full Fourier interval. Alternatively, one can define a function $\tilde{f}(x)$ which is the periodic extension of $f(x)$ over the full Fourier interval. Thus, $\tilde{f}(x)$ is the periodic extension of $f(x)$, $-L \leq x \leq L$, and has the property $\tilde{f}(x+2L) = \tilde{f}(x)$ whereas the function $f(x)$, for all x, need not be periodic (See the example 3-1).

5. In order for a function $f(x)$ to have a Fourier series representation one must be able to calculate the Fourier coefficients a_0, a_n, b_n given by the equations (3.2), (3.3), and (3.4). Consequently, some functions will not have a Fourier series. For example, the functions $\frac{1}{x}$, $\frac{1}{x^2}$ are examples of functions which do not have a Fourier trigonometric series representation over the interval $(-L, L)$. Note that these functions are unbounded over the interval.

6. A function $f(x)$ is said to have a jump discontinuity at a point x_0 if

$$f(x_0^-) = \lim_{\substack{\epsilon \to 0 \\ \epsilon > 0}} f(x_0 - \epsilon) \neq f(x_0^+) = \lim_{\substack{\epsilon \to 0 \\ \epsilon > 0}} f(x_0 + \epsilon).$$

The situation is illustrated in the figure 3-2.

[†] See Appendix C

116

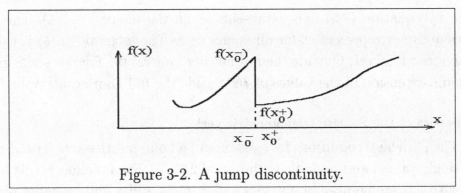

Figure 3-2. A jump discontinuity.

7. If the functions $f(x)$ and $f'(x)$ are piecewise continuous over the interval $(-L, L)$ then the Fourier series representation for $f(x)$: (a) Converges to $f(x)$ at points where $f(x)$ is continuous. (b) Converges to the periodic extension of $f(x)$ if x is outside the full Fourier interval. (c) At points x_0 where there is a finite jump discontinuity, the Fourier trigonometric series converges to $\frac{1}{2}\left[f(x_0^+) + f(x_0^-)\right]$ which represents the average of the left and right-hand limits associated with the jump discontinuity.

8. The function $S_N(x) = a_0 + \sum_{n=1}^{N}\left(a_n \cos \frac{n\pi x}{L} + b_n \sin \frac{n\pi x}{L}\right)$ is called the Nth partial sum and represents a truncation of the series after N terms. One usually plots the approximating function $S_N(x)$ when representing the Fourier series graphically. Whenever the function $f(x)$ being approximated has a point where a jump discontinuity occurs, then the approximating function $S_N(x)$ has oscillations in the neighborhood of the jump discontinuity as well as an "overshoot" of the jump in the function. These effects are known as the Gibb's[†] phenomenon. The Gibb's phenomenon always occurs whenever one attempts to use a series of continuous functions to represent a discontinuous function. The Gibb's phenomenon can be seen in the figures 3-1, 3-5 and 3-9. These effects are not eliminated by increasing the value of N in the partial sum. See for example figure 3-5.

9. A series of the form $a_0 + \sum_{n=1}^{\infty}\left(a_n \cos \frac{n\pi x}{L} + b_n \sin \frac{n\pi x}{L}\right)$ can also be written in the form $a_0 + \sum_{n=1}^{\infty} C_n \sin\left(\frac{n\pi x}{L} + \phi_n\right)$ where $C_n = \sqrt{a_n^2 + b_n^2}$ are called amplitudes and $\phi_n = \arctan(a_n/b_n)$ are called phase angles. The nth term $C_n \sin\left(\frac{n\pi x}{L} + \phi_n\right)$ of the series is called the nth harmonic. The first harmonic is also referred to as the fundamental harmonic.

[†] See Appendix C

Even and Odd functions

A function $f(x)$ is an even function of x if $f(-x) = f(x)$ for all x. Such a function is symmetric about the y-axis. A function $f(x)$ is an odd function of x, if for all values of x, we have $f(-x) = -f(x)$. In this case the function is symmetric about the origin. Three important properties of even and odd functions will enable us to simplify the Fourier series representation given above.

1. If $f(x)$ is an even function of x, then $\int_{-L}^{L} f(x)\,dx = 2\int_{0}^{L} f(x)\,dx$.

2. If $f(x)$ is an odd function of x, then $\int_{-L}^{L} f(x)\,dx = 0$.

3. If $f_e(x), g_e(x)$ are even functions and $f_o(x), g_o(x)$ are odd functions, then an even function times an even function $f_e(x)g_e(x) = h(x)$ is an even function. The product of an even function times an odd function $f_e(x)g_o(x) = h(x)$ is an odd function. An odd function times an odd function $f_o(x)g_o(x) = h(x)$ is an even function.

Example 3-2. (**Symmetry and integral of even function.**)
Show that if $f(x)$ is even, then $\int_{-L}^{L} f(x)\,dx = 2\int_{0}^{L} f(x)\,dx$.
Solution: By hypothesis $f(-x) = f(x)$ and so we can write

$$\int_{-L}^{L} f(\xi)\,d\xi = \int_{-L}^{0} f(\xi)\,d\xi + \int_{0}^{L} f(\xi)\,d\xi.$$

In the integral from $-L$ to 0 we make the change of variable $\xi = -x$ with $d\xi = -dx$ to obtain

$$\int_{-L}^{L} f(\xi)\,d\xi = \int_{L}^{0} f(-x)\,(-dx) + \int_{0}^{L} f(\xi)\,d\xi$$

$$= \int_{0}^{L} f(x)\,dx + \int_{0}^{L} f(x)\,dx = 2\int_{0}^{L} f(x)\,dx.$$

∎

Example 3-3. (**Product of odd functions.**)
Show that if $f_o(x), g_o(x)$ are odd functions, then $h(x) = f_o(x)g_o(x)$ is an even function.
Solution: By hypothesis $f_o(-x) = -f_o(x)$ and $g_o(-x) = -g_o(x)$ so that

$$h(-x) = f_o(-x)g_o(-x) = (-f_o(x))(-g_o(x)) = f_o(x)g_o(x) = h(x)$$

so that by definition $h(x)$ is an even function.

∎

The Fourier trigonometric series representation of a piecewise smooth function $f(x)$ is given by

$$f(x) = a_0 + \sum_{n=1}^{\infty} \left(a_n \cos \frac{n\pi x}{L} + b_n \sin \frac{n\pi x}{L} \right) \tag{3.6}$$

with Fourier coefficients

$$a_0 = \frac{(f,1)}{\|1\|^2} = \frac{1}{2L} \int_{-L}^{L} f(x)\, dx$$

$$a_n = \frac{(f, \cos \frac{n\pi x}{L})}{\| \cos \frac{n\pi x}{L} \|^2} = \frac{1}{L} \int_{L}^{L} f(x) \cos \frac{n\pi x}{L}\, dx, \quad n = 1, 2, 3, \ldots \tag{3.7}$$

$$b_n = \frac{(f, \sin \frac{n\pi x}{L})}{\| \sin \frac{n\pi x}{L} \|^2} = \frac{1}{L} \int_{-L}^{L} f(x) \sin \frac{n\pi x}{L}\, dx, \quad n = 1, 2, 3, \ldots$$

In the special case where $f(x)$ is an odd function the Fourier series representation given by equations (3.6), (3.7) reduces to a Fourier sine series representation

$$f(x) = \sum_{n=1}^{\infty} b_n \sin \frac{n\pi x}{L}, \quad 0 < x < L \tag{3.8}$$

$$\text{with} \qquad b_n = \frac{2}{L} \int_{0}^{L} f(x) \sin \frac{n\pi x}{L}\, dx, \quad n = 1, 2, 3, \ldots$$

In the special case where $f(x)$ is an even function the Fourier series representation given by equations (3.6), (3.7) reduces to the Fourier cosine series representation

$$f(x) = a_0 + \sum_{n=1}^{\infty} a_n \cos \frac{n\pi x}{L}, \quad 0 < x < L$$

$$a_0 = \frac{1}{L} \int_{0}^{L} f(x)\, dx \tag{3.9}$$

$$a_n = \frac{2}{L} \int_{0}^{L} f(x) \cos \frac{n\pi x}{L}\, dx, \quad n = 1, 2, 3, \ldots$$

Example 3-4. (**Fourier series.**)

Find the Fourier series for the function $f(x) = \begin{cases} x, & 0 < x < L \\ -x, & -L < x < 0 \end{cases}$

Solution: Here $f(x)$ is an even function of x and so the Fourier series is a cosine series

$$f(x) = a_0 + \sum_{n=1}^{\infty} a_n \cos \frac{n\pi x}{L}$$

with coefficients

$$a_0 = \frac{1}{L} \int_0^L x \, dx = \frac{L}{2}$$

$$a_n = \frac{2}{L} \int_0^L x \cos \frac{n\pi x}{L} \, dx = \frac{2L}{n^2\pi^2}[-1 + (-1)^n].$$

This gives the Fourier cosine series representation

$$x = \frac{L}{2} - \sum_{n=1}^{\infty} [1 - (-1)^n] \cos \frac{n\pi x}{L}.$$

∎

Suppose we are only interested in a function $f(x)$ which is defined on the interval $(0, L)$. In this case we will not be concerned about how the function looks outside of the interval $(0, L)$. If the given function $f(x)$ is not defined on the full Fourier interval $(-L, L)$, then we can expand this function in a Fourier trigonometric series in three possible ways. (i) We can extend $f(x)$ as an odd function to get a Fourier sine series. (ii) We can extend $f(x)$ as an even function to get a Fourier cosine series. (iii) We can extend $f(x)$ some other way to get the full Fourier trigonometric series having both sine and cosine terms. The situation is illustrated in the figure 3-3.

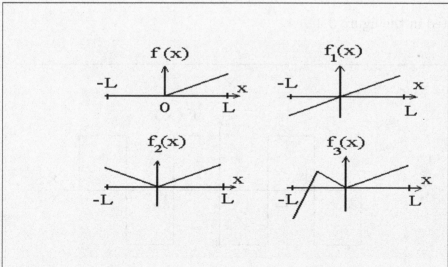

Figure 3-3. The function $f(x)$ extended to full Fourier interval.

120

Example 3-5. (**Fourier sine series.**)

Find the Fourier sine series expansion for the function which is given by $f(x) = x$ over the interval $0 < x < L$.

Solution: Here $f(x) = x$ is only defined on the interval $0 < x < L$ and so if it is extended to the full Fourier interval as an odd function, we obtain the Fourier sine series

$$f(x) = \sum_{n=1}^{\infty} b_n \sin \frac{n\pi x}{L} \quad \text{with} \quad b_n = \frac{2}{L} \int_0^L x \sin \frac{n\pi x}{L} \, dx = -\frac{2(-1)^n L}{n\pi}.$$

This gives the Fourier sine series expansion

$$x = \frac{2L}{\pi} \sum_{n=1}^{\infty} \frac{(-1)^{n+1}}{n} \sin \frac{n\pi x}{L}$$

If $f(x) = x$ is extended as an even function over the full Fourier interval, then it is left as an exercise to show there results the Fourier cosine series expansion

$$x = \frac{L}{2} + \frac{2}{L} \sum_{n=1}^{\infty} \left(\frac{L}{n\pi}\right)^2 [(-1)^n - 1] \cos \frac{n\pi x}{L}.$$

■

Example 3-6. (**Square wave.**)

Find the Fourier series expansion for the square wave

$$f(x) = \begin{cases} 1, & 0 < x < L \\ -1, & -L < x < 0 \end{cases}, \qquad f(x + 2L) = f(x).$$

illustrated in the figure 3-4.

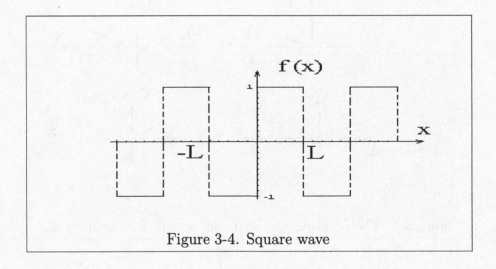

Figure 3-4. Square wave

Solution: Here $f(x)$ is an odd function of x and so can be expanded in a Fourier sine series

$$f(x) = \sum_{n=1}^{\infty} b_n \sin \frac{n\pi x}{L}$$

with $b_n = 2 \int_0^L (1) \sin \frac{n\pi x}{L}\, dx = \frac{2}{n\pi}[1 - \cos n\pi] = \frac{2}{n\pi}[1 - (-1)^n]$. The resulting Fourier sine series is

$$f(x) = \sum_{n=1}^{\infty} \frac{2}{n\pi}[1 - (-1)^n] \sin \frac{n\pi x}{L}.$$

Another form for this Fourier series is obtained by changing the summation index. Notice the term $1 - (-1)^n$ is zero for all even values of n so that only odd values of n give a nonzero result. If we let $n = 2m+1$ for $m = 0, 1, 2, 3, \ldots$, then the resulting form for the Fourier series is

$$f(x) = \frac{4}{\pi} \sum_{m=0}^{\infty} \frac{1}{2m+1} \sin \frac{(2m+1)\pi x}{L}.$$

In expanded form this series can be represented

$$f(x) = \frac{4}{\pi} \left[\sin \frac{\pi x}{L} + \frac{1}{3} \sin \frac{3\pi x}{L} + \frac{1}{5} \sin \frac{5\pi x}{L} + \frac{1}{7} \sin \frac{7\pi x}{L} + \cdots \right].$$

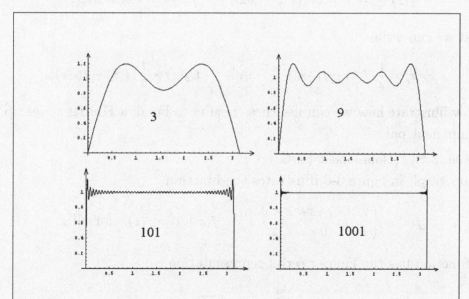

Figure 3-5. Selected partial sums for square wave representation.

Let

$$S_n(x) = \frac{4}{\pi} \sum_{m=0}^{n} \frac{1}{2m+1} \sin \frac{(2m+1)\pi x}{L}$$

denote the nth partial sum of the series. The figure 3-5 illustrates some graphs, plotted over the interval $(0, \pi)$, of some selected partial sums in the special case where $L = \pi$. The number on each graph represents the last harmonic in the series of partial sums. In this figure observe the undershoot and overshoot occurring near the points of discontinuity. Note also that the value of the partial sums at $x = 0$ and $x = \pi$ are zero as this represents the average value of the left and right-hand values for the function at these points.

■

Every function $f(x)$ can be represented as the sum of an even function $F_e(x)$ and an odd function $G_o(x)$. Using the properties of even and odd functions we can write

$$f(x) = F_e(x) + G_o(x)$$

$$f(-x) = F_e(-x) + G_o(-x)$$

$$f(-x) = F_e(x) - G_o(x)$$

By adding and then subtracting the first and last of the above equations we find

$$f(x) + f(-x) = 2F_e(x) \qquad \text{and} \qquad f(x) - f(-x) = 2G_o(x)$$

so that we can write

$$F_e(x) = \frac{1}{2}\left(f(x) + f(-x)\right) \qquad \text{and} \qquad G_o(x) = \frac{1}{2}\left(f(x) - f(-x)\right).$$

We now illustrate how we can use these results to break a Fourier series up into its component parts.

Example 3-7. (**Component parts.**)
The top graph in figure 3-6 illustrates the function

$$f(x) = \begin{cases} 2, & -2 < x < 0 \\ x, & 0 < x < 2 \end{cases}, \qquad f(x+4) = f(x) \quad \text{for all } x.$$

This function has the Fourier series representation

$$f(x) = a_0 + \sum_{n=1}^{\infty} \left(a_n \cos \frac{n\pi x}{2} + b_n \sin \frac{n\pi x}{2}\right)$$

where

$$a_0 = \frac{1}{2L} \int_{-L}^{L} f(x)\, dx = \frac{1}{4} \int_{-2}^{2} f(x)\, dx = \frac{1}{4} \left[\int_{-2}^{0} 2\, dx + \int_{0}^{2} x\, dx \right] = \frac{3}{2}$$

$$a_n = \frac{1}{L} \int_{-L}^{L} f(x) \cos \frac{n\pi x}{L}\, dx$$

$$a_n = \frac{1}{2} \left[\int_{-2}^{0} 2 \cos \frac{n\pi x}{2}\, dx + \int_{0}^{2} x \cos \frac{n\pi x}{2}\, dx \right] = \frac{-2}{n^2 \pi^2} \left[1 - (-1)^n \right]$$

$$b_n = \frac{1}{L} \int_{-L}^{L} f(x) \sin \frac{n\pi x}{L}\, dx = \frac{1}{2} \left[\int_{-2}^{0} 2 \sin \frac{n\pi x}{2}\, dx + \int_{0}^{2} x \sin \frac{n\pi x}{2}\, dx \right] = \frac{-2}{n\pi}$$

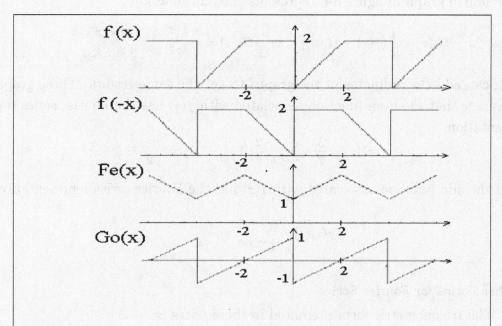

Figure 3-6. Even and odd functions associated with Fourier Series.

This gives the Fourier series representation

$$f(x) = \frac{3}{2} + \sum_{n=1}^{\infty} \frac{-2}{n^2 \pi^2} \left[1 - (-1)^n \right] \cos \frac{n\pi x}{2} + \sum_{n=1}^{\infty} \frac{-2}{n\pi} \sin \frac{n\pi x}{2}.$$

The second graph in figure 3-6 illustrates the function $f(-x)$ vs x over the same range as the first graph. To obtain this graph you can rotate the $f(x)$ curve about the y-axis or you can calculate $f(-x)$ from the given function $f(x)$. We have

$$f(-x) = \begin{cases} 2 & -2 < -x < 0 \\ -x & 0 < -x < 2 \end{cases} \quad \text{which can be} \atop \text{rewritten as} \quad f(-x) = \begin{cases} -x & -2 < x < 0 \\ 2 & 0 < x < 2 \end{cases}$$

The third figure in figure 3-6 represents the even function

$$F_e(x) = \frac{1}{2}\left(f(x) + f(-x)\right) \quad \text{or} \quad F_e(x) = \begin{cases} \frac{2-x}{2} & -2 < x < 0 \\ \frac{x+2}{2} & 0 < x < 2 \end{cases}.$$

The fourth graph in figure 3-6 represents the odd function

$$G_o(x) = \frac{1}{2}\left(f(x) - f(-x)\right) \quad \text{or} \quad G_o(x) = \begin{cases} \frac{2+x}{2} & -2 < x < 0 \\ \frac{x-2}{2} & 0 < x < 2 \end{cases}.$$

You can add the ordinates of the graphs to get the same results. These graphs illustrate that the even function associated with $f(x)$ has the Fourier series representation

$$F_e(x) = \frac{3}{2} + \sum_{n=1}^{\infty} \frac{-2}{n^2\pi^2}[1 - (-1)^n]\cos\frac{n\pi x}{2}$$

and the odd function associated with $f(x)$ has the Fourier series representation

$$G_o(x) = \sum_{n=1}^{\infty} \frac{-2}{n\pi}\sin\frac{n\pi x}{2}.$$

$\blacksquare$

Other Forms for Fourier Series

The trigonometric form presented in these notes is

$$f(x) = a_0 + \sum_{n=1}^{\infty}\left(a_n\cos\frac{n\pi x}{L} + b_n\sin\frac{n\pi x}{L}\right) \qquad -L < x < L$$

with $\quad a_0 = \dfrac{1}{2L}\displaystyle\int_{-L}^{L} f(x)\,dx$

$$a_n = \frac{1}{L}\int_{-L}^{L} f(x)\cos\frac{n\pi x}{L}\,dx, \quad n = 1,2,3,\ldots$$

$$b_n = \frac{1}{L}\int_{-L}^{L} f(x)\sin\frac{n\pi x}{L}\,dx, \quad n = 1,2,3,\ldots$$

(3.10)

The symmetric form of the Fourier trigonometric series presented in many text-books is

$$f(x) = \frac{A_0}{2} + \sum_{n=1}^{\infty} \left(A_n \cos \frac{n\pi x}{L} + B_n \sin \frac{n\pi x}{L} \right)$$

with $\quad A_n = \frac{1}{L} \int_{-L}^{L} f(x) \cos \frac{n\pi x}{L}\, dx, \qquad n = 0, 1, 2, 3, \ldots \qquad (3.11)$

$$B_n = \frac{1}{L} \int_{-L}^{L} f(x) \sin \frac{n\pi x}{L}\, dx \qquad n = 1, 2, 3, \ldots$$

Observe the factor of one-half in the a_0 term of equation (3.10) has been shifted to the constant term of the series in equation (3.11).

The function $f(x)$ need not be defined only on the interval $(-L, L)$. For $f(x)$ defined on the interval $\alpha \le x \le \alpha + 2L$ with $f(x + 2L) = f(x)$ there is the Fourier trigonometric series representation

$$f(x) = a_0 + \sum_{n=1}^{\infty} \left(a_n \cos \frac{n\pi x}{L} + b_n \sin \frac{n\pi x}{L} \right)$$

with $\quad a_0 = \frac{1}{2L} \int_{\alpha}^{\alpha+2L} f(x)\, dx$

$$a_n = \frac{1}{L} \int_{\alpha}^{\alpha+2L} f(x) \cos \frac{n\pi x}{L}\, dx \qquad (3.12)$$

$$b_n = \frac{1}{L} \int_{\alpha}^{\alpha+2L} f(x) \sin \frac{n\pi x}{L}\, dx.$$

The complex form of the Fourier series is obtained by using the Euler identity $e^{in\pi x/L} = \cos \frac{n\pi x}{L} + i \sin \frac{n\pi x}{L}$, where $i^2 = -1$ and then substituting

$$\cos \frac{n\pi x}{L} = \frac{1}{2} \left(e^{in\pi x/L} + e^{-in\pi x/L} \right) \quad \text{and} \quad \sin \frac{n\pi x}{L} = \frac{-i}{2} \left(e^{in\pi x/L} - e^{-in\pi x/L} \right)$$

into the symmetric form of the Fourier series given by equation (3.11). This produces the series

$$f(x) = \frac{A_0}{2} + \sum_{n=1}^{\infty} \frac{A_n}{2} \left(e^{in\pi x/L} + e^{-in\pi x/L} \right) + \frac{-iB_n}{2} \left(e^{in\pi x/L} - e^{-in\pi x/L} \right)$$

$$f(x) = \frac{A_0}{2} + \frac{1}{2} \sum_{n=1}^{\infty} (A_n - iB_n) e^{in\pi x/L} + \frac{1}{2} \sum_{n=1}^{\infty} (A_n + iB_n) e^{-in\pi x/L}$$

Define the constants

$$C_0 = \frac{A_0}{2} = \frac{1}{2L} \int_{-L}^{L} f(x)\,dx$$

$$C_n = \frac{1}{2}(A_n - iB_n) = \frac{1}{2L} \int_{-L}^{L} f(x) \left(\cos\frac{n\pi x}{L} - i\sin\frac{n\pi x}{L}\right) dx = \frac{1}{2L} \int_{-L}^{L} f(x) e^{-i\,n\pi x/L}\,dx$$

$$C_{-n} = \frac{1}{2}(A_n + iB_n) = \frac{1}{2L} \int_{-L}^{L} f(x) \left(\cos\frac{n\pi x}{L} + i\sin\frac{n\pi x}{L}\right) dx = \frac{1}{2L} \int_{-L}^{L} f(x) e^{i\,n\pi x/L}\,dx$$

and write

$$f(x) = C_0 + \sum_{n=1}^{\infty} C_n e^{i\,n\pi x/L} + \sum_{n=1}^{\infty} C_{-n} e^{-i\,n\pi x/L}.$$

In the last summation replace n by $-n$ to obtain

$$f(x) = C_0 + \sum_{n=1}^{\infty} C_n e^{i\,n\pi x/L} + \sum_{n=-1}^{-\infty} C_n e^{i\,n\pi x/L}.$$

This produces the complex form of the Fourier series

$$f(x) = \sum_{n=-\infty}^{\infty} C_n e^{i\,n\pi x/L}$$

(3.13)

$$\text{where} \qquad C_n = \frac{1}{2L} \int_{-L}^{L} f(x) e^{-i\,n\pi x/L}.$$

Differentiation and Integration of Fourier Series

The Fourier series is an infinite series and so we ask the question "Under what conditions can you differentiate and integrate this infinite series term by term?" Consider a more general situation. Assume we are given a set of orthogonal functions $\{y_n(x)\}$ defined over an interval (a,b) with inner product

$$(y_n, y_m) = \int_a^b r(x) y_n(x) y_m(x)\,dx = \| y_n \|^2 \delta_{nm}$$

with $r(x) > 0$. We represent a function $f(x)$ in the form of an infinite series of these orthogonal functions

$$f(x) = \sum_{m=0}^{\infty} a_m y_m(x) = \sum_{m=0}^{\infty} u_m(x), \quad a \le x \le b$$

(3.14)

having constant coefficients a_m and we form the sequence of partial sums

$$S_n(x) = \sum_{m=0}^{n} a_m y_m(x) = \sum_{m=0}^{n} u_m(x),$$

(3.15)

where $f(x) = \lim_{n \to \infty} S_n(x)$. The series defined by (3.15) is said to converge uniformly with respect to x, over an interval (a, b), if for every small positive number ϵ, there exists an integer n_ϵ, independent of x, such that $|f(x) - S_N(x)| < \epsilon$ for $N > n_\epsilon$ and x satisfying $a \leq x \leq b$. Recall the Weierstrass[†] M-test, which states that if we can construct positive constants M_n satisfying $|a_n y_n(x)| < M_n$ for $a \leq x \leq b$, and if the series $\sum_{n=0}^{\infty} M_n$ converges, then the series (3.14) is uniformly convergent over the interval (a, b). Uniformly convergent series have the following properties which can be found in most advanced calculus books.

(a) If each of the functions $u_m(x)$ is continuous over the interval (a, b), and the series given by equation (3.14) converges uniformly, then the sum $f(x)$ is also continuous.

(b) If each $u_m(x)$ in the series (3.14) is integrable and the series (3.14) converges uniformly in (a, b), then the series (3.14) can be integrated term by term and

$$\sum_{n=1}^{\infty} \int_a^x u_n(x)\,dx = \int_a^x \left[\sum_{n=0}^{\infty} u_n(x) \right] dx.$$

The series of integrated functions also converges uniformly in (a, b).

(c) If each function $u_m(x)$ is differentiable and the series of the derivatives $\sum_{m=0}^{\infty} u'_m(x)$ converges uniformly in (a, b), then we can write

$$\frac{d}{dx} \sum_{n=0}^{\infty} u_n(x) = \sum_{n=0}^{\infty} \frac{d}{dx}(u_n(x))$$

In simpler terms, just look at a graph of the function represented by the Fourier series function $\tilde{f}(x)$ (See page 115 for definition of $\tilde{f}(x)$). If this graph is continuous and its derivative curve exists and is sectionally continuous, having only a finite number of jump discontinuities, then the Fourier series can be integrated term by term. Special conditions have to exist in order to differentiate a Fourier series term by term. If $f(x)$ is continuous over the interval $(-L, L)$ and periodic with $f'(x)$ satisfying the Dirichlet conditions, then the Fourier series can be differentiated term by term. If the Fourier series function $\tilde{f}(x)$ and n of its derivatives are continuous with the $(n+1)$st derivative sectionally continuous, then the Fourier series can be differentiated n-times.

[†] See Appendix C

Consider the following special case. The Fourier sine series

$$g(x) = \sum_{n=1}^{\infty} b_n \sin \frac{n\pi x}{L}, \quad \text{with} \quad b_n = \frac{2}{L} \int_0^L g(x) \sin \frac{n\pi x}{L} \, dx \qquad (3.16)$$

on the interval $-L < x < L$, can be differentiated to obtain

$$g'(x) = \sum_{n=1}^{\infty} \frac{n\pi}{L} b_n \cos \frac{n\pi x}{L}. \qquad (3.17)$$

This is not a full Fourier cosine series. If the function $g'(x)$ is continuous and its derivative $g''(x)$ is piecewise continuous we would expect to be able to represent this derivative function by a Fourier cosine series having the form

$$g'(x) = a_0 + \sum_{n=1}^{\infty} a_n \cos \frac{n\pi x}{L}$$

$$\text{where} \quad a_0 = \frac{1}{L} \int_0^L g'(x) \, dx = \frac{1}{L} [g(L) - g(0)] \qquad (3.18)$$

$$\text{and} \quad a_n = \frac{2}{L} \int_0^L g'(x) \cos \frac{n\pi x}{L} \, dx$$

This last integral can be integrated by parts to obtain

$$a_n = \frac{n\pi}{L} \left[\frac{2}{L} \int_0^L g(x) \sin \frac{n\pi x}{L} \, dx \right] + \frac{2}{L} \left[(-1)^n g(L) - g(0) \right]. \qquad (3.19)$$

This tells us $g'(x)$ should have the Fourier cosine series representation

$$g'(x) = \frac{1}{L} [g(L) - g(0)] + \sum_{n=1}^{\infty} \left[\frac{n\pi}{L} b_n + \frac{2}{L} [(-1)^n g(L) - g(0)] \right] \cos \frac{n\pi x}{L}. \qquad (3.20)$$

We can now compare the equations (3.17) and (3.20) . We note these equations will be identical only if $a_0 = 0$ and $a_n = \frac{n\pi}{L} b_n$. This requires that $g(L) - g(0) = 0$ and simultaneously the equation $(-1)^n g(L) - g(0) = 0$ is satisfied. Observe also the conditions $g(0) = 0$ and $g(L) = 0$ are necessary conditions for the Fourier sine series representation of $g(x)$ to be a continuous function. These conditions also allow the equations (3.17) and (3.20) to be identical. Therefore, we cannot differentiate a Fourier sine series unless the derivative function is piecewise smooth and the conditions $g(0) = 0$ and $g(L) = 0$ are satisfied. However, if $g'(x)$ is piecewise smooth we can always use the equation (3.20) instead of differentiating the sine series.

The rules for differentiation and integration of a Fourier series can be illustrated by the following examples.

Example 3-8. (**Fourier cosine series.**)
Consider the Fourier series expansion of the function $f(x) = x^2$, over the interval $(-L, L)$. Here $f(x)$ is an even function and so it has a Fourier cosine series representation given by

$$f(x) = a_0 + \sum_{n=1}^{\infty} a_n \cos \frac{n\pi x}{L}$$

where

$$a_0 = \frac{1}{L} \int_0^L x^2 \, dx = \frac{L^2}{3} \tag{3.21}$$

$$a_n = \frac{2}{L} \int_0^L x^2 \cos \frac{n\pi x}{L} \, dx = \frac{4L^2(-1)^n}{n^2\pi^2}$$

which produces the Fourier cosine series

$$f(x) = x^2 = \frac{L^2}{3} + \frac{4L^2}{\pi^2} \sum_{n=1}^{\infty} \frac{(-1)^n}{n^2} \cos \frac{n\pi x}{L} \tag{3.22}$$

the graph of which is illustrated in the figure 3-7.

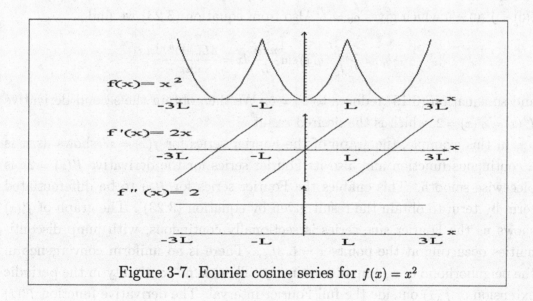

Figure 3-7. Fourier cosine series for $f(x) = x^2$

If we differentiate this series we obtain the Fourier sine series

$$f'(x) = 2x = \frac{4L}{\pi} \sum_{n=1}^{\infty} \frac{(-1)^{n+1}}{n} \sin \frac{n\pi x}{L} \tag{3.23}$$

which is also illustrated in figure 3-7. Compare this result with example 3-5 to see it is correct. One can verify that the integration of the equation (3.23) from 0 to x produces the result

$$\int_0^x 2x \, dx = x^2 = \frac{4L^2}{\pi^2} \sum_{n=1}^{\infty} \frac{(-1)^{n+1}}{n^2} + \frac{4L^2}{\pi^2} \sum_{n=1}^{\infty} \frac{(-1)^n}{n^2} \cos \frac{n\pi x}{L}. \tag{3.24}$$

From exercise number 12 at the end of this chapter we find that $\dfrac{1}{3} = \dfrac{4}{\pi^2} \sum_{n=1}^{\infty} \dfrac{(-1)^{n+1}}{n^2}$ and so the equation (3.24) reduces to the original Fourier series for x^2.

If we differentiate the equation (3.23) we obtain

$$f''(x) = 2 = 4 \sum_{n=1}^{\infty} (-1)^{n+1} \cos \frac{n\pi x}{L}$$

which is clearly not true since the series on the right-hand side does not converge. Here $f'(x)$ is sectionally continuous as can be seen from the graph and so we can use the equation (3.20) to obtain $g'(x) = f''(x)$. We use $g(L) = f'(L) = 2L$, $g(0) = f'(0) = 0$ which gives $a_0 = 2$. Also from equation (3.23) we find

$$b_n = \frac{2}{L} \int_0^L g(x) \sin \frac{n\pi x}{L} \, dx = \frac{4L(-1)^{n+1}}{n\pi}$$

and so equation (3.19) reduces to $a_n = 0$. We then obtain the second derivative $f''(x) = g'(x) = 2$. which is the desired result.

In this example, the graph of the Fourier series for $f(x) = x^2$ shows us it is a continuous function and also its Fourier series for the derivative $f'(x) = 2x$ is piecewise smooth. This enables the Fourier series for $f(x)$ to be differentiated term by term to obtain the result given by equation (3.23) . The graph of $f'(x)$ shows us the Fourier sine series is sectionally continuous, with jump discontinuities occurring at the points $x = L, 3L, \ldots$ There is no uniform convergence in the neighborhood of a jump discontinuity. There is no continuity in the periodic extension of $f'(x)$ outside the full Fourier interval. The derivative function $f''(x)$ did not produce a Fourier cosine series which converged. Also $f'(x) = 2x$ does not equal zero at both end points $x = 0$ and $x = L$. The special formula given by equation (3.20) for the differentiation of a Fourier sine series can be applied because the function the series represents, which is $f'(x) = 2x$, is continuous over the full Fourier interval and its derivative is piecewise smooth.

■

Example 3-9. (**Fourier sine series.**)

If we expand the function $f(x) = x^2$ on $0 < x < L$ as a Fourier sine series we obtain

$$f(x) = x^2 = \sum_{n=1}^{\infty} \left[-\frac{2L^2}{n\pi} - \frac{4L^2}{n^3\pi^3}[1 - (-1)^n] \right] \sin \frac{n\pi x}{L}$$

the graph of which is illustrated in the figure 3-8. Notice the graph for $f(x)$ is not continuous, but sectionally continuous. Therefore we cannot differentiate this series term by term. (Try it and test for convergence.) However, we can use the equation (3.20) for the differentiate series. Here $f(0) = 0$ and $f(L) = L^2$ so the equation (3.20) gives us

$$f'(x) = 2x = L - \sum_{n=1}^{\infty} \frac{4L}{n^2\pi^2}[1 - (-1)^n] \cos \frac{n\pi x}{L}$$

which agrees with the results from example 3-4. The graph of this function is the continuous curve in figure 3-8. Consequently, we can differentiate this series to obtain

$$f''(x) = 2 = \sum_{n=1}^{\infty} \frac{4}{n\pi}[1 - (-1)^n] \sin \frac{n\pi x}{L}$$

which agrees with the results from example 3-6.

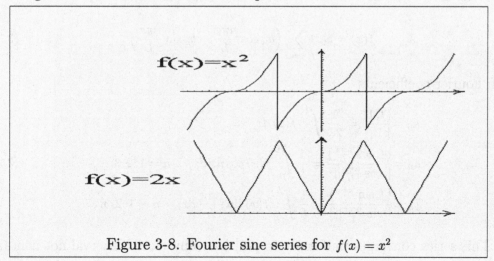

Figure 3-8. Fourier sine series for $f(x) = x^2$

■

To summarize we have

The Fourier sine series representation of $f(x)$ is given by

$$f(x) = \sum_{n=1}^{\infty} b_n \sin \frac{n\pi x}{L}.$$

This series can only be differentiated if (i) $f(0) = f(L) = 0$ and (ii) $f'(x)$ is a smooth function. If these conditions are not satisfied, then one can use the formula

$$f'(x) = \frac{1}{L}[f(L) - f(0)] + \sum_{n=1}^{\infty} \left[\frac{n\pi}{L} b_n + \frac{2}{L} \left((-1)^n f(L) - f(0) \right) \right] \cos \frac{n\pi x}{L} \quad (3.25)$$

whenever $f(x)$ is continuous with derivative $f'(x)$ which is piecewise smooth.

Consider a function $f(x)$ which is defined everywhere on the interval $(-L, L)$ and satisfies the conditions: (i) $f(-L) = f(L)$. (ii) The function is continuous over the interval $(-L, L)$. (iii) The derivative $f'(x)$ is piecewise continuous over the interval $(-L, L)$ with a finite number of jump discontinuities. (iv) Both $f(x)$ and $f'(x)$ are bounded over the interval $(-L, L)$. These conditions allow $f(x)$ to be represented by a Fourier trigonometric series expansion given by

$$f(x) = a_0 + \sum_{n=1}^{\infty} \left(a_n \cos \frac{n\pi x}{L} + b_n \sin \frac{n\pi x}{L} \right) \quad (3.26)$$

with Fourier coefficients

$$a_0 = \frac{(f, 1)}{\|1\|^2} = \frac{1}{2L} \int_{-L}^{L} f(x)\, dx$$

$$a_n = \frac{(f, \cos \frac{n\pi x}{L})}{\|\cos \frac{n\pi x}{L}\|^2} = \frac{1}{L} \int_{L}^{L} f(x) \cos nx\, dx, \quad n = 1, 2, 3, \ldots \quad (3.27)$$

$$b_n = \frac{(f, \sin \frac{n\pi x}{L})}{\|\sin \frac{n\pi x}{L}\|^2} = \frac{1}{L} \int_{-L}^{L} f(x) \sin \frac{n\pi x}{L}\, dx, \quad n = 1, 2, 3, \ldots$$

This series converges uniformly to $f(x)$ over any closed interval not containing a jump discontinuity of $f(x)$. At points ξ of discontinuity the Fourier series converges to $\dfrac{1}{2}\left[\lim\limits_{x\to\xi^+}f(x)+\lim\limits_{x\to\xi^-}f(x)\right]$. The series can be differentiated term by term at each point x where $f'(x)$ is uniformly convergent.

When $f(x)$ is piecewise continuous over an interval $a\le x\le b$ and $g(x)$ is also a piecewise continuous function over this same interval, then

$$\int_a^b g(x)f(x)\,dx=\int_a^b g(x)\left[a_0+\sum_{n=1}^{\infty}\left(a_n\cos\frac{n\pi x}{L}+b_n\sin\frac{n\pi x}{L}\right)\right]dx.$$

Note that differentiation is a roughening process associated with the curve $f(x)$ while integration is a smoothing process. That is each time the Fourier series is differentiated the nth term of the series is multiplied by a factor of n which does help the convergence process. An integration of $f(x)$ causes the nth term of the series to get divided by a factor of n which helps the convergence process. If the nonzero term a_0x appears in the integrated series, then the series is no longer called a Fourier series.

Caution should be taken when differentiating a Fourier series. The differentiation is allowed provided that the resulting series converges uniformly. Thus convergence and divergence of series is an important topic in mathematics. More information on uniform convergence of series can be found in most advanced calculus textbooks.

Fourier Coefficients

Assume we are given a set of orthogonal functions $\{y_n(x)\}$, $a\le x\le b$ and we want to represent an arbitrary piecewise smooth function $F(x)$ on the interval (a,b) by a series having the form $\sum K_m y_m(x)$ where K_m are constants. How do we go about selecting the constants K_m such that the finite sum

$$S_n(x)=\sum_{m=0}^{n}K_m y_m(x)\tag{3.28}$$

approximates $F(x)$ in some sense. Let us choose the coefficients K_m, $m=0,1,2,\ldots$ such that the weighted mean square error E_n defined by

$$E_n=\int_a^b r(x)\left[F(x)-S_n(x)\right]^2dx\tag{3.29}$$

is a minimum. Here our measure of the 'goodness of approximation' is the norm defined by the inner product

$$(F-S_n,F-S_n)=\parallel F(x)-S_n(x)\parallel^2=E_n$$

which is the same inner product associated with the orthogonal functions $\{y_n(x)\}$. Substituting (3.28) into (3.29) we obtain

$$E_n = E_n(K_0, K_1, \ldots, K_n) = \int_a^b r(x) \left[F(x) - \sum_{m=0}^n K_m y_m(x) \right]^2 dx. \qquad (3.30)$$

In equation (3.30) we hold all the K_m constant except K_i and let K_i vary. Then the mean square error E_n is a minimum when the K_i are chosen such that $\frac{\partial E_n}{\partial K_i} = 0$ for $i = 1, 2, \ldots, n$. Differentiating equation (3.30) with respect to K_i and setting the result equal to zero produces the equation

$$\frac{\partial E_n}{\partial K_i} = \int_a^b 2r(x) \left[F(x) - \sum_{m=0}^n K_m y_m(x) \right] (-y_i(x)) \, dx = 0 \qquad (3.31)$$

which can be written in terms of inner products and norms as

$$-2(F, y_i) + 2K_i \parallel y_i \parallel^2 = 0. \qquad (3.32)$$

This implies K_i is given by an inner product divided by a norm squared

$$K_i = \frac{(F, y_i)}{\parallel y_i \parallel^2} = \frac{\int_a^b r(x) F(x) y_i(x) \, dx}{\parallel y_i \parallel^2}, \quad i = 0, 1, 2, \ldots, n. \qquad (3.33)$$

The coefficients given by equation (3.33) are called the Fourier coefficients of $F(x)$ relative to the orthogonal set $\{y_n(x)\}$. For each Fourier coefficient K_i given by equation (3.33) the minimum value of the weighted mean square error E_n is found to be

$$\min E_n = \parallel F \parallel^2 - \sum_{m=0}^n K_m^2 \parallel y_m \parallel^2 \geq 0. \qquad (3.34)$$

The inequality in equation (3.34) implies

$$\sum_{m=0}^\infty K_m^2 \parallel y_m \parallel^2 \leq \parallel F \parallel^2 \qquad (3.35)$$

which is known as Bessel's inequality and tells us if the norm squared of F exists, then as n increases without bound the series in equation (3.34) converges.

If as n increases without bound the error E_n approaches zero, then equation (3.34) becomes Parseval's[†] equation

$$\parallel F \parallel^2 = \sum_{m=0}^\infty K_m^2 \parallel y_m \parallel^2 \qquad (3.36)$$

---

†   See Appendix C

which for the special case where $\| y_m \|^2 = 1$ for all m, becomes

$$\| F \|^2 = \sum_{m=0}^{\infty} K_m^2. \tag{3.37}$$

Definition: (**Complete set**) An orthogonal set $\{y_n(x)\}$ is called complete if $\lim_{n \to \infty} E_n = 0$ for every function $F(x)$ which is square integrable over (a, b).

The equation (3.36) provides another way of expressing completeness. A set $\{y_n(x)\}$ is called complete if all square integrable functions $F(x)$ satisfy equation (3.36), where K_m are the Fourier coefficients given by equation (3.33).

If a function is represented by a convergent series of orthogonal functions $\{y_n(x)\}$ in the form

$$F(x) = \sum_{n=0}^{\infty} a_n y_n(x), \quad a \leq x \leq b, \tag{3.38}$$

then the series is called a generalized Fourier series and the coefficients a_n are given by the inner product divided by a norm squared relation

$$a_n = \frac{(F, y_n)}{\| y_n \|^2} = \frac{\int_a^b r(x) F(x) y_n(x)\, dx}{\int_a^b r(x) y_n^2(x)\, dx}, \quad n = 0, 1, 2, \ldots$$

which are called Fourier coefficients. If the series given by equation (3.38) is uniformly convergent we can take the inner product of both sides of the equation with respect to y_m, where m is constant but arbitrary. This gives the relations

$$(F, y_m) = \left(y_m, \sum_{n=0}^{\infty} a_n y_n\right) = \sum_{n=0}^{\infty} (y_m, a_n y_n) = \sum_{n=0}^{\infty} a_n (y_m, y_n)$$

$$= a_0 (y_m, y_0) + a_1 (y_m, y_1) + \cdots + a_m (y_m, y_m) + \cdots$$

$$= a_m \| y_m \|^2.$$

Solving for a_m we find it is given by an inner product divided by a norm squared

$$a_m = \frac{(F, y_m)}{\| y_m \|^2}, \quad m = 0, 1, 2, \ldots \tag{3.39}$$

which are the Fourier coefficients.

Throughout the remainder of this text we make use of the following theorem which is stated without proof.

> **THEOREM** If the Sturm-Liouville system
>
> $$\frac{d}{dx}[p(x)y']+[q(x)+\lambda r(x)]y=0, \quad a \le x \le b$$
> $$Ay(a)+By'(a)=0, \qquad Ey(b)+Fy'(b)=0$$
>
> is such that $p(x)$, $p'(x)$, $q(x)$, $r(x)$, and $[p(x)r(x)]''$ are real and continuous on the interval (a,b) with $r(x) > 0$ and $p(x) > 0$ and the constants A, B, E, F are real and independent of λ, then there exists an infinite set of real eigenvalues λ_n and corresponding real eigenfunctions $\{y_n(x)\}$ which are orthogonal on (a,b) with respect to the weight function $r(x)$. For $f(x)$ and $f'(x)$ piecewise smooth functions the series $\sum_{n=0}^{\infty} a_n y_n(x) = f(x)$, where a_n are the Fourier coefficients, converges to $f(x)$ at each point x where $f(x)$ is continuous. The series converges to $\frac{1}{2}[f(x^+) + f(x^-)]$ at points x where $f(x)$ has a jump discontinuity.

Fourier Bessel Series

Much of what has been done for the Fourier trigonometric series can also be done for other sets of orthogonal functions. For example, consider the orthogonal sets of Bessel functions from the examples 2-7, 2-8 and 2-9 which are summarize in the Table 3-1.

Table 3-1. Orthogonality of Bessel functions.			
Orthogonal functions	Interval weight function	Boundary conditions	Eigenvalues
$y_n(x) = J_k(\lambda_n x)$	$0 \le x \le b$ $r(x) = x$	$y(b) = 0$	$\lambda_n = \xi_{kn}/b$ for $n = 1,2,3,\ldots$ $J_k(\xi_{kn}) = 0$
$y_n(x) = J_k(\lambda_n x)$	$0 \le x \le b$ $r(x) = x$	$by'(b) + \beta y(b) = 0$	$\lambda_n = \alpha_{kn}/b$ for $n = 1,2,3,\ldots$ $\alpha_{kn} J_k'(\alpha_{kn}) + \beta J_k(\alpha_{kn}) = 0$
$y_n(x) = U_k(\lambda_n x)$	$a \le x \le b$ $r(x) = x$	$y_n(a) = 0$ $y_n(b) = 0$	$\lambda_n = \gamma_{kn}/a$ for $n = 1,2,3,\ldots$ $U_k(\gamma_{kn}) = 0$ $Y_k(\gamma_{kn}\frac{b}{a})J_k(\gamma_{kn}) - J_k(\gamma_{kn}\frac{b}{a})Y_k(\gamma_{kn}) = 0$

For the boundary condition $y(b) = 0$ we have the following result.

The Fourier Bessel series representation for a piecewise continuous function $f(x)$ defined on the interval $(0, b)$ is given by

$$f(x) = \sum_{n=1}^{\infty} c_n J_k(\lambda_n x) \tag{3.40}$$

with Fourier coefficients calculated from the ratio of inner product divided by a norm squared

$$c_n = \frac{(f, J_k(\lambda_n x))}{||J_k(\lambda_n x)||^2} = \frac{2}{b^2 J_{k+1}(\lambda_n b)} \int_0^b x f(x) J_k(\lambda_n x)\, dx, \qquad n = 1, 2, 3, \ldots \tag{3.41}$$

where the eigenvalues are determined from the positive roots of the equation $J_k(\lambda_n b) = 0$.

For the boundary condition $by'(b) + \beta y(b) = 0$ represented by a linear combination of the function and derivative at $x = b$ we have the result.

The Fourier Bessel series representation for a piecewise continuous function $f(x)$ defined on the interval $(0, b)$ is given by

$$f(x) = \sum_{n=1}^{\infty} c_n J_k(\lambda_n x) \tag{3.42}$$

with Fourier coefficients

$$c_n = \frac{(f, J_k(\lambda_n x))}{||J_k(\lambda_n x)||^2} = \frac{2\lambda_n^2}{(\beta^2 + \lambda_n^2 b^2 - k^2) J_k^2(\lambda_n b)} \int_0^b x f(x) J_k(\lambda_n x)\, dx, \quad n = 1, 2, 3, \ldots \tag{3.43}$$

where the eigenvalues are determined from the positive roots of the equation

$$b\lambda_n J_k'(\lambda_n b) + \beta J_k(\lambda_n b) = 0, \quad n = 1, 2, 3, \ldots \tag{3.44}$$

A special case of the above is the following.

In the special case where $\beta = 0$ and $k = 0$, the function $f(x)$ is represented by the Fourier Bessel series

$$f(x) = c_0 + \sum_{n=1}^{\infty} c_n J_0(\lambda_n x) \tag{3.45}$$

with Fourier coefficients

$$c_0 = \frac{2}{b^2} \int_0^b x f(x)\, dx$$
$$c_n = \frac{2}{b^2 J_0^2(\lambda_n b)} \int_0^b x f(x) J_0(\lambda_n x)\, dx \tag{3.46}$$

where the eigenvalues are the positive roots of the equation $J_1(\lambda_n b) = 0$.

For the zero boundary conditions at $x = a$ and $x = b$ we have the result.

The Fourier Bessel series representation for a piecewise continuous function $f(x)$ defined over the interval (a, b) is given by

$$f(x) = \sum_{n=1}^{\infty} c_n U_k(\lambda_n x) \tag{3.47}$$

with Fourier coefficients given by

$$c_n = \frac{(f, U_k(\lambda_n x))}{\|U_k(\lambda_n x)\|^2} = \frac{\pi^2 \lambda_n^2 J_k^2(\lambda_n b)}{2(J_k^2(\lambda_n a) - J_k^2(\lambda_n b))} \int_a^b x f(x) U_k(\lambda_n x)\, dx \tag{3.48}$$

where the eigenvalues are the positive roots of the equation

$$U_k(\lambda_n a) = Y_k(\lambda_n b) J_k(\lambda_n a) - J_k(\lambda_n b) Y_k(\lambda_n a) = 0. \tag{3.49}$$

Example 3-10.   (Fourier Bessel series.)

Represent the function $f(x) = x^2$, over the interval $0 \le x \le b$ as a Fourier Bessel series in terms of the orthogonal functions $\{J_0(\lambda_n x)\}$, where $J_0(\lambda_n b) = 0$.

Solution: It is desired to represent $f(x)$ in the form

$$f(x) = x^2 = \sum_{n=1}^{\infty} a_n J_0(\lambda_n x),$$

where $y_n(x) = J_0(\lambda_n x)$, $n = 1, 2, 3, \ldots$ are the zeroth order Bessel functions. The Fourier coefficients are given by equation (3.41) and are found to be

$$a_n = \frac{(x^2, J_0(\lambda_n x))}{\| J_0(\lambda_n x) \|^2} = \frac{\int_0^b x^3 J_0(\lambda_n x)\, dx}{\frac{b^2}{2} J_1^2(\lambda_n b)}.$$

Using results from table 1 of chapter 1, the numerator can be evaluated and after simplification there results

$$a_n = \frac{2(\lambda_n^2 b^2 - 4)}{b \lambda_n^3 J_1^2(\lambda_n b)}$$

and hence

$$f(x) = \frac{2}{b} \sum_{n=0}^{\infty} \frac{(\lambda_n^2 b^2 - 4) J_0(\lambda_n x)}{\lambda_n^3 J_1(\lambda_n x)}.$$

∎

Fourier Legendre series

The Fourier Legendre series results when the set of Legendre functions are employed as the orthogonal set. The Fourier Legendre series representation for a piecewise continuous function $f(x)$ defined over the interval $(-1, 1)$ is given by

$$f(x) = \sum_{n=0}^{\infty} c_n P_n(x) \tag{3.50}$$

with Fourier coefficients given by

$$c_n = \frac{(f, P_n(x))}{\|P_n(x)\|^2} = \frac{2n+1}{2} \int_{-1}^{1} f(x) P_n(x)\, dx \tag{3.51}$$

The Fourier Legendre series representation for a piecewise continuous function $f(\theta)$ defined over the interval $(0, \pi)$ is given by

$$f(x) = \sum_{n=0}^{\infty} c_n P_n(\cos\theta), \qquad 0 < \theta < \pi \tag{3.52}$$

with Fourier coefficients given by

$$c_n = \frac{(f, P_n(\cos\theta))}{\|P_n(\cos\theta)\|^2} = \frac{2n+1}{2} \int_{0}^{\pi} f(\theta) P_n(\cos\theta) \sin\theta\, d\theta \tag{3.53}$$

Fourier series in two-dimensions

If for $n = 1, 2, 3, \ldots$, we have a set of functions $p_n(x)$ over an interval $a < x < b$ which are orthogonal with respect a weight function $w_p(x) > 0$ and another set of functions $q_n(y)$ which are orthogonal over an interval $c < y < d$, with respect to a weight function $w_q(y) > 0$ then one can represent a smooth function of two variables $f(x, y)$, $a < x < b$, $c < x < d$, by the double Fourier series

$$f(x, y) = \sum_{n=1}^{\infty} \sum_{m=1}^{\infty} C_{nm} p_n(x) q_m(y) \tag{3.54}$$

where C_{nm} are the Fourier coefficients. We make use of the orthogonality property of the given set of functions and solve for the Fourier coefficients in exactly the same way as we did for functions of one variable, except now we must do it twice, once for each variable. For example, if we multiply both sides of equation (3.54) by $w_q(y) q_i(y) \, dy$ and then integrate both sides of the equation from c to d we obtain the inner products

$$(f, q_i) = \sum_{n=1}^{\infty} \sum_{m=1}^{\infty} C_{nm} p_n(x) (q_m, q_i) \tag{3.55}$$

Note as you sum on the index m all terms in the inner product (q_m, q_i) are zero except when m takes on the value $m = i$. Therefore, the double Fourier series reduces to

$$(f, q_i) = \|q_i\|^2 \sum_{n=1}^{\infty} C_{ni} p_n(x). \tag{3.56}$$

Now multiply both sides of this equation by $w_p(x) p_j(x) \, dx$ and integrate both sides from a to b to obtain the series of inner products

$$((f, q_i), p_j) = \|q_i\|^2 \sum_{n=1}^{\infty} C_{ni} (p_n, p_j) \tag{3.57}$$

Again, note all inner products on the right-hand side are zero except when the index n takes on the value $n = j$. This reduces the equation (3.57) to the form

$$((f, q_i), p_j) = \|q_i\|^2 \|p_j\|^2 C_{ji} \tag{3.58}$$

from which we can solve for the Fourier coefficients. These coefficients are given by

$$C_{ji} = \frac{((f, q_i), p_j)}{\|q_i\|^2 \|p_j\|^2} \tag{3.59}$$

By changing j to n and i to m we write these coefficients as an inner product of an inner product divide by the norm squared from each orthogonal set. In terms of integrals

$$C_{nm} = \frac{((f,q_m),p_n)}{||q_n||^2||p_n||^2} = \frac{\int_a^b \int_c^d w_p(x)w_q(y)p_n(x)q_m(y)\,dydx}{\left(\int_c^d w_q(y)q_m^2(y)\,dy\right)\left(\int_a^b w_p(x)p_n^2(x)\,dx\right)} \qquad (3.60)$$

In a similar fashion Fourier series representation for smooth functions of more than two variables can be constructed. For example, if $p_n(x), q_n(y), r_n(z)$ are orthogonal sets over the intervals $a < x < b$, $c < x < d$, and $e < z < f$ with respect to positive weight functions $w_p(x), w_q(y), w_r(z)$ respectively, then one can readily verify the triple Fourier series representation

$$f(x,y,z) = \sum_{\ell=1}^{\infty} \sum_{m=1}^{\infty} \sum_{n=1}^{\infty} C_{\ell mn} p_\ell(x) q_m(y) r_n(z)$$

with Fourier coefficients $C_{\ell mn}$ given by a triple inner product divided by norm squared terms. This is written

$$C_{\ell mn} = \frac{(((f,r_n),q_m),p_\ell)}{||r_n||^2||q_m||^2||p_\ell||^2}$$

which has the expanded form

$$C_{\ell mn} = \frac{\int_a^b \int_c^d \int_e^f w_p(x)w_q(y)q_r(z)p_\ell(x)q_m(y)r_n(z)\,dzdydx}{||r_n||^2||q_m||^2||p_\ell||^2}. \qquad (3.61)$$

In those cases where the norm squared is not the same for all functions of the set, as for example the cosine series terms $\{1, \cos\frac{n\pi x}{L}\}$, then one must isolated those distinct cases and treat the above inner products divided by a norm squared on a term by term basis.

Fourier integrals

The Fourier trigonometric series representation of a piecewise continuous function defined on the interval $(-L, L)$ is given by

$$f(x) = a_0 + \sum_{n=1}^{\infty} \left(a_n \cos\frac{n\pi x}{L} + b_n \sin\frac{n\pi x}{L} \right)$$

$$\text{where} \quad a_0 = \frac{1}{2L} \int_{-L}^{L} f(\xi)\,d\xi$$

$$a_n = \frac{1}{L} \int_{-L}^{L} f(\xi) \cos\frac{n\pi\xi}{L}\,d\xi$$

$$b_n = \frac{1}{L} \int_{-L}^{L} f(\xi) \sin\frac{n\pi\xi}{L}\,d\xi$$

We investigate the behavior of the above series in the limit as L increases without bound. Substituting the coefficients into the series we find one must evaluate the limits

$$f(x) = \lim_{L \to \infty} \frac{1}{2L} \int_{-L}^{L} f(\xi) d\xi$$
$$+ \lim_{L \to \infty} \sum_{n=1}^{\infty} \frac{1}{L} \int_{-L}^{L} f(\xi) \left[\cos \frac{n\pi\xi}{L} \cos \frac{n\pi x}{L} + \sin \frac{n\pi\xi}{L} \sin \frac{n\pi x}{L} \right] d\xi \tag{3.62}$$

In order for the equation (3.62) to have a limit as L increases without bound we require

$$\lim_{L \to \infty} \int_{-L}^{L} |f(\xi)| \, d\xi < M = \text{Constant}. \tag{3.63}$$

If this condition is true we say $f(x)$ is an absolute integrable function. Under these conditions the first term in equation (3.62) tends to zero as L increases without bound.

Define the quantities $\omega_n = \frac{n\pi}{L}$ and $\Delta\omega_n = \frac{\pi}{L}$ and write the equation (3.62) in the form

$$f(x) = \lim_{L \to \infty} \sum_{n=1}^{\infty} [A(\omega_n) \cos \omega_n x + B(\omega_n) \sin \omega_n x] \, \Delta\omega_n \tag{3.64}$$

where

$$A(\omega_n) = \frac{1}{\pi} \int_{-L}^{L} f(\xi) \cos \omega_n \xi \, d\xi$$
$$B(\omega_n) = \frac{1}{\pi} \int_{-L}^{L} f(\xi) \sin \omega_n \xi \, d\xi. \tag{3.65}$$

The summation given by equation (3.64) is like a Riemannian sum which turns into an integral as $\Delta\omega_n \to 0$. It can be shown,[1] that in the limit as $L \to \infty$, $(\Delta\omega_n \to 0)$, the summation given by equation (3.64) reduces to the Fourier integral formula

$$f(x) = \int_0^{\infty} [A(\omega) \cos \omega x + B(\omega) \sin \omega x] \, d\omega$$

where

$$A(\omega) = \frac{1}{\pi} \int_{-\infty}^{\infty} f(\xi) \cos \omega \xi \, d\xi \tag{3.66}$$

$$B(\omega) = \frac{1}{\pi} \int_{-\infty}^{\infty} f(\xi) \sin \omega \xi \, d\xi.$$

---

[1] W. Rogosinski, Fourier Series, Chelsea Publishing Co., 1959.

are called the Fourier integral coefficients. This represents the continuum analog of the discrete Fourier series representation of a function. The above result can also be written in many different forms. One form is

$$f(x) = \frac{1}{\pi} \int_0^\infty \int_{-\infty}^\infty f(\xi) \cos \omega(\xi - x) \, d\xi d\omega. \tag{3.67}$$

We make use of the fact that $\cos \omega(\xi - x)$ is an even function of ω and write the integral (3.67) as

$$f(x) = \frac{1}{2\pi} \int_{-\infty}^\infty \int_{-\infty}^\infty f(\xi) \cos \omega(\xi - x) \, d\xi d\omega. \tag{3.68}$$

Observe also $\sin \omega(\xi - x)$ is an odd function of ω and so we can write

$$0 = \frac{i}{2\pi} \int_{-\infty}^\infty \int_{-\infty}^\infty f(\xi) \sin \omega(\xi - x) \, d\xi d\omega \tag{3.69}$$

where i is an imaginary unit satisfying $i^2 = -1$. By adding the equations (3.69) and (3.68) we obtain still another form for the Fourier integral formula

$$f(x) = \frac{1}{2\pi} \int_{-\infty}^\infty \int_{-\infty}^\infty f(\xi) \left[\cos \omega(\xi - x) + i \sin \omega(\xi - x) \right] \, d\xi d\omega. \tag{3.70}$$

We use Euler's identity $e^{i\theta} = \cos \theta + i \sin \theta$ and write the Fourier integral formula in the exponential form

$$f(x) = \frac{1}{2\pi} \int_{-\infty}^\infty \int_{-\infty}^\infty f(\xi) e^{i\omega(\xi - x)} \, d\xi d\omega. \tag{3.71}$$

When the Fourier integral formula converges it converges to points $f(x)$ where $f(x)$ is continuous. At points x where $f(x)$ has a jump discontinuity, the Fourier integral formula converges to the average of the left and right-hand limits associated with the jump discontinuity or $\frac{1}{2}(f(x^+) + f(x^-))$. These results produce the following Fourier integral formulas.

The Fourier integral representation of a piecewise smooth function $f(x)$ which is bounded and absolutely integrable over the x-axis is given by

$$f(x) = \int_0^\infty (A(\omega) \cos \omega x + B(\omega) \sin \omega x) \, d\omega$$

where
$$A(\omega) = \frac{1}{\pi} \int_{-\infty}^\infty f(\xi) \cos \omega \xi \, d\xi \tag{3.72}$$

$$B(\omega) = \frac{1}{\pi} \int_{-\infty}^\infty f(\xi) \sin \omega \xi \, d\xi$$

are the Fourier integral coefficients.

If $f(x)$ is an odd function of x the Fourier integral representation given by equation (3.72) reduces to the Fourier sine integral representation

$$f(x) = \int_0^\infty B(\omega) \sin \omega x \, d\omega$$

where $\quad B(\omega) = \frac{2}{\pi} \int_0^\infty f(\xi) \sin \omega \xi \, d\xi$

(3.73)

If $f(x)$ is an even function of x the Fourier integral representation given by equation (3.72) reduces to the Fourier cosine integral representation

$$f(x) = \int_0^\infty A(\omega) \cos \omega x \, d\omega$$

where $\quad A(\omega) = \frac{2}{\pi} \int_0^\infty f(\xi) \cos \omega \xi \, d\xi$

(3.74)

Example 3-11. (**Fourier integral.**)
Find the Fourier integral representation of the function

$$f(x) = \begin{cases} 1, & -1 < x < 1 \\ 0, & \text{otherwise} \end{cases}$$

Solution: The Fourier integral coefficients are given by

$$A(\omega) = \frac{1}{\pi} \int_{-\infty}^\infty f(\xi) \cos \omega \xi \, d\xi = \frac{1}{\pi} \int_{-1}^1 \cos \omega \xi \, d\xi = \frac{2}{\pi} \frac{\sin \omega}{\omega}$$

$$B(\omega) = \frac{1}{\pi} \int_{-\infty}^\infty f(\xi) \sin \omega \xi \, d\xi = \frac{1}{\pi} \int_{-1}^1 \sin \omega \xi \, d\xi = 0$$

Note $B(\omega) = 0$ because $f(x)$ is an even function of x.

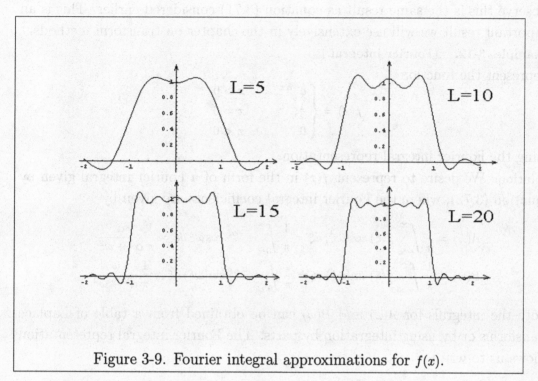

Figure 3-9. Fourier integral approximations for $f(x)$.

The Fourier integral representation of $f(x)$ is given by the Fourier cosine integral representation

$$f(x) = \frac{2}{\pi} \int_0^\infty \frac{\sin \omega}{\omega} \cos \omega x \, d\omega \qquad (3.75)$$

Analogous to the partial sums used to approximate a Fourier series, the integral

$$f_L(x) = \frac{2}{\pi} \int_0^L \frac{\sin \omega}{\omega} \cos \omega x \, d\omega \qquad (3.76)$$

is used to approximate the integral given by equation (3.75). The figure 3-9 illustrates the Fourier integral approximations for selected values of the upper limit L.

■

It can also be demonstrated that the limit as $L \to \infty$ of the complex exponential form of the Fourier series, given by equation (3.13), produces the Fourier integral formula

$$f(x) = \frac{1}{2\pi} \int_{-\infty}^{\infty} \int_{-\infty}^{\infty} f(\xi) e^{i\omega(\xi - x)} \, d\xi d\omega. \qquad (3.77)$$

Observe this is the same result as equation (3.71) considered earlier. This is an important result we will use extensively in the chapter on transform methods.

Example 3-12. (Fourier integral.)

Represent the function

$$f(x) = \begin{cases} e^{-\alpha x}, & x > 0 \\ \frac{1}{2}, & x = 0 \\ 0, & x < 0 \end{cases}$$

using the Fourier integral representation.

Solution: We desire to represent $f(x)$ in the form of a Fourier integral given by equation (3.72), where the Fourier integral coefficients are given by

$$A(\omega) = \frac{1}{\pi} \int_{-\infty}^{\infty} f(\xi) \cos(\omega \xi) \, d\xi = \frac{1}{\pi} \int_{0}^{\infty} e^{-\alpha \xi} \cos \omega \xi \, d\xi = \frac{1}{\pi} \frac{\alpha}{\alpha^2 + \omega^2}$$

$$B(\omega) = \frac{1}{\pi} \int_{-\infty}^{\infty} f(\xi) \sin(\omega \xi) \, d\xi = \frac{1}{\pi} \int_{0}^{\infty} e^{-\alpha \xi} \sin(\omega \xi) \, d\xi = \frac{1}{\pi} \frac{\omega}{\alpha^2 + \omega^2}.$$

Note the integrals for $A(\omega)$ and $B(\omega)$ can be obtained from a table of Laplace transforms or by using integration by parts. The Fourier integral representation allows us to write

$$e^{-\alpha x} = \frac{1}{\pi} \int_{0}^{\infty} \left[\frac{\alpha \cos \omega x + \omega \sin \omega x}{\alpha^2 + \omega^2} \right] d\omega, \quad x > 0.$$

In the special case when $\alpha = 1$ the above can be expressed

$$e^{-x} = f_1(x) + f_2(x), \quad x > 0,$$

where

$$f_1(x) = \frac{1}{\pi} \int_{0}^{\infty} \frac{\cos \omega x}{1 + \omega^2} \, d\omega \quad \text{and} \quad f_2(x) = \frac{1}{\pi} \int_{0}^{\infty} \frac{\omega \sin \omega x}{1 + \omega^2} \, d\omega.$$

By differentiating under the integral sign we obtain

$$\frac{df_1}{dx} = \frac{1}{\pi} \int_{0}^{\infty} -\frac{\omega \sin \omega x}{1 + \omega^2} \, d\omega = -f_2(x) = f_1(x) - e^{-x}.$$

The differential equation

$$\frac{df_1}{dx} - f_1 = -e^{-x}, \quad x > 0$$

is subject to the initial condition

$$f_1(0) = \frac{1}{\pi} \int_{0}^{\infty} \frac{d\omega}{1 + \omega^2} = \frac{1}{2}.$$

This initial value problem is easily solved and we find the solution $f_1(x) = \frac{1}{2}e^{-x}$ and consequently we have

$$\frac{1}{\pi}\int_0^\infty \frac{\cos\omega x}{1+\omega^2}\, d\omega = \frac{1}{2}e^{-x}, \qquad x > 0$$

$$\frac{1}{\pi}\int_0^\infty \frac{\omega\sin\omega x}{1+\omega^2}\, d\omega = \frac{1}{2}e^{-x}, \qquad x > 0.$$

Define the functions

$$F_L(x) = \frac{1}{\pi}\int_0^L \frac{\cos\omega x}{1+\omega^2}\, d\omega \qquad \text{and} \qquad G_L(x) = \frac{1}{\pi}\int_0^L \frac{\omega\sin\omega x}{1+\omega^2}\, d\omega$$

representing partial sums of the above integrals. Graphs of these partial sums for selected L values appear in the figure 3-10.

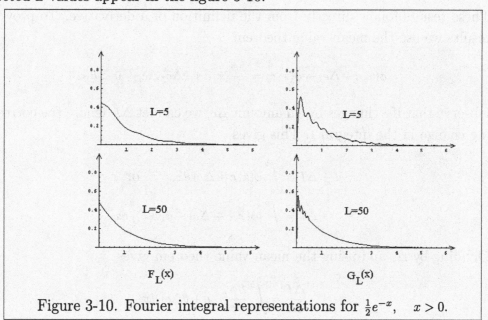

Figure 3-10. Fourier integral representations for $\frac{1}{2}e^{-x}$, $\quad x > 0$.

$\blacksquare$

Differentiation and Integration of Integrals

The previous example illustrates that once the integral representation of a function is known, it is possible other integral representations can be obtained by differentiation or integration with respect to a parameter. Consider the definite integral

$$I = \int_a^b \phi(x,c)\, dx \tag{3.78}$$

which contains a parameter c and where the limits of integration a, b are finite and independent of c. The value of this integral is independent of the variable of integration. Its value depends only upon the parameter c and the limits of integration. For $\phi(x, c)$ and $\frac{\partial \phi(x,c)}{\partial c}$ finite, single valued and continuous functions over the range of integration $a \leq x \leq b$, and for a and b independent of c we have

$$\frac{\partial I}{\partial c} = \int_a^b \frac{\partial \phi(x,c)}{\partial c}\, dx. \qquad (3.79)$$

If the limits of integration are dependent upon c we may write

$$\frac{dI}{dc} = \int_a^b \frac{\partial \phi(x,c)}{\partial c}\, dx + \phi(b,c)\frac{db}{dc} - \phi(a,c)\frac{da}{dc}. \qquad (3.80)$$

These results follow directly from the definition of a derivative. To prove these results we use the mean value theorem

$$\phi(x, c + \Delta c) - \phi(x, c) = \frac{\partial \phi}{\partial c}(x, c + \theta \Delta c)\, \Delta c, \quad 0 < \theta < 1.$$

Observe that if c changes by an amount Δc, we can let ΔI denote the corresponding change in the integral I. This gives

$$I + \Delta I = \int_a^b \phi(x, c + \Delta c)\, dx, \qquad \text{or}$$

$$\Delta I = \int_a^b [\phi(x, c + \Delta c) - \phi(x, c)]\, dx.$$

Dividing by Δc and using the mean value theorem gives

$$\frac{\Delta I}{\Delta c} = \int_a^b \frac{\partial \phi}{\partial c}(x, c + \theta \Delta c)\, dx$$

and

$$\frac{\partial I}{\partial c} = \lim_{\Delta c \to 0} \int_a^b \frac{\partial \phi}{\partial c}(x, c + \theta \Delta c)\, dx = \int_a^b \frac{\partial \phi}{\partial c}(x, c)\, dx$$

which establishes the result given by equation (3.79). Now if a and b are functions of c, then the result given by equation (3.80) follows from the chain rule differentiation and

$$\frac{dI}{dc} = \frac{\partial I}{\partial c} + \frac{\partial I}{\partial a}\frac{da}{dc} + \frac{\partial I}{\partial b}\frac{db}{dc}.$$

Not only can we differentiate with respect to the parameter c in equation (3.78), we can also integrate this integral with respect to the parameter c, with limits α to β, where α and β are independent of a and b. In general, we can write

$$I_1 = \int_\alpha^\beta \left[\int_a^b \phi(x,c)\,dx \right] dc = \int_a^b \left[\int_\alpha^\beta \phi(x,c)\,dc \right] dx = I_2$$

Here the order of integration is immaterial so long as $\phi(x,c)$ remains single valued, finite and continuous for all values of x and c between the limits of integration. To prove this result, let α be a constant and let β vary. By defining the integral

$$f(x,\beta) = \int_\alpha^\beta \phi(x,c)\,dc, \quad \text{with derivative} \quad \frac{\partial f}{\partial \beta} = \phi(x,\beta),$$

we demonstrate $I_1 = I_2$ by showing $\frac{\partial I_1}{\partial \beta} = \frac{\partial I_2}{\partial \beta}$. That is, two functions which have the same derivative differ by at most a constant and in this case the constant is zero. We differentiate I_2 to obtain

$$\frac{\partial I_2}{\partial \beta} = \frac{\partial}{\partial \beta} \int_a^b \left[\int_\alpha^\beta \phi(x,c)\,dc \right] dx = \frac{\partial}{\partial \beta} \int_a^b f(x,\beta)\,dx = \int_a^b \frac{\partial f(x,\beta)}{\partial \beta}\,dx = \int_a^b \phi(x,\beta)\,dx.$$

Also we have,

$$\frac{\partial I_1}{\partial \beta} = \frac{\partial}{\partial \beta} \int_\alpha^\beta \left[\int_a^b \phi(x,\beta)\,dx \right] dc = \int_a^b \phi(x,\beta)\,dx$$

and hence the two partial derivatives are equal. Therefore I_1 and I_2 differ by at most a constant and $I_1 - I_2 = k$. But when $\beta = \alpha$ we have $I_1 = I_2$ and hence $k = 0$ and consequently $I_1 = I_2$.

In order to calculate the Fourier integral representation of a function we must know how to evaluate integrals of the form

$$f(\omega) = \int_{-\infty}^\infty G(\xi,\omega)\,d\xi, \quad \text{or} \quad g(\omega) = \int_0^\infty G(\xi,\omega)d\xi$$

which contain a parameter ω. If the integrals defining $f(\omega)$, $f'(\omega)$, $g(\omega)$, $g'(\omega)$ are uniformly convergent and $G(\xi,\omega)$, $\frac{\partial G}{\partial \omega}(\xi,\omega)$ are continuous functions of ξ and ω for all values of ξ over the interval (a,b) and for all values of ω in the interval (α,β), then it can be established that these integrals with infinite limits may also be differentiated and integrated with respect to the parameter ω. The uniform convergence of an improper integral can be determined by the M test for improper integrals.

M-Test for Improper Integrals

If $g(x, \xi)$ is continuous for $a \leq x < \infty$ and $|g(x, \xi)| \leq M(\xi)$ for all x, where M is continuous for $a \leq \xi < \infty$, and if $\int_a^\infty M(\xi)\, d\xi$ converges, then the integral $\int_a^\infty g(x, \xi)\, d\xi$ is uniformly convergent and absolutely convergent.

Example 3-13. (**Differentiation with respect to parameter.**)

Define the integral

$$I(\alpha) = \int_0^\infty e^{-\alpha x} \frac{\sin x}{x}\, dx, \quad \alpha > 0.$$

Differentiate this integral with respect to the parameter α and obtain

$$\frac{dI}{d\alpha} = -\int_0^\infty e^{-\alpha x} \sin x\, dx,$$

where the integral on the right is the Laplace transform of $\sin x$ with α as parameter. Therefore, we can write

$$\frac{dI}{d\alpha} = -\frac{1}{1 + \alpha^2}, \quad \text{or} \quad dI = \frac{-d\alpha}{1 + \alpha^2}.$$

Integrating this equation with respect to α gives $I = -\arctan(\alpha) + c$. Now as $\alpha \to \infty$ we have $I(\alpha) \to 0$ and this implies $c = \frac{\pi}{2}$ and

$$I = I(\alpha) = \frac{\pi}{2} - \arctan(\alpha) = \text{arccot}(\alpha) = \arctan\left(\frac{1}{\alpha}\right).$$

As $\alpha \to 0$ we have $I(0) = \int_0^\infty \frac{\sin x}{x}\, dx = \lim_{\alpha \to 0} \arctan\left(\frac{1}{\alpha}\right) = \frac{\pi}{2}$. This integral is the limit of the sine integral

$$Si(x) = \int_0^x \frac{\sin u}{u}\, du$$

as x increases without bound.

■

Many new integration formulas can be obtained by differentiating or integrating with regard to parameters which have been considered constants during previous integrations.

Example 3-14. (**Differentiation with respect to parameter.**)

If $I = \int_0^\pi \frac{d\xi}{\alpha + \beta \cos \xi} = \frac{\pi}{(\alpha^2 - \beta^2)^{1/2}}$, $\alpha > \beta$, then

$$\frac{dI}{d\alpha} = \int_0^\pi \frac{-d\xi}{(\alpha + \beta \cos \xi)^2} = \frac{-\pi \alpha}{(\alpha^2 - \beta^2)^{3/2}} \quad \text{and} \quad \frac{dI}{d\beta} = \int_0^\pi \frac{-\cos \xi\, d\xi}{(\alpha + \beta \cos \xi)^2} = \frac{\pi \beta}{(\alpha^2 - \beta^2)^{3/2}}$$

■

Exercises 3

▶ **1.**

Certain Fourier series are easy to recognize. In equation (3.10) let $L = \pi$.

(a) Find the Fourier cosine series of $f(x) = 1$, $\quad -\pi \le x \le \pi$
(b) Find the Fourier sine series of $f(x) = \sin x$, $\quad -\pi \le x \le \pi$.
(c) Find the Fourier sine series of $f(x) = \sin 5x$, $\quad -\pi \le x \le \pi$.

▶ **2.**

Verify the given Fourier sine series expansions and then graph the Fourier series representation over the interval $-3L \le x \le 3L$.

(a) $1 \sim \dfrac{4}{\pi} \left(\sin \dfrac{\pi x}{L} + \dfrac{1}{3} \sin \dfrac{3\pi x}{L} + \dfrac{1}{5} \sin \dfrac{5\pi x}{L} + \cdots \right)$

(b) $x \sim \dfrac{2L}{\pi} \left(\sin \dfrac{\pi x}{L} - \dfrac{1}{2} \sin \dfrac{2\pi x}{L} + \dfrac{1}{3} \sin \dfrac{3\pi x}{L} - \cdots \right)$

▶ **3.**

Verify the given Fourier cosine series expansions and then graph the Fourier series representation over the interval $-3L \le x \le 3L$.

(a) $x \sim \dfrac{L}{2} - \dfrac{4L}{\pi^2} \left(\cos \dfrac{\pi x}{L} + \dfrac{1}{9} \cos \dfrac{3\pi x}{L} + \dfrac{1}{25} \cos \dfrac{5\pi x}{L} + \cdots \right)$

(b) $x^2 \sim \dfrac{L^2}{3} - \dfrac{4L^2}{\pi^2} \left(\cos \dfrac{\pi x}{L} - \dfrac{1}{4} \cos \dfrac{2\pi x}{L} + \dfrac{1}{9} \cos \dfrac{3\pi x}{L} - \cdots \right)$

▶ **4.**

Find the Fourier trigonometric series of the function given and illustrate with a sketch the function represented by the Fourier series.

(a) $\quad F(x) = x$, $\quad -1 \le x \le 1$, $\quad F(x+2) = F(x)$

(b) $\quad F(x) = \begin{cases} -1, & 0 < x < 1 \\ 1, & -1 < x < 0 \end{cases}$, $\quad F(x+2) = F(x)$

(c) $\quad F(x) = \begin{cases} x, & 0 < x < 2\pi \\ x + 2\pi, & -2\pi < x < 0 \end{cases}$, $\quad F(x+4\pi) = F(x)$

(d) $\quad F(x) = \begin{cases} x, & 0 < x < \pi \\ -x, & -\pi < x < 0 \end{cases}$, $\quad F(x+2\pi) = F(x)$

(e) $\quad F(x) = |\sin x|$, $\quad -\pi < x < \pi$, $\quad F(x+2\pi) = F(x)$

(f) $\quad F(x) = \begin{cases} \sin x, & 0 < x < \pi \\ 0, & -\pi < x < 0 \end{cases}$, $\quad F(x+2\pi) = F(x)$

▶ **5.**

Let $F_e(x)$, $G_e(x)$ denote even functions of x and let $F_o(x), G_o(x)$ denote odd functions of x.

(a) Show $h(x) = F_e(x)G_e(x)$ is an even function.

(b) Show $h(x) = F_e(x)F_o(x)$ is an odd function.

(c) Show $h(x) = F_o(x)G_o(x)$ is an even function.

▶ **6.**

Let $F(x)$ denote a smooth function defined over the interval $(-L, L)$.

(a) Show $G(x) = \frac{F(x)+F(-x)}{2}$ is an even function.

(b) Show $H(x) = \frac{F(x)-F(-x)}{2}$ is an odd function.

(c) Show $F(x)$ can be represented as the sum of an even and an odd function. What does this imply about the Fourier series representation of $F(x)$?

▶ **7.**

(a) Prove that if $F(x)$ is an odd function of x, then $\displaystyle\int_{-L}^{L} F(x)\, dx = 0$

(b) Prove that if $F(x)$ is an even function of x, then $\displaystyle\int_{-L}^{L} F(x)\, dx = 2\int_{0}^{L} F(x)\, dx$

▶ **8.**

(a) If $F(x+T) = F(x)$ for all x, with T constant, then show $F(x+nT) = F(x)$ for $n = 1, 2, \ldots$

(b) For $F(x)$ a smooth function on $(-L, L)$ and periodic with period $2L$ such that $F(x+2L) = F(x)$ for all x, let

$$I_1 = \int_{-L}^{L} F(x)\, dx, \qquad I_2 = \int_{0}^{2L} F(x)\, dx, \qquad I_3 = \int_{\alpha}^{\alpha+2L} F(x)\, dx.$$

Show $I_1 = I_2 = I_3$.

(c) Use the result in part (b) applied to the Euler equations defining the Fourier coefficients and show how to calculate the Fourier coefficients if $F(x)$ is defined over an interval $(\alpha, \alpha + 2L)$.

▶ **9.**

(a) Solve the differential equation $\dfrac{d^2y}{dt^2} + \omega^2 y = S(t)$, $\quad \omega \neq n\pi, \quad n = 1, 2, \ldots$, where $S(t)$ is the square wave $S(t) = \begin{cases} 1, & 0 < t < 1 \\ -1, & -1 < t < 0 \end{cases}$, $\quad S(t+2) = S(t)$ by first expanding $S(t)$ in a Fourier series.

(b) From your results for part (a), what restrictions must be placed upon ω?

▶ 10.

Show how to calculate the Fourier coefficients if $f = f(x, y)$, $0 \le x \le a$, $0 \le y \le b$ is to be expanded in a double Fourier series of the form:

$$(a) \quad f(x, y) = \sum_{n=1}^{\infty} \sum_{m=1}^{\infty} a_{nm} \sin(\frac{n\pi x}{a}) \sin(\frac{m\pi y}{b})$$

$$(b) \quad f(x, y) = \sum_{m=1}^{\infty} \sum_{n=1}^{\infty} a_{nm} \sin(\frac{n\pi x}{a}) J_k(\lambda_m y)$$

where $J_k(\lambda_m b) = 0$

$$(c) \quad f(x, y) = \sum_{m=0}^{\infty} \sum_{n=0}^{\infty} a_{nm} \cos(\frac{m\pi x}{a}) \cos(\frac{n\pi y}{b})$$

▶ 11.

(a) Expand the function e^x as a Fourier sine series over the interval $0 < x < \pi$ and show

$$e^x = \frac{2}{\pi} \sum_{n=1}^{\infty} \frac{n}{n^2 + 1} [1 - (-1)^n e^{\pi}] \sin(nx)$$

(b) Sketch the function e^x, all x, and sketch the Fourier series representation of the function e^x, all x. Discuss the use of the equal sign in part (a).

▶ 12.

(a) Represent the function $f(x) = x^2$, $0 \le x \le L$ as a Fourier cosine series.

(b) Evaluate the series at $x = 0$ and at $x = L$ to show

$$\sum_{n=1}^{\infty} \frac{(-1)^{n+1}}{n^2} = \frac{\pi^2}{12} \quad \text{and} \quad \sum_{n=1}^{\infty} \frac{1}{n^2} = \frac{\pi^2}{6}$$

What property of Fourier series gives the above results?

▶ 13.

Represent the function illustrated in the figure 3-11 as:

(a)   A Fourier cosine series.

(b)   A Fourier sine series.

(c)   Sketch the functions represented by the Fourier series in the cases (a) and (b) above over the interval $-3L \le x \le 4L$.

(d) Add the results in (a) and (b) above and divide by 2. Sketch over the interval $-3L \leq x \leq 4L$ the function represented by the resulting Fourier series.

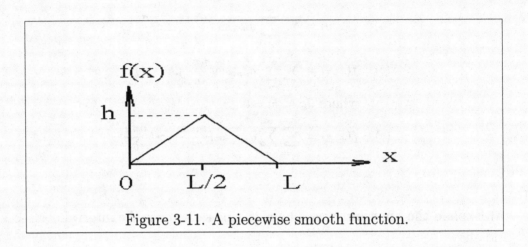

Figure 3-11. A piecewise smooth function.

▶ 14.

Let x^n be expanded in a Fourier-Bessel series

$$x^n = \sum_{m=1}^{\infty} A_m J_n(\lambda_m x), \quad \text{for} \quad 0 \leq x \leq a \quad \text{where } J_n(\lambda_m a) = 0.$$

Show the Fourier coefficients are given by an inner product divided by a norm squared which gives $A_m = \dfrac{2a^{n-1}}{\lambda_m J_{n+1}(\lambda_m a)}$ for $m = 1, 2, \ldots$.
Hint: Don't forget the weight function.

▶ 15.

(a) Define $f(x) = x + 1$ over the interval $1 < x < 2$, satisfying the periodicity condition $f(x + 1) = f(x)$ for all x. Expand the function $f(x)$ in a Fourier series which is valid for $1 < x < 2$.

(b) Graph the function represented by the Fourier series in part (a) over the interval $-3 < x < 3$.

▶ 16.

Represent the function $\sin x$, $0 \leq x \leq \pi$ as a Fourier cosine series and sketch a graph of the representation over the interval $-3\pi < x < 3\pi$. Is it permissible to differentiate this Fourier cosine series term by term? If so what is your result?

▶ **17.**

Represent the function $\cos x$, $0 \leq x \leq \pi$ as a Fourier sine series and sketch a graph of the representation over the interval $-3\pi < x < 3\pi$. Is it permissible to differentiate this Fourier sine series term by term? If so what is your result?

▶ **18.**

Represent the function $f(x) = 3\sin\dfrac{5\pi x}{L} + 4\cos\dfrac{5\pi x}{L}$ by a Fourier trigonometric series on the interval $(-L, L)$. Find the Fourier coefficients a_0, a_n, b_n.

▶ **19.**

Show the function

$$f(x) = \begin{cases} x+1, & 0 < x < 1 \\ -(x+1), & -1 < x < 0 \end{cases}$$

has the Fourier series representation

$$f(x) = \frac{1}{2} - \frac{4}{\pi^2} \sum_{k=0}^{\infty} \frac{1}{(2k+1)^2} \cos[(2k+1)\pi x] + \frac{4}{\pi} \sum_{k=0}^{\infty} \frac{1}{(2k+1)} \sin[(2k+1)\pi x].$$

(a) Sketch a graph of the Fourier series representation for $-3 \leq x \leq 3$.

(b) Sketch a graph of the Fourier series representation

$$H(x) = \frac{4}{\pi} \sum_{k=0}^{\infty} \frac{1}{(2k+1)} \sin[(2k+1)\pi x], \quad -3 \leq x \leq 3$$

(c) Sketch a graph of the Fourier series representation of the function

$$G(x) = \frac{1}{2} - \frac{4}{\pi^2} \sum_{k=0}^{\infty} \frac{1}{(2k+1)^2} \cos[(2k+1)\pi x], \quad -3 \leq x \leq 3$$

(d) Which of the above Fourier series can be differentiated term by term?

▶ **20.**

Expand the function $f(x) = \begin{cases} 0, & -1 < x < 0 \\ 1, & 0 < x < 1 \end{cases}$ in a Fourier Legendre series

and find the first four terms of the Fourier series expansion $f(x) = \displaystyle\sum_{m=0}^{\infty} c_m P_m(x)$.

▶ **21.**

Verify the Fourier series representation for the functions illustrated. In each graph assume the maximum amplitude of each function is +1 and the minimum amplitude of each function is either zero or -1 depending upon the graph.

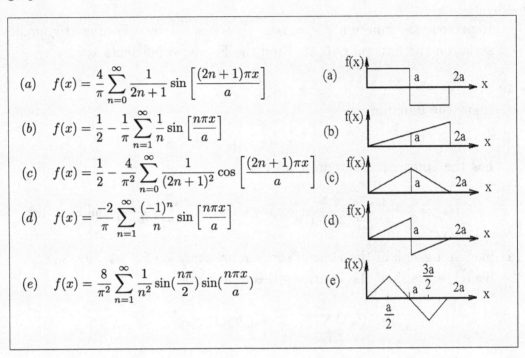

(a) $f(x) = \dfrac{4}{\pi} \displaystyle\sum_{n=0}^{\infty} \dfrac{1}{2n+1} \sin\left[\dfrac{(2n+1)\pi x}{a}\right]$

(b) $f(x) = \dfrac{1}{2} - \dfrac{1}{\pi} \displaystyle\sum_{n=1}^{\infty} \dfrac{1}{n} \sin\left[\dfrac{n\pi x}{a}\right]$

(c) $f(x) = \dfrac{1}{2} - \dfrac{4}{\pi^2} \displaystyle\sum_{n=0}^{\infty} \dfrac{1}{(2n+1)^2} \cos\left[\dfrac{(2n+1)\pi x}{a}\right]$

(d) $f(x) = \dfrac{-2}{\pi} \displaystyle\sum_{n=1}^{\infty} \dfrac{(-1)^n}{n} \sin\left[\dfrac{n\pi x}{a}\right]$

(e) $f(x) = \dfrac{8}{\pi^2} \displaystyle\sum_{n=1}^{\infty} \dfrac{1}{n^2} \sin(\dfrac{n\pi}{2}) \sin(\dfrac{n\pi x}{a})$

▶ **22.**

(a) Verify that the function $f(x) = \begin{cases} 0, & -\pi < x < 0 \\ \pi - x, & 0 < x < \pi \end{cases}$ has the Fourier trigonometric series $f(x) = \dfrac{\pi}{4} + \displaystyle\sum_{n=1}^{\infty} \dfrac{1-(-1)^n}{n^2 \pi} \cos nx + \displaystyle\sum_{n=1}^{\infty} \dfrac{1}{n} \sin nx, \quad -\pi < x < \pi.$

(b) Sketch a graph of this Fourier series over the interval $-3\pi < x < 3\pi$.

(c) Sketch a graph over the interval $-3\pi < x < 3\pi$ of the function represented by the Fourier sine series $G(x) = \displaystyle\sum_{n=1}^{\infty} \dfrac{1}{n} \sin nx.$

(d) Sketch a graph over the interval $-3\pi < x < 3\pi$ of the function represented by the Fourier cosine series $H(x) = \dfrac{\pi}{4} + \displaystyle\sum_{n=1}^{\infty} \dfrac{1-(-1)^n}{n^2 \pi} \cos nx.$

▶ **23.**

(a) Verify that the function $f(x) = (x + 2\pi)^2$, $-\pi < x < \pi$ has the Fourier series expansion $f(x) = \dfrac{13\pi^2}{3} + \sum\limits_{n=1}^{\infty} 4\dfrac{(-1)^n}{n^2} \cos nx + \sum\limits_{n=1}^{\infty} 8\pi\dfrac{(-1)^{n+1}}{n} \sin nx$.

(b) What function has the Fourier cosine series $f_e(x) = \dfrac{13\pi^2}{3} + \sum\limits_{n=1}^{\infty} 4\dfrac{(-1)^n}{n^2} \cos nx$? Sketch this function over the interval $-3\pi < x < 3\pi$.

(c) What function has the Fourier sine series $f_o(x) = \sum\limits_{n=1}^{\infty} 8\pi\dfrac{(-1)^{n+1}}{n} \sin nx$? Sketch this function over the interval $-3\pi < x < 3\pi$.

▶ **24.**

Denote the positive roots of the equation $\lambda J_1(\lambda) - h J_0(\lambda) = 0$, where h is a positive constant, by λ_m for $m = 1, 2, 3, \ldots$. Expand the function $f(x) = 1$ over the interval $0 \leq x \leq 1$ in a Fourier Bessel series involving the set of functions $\{J_0(\lambda_n x)\}$, where λ_n is a root of the above equation.

▶ **25.**

(a) Show $1 - \dfrac{x}{L} = 2\sum\limits_{n=0}^{\infty} \dfrac{1}{n\pi} \sin\left(\dfrac{n\pi x}{L}\right)$ and sketch a graph of the Fourier series representation over the interval $(-3L, 3L)$.

(b) Show $\dfrac{x}{L} = 2\sum\limits_{n=1}^{\infty} \dfrac{(-1)^{n+1}}{n\pi} \sin\left(\dfrac{n\pi x}{L}\right)$ and sketch a graph of the Fourier series representation over the interval $(-3L, 3L)$.

▶ **26.**

Let $\phi_n(x)$ denote an orthogonal set over the interval $a \leq x \leq b$ with respect to a weight function $\omega_1(x)$. Let $\psi_m(y)$ denote an orthogonal set of functions over the interval $\alpha \leq y \leq \beta$ with respect to the weight function $\omega_2(y)$. Let $H_{mn}(x, y) = \phi_n(x)\psi_m(y)$. If $f(x, y)$, $a \leq x \leq b$, $\alpha \leq y \leq \beta$ is expanded as a double Fourier series of the form $f = \sum\sum c_{nm} H_{nm}$, then show the Fourier coefficients are given by

$$c_{nm} = \frac{\int_{\alpha}^{\beta} \int_{a}^{b} \omega_1(x)\omega_2(y)f(x,y)\phi_n(x)\psi_m(y)\,dx\,dy}{\|\phi_n\|^2 \cdot \|\psi_m\|^2}.$$

▶ **27.**

(a) Expand the function $f(x) = x^2$, $1 < x < 2$ in a Fourier series which is valid over the interval $1 < x < 2$ and such that $f(x+1) = f(x)$ for all x.

(b) Graph the function, which is represented by the Fourier series in part (a), over the interval $-3 < x < 3$.

▶ **28.**

If $I_1 = \int_0^\infty e^{-x^2}\, dx = \dfrac{\sqrt{\pi}}{2}$, then find, $I_2 = \int_0^\infty e^{-\beta x^2}\, dx$

▶ **29.**

Let $I = \int_0^\infty e^{-\alpha x^2} \cos(\beta x)\, dx$

(a) Show $\dfrac{dI}{d\beta} = \dfrac{-\beta}{2\alpha} I$ Hint: Integrate by parts after differentiation.

(b) Solve the differential equation in (a) after determining the initial condition and hence determine I.

▶ **30.**

For $x > 0$ and $\alpha > 0$ use the given integrals

$$\int_{-\infty}^\infty \frac{\sin \omega x}{(\omega - \beta)^2 + \alpha^2}\, d\omega = \frac{\pi}{2} e^{-\alpha x} \sin \beta x, \qquad \int_{-\infty}^\infty \frac{\cos \omega x}{(\omega - \beta)^2 + \alpha^2}\, d\omega = \frac{\pi}{\alpha} e^{-\alpha x} \cos \beta x$$

to find the Fourier integral representation of the following functions.

(a) $f(x) = \dfrac{1}{x^2 + 4}$ (b) $f(x) = \dfrac{1}{(x-2)^2 + 4}$

▶ **31.**

Find the Fourier integral representation of the given functions

(a) $f(x) = \begin{cases} \cos x, & |x| < \frac{\pi}{2} \\ 0, & |x| > \frac{\pi}{2} \end{cases}$ (b) $f(x) = \begin{cases} \sin x, & |x| < \pi \\ 0, & |x| > \pi \end{cases}$

▶ **32.**

Find the Fourier integral representation of the given functions and sketch the functions.

(a) $f(x) = e^{-kx}$, $k > 0$, $x \geq 0$, $f(-x) = f(x)$

(b) $f(x) = e^{-kx}$, $k > 0$, $x > 0$, $f(-x) = -f(x)$

▶ **33.**

Show the Fourier integral representation of the function

$$f(x) = \begin{cases} 0, & x > 1 \\ 1 - x, & 0 \le x \le 1 \end{cases}, \quad f(-x) = f(x)$$

is given by $f(x) = \dfrac{4}{\pi} \displaystyle\int_0^\infty \dfrac{\sin^2(\frac{\omega}{2}) \cos(\omega x)}{\omega^2} \, d\omega$.

▶ **34.**

Find the Fourier-Bessel coefficients a_i if $f(x) = 1 = \displaystyle\sum_{i=1}^\infty a_i J_0(\xi_{0i} \dfrac{x}{b})$, $\quad 0 < x < b$

where $J_0(\xi_{0i}) = 0$ for $i = 1, 2, 3, \ldots$.

▶ **35.**

(a) Show the function $f(x) = \begin{cases} 2 + 2x, & 0 < x < 1 \\ 0, & -1 < x < 0 \end{cases}$ has the Fourier series expansion $f(x) = f_e(x) + f_o(x)$ where

$$f_e(x) = \frac{3}{2} + \sum_{n=1}^\infty \frac{2}{\pi^2} \frac{[(-1)^n - 1]}{n^2} \cos nx, \qquad f_o(x) = \frac{2}{\pi} \sum_{n=1}^\infty \frac{[1 - 2(-1)^n]}{n} \sin n\pi x$$

(b) Sketch the function $f_e(x)$ over the interval $-3 < x < 3$.

(c) Sketch the function $f_o(x)$ over the interval $-3 < x < 3$.

▶ **36.**

(a) Show that a function which is expanded in a Fourier-Legendre series of the form

$$f(x) = \sum_{n=0}^\infty c_n P_n(x), \quad -1 < x < 1$$

has the Fourier-Legendre coefficients

$$c_n = \frac{(f(x), P_n(x))}{\| P_n \|^2} = \frac{2n+1}{2} \int_{-1}^1 f(x) P_n(x) \, dx.$$

(b) Expand the function $f(x) = x^2$ in a Fourier-Legendre series.

(c) Expand the function $f(x) = x^4$ in a Fourier-Legendre series.

▶ **37.**

(a) Show the function $f(\omega) = \begin{cases} -\frac{\pi}{2}, & \omega < 0 \\ 0, & \omega = 0 \\ \frac{\pi}{2}, & \omega > 0 \end{cases}$

does not have a Fourier integral representation in the usual sense.

(b) Show the function $f(\omega) = \begin{cases} -\frac{\pi}{2}e^{\alpha\omega}, & \omega < 0 \\ 0, & \omega = 0 \\ \frac{\pi}{2}e^{-\alpha\omega}, & \omega > 0 \end{cases}$

where $\alpha > 0$, has the Fourier integral representation

$$f(\omega) = \int_0^\infty \frac{\xi}{\alpha^2 + \xi^2} \sin \xi\omega \, d\xi$$

(c) Show in the limit as $\alpha \to 0$ there results the limiting sense Fourier integral representation

$$f(\omega) = \int_0^\infty \frac{\sin \omega\xi}{\xi} \, d\xi = \begin{cases} -\frac{\pi}{2}, & \omega < 0 \\ 0, & \omega = 0 \\ \frac{\pi}{2}, & \omega > 0 \end{cases}.$$

(d) Show the Fourier integral representation given by equation (3.75) can also be written in the form

$$f(x) = \frac{1}{\pi} \int_0^\infty \frac{1}{\omega} \left(\sin[(1+x)\omega] + \sin[(1-x)\omega] \right) d\omega$$

Now use the results from part (c) to evaluate this integral.

▶ **38.**

Expand the function $f(x) = e^{-|x|}$ in a Fourier integral representation and show

$$e^{-|x|} = \int_0^\infty \frac{\cos \omega x}{1 + \omega^2} \, d\omega$$

▶ **39.**

Expand the following functions in a Fourier integral representation.

(a) $f(x) = \begin{cases} x, & 0 < x < 1 \\ 0, & \text{otherwise} \end{cases}$

(b) $f(x) = \begin{cases} 1, & 0 < x < 2 \\ 0, & \text{otherwise} \end{cases}$

Then evaluate $f(x)$ at $x = 1$ to show $\int_0^\infty \frac{\sin \omega}{\omega} \, d\omega = \frac{\pi}{2}$.

(c) $f(x) = \begin{cases} e^{-x}, & 0 < x < \infty \\ 0, & \text{otherwise} \end{cases}$

Chapter 4
Parabolic Equations

Partial differential equations occur in abundance in a variety of areas from engineering, mathematical biology and physics. In this chapter we will concentrate upon the partial differential equations representing heat flow and diffusion which are parabolic equations.

Heat Equation

Heat can flow from regions of high temperature to regions of low temperature in essentially three different ways. These three ways are conduction, radiation and convection. Internal conduction occurs because of molecular motion within the body. That is, within solid bodies we have heat flowing from regions of high temperature, which have a large amount of molecular motion, (large velocities, large kinetic energy of the molecules) to regions of low temperature (smaller velocities, less kinetic energy of the molecules). Radiation heat transfer between two objects occurs when heat passes through space from the hotter object to the cooler object without heating the space between the objects. It is due to wave motion. A common example is the Sun heating the Earth. Heat transfer by convection occurs when some type of motion moves heat from one place to another. Forced convection occurs when a blower blows heat from one area to another. There are other types of heat transfer such as that which occurs during evaporation and condensation processes. All of these heat transfer methods are best studied in a course on the subject. For our purposes we will concentrate on heat transfer by conduction. Recall the Gauss[†] divergence theorem

---

For $\vec{F} = \vec{F}(x, y, z, t)$ a continuous vector field defined everywhere within a simply connected volume V and surrounded by a closed surface S we have

$$\iiint_V \operatorname{div} \vec{V}\, d\tau = \iint_S \vec{F} \cdot \hat{n}\, d\sigma \qquad (4.1)$$

where $d\tau$ is a volume element, $d\sigma$ is an element of surface area and $\hat{n}$ is a unit exterior normal to the surface.

[†] See Appendix C

which can be used to convert surface integrals to volume integrals. We will use this theorem together with the Fourier law of heat conduction and a conservation law for energy transfer to derive the heat conduction equation. The Fourier law of heat conduction is given by

$$\vec{q} = -\kappa \operatorname{grad} u = -\kappa \nabla u = -\kappa \left(\frac{\partial u}{\partial x}\hat{i} + \frac{\partial u}{\partial y}\hat{j} + \frac{\partial u}{\partial z}\hat{k} \right) \tag{4.2}$$

where $\vec{q}$ (cal/cm^{2}sec) is the rate of heat flow per unit area per unit of time, κ (cal/sec cm^2 °C/cm) is the thermal conductivity of the region where heat flows and depends upon the material properties in which heat is flowing. The quantity $u = u(x,y,z)$ (°C) represents the temperature. The surfaces $u(x,y,z) = c$, where c is constant, are called isothermal surfaces or surfaces of constant temperature. Observe the gradient vector ∇u is always normal to any point on an isothermal surface $u(x,y,z) =$ constant and points in the direction of greatest increase of temperature. Because heat flows from hot to cold regions we need the above negative sign in Fourier's law. Thus, Fourier's law for heat conduction can be interpreted as representing heat flow in the direction which temperature decreases. The quantity $\vec{q}$ is called the thermal current vector and represents the rate of heat flow per unit of area.

Introduce the following notation and units of measurements:

$c = c(x,y,z)$ is the specific heat of the solid, (cal/gm °C)

$\varrho = \varrho(x,y,z)$ is the volume density of the solid, (gm/cm^3)

$\kappa = \kappa(x,y,z)$ is the thermal conductivity of the solid, (cal/sec cm^2 °C/cm)

$\vec{q} = \vec{q}(x,y,z,t)$ is the rate of heat flow per unit area, (cal/cm^{2}sec)

$H = H(x,y,z,t)$ is the rate of heat generation per unit volume, (cal/sec cm^3)

$u = u(x,y,z,t)$ denotes temperature, (°C)

We employ the above symbols and write out the law of conservation of energy for an arbitrary simply-connected region[‡] V with closed surface S. Let H_s denote the change in the amount of heat stored in the region V during a time interval Δt. We write H_s as the amount of heat H_c which crosses the surface S into or leaving the region during the time interval Δt plus any heat generated H_g within

---

[‡] A simply-connected region is such that any simple closed curve within the region can be continuously shrunk to a point without leaving the region. Regions which are not simply-connected are called multiply-connected.

the region V during this time interval. The conservation law is then written in either of the forms

$$H_s = H_c + H_g \qquad \text{or} \qquad H_c + H_g - H_s = 0. \qquad (4.3)$$

The heat stored in a volume element $d\tau$ of V is given by $c\varrho u\, d\tau$ in units of calories. The quantity H_s is the rate of change of heat stored within the volume element and is given by

$$H_s = \frac{\partial}{\partial t} \iiint_V c\varrho u\, d\tau \qquad \left(\frac{\text{cal}}{\text{sec}}\right). \qquad (4.4)$$

The amount of heat which crosses into the region during a time interval Δt, or flux of heat into the region, is given by

$$H_c = \iint_S -\vec{q}\cdot\hat{n}\, d\sigma \qquad \left(\frac{\text{cal}}{\text{sec}}\right). \qquad (4.5)$$

where the negative sign is used to change the sign of the exterior unit normal vector $\hat{n}$. This surface integral is converted to a volume integral using the above Gauss divergence theorem to obtain

$$H_c = \iiint_V -\operatorname{div}\vec{q}\, d\tau \qquad \left(\frac{\text{cal}}{\text{sec}}\right). \qquad (4.6)$$

The heat generated within the region is given by

$$H_g = \iiint_V H\, d\tau \qquad \left(\frac{\text{cal}}{\text{sec}}\right). \qquad (4.7)$$

The results from equations (4.4), (4.6), and (4.7) enable us to write the conservation law given by equation (4.3) in the form

$$\iiint_V \left\{-\operatorname{div}\vec{q} + H - \frac{\partial}{\partial t}(c\varrho u)\right\} d\tau = 0 \qquad (4.8)$$

which implies, for an arbitrary volume and arbitrary time Δt, the term inside the brackets must equal zero. Also note each term in equation (4.8) has the same units of $(\text{cal/cm}^3\,\text{sec})$. Now substitute for $\vec{q}$ from equation (4.2), representing Fourier's law of heat conduction, to obtain the heat conduction equation

$$\operatorname{div}(\kappa\operatorname{grad} u) + H = \frac{\partial}{\partial t}(c\varrho u). \qquad (4.9)$$

which has the expanded form

$$\frac{\partial}{\partial x}\left(\kappa\frac{\partial u}{\partial x}\right) + \frac{\partial}{\partial y}\left(\kappa\frac{\partial u}{\partial y}\right) + \frac{\partial}{\partial z}\left(\kappa\frac{\partial u}{\partial z}\right) + H = \frac{\partial}{\partial t}(c\,\varrho\,u). \tag{4.10}$$

In the special case the thermal conductivity κ is constant, and does not vary with position, we can write

$$\operatorname{div}(\kappa\operatorname{grad}u) = \nabla(\kappa\,\nabla u) = \kappa\,\nabla^2 u. \tag{4.11}$$

where $\nabla^2 u$ is called the Laplacian operator. In the case where all coefficients are constants we can write the heat equation in the form

$$\frac{\partial u}{\partial t} = K\nabla^2 u + Q, \qquad K = \frac{\kappa}{c\,\varrho}, \qquad Q = \frac{H}{c\,\varrho} \tag{4.12}$$

where K is called the diffusivity of the material. If $\lim_{t\to\infty}\frac{\partial u}{\partial t} = 0$ the temperature is said to have reached steady state conditions. For steady state conditions we set $\frac{\partial u}{\partial t} = 0$ in equation (4.12) and assume the temperature u depends only upon position. If there are no sources or sinks we set $Q = 0$ and under these conditions the heat equation becomes homogeneous. The tables 4-1 and 4-2 show various representations of the heat equation in Cartesian coordinates. The construction of similar tables for cylindrical and spherical coordinates is left as an exercise.

Note in the table 4-1 that the exact differential forms are easily integrated. Also note the different forms of the Laplacian operator in one, two and three dimensions.

The Laplacian operator $\nabla^2 u$ is sometimes written using the notation Δu. The Laplacian takes on different forms in different coordinate systems. Given a set of transformation equations $x = x(\xi,\eta,\zeta)$ $\quad y = y(\xi,\eta,\zeta)$ $\quad z = z(\xi,\eta,\zeta)$ we form the position vector $\vec{r} = \vec{r}(\xi,\eta,\zeta) = x(\xi,\eta,\zeta)\hat{i} + y(\xi,\eta,\zeta)\hat{j} + z(\xi,\eta,\zeta)\hat{k}$ and calculate the derivatives $\frac{\partial\vec{r}}{\partial\xi}$, $\frac{\partial\vec{r}}{\partial\eta}$, $\frac{\partial\vec{r}}{\partial\zeta}$ which represent tangent vectors to the coordinate curves defining the (ξ,η,ζ) coordinates.

From these tangent vectors one can calculate the magnitudes

$$h_\xi = \left|\frac{\partial\vec{r}}{\partial\xi}\right|, \quad h_\eta = \left|\frac{\partial\vec{r}}{\partial\eta}\right|, \quad h_\zeta = \left|\frac{\partial\vec{r}}{\partial\zeta}\right|$$

and the dot products $\frac{\partial\vec{r}}{\partial\xi}\cdot\frac{\partial\vec{r}}{\partial\eta}$, $\frac{\partial\vec{r}}{\partial\xi}\cdot\frac{\partial\vec{r}}{\partial\zeta}$, $\frac{\partial\vec{r}}{\partial\eta}\cdot\frac{\partial\vec{r}}{\partial\zeta}$. If these dot products are zero the coordinate curves have orthogonal intersections and under these conditions

Table 4-1 Various Forms for Heat Equation Cartesian Coordinates

Special Cases	Operator Form	1-Dimensional Form
General	$\nabla\left(\kappa\nabla u\right) + H = \varrho c\dfrac{\partial u}{\partial t}$	$\dfrac{\partial}{\partial x}\left(\kappa\dfrac{\partial u}{\partial x}\right) + H = \varrho c\dfrac{\partial u}{\partial t}$
Homogeneous Material	$\nabla^2 u + \dfrac{H}{\kappa} = \dfrac{\varrho c}{\kappa}\dfrac{\partial u}{\partial t}$	$\dfrac{\partial^2 u}{\partial x^2} + \dfrac{H}{\kappa} = \dfrac{\varrho c}{\kappa}\dfrac{\partial u}{\partial t}$
Steady State	$\nabla\left(\kappa\nabla u\right) + H = 0$	$\dfrac{\partial}{\partial x}\left(\kappa\dfrac{\partial u}{\partial x}\right) + H = 0$
Source Free	$\nabla\left(\kappa\nabla u\right) = \varrho c\dfrac{\partial u}{\partial t}$	$\dfrac{\partial}{\partial x}\left(\kappa\dfrac{\partial u}{\partial x}\right) = \varrho c\dfrac{\partial u}{\partial t}$
Steady State Homogeneous Material	$\nabla^2 u + \dfrac{H}{\kappa} = 0$	$\dfrac{d^2 u}{dx^2} + \dfrac{H}{\kappa} = 0$
Steady State Source Free	$\nabla\left(\kappa\nabla u\right) = 0$	$\dfrac{d}{dx}\left(\kappa\dfrac{du}{dx}\right) = 0$
Homogeneous Material Source Free	$\nabla^2 u = \dfrac{\varrho c}{\kappa}\dfrac{\partial u}{\partial t}$	$\dfrac{\partial^2 u}{\partial x^2} = \dfrac{\varrho c}{\kappa}\dfrac{\partial u}{\partial t}$
Steady State Source Free Homogeneous Material	$\nabla^2 u = 0$	$\dfrac{d^2 u}{dx^2} = 0$

Table 4-2 Various Forms for Heat Equation Cartesian Coordinates

Special Cases	2-Dimensional Form	3-Dimensional Form
General	$\dfrac{\partial}{\partial x}\left(\kappa\dfrac{\partial u}{\partial x}\right) + \dfrac{\partial}{\partial y}\left(\kappa\dfrac{\partial u}{\partial y}\right) + H = \varrho c\dfrac{\partial u}{\partial t}$	$\dfrac{\partial}{\partial x}\left(\kappa\dfrac{\partial u}{\partial x}\right) + \dfrac{\partial}{\partial y}\left(\kappa\dfrac{\partial u}{\partial y}\right) + \dfrac{\partial}{\partial z}\left(\kappa\dfrac{\partial u}{\partial z}\right) + H = \varrho c\dfrac{\partial u}{\partial t}$
Homogeneous Material	$\dfrac{\partial^2 u}{\partial x^2} + \dfrac{\partial^2 u}{\partial y^2} + \dfrac{\partial^2 u}{\partial z^2} + \dfrac{H}{\kappa} = \dfrac{\varrho c}{\kappa}\dfrac{\partial u}{\partial t}$	$\dfrac{\partial^2 u}{\partial x^2} + \dfrac{\partial^2 u}{\partial y^2} + \dfrac{\partial^2 u}{\partial z^2} + \dfrac{H}{K} = \dfrac{\varrho c}{K}\dfrac{\partial u}{\partial t}$
Steady State	$\dfrac{\partial}{\partial x}\left(\kappa\dfrac{\partial u}{\partial x}\right) + \dfrac{\partial}{\partial y}\left(\kappa\dfrac{\partial u}{\partial y}\right) + H = 0$	$\dfrac{\partial}{\partial x}\left(\kappa\dfrac{\partial u}{\partial x}\right) + \dfrac{\partial}{\partial y}\left(\kappa\dfrac{\partial u}{\partial y}\right) + \dfrac{\partial}{\partial z}\left(\kappa\dfrac{\partial u}{\partial z}\right) + H = 0$
Source Free	$\dfrac{\partial}{\partial x}\left(\kappa\dfrac{\partial u}{\partial x}\right) + \dfrac{\partial}{\partial y}\left(\kappa\dfrac{\partial u}{\partial y}\right) = \varrho c\dfrac{\partial u}{\partial t}$	$\dfrac{\partial}{\partial x}\left(\kappa\dfrac{\partial u}{\partial x}\right) + \dfrac{\partial}{\partial y}\left(\kappa\dfrac{\partial u}{\partial y}\right) + \dfrac{\partial}{\partial z}\left(\kappa\dfrac{\partial u}{\partial z}\right) = \varrho c\dfrac{\partial u}{\partial t}$
Steady State Homogeneous Material	$\dfrac{\partial^2 u}{\partial x^2} + \dfrac{\partial^2 u}{\partial y^2} + \dfrac{H}{\kappa} = 0$	$\dfrac{\partial^2 u}{\partial x^2} + \dfrac{\partial^2 u}{\partial y^2} + \dfrac{\partial^2 u}{\partial z^2} + \dfrac{H}{\kappa} = 0$
Steady State Source Free	$\dfrac{\partial}{\partial x}\left(\kappa\dfrac{\partial u}{\partial x}\right) + \dfrac{\partial}{\partial y}\left(\kappa\dfrac{\partial u}{\partial y}\right) = 0$	$\dfrac{\partial}{\partial x}\left(\kappa\dfrac{\partial u}{\partial x}\right) + \dfrac{\partial}{\partial y}\left(\kappa\dfrac{\partial u}{\partial y}\right) + \dfrac{\partial}{\partial z}\left(\kappa\dfrac{\partial u}{\partial z}\right) = 0$
Homogeneous Material Source Free	$\dfrac{\partial^2 u}{\partial x^2} + \dfrac{\partial^2 u}{\partial y^2} = \dfrac{\varrho c}{\kappa}\dfrac{\partial u}{\partial t}$	$\dfrac{\partial^2 u}{\partial x^2} + \dfrac{\partial^2 u}{\partial y^2} + \dfrac{\partial^2 u}{\partial z^2} = \dfrac{\varrho c}{\kappa}\dfrac{\partial u}{\partial t}$
Steady State Source Free Homogeneous Material	$\dfrac{\partial^2 u}{\partial x^2} + \dfrac{\partial^2 u}{\partial y^2} = 0$	$\dfrac{\partial^2 u}{\partial x^2} + \dfrac{\partial^2 u}{\partial y^2} + \dfrac{\partial^2 u}{\partial z^2} = 0$

we say the (ξ, η, ζ) coordinate system is an orthogonal system. The Laplacian is calculated in an orthogonal system from the relation

$$\nabla^2 u = \frac{1}{h_\xi h_\eta h_\zeta}\left[\frac{\partial}{\partial \xi}\left(\frac{h_\eta h_\zeta}{h_\xi}\frac{\partial u}{\partial \xi}\right) + \frac{\partial}{\partial \eta}\left(\frac{h_\xi h_\zeta}{h_\eta}\frac{\partial u}{\partial \eta}\right) + \frac{\partial}{\partial \zeta}\left(\frac{h_\xi h_\eta}{h_\zeta}\frac{\partial u}{\partial \zeta}\right)\right]. \tag{4.13}$$

For a derivation of this relation see the reference by Spiegel in the bibliography section. The above representation is only valid in an orthogonal coordinate system. For future reference we list the representation of the Laplacian in Cartesian, cylindrical and spherical coordinate systems.

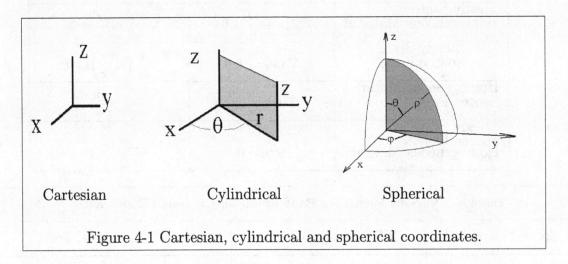

Figure 4-1 Cartesian, cylindrical and spherical coordinates.

In rectangular coordinates (x, y, z) we have

$$\Delta u = \nabla^2 u = \frac{\partial^2 u}{\partial x^2} + \frac{\partial^2 u}{\partial y^2} + \frac{\partial^2 u}{\partial z^2}. \tag{4.14}$$

In cylindrical coordinates (r, θ, z) we have the transformation equations

$$x = r\cos\theta, \qquad y = r\sin\theta, \qquad z = z$$

which produces the Laplacian

$$\Delta u = \nabla^2 u = \frac{1}{r}\frac{\partial}{\partial r}\left(r\frac{\partial u}{\partial r}\right) + \frac{1}{r^2}\frac{\partial^2 u}{\partial \theta^2} + \frac{\partial^2 u}{\partial z^2}$$

$$\text{or} \qquad \nabla^2 u = \frac{\partial^2 u}{\partial r^2} + \frac{1}{r}\frac{\partial u}{\partial r} + \frac{1}{r^2}\frac{\partial^2 u}{\partial \theta^2} + \frac{\partial^2 u}{\partial z^2}. \tag{4.15}$$

In spherical coordinates (ρ, θ, ϕ) we have the transformation equations

$$x = \rho\sin\theta\cos\phi, \qquad y = \rho\sin\theta\sin\phi, \qquad z = \rho\cos\theta$$

which produces the Laplacian

$$\Delta u = \nabla^2 u = \frac{1}{\rho^2}\frac{\partial}{\partial \rho}\left(\rho^2 \frac{\partial u}{\partial \rho}\right) + \frac{1}{\rho^2 \sin\theta}\frac{\partial}{\partial \theta}\left(\sin\theta \frac{\partial u}{\partial \theta}\right) + \frac{1}{\rho^2 \sin^2\theta}\frac{\partial^2 u}{\partial \phi^2}$$

$$\text{or}\qquad \nabla^2 u = \frac{\partial^2 u}{\partial \rho^2} + \frac{2}{\rho}\frac{\partial u}{\partial \rho} + \frac{1}{\rho^2}\frac{\partial^2 u}{\partial \theta^2} + \frac{1}{\rho^2}\cot\theta\frac{\partial u}{\partial \theta} + \frac{1}{\rho^2 \sin^2\theta}\frac{\partial^2 u}{\partial \phi^2}$$

(4.16)

Boundary and initial conditions for the heat equation

Let S denote a closed surface enclosing a volume V where we wish to solve the heat equation. The various boundary conditions that can be assigned are as follows:

Dirichlet[†] boundary conditions or boundary value problem of the first kind requires the temperature be specified on the boundary of the region where the heat equation is solved. This type of boundary condition can be written in the form

$$u(x,y,z,t)\Big|_{(x,y,z)\in S} = f_1(x,y,z,t)$$

(4.17)

where f_1 is a specified temperature.

Neumann[†] boundary conditions or boundary value problem of the second kind requires the heat flux across the boundary be specified on the boundary of the region where the heat equation is solved. This type of boundary condition is expressed

$$\frac{\partial u}{\partial n} = \operatorname{grad} u \cdot \hat{n} = f_2(x,y,z,t)\Big|_{(x,y,z)\in S}$$

(4.18)

where f_2 is a specified heat flux.

Note for an insulated boundary we have $\frac{\partial u}{\partial n} = \operatorname{grad} u \cdot \hat{n} = 0$ and no heat flows across the boundary.

Robin[†] boundary condition or boundary value problem of the third kind requires that the heat loss from the boundary to the surrounding medium be specified. This type of condition can be expressed

$$\frac{\partial u}{\partial n} + hu = f_3(x,y,z,t)\Big|_{(x,y,z)\in S}$$

(4.19)

[†] See Appendix C

where $h > 0$ is a constant and f_3 is a specified function. Here the heat exchange depends upon the temperature difference with the surrounding medium. In this case we can think of u as representing $T - T_0$ where T_0 is the temperature of the surrounding medium.

Mixed boundary conditions or mixed boundary value problems require different types of boundary conditions be specified over different portions of the bounding surface of the region of interest. For example, temperature can be specified over one portion of the surface while the heat flux can be specified over the other part of the surface. This type of mixed boundary condition can be expressed

$$
\begin{aligned}
u(x,y,z,t) = & f_1(x,y,z)\Big|_{(x,y,z)\in S_1} \\
\frac{\partial u}{\partial n} = & f_2(x,y,z)\Big|_{(x,y,z)\in S_2}
\end{aligned}
\quad \text{where} \quad
\begin{aligned}
S_1 \cup S_2 &= S \\
S_1 \cap S_2 &= \emptyset
\end{aligned}. \tag{4.20}
$$

Mixed boundary value problems are in general much harder to solve. In the Neumann and Robin conditions there occurs the normal derivative $\dfrac{\partial u}{\partial n}$ which is calculated at a boundary point and given by $\dfrac{\partial u}{\partial n} = \operatorname{grad} u \cdot \hat{n}$ where $\hat{n}$ is the exterior unit normal to the boundary surface.

The above boundary conditions can be represented in an alternative fashion by using a boundary operator

$$
B(u)\Big|_{x,y,z\in S} = \left[\alpha\frac{\partial u}{\partial n} + \beta u\right]_{x,y,z\in S} = \text{a specified condition.}
$$

where a Dirichlet boundary condition results when $\alpha = 0, \beta = 1$, a Neumann condition results when $\alpha = 1, \beta = 0$, and a Robin condition results for the parameter values $\alpha = 1, \beta = h$.

Initial conditions are expressed

$$
u(x,y,z,0) = f_4(x,y,z) \quad \text{where } (x,y,z) \in V \tag{4.21}
$$

Here f_4 is a specified initial temperature distribution.

Diffusion Equation

The equation (4.10) is sometimes referred to as the diffusion equation. For diffusion processes we replace the temperature $u(x,y,z,t)$ ($°C$) by concentration $C(x,y,z,t)$ (gm/cm^3) which represents the concentration of diffusing material at the point (x, y, z) at time t. The Fourier law of heat conduction is replaced by Fick's[†] law of diffusion which states that the rate of transfer of diffusing material across a unit area of cross section is proportional to the concentration gradient in the direction of diffusion. In terms of symbols one can write

$$\vec{J} = -D \operatorname{grad} C = -D \nabla C \tag{4.22}$$

where $\vec{J}$ (gm/cm²sec) is the mass flow rate per unit area, D (cm²/sec) is the diffusion coefficient or mass diffusivity, and C (gm/cm³) is the concentration. Fick's law is valid for isotropic media of gases, liquids and solids. The diffusion coefficient D depends upon the process being considered. Nonlinear diffusion equations result if D is a function of the concentration. For anisotropic material, Fick's law is written

$$\vec{J} = -J_x \hat{i} - J_y \hat{j} - J_z \hat{k} \tag{4.23}$$

where

$$\begin{aligned}
-J_x &= D_{11} \frac{\partial C}{\partial x} + D_{12} \frac{\partial C}{\partial y} + D_{13} \frac{\partial C}{\partial z} \\
-J_y &= D_{21} \frac{\partial C}{\partial x} + D_{22} \frac{\partial C}{\partial y} + D_{23} \frac{\partial C}{\partial z} \\
-J_z &= D_{31} \frac{\partial C}{\partial x} + D_{32} \frac{\partial C}{\partial y} + D_{33} \frac{\partial C}{\partial z}
\end{aligned} \tag{4.24}$$

and the D_{ij}, $i, j = 1, 2, 3$, are the diffusion coefficients reflecting the different diffusion properties in different directions. If all the diffusion coefficients D_{ij} are constants, then the analog of the heat equation becomes

$$\begin{aligned}
\frac{\partial C}{\partial t} = & D_{11} \frac{\partial^2 C}{\partial x^2} + D_{12} \frac{\partial^2 C}{\partial x \partial y} + D_{13} \frac{\partial^2 C}{\partial x \partial z} \\
& + D_{21} \frac{\partial^2 C}{\partial x \partial y} + D_{22} \frac{\partial^2 C}{\partial y^2} + D_{23} \frac{\partial^2 C}{\partial z \partial y} \\
& + D_{31} \frac{\partial^2 C}{\partial x \partial z} + D_{32} \frac{\partial^2 C}{\partial y \partial z} + D_{33} \frac{\partial^2 C}{\partial z^2}.
\end{aligned} \tag{4.25}$$

In the special case $D_{ij} = 0$ for $i \neq j$ and $D_{11} = D_{22} = D_{33} = D$ the equation (4.25) reduces to

$$\frac{\partial C}{\partial t} = D \left(\frac{\partial^2 C}{\partial x^2} + \frac{\partial^2 C}{\partial y^2} + \frac{\partial^2 C}{\partial z^2} \right) = D \nabla^2 C \tag{4.26}$$

which has the exact same form as the heat equation given by equation (4.9) when the coefficients are constant and the heat source term is zero.

[†] See Appendix C

170

Boundary conditions associated with the diffusion equation are just like those of the heat equation. Dirichlet conditions require a specified concentration at the bounding surface. Neumann conditions require the flux of the diffusants be specified. A Robin condition requires a surface evaporation be proportional to the difference in concentration between the surface and surrounding medium.

Example 4-1. (Heat flow one-dimension.)

The previous derivation of the heat equation was for a general region of space. Sometimes the concepts are easier if one is not so general. Therefore, we will derive the one-dimensional heat equation for heat flow in a rod. With reference to figure 4-2 we formulate the equations describing the temperature distribution in a finite rod of length L which is insulated along its length and subject to end conditions. Imagine a thin rod made of a homogeneous material where the temperature remains constant at any cross section. Also assume the surface of the rod is insulated so heat flows only in the x-direction. We examine the law of conservation of energy for an arbitrary Δx section of the rod.

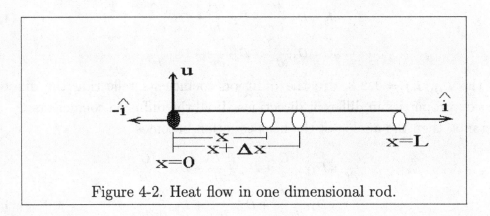

Figure 4-2. Heat flow in one dimensional rod.

We use the notations

c specific heat   (cal/gm°C)

ϱ density (gm/cm^3)

A cross sectional area   (cm^2)

$u = u(x,t)$ temperature   (°C)

t time   (sec)

Δx distance   (cm)

κ thermal conductivity   (cal/cm^2 sec °C/cm)

H heat generation source   (cal/cm^3)sec.

and assume c, ϱ and κ are constants. The element of volume of the Δx section is given by $A\Delta x$ and the conservation of energy requires the rate of change of heat energy for this volume element must equal the rate of change of heat energy flowing across the ends plus any heat energy produced inside the volume element. We assume the sides are insulated. The rate of change of heat stored in the volume element between x and $x + \Delta x$ is given by

$$H_s = \frac{\partial}{\partial t} \int_x^{x+\Delta x} c\varrho A u(x,t)\,dx = \int_x^{x+\Delta x} c\varrho A \frac{\partial u(x,t)}{\partial t}\,dx$$

where $c\varrho u = e$ represents the thermal energy density with units of (cal/cm^3). The heat loss from the left and right ends are determined from the relations

$$\frac{\partial u}{\partial n} = \operatorname{grad} u \cdot \hat{n} = \frac{\partial u}{\partial x}\hat{i} \cdot (-\hat{i}) \qquad\qquad \frac{\partial u}{\partial n} = \operatorname{grad} u \cdot \hat{n} = \frac{\partial u}{\partial x}\hat{i} \cdot (\hat{i})$$

so that

$$H_c = \kappa A \left[\frac{\partial u(x+\Delta x,t)}{\partial x} - \frac{\partial u(x,t)}{\partial x} \right].$$

The heat generated by a source within the volume element is given by

$$H_g = A \int_x^{x+\Delta x} H(x,t)\,dx.$$

We write the conservation of energy in the form

$$H_c + H_g - H_s = 0 \tag{4.27}$$

and then use the mean value theorem for integrals

$$\int_x^{x+\Delta x} f(x)\,dx = f(x+\theta\Delta x)\,\Delta x, \qquad 0 < \theta < 1$$

on the H_g and H_s terms and write

$$\kappa A \left[\frac{\partial u(x+\Delta x,t)}{\partial x} - \frac{\partial u(x,t)}{\partial x} \right] + A H(x+\theta_1\Delta x,t)\,\Delta x - c\varrho A \frac{\partial u(x+\theta_2\Delta x,t)}{\partial t}\,\Delta x = 0.$$

Now divide by $A\Delta x$ and take the limit as $\Delta x \to 0$ to obtain the one-dimensional heat equation

$$\kappa \frac{\partial^2 u}{\partial x^2} + H(x,t) = c\varrho \frac{\partial u}{\partial t}, \qquad u = u(x,t), \quad 0 < x < L \tag{4.28}$$

or

$$\frac{\partial u}{\partial t} = \alpha^2 \frac{\partial^2 u}{\partial x^2} + Q(x,t), \qquad Q = \frac{H}{c\varrho}\,(°C/\text{sec}), \qquad \alpha^2 = K = \frac{\kappa}{c\varrho}\,(\text{cm}^2/\text{sec}) \tag{4.29}$$

where $\alpha^2 = K$ is called the diffusivity of the rod material. Note that this result is a special case of our previous result.

By changing the assumptions used to derive the above heat equation one can produce various modifications of the heat equation. For example, if we remove the lateral insulation we will have heat loss from the surface of the rod. Using Newton's law of cooling, we write this heat loss as being proportional to the temperature difference between the rod temperature $u(x,t)$ and the surrounding medium temperature u_0 which is assumed constant. This lateral heat loss from the Δx volume element is written as

$$H_L = -\beta_1(u - u_0)A\Delta x$$

where $\beta_1 > 0$ is the proportionality heat transfer constant. This is an additional term which must be added to the energy balance equation (4.27). This additional term changes the one-dimensional heat equation to the form

$$\frac{\partial u}{\partial t} = \alpha^2 \frac{\partial^2 u}{\partial x^2} - \beta(u - u_0) + Q(x,t) \qquad (4.30)$$

where β (sec^{-1}) is a new heat loss rate constant obtained from $\beta_1/c\varrho$. Letting $U = u - u_0$ denote the temperature difference with the surrounding medium the one-dimensional heat equation can be written in the form

$$\frac{\partial U}{\partial t} = \alpha^2 \frac{\partial^2 U}{\partial x^2} - \beta U + Q(x,t) \qquad (4.31)$$

Another modification to the heat equation is to consider convection. If we assume the molecules within the material move with a velocity $\vec{V}$, then the Fourier law of heat conduction can be modified to read $\vec{q} = -\kappa \operatorname{grad} u + c\varrho u\vec{V}$. When $\vec{q}$ is substituted into equation (4.8) there results the heat equation with convection

$$\operatorname{div}\left(\kappa\nabla u - c\varrho u\vec{V}\right) + H = \frac{\partial}{\partial t}(c\varrho u)$$

which for c, ϱ and κ constants reduces to

$$\frac{\partial u}{\partial t} = K\left(\frac{\partial^2 u}{\partial x^2} - u\nabla \cdot \vec{V} - \vec{V} \cdot \nabla u\right) + Q(x,t) \qquad (4.32)$$

where $K = \frac{\kappa}{c\varrho}$, $Q = \frac{H}{c\varrho}$. A special case is the one-dimensional version of equation (4.32) where V is a constant representing the average velocity of the molecules

$$\frac{\partial u}{\partial t} = K\frac{\partial^2 u}{\partial x^2} - V\frac{\partial u}{\partial x} + Q(x,t). \qquad (4.33)$$

Method of separation of variables.

The method of separation of variables is a way of reducing linear partial differential equations with homogeneous boundary conditions to a study of independent ordinary differential equations. Some of the resulting ordinary differential equations together with the homogeneous boundary conditions produce Sturm-Liouville problems and associated orthogonal functions. The solution of the partial differential equation can then be represented using some kind of series of these orthogonal functions.

In the following problems and examples the method of separation of variables will be illustrated. In these examples note the initial conditions (IC) will always be the last thing we consider. Also look for Fourier series representations of the initial conditions.

Example 4-2. (Heat equation finite domain.)
Solve the following boundary value problem for the temperature distribution $u = u(x,t)$ in a one-dimensional insulated rod.

$$\begin{aligned}
\text{PDE:} \qquad & L(u) = \frac{\partial u}{\partial t} - K\frac{\partial^2 u}{\partial x^2} = 0, \quad 0 < x < L \\
\text{BC:} \qquad & B(u)|_{x=0} = u(0,t) = 0 \\
& B(u)|_{x=L} = u(L,t) = 0 \\
\text{IC:} \qquad & u(x,0) = f(x)
\end{aligned}$$
(4.34)

Solution: Here Dirichlet conditions specify the temperature at the ends of the rod and $f(x)$ is an assigned initial temperature distribution. Observe the boundary conditions are homogeneous which is necessary for the method of separation of variables to be applicable. We desire to find a nonzero solution to equation (4.34) which satisfies the PDE, the BC's and the IC. The method of separation of variables begins by assuming a product solution where the independent variables are separated. We desire a nonzero solution to equation (4.34) of the form

$$u = u(x,t) = F(x)G(t) \tag{4.35}$$

where $F(x)$ and $G(t)$ are functions to be determined. We differentiate the assumed solution

$$\frac{\partial u}{\partial t} = F(x)G'(t), \qquad \frac{\partial u}{\partial x} = F'(x)G(t), \qquad \frac{\partial^2 u}{\partial x^2} = F''(x)G(t) \tag{4.36}$$

where primes denote differentiation with respect to the argument of the function. (i.e. $F'(x) = \frac{dF}{dx}$ $F''(x) = \frac{d^2 F}{dx^2}$.) It is necessary that the assumed solution satisfy both the PDE and BC's so we substitute the derivatives from equation (4.36) into the PDE and BC's to obtain

$$F(x)\frac{dG}{dt} - K\frac{d^2 F}{dx^2}G(t) = 0 \tag{4.37}$$

and

$$u(0,t) = F(0)G(t) = 0 \quad \text{which implies} \quad F(0) = 0$$
$$u(L,t) = F(L)G(t) = 0 \quad \text{which implies} \quad F(L) = 0.$$

Here $G(t) \neq 0$ because u is to be a nonzero solution. In equation (4.37) we separate the variables and write

$$\frac{1}{KG}\frac{dG}{dt} = \frac{1}{F}\frac{d^2 F}{dx^2} = \lambda \tag{4.38}$$

where λ is a constant called the separation constant. Note x and t are assumed to be independent variables and so we could hold x fixed and vary t so that the left-hand side of equation (4.38) would have to equal a constant. Similarly, we could hold t fixed and vary x so that the right-hand side of equation (4.37) would equal a constant. The equality sign of the ratios hold only if both ratios are constants. We remark in passing that we could have placed the constant K on either the right or left-hand side of equation (4.38) . We selected the side for the constant K which did not involve the boundary conditions. That is, the boundary conditions were on the F equation and so we want to keep that equation as simple as possible. The equation (4.37) produces two ordinary differential equations

$$\frac{dG}{dt} = \lambda KG, \qquad \begin{aligned} \frac{d^2 F}{dx^2} - \lambda F &= 0 \\ F(0) = 0, \qquad F(L) &= 0 \end{aligned} \tag{4.39}$$

Observe the F equation is a Sturm-Liouville problem we have previously solved. In general, we would investigate the choices of $\lambda = \omega^2$, $\lambda = 0$ and $\lambda = -\omega^2$ of positive, zero and negative separation constants. We find only $\lambda = -\omega^2$ leads to nonzero solutions for $F = F(x)$. The eigenvalue set of values

$$\lambda = \lambda_n = -\omega_n^2 = -\frac{n^2\pi^2}{L^2} \quad \text{for} \quad n = 1, 2, 3, \ldots \tag{4.40}$$

is called the discrete spectrum of values which produces a solution set. In later sections we will find that some problems have a discrete spectrum of values while other problems have a continuous spectrum of values which produce solutions in the limit as $L \to \infty$. Associate with each eigenvalue there is a corresponding eigenfunction. These are written

$$F(x) = F_n(x) = \sin \omega_n x = \sin \left(\frac{n\pi x}{L} \right), \quad n = 1, 2, 3, \ldots \tag{4.41}$$

and further these eigenfunctions are a set of orthogonal functions over the interval $(0, L)$. For each separation constant λ_n we must also solve the equation for G that resulted from our separation of variables. This equation is

$$\frac{dG}{dt} = K\lambda G = -K\omega_n^2 G \tag{4.42}$$

and has solutions

$$G(t) = G_n(t) = e^{-K\omega_n^2 t} = e^{-\frac{Kn^2\pi^2}{L^2}t}, \quad n = 1, 2, 3, \ldots. \tag{4.43}$$

We find there is not just one solution to the given partial differential equation, but an infinite number of solutions given by

$$u(x, t) = u_n(x, t) = F_n(x)G_n(t) = \sin \left(\frac{n\pi x}{L} \right) e^{-\frac{Kn^2\pi^2}{L^2}t}, \quad n = 1, 2, 3, \ldots \tag{4.44}$$

By the principal of superposition we can multiply these solutions by arbitrary constants and add them to form more general type solutions. The more general solution is then

$$u(x, t) = \sum_{n=1}^{\infty} B_n \sin \left(\frac{n\pi x}{L} \right) e^{-\frac{Kn^2\pi^2}{L^2}t}. \tag{4.45}$$

The initial condition requires the coefficients B_n to be selected such that

$$u(x, 0) = f(x) = \sum_{n=1}^{\infty} B_n F_n(x) = \sum_{n=1}^{\infty} B_n \sin \left(\frac{n\pi x}{L} \right) \tag{4.46}$$

This is nothing more than a Fourier sine series representation for $f(x)$ with Fourier coefficients B_n. These coefficients are found from an inner product divided by a norm squared

$$B_n = \frac{(f, F_n)}{||F_n||^2} = \frac{2}{L} \int_0^L f(x) \sin \left(\frac{n\pi x}{L} \right) dx, \quad n = 1, 2, 3, \ldots. \tag{4.47}$$

The solution to the one-dimensional heat equation over the interval $(0, L)$ is thus represented

$$u(x, t) = \sum_{n=1}^{\infty} B_n \sin \left(\frac{n\pi x}{L} \right) e^{-\frac{Kn^2\pi^2}{L^2}t}$$

$$\text{where} \quad B_n = \frac{(f, F_n)}{||F_n||^2} = \frac{2}{L} \int_0^L f(x) \sin \left(\frac{n\pi x}{L} \right) dx \quad n = 1, 2, 3, \ldots. \tag{4.48}$$

Remark 1: We have set all constants in the solutions of the differential equation for F and G equal to unity. Suppose we had used arbitrary constants and written the solutions for F and G in the forms

$$F_n(x) = \beta_n \sin\left(\frac{n\pi x}{L}\right), \qquad G_n(t) = \gamma_n e^{-\frac{Kn^2\pi^2}{L^2}t}, \qquad n = 1, 2, 3, \ldots$$

where β_n and γ_n are arbitrary constants. Then when we multiplied the solutions together to form $u_n(x,t) = F_n(x)G_n(t)$ we would have to define new arbitrary constants $\delta_n = \beta_n\gamma_n$. Also when we used the principal of superposition, the multiplication of the solutions by constants would require the naming of new constants. All this renaming of new constants is avoided by letting the general constants in the solutions of the F and G equation be unity. The last step of superposition will bring back the arbitrary constants we need.

Remark 2: To emphasize the fact there is not just one solution but an infinite number of solutions for the F and G equations we have added subscripts and have written F_n and G_n for $n = 1, 2, 3, \ldots$. This requires $u = u(x,t) = F(x)G(t)$ also be adjusted in notation to $u_n = u_n(x,t)$ to show there is an infinite number of solutions. Watch for this notation change in other problems.

Remark 3: In the special case of a gold bar we find from a handbook the approximate values of thermal conductivity, specific heat and density given by $\kappa \approx 0.70$ cal/cm^2 sec C/cm, $c \approx 0.03$ cal/gm C, $\varrho \approx 19.3$ gm/cm^3. These values produce the diffusivity $K \approx 1.209$ cm^2/sec. Using this value for the diffusivity and letting $L = \pi$ we select the initial condition $f(x) = T_0 x(\pi - x)$. This gives the Fourier coefficients $B_n = \dfrac{(f, F_n)}{\| F_n \|^2} = \dfrac{2T_0}{\pi} \displaystyle\int_0^\pi x(\pi - x) \sin nx \, dx = \dfrac{4T_0}{\pi}\dfrac{1 - (-1)^n}{n^3}$ so that the solution can be represented

$$u(x,t) = \frac{4T_0}{\pi} \sum_{n=1}^{\infty} \frac{1 - (-1)^n}{n^3} \exp\left(-1.209\, n^2 t\right) \sin nx.$$

Graphs of this solution for selected fixed times are illustrated in the figure 4-3 along with a three-dimensional plot of temperature vs time and distance.

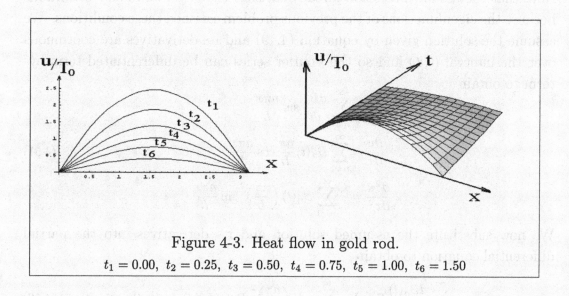

Figure 4-3. Heat flow in gold rod.

$t_1 = 0.00$, $t_2 = 0.25$, $t_3 = 0.50$, $t_4 = 0.75$, $t_5 = 1.00$, $t_6 = 1.50$

∎

Example 4-3. (**Heat equation with source term.**)
Solve the following boundary value problem for the temperature $u = u(x,t)$ in a one-dimensional rod which includes a source term.

PDE: $\quad L(u) = \dfrac{\partial u}{\partial t} - K \dfrac{\partial^2 u}{\partial x^2} = q(x,t), \qquad 0 < x < L$

BC: $\qquad\qquad u(0,t) = 0$

$\qquad\qquad\qquad u(L,t) = 0$

IC: $\qquad\qquad u(x,0) = f(x)$

Solution: This problem is the same as the previous problem except for the source term on the right-hand side of the equation. We make use of the results from the previous problem and use the same set of orthogonal functions $\{F_n(x) = \sin \frac{n\pi x}{L}\}$ over the interval $(0, L)$ and assume a solution of the form

$$u = u(x,t) = \sum_{n=1}^{\infty} B_n(t) \sin \frac{n\pi x}{L} = \sum_{n=1}^{\infty} B_n(t) F_n(x). \qquad (4.49)$$

Recall the functions $F_n(x)$ satisfy the orthogonality property

$$(F_n, F_m) = \int_0^L F_n(x) F_m(x)\, dx = ||F_n||^2 \delta_{mn}, \qquad ||F_n||^2 = \frac{L}{2}. \qquad (4.50)$$

Note that the assumed solution already satisfies the given boundary conditions because the eigenfunctions of the previous problem satisfied these conditions. We assume the solution given by equation (4.49) and its derivatives are continuous over the interval $(0, L)$ and so the Fourier series can be differentiated term-by-term to obtain

$$\frac{\partial u}{\partial t} = \sum_{n=1}^{\infty} \frac{dB_n}{dt} \sin \frac{n\pi x}{L}$$

$$\frac{\partial u}{\partial x} = \sum_{n=1}^{\infty} B_n(t) \frac{n\pi}{L} \cos \frac{n\pi x}{L} \quad (4.51)$$

$$\frac{\partial^2 u}{\partial x^2} = - \sum_{n=1}^{\infty} B_n(t) \left(\frac{n\pi}{L}\right)^2 \sin \frac{n\pi x}{L}.$$

We now substitute the assumed solution and its derivatives into the partial differential equation to obtain

$$\sum_{n=1}^{\infty} \frac{dB_n(t)}{dt} F_n(x) = \sum_{n=1}^{\infty} -K \left(\frac{n\pi}{L}\right)^2 B_n(t) F_n(x) + q(x,t). \quad (4.52)$$

In order to solve for the unknown functions $B_n(t)$, we use the orthogonality of the functions $\{F_n(x)\}$ in the following manner. We multiply both sides of the equation (4.52) by $F_m(x)\, dx$, for some fixed value of the index m, and then integrate the equation from 0 to L to obtain

$$\sum_{n=1}^{\infty} \frac{dB_n(t)}{dt} (F_n, F_m) + \sum_{n=1}^{\infty} K \left(\frac{n\pi}{L}\right)^2 B_n(t)(F_n, F_m) = \int_0^L q(x,t) F_m(x)\, dx. \quad (4.53)$$

Because of the orthogonality of the set of functions $\{F_n(x)\}$ the only nonzero terms on the left-hand side of equation (4.53) occur when the summation index n has the value of m. When $n = m$ we obtain a nonzero inner product and equation (4.53) reduces to the ordinary differential equation

$$\frac{dB_m(t)}{dt} + K \left(\frac{m\pi}{L}\right)^2 B_m(t) = q_m(t) \quad (4.54)$$

where we have defined

$$q_n(t) = \frac{2}{L} \int_0^L q(x,t) \sin \frac{n\pi x}{L}\, dx, \quad n = 1, 2, 3, \ldots \quad (4.55)$$

The equation (4.54) is a first order linear differential equation which is easily solved. Changing the index m back to n we solve this equation for $B_n(t)$. We find the resulting equation has the integrating factor $\exp\left(K\left(\frac{n\pi}{L}\right)^2 t\right)$. This means that if we multiply equation (4.54) by this integrating factor, and replace m by n, we obtain the exact differential equation

$$\frac{d}{dt}\left[B_n(t)e^{K\left(\frac{n\pi}{L}\right)^2 t}\right] = q_n(t)e^{K\left(\frac{n\pi}{L}\right)^2 t} \tag{4.56}$$

which can easily be integrated

$$\int_0^t d\left[B_n(t)e^{K\left(\frac{n\pi}{L}\right)^2 t}\right] = \int_0^t q_n(\tau)e^{K\left(\frac{n\pi}{L}\right)^2 \tau}\, d\tau \tag{4.57}$$

to obtain the solution

$$B_n(t) = B_n(0)e^{-K\left(\frac{n\pi}{L}\right)^2 t} + e^{-K\left(\frac{n\pi}{L}\right)^2 t}\int_0^t q_n(\tau)e^{K\left(\frac{n\pi}{L}\right)^2 \tau}\, d\tau. \tag{4.58}$$

Now we can use the initial condition information to solve for the coefficients $B_n(0)$. These coefficients must satisfy

$$u(x,0) = f(x) = \sum_{n=1}^{\infty} B_n(0)F_n(x) = \sum_{n=1}^{\infty} B_n(0)\sin\frac{n\pi x}{L}. \tag{4.59}$$

This says $f(x)$ is to be represented as a Fourier sine series expansion with Fourier coefficients $B_n(0)$ which represent the initial conditions needed for the differential equation given by equation (4.54). Again we make use of the orthogonality of the functions $F_n(x)$ and obtain the necessary Fourier coefficients $B_n(0)$ which we find are represented by an inner product divided by a norm squared

$$B_n(0) = \frac{(f, F_n)}{\|F_n\|^2} = \frac{2}{L}\int_0^L f(x)\sin\frac{n\pi x}{L}\, dx. \tag{4.60}$$

The solution to the original problem can now be represented in the form

$$u = u(x,t) = \sum_{n=1}^{\infty} B_n(t)\sin\frac{n\pi x}{L} = \sum_{n=1}^{\infty} B_n(t)F_n(x)$$

where $\quad B_n(t) = B_n(0)e^{-K\left(\frac{n\pi}{L}\right)^2 t} + e^{-K\left(\frac{n\pi}{L}\right)^2 t}\int_0^t q_n(\tau)e^{K\left(\frac{n\pi}{L}\right)^2 \tau}\, d\tau$

$$\tag{4.61}$$

with $\quad q_n(t) = \frac{2}{L}\int_0^L q(x,t)\sin\frac{n\pi x}{L}\, dx$

and $\quad B_n(0) = \frac{(f, F_n)}{\|F_n\|^2} = \frac{2}{L}\int_0^L f(x)\sin\frac{n\pi x}{L}\, dx.$

Example 4-4. (Nonhomogeneous boundary conditions.)

Solve the boundary value problem for the temperature $u = u(x, t)$ in a one-dimensional rod which includes a source term and in addition there are nonhomogeneous boundary conditions.

$$\text{PDE:} \quad L(u) = \frac{\partial u}{\partial t} - K\frac{\partial^2 u}{\partial x^2} = q(x, t), \qquad 0 < x < L$$

$$\text{BC:} \quad u(0, t) = A(t)$$

$$u(L, t) = B(t)$$

$$\text{IC:} \quad u(x, 0) = f(x)$$

Solution: Use superposition of solutions to break the given problem up into two problems. One problem has homogeneous boundary conditions and is exactly the same form as the previous example and the other problem is an easy to solve partial differential equation which has the desired boundary conditions. We begin by assuming a solution of the form

$$u(x, t) = w(x, t) + v(x, t) \tag{4.62}$$

and substitute this solution into the PDE, BC and IC to obtain

$$\text{PDE:} \quad \frac{\partial w}{\partial t} + \frac{\partial v}{\partial t} - K\left(\frac{\partial^2 w}{\partial x^2} + \frac{\partial^2 v}{\partial x^2}\right) = q(x, t)$$

$$\text{BC:} \quad u(0, t) = w(0, t) + v(0, t) = A(t)$$

$$u(L, t) = w(L, t) + v(L, t) = B(t) \tag{4.63}$$

$$\text{IC:} \quad u(x, 0) = w(x, 0) + v(x, 0) = f(x)$$

Let w be the function which satisfies the same type of problem as the previous example, except for a modified source term. The important property to observe is that we want w to satisfy homogeneous boundary conditions. Therefore, we require w to satisfy

$$\text{PDE:} \quad \frac{\partial w}{\partial t} - K\frac{\partial^2 w}{\partial x^2} = q(x, t) - \frac{\partial v}{\partial t}$$

$$\text{BC:} \quad w(0, t) = 0$$

$$w(L, t) = 0 \tag{4.64}$$

$$\text{IC:} \quad w(x, 0) = f(x) - v(x, 0).$$

These conditions simplify the equations (4.63) to the form

$$\text{PDE:} \qquad \frac{\partial^2 v}{\partial x^2} = 0$$

$$\text{BC:} \qquad v(0, t) = A(t) \qquad\qquad (4.65)$$

$$v(L, t) = B(t)$$

where the nonhomogeneous boundary conditions have been assigned to the $v(x, t)$ term. The resulting partial differential equation is now an easy to solve problem. Integrating with respect to x twice gives the solution

$$v = v(x, t) = c_1 x + c_2 \qquad\qquad (4.66)$$

where c_1 and c_2 are functions of t and are selected to satisfy the boundary conditions. We have

$$v(0, t) = A(t) = c_2$$

$$\text{and} \qquad v(L, t) = B(t) = c_1 L + c_2. \qquad\qquad (4.67)$$

Solving for c_1 and c_2 gives the solution

$$v = v(x, t) = [B(t) - A(t)]\frac{x}{L} + A(t). \qquad\qquad (4.68)$$

The equation (4.64) then gets the new source term

$$Q(x, t) = q(x, t) - \frac{\partial v}{\partial t} = q(x, t) - [B'(t) - A'(t)]\frac{x}{L} - A'(t) \qquad\qquad (4.69)$$

and new initial condition

$$w(x, 0) = f(x) - v(x, 0) = f(x) - [B(0) - A(0)]\frac{x}{L} - A(0) \qquad\qquad (4.70)$$

and is solved in the same way that example 4-3 was solved.

Another way of viewing the original problem is as follows. One can construct a function

$$v(x, t) = \frac{L - x}{L} A(t) + \frac{x}{L} B(t)$$

which satisfies the boundary conditions, then the function $w(x, t) = u(x, t) - v(x, t)$ defines a new function which is a solution of the partial differential equation

$$\frac{\partial w}{\partial t} = K\frac{\partial^2 w}{\partial x^2} + Q(x, t),$$

where Q is defined by equation (4.69), and w is to satisfy the homogeneous boundary conditions

$$w(0, t) = 0, \qquad w(L, t) = 0$$

together with the initial condition $w(x, 0) = u(x, 0) - v(x, 0)$. The resulting problem in w can now be solved by the method of separation of variables.

■

Example 4-5. **(General heat conduction problem.)**
Determine a solution technique for the general boundary value problem for temperature $u = u(x,t)$ along a thin rod with heat source term and nonhomogeneous boundary conditions. This type of problem is modeled as

$$PDE: \quad \varrho(x)c(x)\frac{\partial u}{\partial t} = \frac{\partial}{\partial x}\left(\kappa(x)\frac{\partial u}{\partial x}\right) + Q(x,t), \quad a \le x \le b, \ t > 0$$

$$BC: \quad \alpha_1 u(a,t) - \alpha_2 \frac{\partial u(a,t)}{\partial x} = A(t)$$

$$\beta_1 u(b,t) + \beta_2 \frac{\partial u(b,t)}{\partial x} = B(t)$$

$$IC: \quad u(x,0) = f(x)$$

where $\alpha_1, \alpha_2, \beta_1, \beta_2$ are constants and the density, specific heat and thermal conductivity are treated as functions of position.

Solution: Let $u = u(x,t) = w(x,t) + v(x,t)$ and substitute into the given partial differential equation, boundary conditions and initial condition to obtain

$$\varrho c\left(\frac{\partial w}{\partial t} + \frac{\partial v}{\partial t}\right) = \frac{\partial}{\partial x}\left(\kappa\left[\frac{\partial w}{\partial x} + \frac{\partial v}{\partial x}\right]\right) + Q(x,t)$$

$$\alpha_1\left[w(a,t) + v(a,t)\right] - \alpha_2\left[\frac{\partial w(a,t)}{\partial x} + \frac{\partial v(a,t)}{\partial x}\right] = A(t) \qquad (4.71)$$

$$\beta_1\left[w(b,t) + v(b,t)\right] + \beta_2\left[\frac{\partial w(b,t)}{\partial x} + \frac{\partial v(b,t)}{\partial x}\right] = B(t)$$

$$u(x,0) = w(x,0) + v(x,0) = f(x)$$

We break the original problem up into two separate problems so the solution to the original problem is the sum of the solutions from these two problems.

Problem 1: We assign nonhomogeneous boundary conditions to the $v(x,t)$ term. Let $v(x,t)$ satisfy the following conditions

$$\frac{\partial}{\partial x}\left(\kappa(x)\frac{\partial v}{\partial x}\right) = 0$$

$$\alpha_1 v(a,t) - \alpha_2 \frac{\partial v(a,t)}{\partial x} = A(t) \qquad (4.72)$$

$$\beta_1 v(b,t) + \beta_2 \frac{\partial v(b,t)}{\partial x} = B(t)$$

where the nonhomogeneous boundary conditions are now shifted to the v-equation. This equation is easily solved. An integration produces

$$\kappa(x)\frac{\partial v}{\partial x} = c_1(t) \quad \text{or} \quad \partial v = \frac{c_1(t)}{\kappa(x)}\partial x.$$

A second integration produces

$$v = v(x,t) = c_1(t) \int_a^x \frac{dx}{\kappa(x)} + c_2(t)$$

where $c_1(t)$ and $c_2(t)$ are arbitrary functions of t which are selected to satisfy the given boundary conditions. This requires

$$\alpha_1 c_1(t) - \alpha_2 \frac{c_1(t)}{\kappa(a)} = A(t)$$

$$\beta_1 \left[c_1(t) \int_a^b \frac{dx}{\kappa(x)} + c_2(t) \right] + \beta_2 \frac{c_1(t)}{\kappa(b)} = B(t)$$

from which one can solve for $c_1(t)$ and $c_2(t)$.

Problem 2: The selection of conditions given by equation (4.72) simplifies the equation (4.71) to a heat equation with source term and homogeneous boundary conditions given by

$$\varrho c \frac{\partial w}{\partial t} = \frac{\partial}{\partial x} \left(\kappa \frac{\partial w}{\partial x} \right) + Q(x,t) - \varrho c \frac{\partial v}{\partial t}$$

$$\alpha_1 w(a,t) - \alpha_2 \frac{\partial w(a,t)}{\partial x} = 0 \tag{4.73}$$

$$\beta_1 w(b,t) + \beta_2 \frac{\partial w(b,t)}{\partial x} = 0$$

$$w(x,0) = f(x) - v(x,0).$$

We let $q(x,t) = Q(x,t) - \varrho c \frac{\partial v}{\partial t}$ and let $h(x) = f(x) - v(x,0)$. To solve the problem 2, we first solve the homogeneous equation associated with problem 2. We solve for $w = w(x,t)$ which satisfies the homogeneous heat equation

$$\varrho c \frac{\partial w}{\partial t} = \frac{\partial}{\partial x} \left(\kappa \frac{\partial w}{\partial x} \right)$$

$$\alpha_1 w(a,t) - \alpha_2 \frac{\partial w(a,t)}{\partial x} = 0$$

$$\beta_1 w(b,t) + \beta_2 \frac{\partial w(b,t)}{\partial x} = 0$$

We can assume a solution to this equation in the form $w(x,t) = F(x)G(t)$ where the variables are separated. Upon substituting this assumed solution into the partial differential equation and boundary conditions, there results the differential equations

$$\frac{d}{dx}\left(\kappa(x)\frac{dF}{dx}\right) + \lambda^2\varrho(x)c(x)F(x) = 0$$

$$\alpha_1 F(a) - \alpha_2 F'(a) = 0, \qquad G'(t) + \lambda^2 G(t) = 0$$

$$\beta_1 F(b) + \beta_2 F'(b) = 0, \qquad G(t) = e^{-\lambda^2 t}$$

$$a < x < b$$

where we have selected $-\lambda^2$ as the separation constant. The equation for the function $F = F(x)$ is a Sturm-Liouville problem which has eigenvalues λ_n for $n = 1, 2, 3, \ldots$ and corresponding eigenfunctions $F = F_n(x)$ which are orthogonal over the interval (a, b) satisfying

$$(F_n, F_m) = \int_a^b r(x)F_n(x)F_m(x)\,dx = \begin{cases} 0, & n \neq m \\ \|F_n\|^2, & n = m \end{cases}$$

where $r(x) = \varrho(x)c(x)$ is the weight function.

We are now in a position to solve the problem 2 above. Since each $F_n(x)$ has been selected to satisfy the boundary conditions, we can assume a solution to the problem 2 in the form

$$w = w(x, t) = \sum_{n=1}^{\infty} B_n(t)F_n(x).$$

We substitute this assumed solution into the problem 2 to obtain

$$\sum_{n=1}^{\infty} \varrho(x)c(x)\frac{dB_n(t)}{dt}F_n(x) = \sum_{n=1}^{\infty} B_n(t)\frac{d}{dx}\left(\kappa(x)\frac{dF_n(x)}{dx}\right) + q(x, t)$$

which simplifies to

$$\sum_{n=1}^{\infty} r(x)F_n(x)\frac{dB_n}{dt} = \sum_{n=1}^{\infty} -B_n(t)\lambda_n^2 r(x)F_n(x) + q(x, t)$$

because each F_n is a solution of the above Sturm-Liouville problem. Now multiply both sides of this equation by $F_m(x)\,dx$ and integrate each term from a to b. The orthogonality of the eigenfunctions simplifies the result to an ordinary differential equation

$$\frac{dB_m}{dt} + \lambda_m^2 B_m = Q_m(t) \tag{4.74}$$

where

$$Q_m(t) = \frac{1}{\|F_m\|^2}\int_a^b q(x, t)F_m(x)\,dx.$$

The solution of the equation (4.74) is found to be

$$B_m(t) = B_m(0)e^{-\lambda_m^2 t} + e^{-\lambda_m^2 t} \int_0^t Q_m(\xi)e^{-\lambda_m^2 \xi}\, d\xi.$$

We can now change m back to n and determine $B_n(0)$ from the given initial conditions. We require

$$u(x,0) = \sum_{n=1}^{\infty} B_n(0)F_n(x) = h(x).$$

Here $B_n(0)$ turns out to equal the Fourier coefficients associated with the eigenfunction expansion for the representation of the function $h(x)$. The coefficients of this expansion are given by an inner product divided by a norm squared

$$B_n(0) = \frac{(h, F_n)}{\|F_n\|^2} = \frac{1}{\|F_n\|^2}\int_a^b r(x)h(x)F_n(x)\, dx.$$

By adding the solutions from problems 1 and 2 we obtain the solution to the original problem.

$\blacksquare$

Example 4-6. (Steady state heat equation.)

Solve the steady state heat equation

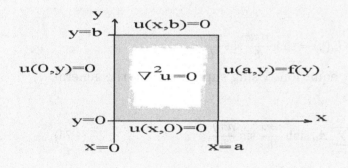

$$\nabla^2 u = \frac{\partial^2 u}{\partial x^2} + \frac{\partial^2 u}{\partial y^2} = 0$$

over the rectangular region $0 < x < a$, $0 < y < b$ subject to the Dirichlet boundary conditions

$$u(0,y) = 0, \quad u(a,y) = f(y)$$
$$u(x,0) = 0, \quad u(x,b) = 0$$

Solution: Assume a solution $u = u(x,y) = F(x)G(y)$ and substitute this solution into the partial differential equation to obtain

$$F''(x)G(y) + F(x)G''(y) = 0.$$

Also substitute the assumed solution into the homogeneous boundary conditions to obtain

$$F(0)G(y) = 0, \qquad F(x)G(0) = 0, \qquad F(x)G(b) = 0.$$

We separate the variables and obtain

$$\frac{F''(x)}{F(x)} = -\frac{G''(y)}{G(y)} = \lambda$$

where λ is a separation constant. This leads to two ordinary differential equations with boundary conditions

$$F''(x) - \lambda F(x) = 0, \qquad\qquad G''(y) + \lambda G(y) = 0$$

$$0 < x < a, \qquad\qquad\qquad 0 < y < b$$

$$F(0) = 0, \qquad\qquad G(0) = 0, \qquad G(b) = 0$$

The differential equation for $G = G(y)$ is a Sturm-Liouville problem. We let $\lambda = -\omega^2$, $\lambda = 0$, $\lambda = \omega^2$ and find the eigenvalues $\lambda = \lambda_n = \omega_n^2 = \frac{n^2\pi^2}{b^2}$ for $n = 1, 2, \ldots$ and corresponding eigenfunctions $G = G_n(y) = \sin\frac{n\pi y}{b}$ for $n = 1, 2, \ldots$. For each eigenvalue λ_n the general solution for $F = F(x)$ can be represented

$$F = F_n(x) = A \sinh\frac{n\pi x}{b} + B \cosh\frac{n\pi x}{b}.$$

The condition $F(0) = 0$ requires $B = 0$. For the simplest form of the solution we let $A = 1$. We thus obtain the solutions

$$G = G_n(y) = \sin\frac{n\pi y}{b}, \qquad F = F_n(x) = \sinh\frac{n\pi x}{b}$$

and therefore for each $n = 1, 2, \ldots$

$$u_n(x, y) = F_n(x)G_n(y) = \sinh\frac{n\pi x}{b} \sin\frac{n\pi y}{b}$$

are solutions of the steady state heat equation. Using superposition, the general solution can be represented

$$u = u(x, y) = \sum_{n=1}^{\infty} A_n \sinh\frac{n\pi x}{b} \sin\frac{n\pi y}{b} \tag{4.75}$$

where the constants A_n are selected to satisfy the nonhomogeneous boundary condition

$$u(a, y) = f(y) = \sum_{n=1}^{\infty} A_n \sinh\frac{n\pi a}{b} \sin\frac{n\pi y}{b}. \tag{4.76}$$

The equation (4.76) is a Fourier sine series representation of the function $f(y)$ with Fourier coefficients $A_n \sinh\frac{n\pi a}{b}$. These Fourier coefficients are given by an inner product divided by a norm squared or

$$A_n \sinh\frac{n\pi a}{b} = \frac{(f, \sin\frac{n\pi y}{b})}{||\sin\frac{n\pi y}{b}||^2} = \frac{2}{b}\int_0^b f(y) \sin\frac{n\pi y}{b}\, dy$$

or

$$A_n = \frac{2}{b \sinh \frac{n\pi a}{b}} \int_0^b f(y) \sin \frac{n\pi y}{b}\, dy.$$

By shifting terms, the solution (4.75) can be expressed in the form

$$u = u(x,y) = \sum_{n=1}^{\infty} \frac{2a_n}{b} \frac{\sinh \frac{n\pi x}{b}}{\sinh \frac{n\pi a}{b}} \sin \frac{n\pi y}{b}$$

where

$$a_n = \int_0^b f(y) \sin \frac{n\pi y}{b}\, dy.$$

■

Example 4-7. (Cylindrical coordinates.)
Consider the Dirichlet problem for temperature in a long cylindrical rod with circular cross section. We assume the temperature $u = u(r,t)$ is a function of the radial distance and time.

$$\text{PDE:} \quad \frac{\partial u}{\partial t} = K\left(\frac{\partial^2 u}{\partial r^2} + \frac{1}{r}\frac{\partial u}{\partial r}\right), \quad 0 < r < b$$
$$\text{BC:} \quad u(b,t) = 0 \qquad\qquad\qquad (4.77)$$
$$\text{IC:} \quad u(r,0) = f(r)$$

Solution: Here the boundary condition is homogeneous and so we assume a solution

$$u = u(r,t) = F(r)G(t)$$

and substitute into both the PDE and BC to obtain

$$F(r)G'(t) = K\left(F''(r)G(t) + \frac{1}{r}F'(r)G(t)\right)$$
$$F(b)G(t) = 0$$

The boundary condition implies $F(b) = 0$ and separation of the variables gives

$$\frac{G'(t)}{KG(t)} = \frac{F''(r) + \frac{1}{r}F'(r)}{F(r)} = \lambda_1$$

where λ_1 is a separation constant. This results in two ordinary differential equations

$$G'(t) - \lambda_1 KG(t) = 0 \qquad \text{and} \qquad F''(r) + \frac{1}{r}F'(r) - \lambda_1 F(r) = 0, \quad 0 < r < b$$
$$F(b) = 0.$$

The second equation for $F = F(r)$ is a singular Sturm-Liouville problem. We examine the cases $\lambda_1 = \omega^2$, $\lambda_1 = 0$ and find that no solution exists. The case where $\lambda_1 = -\omega^2$ results in Bessel's differential equation of order zero. The general solution of this Bessel differential equation is given by

$$F = F(r) = C_1 J_0(\omega r) + C_2 Y_0(\omega r), \quad \text{for} \quad 0 \leq r \leq b, \quad C_1, C_2 \text{ constants.}$$

Note that bounded solutions are always implied. To obtain a bounded solution we set $C_2 = 0$ and for convenience we set $C_1 = 1$. This gives the bounded solution on the interval $0 < r < b$ as

$$F = F(r) = J_0(\omega r)$$

where ω is selected such that the boundary condition is satisfied. This requires

$$F(\omega b) = J_0(\omega b) = 0 \quad \text{or} \quad \omega = \omega_i = \frac{\xi_{0i}}{b} \quad i = 1, 2, 3, \ldots$$

where ξ_{0i} is the ith zero of the zeroth order Bessel function of the first kind. The eigenvalues are $\lambda_i = -\omega_i^2$ and the eigenfunctions are

$$F_i = F_i(r) = J_0(\omega_i r) = J_0(\xi_{0i}\frac{r}{b}) \quad i = 1, 2, 3, \ldots$$

These eigenfunctions are orthogonal over the interval $(0, b)$ with respect to the weight function r. For each value of the index i we must solve for $G = G(t)$. This gives the solution set

$$G_i = G_i(t) = e^{-\omega_i^2 K t} \quad \text{for} \quad i = 1, 2, 3, \ldots$$

so that the original partial differential equation has solutions of the form

$$u_i = u_i(r, t) = F_i(r)G_i(t) = J_0(\omega_i r)e^{-\omega_i^2 K t}$$

for $i = 1, 2, 3, \ldots$. By superposition, the general solution can be represented

$$u(r, t) = \sum_{i=1}^{\infty} a_i J_0(\omega_i r)e^{-\omega_i^2 K t}.$$

The initial condition requires $f(r)$ to be represented by a Fourier-Bessel series.

$$u(r, 0) = f(r) = \sum_{i=1}^{\infty} a_i J_0(\omega_i r).$$

We make use of the orthogonality property of the eigenfunctions and find the coefficients a_i are given by an inner product divided by a norm squared or

$$a_i = \frac{(f, J_0(\omega_i r))}{||J_0(\omega_i r)||^2} = \frac{2}{b^2 J_1^2(\omega_i b)} \int_0^b r f(r) J_0(\omega_i r)\, dr.$$

This gives the solution

$$u = u(r, t) = \frac{2}{b^2} \sum_{i=1}^{\infty} \frac{J_0(\omega_i r)}{J_1^2(\omega_i b)} e^{-\omega_i^2 K t} \int_0^b \eta f(\eta) J_0(\omega_i \eta)\, d\eta$$

where η is a dummy variable of integration.

$\blacksquare$

Example 4-8. (Spherical coordinates.)

Solve the Dirichlet problem for the steady state temperature $u = u(\rho, \theta)$ in spherical coordinates where there is symmetry with respect to ϕ.

PDE: $\quad \rho^2 \dfrac{\partial^2 u}{\partial \rho^2} + 2\rho \dfrac{\partial u}{\partial \rho} + \dfrac{\partial^2 u}{\partial \theta^2} + \cot\theta \dfrac{\partial u}{\partial \theta} = 0, \quad 0 \le \rho \le b, \quad 0 \le \theta \le \pi$

BC: $\qquad\qquad\qquad\qquad\qquad u(b, \theta) = f(\theta)$

Solution: Using separation of variables we assume a product solution of the form
$u = u(\rho, \theta) = F(\rho)G(\theta)$. This solution is substituted into the partial differential equation to obtain

$$\rho^2 F''(\rho)G(\theta) + 2\rho F'(\rho)G(\theta) + F(\rho)G''(\theta) + \cot\theta F(\rho)G'(\theta) = 0.$$

Now divide each term by $F(\rho)G(\theta)$ and separate the variables to obtain

$$\frac{\rho^2 F'' + 2\rho F'}{F} = -\frac{(G'' + \cot\theta G')}{G} = \lambda$$

where λ is a separation constant. This produces two ordinary differential equations

$$\rho^2 \frac{d^2 F}{d\rho^2} + 2\rho \frac{df}{d\rho} - \lambda F = 0 \qquad \text{and} \qquad \sin\theta \frac{d^2 G}{d\theta^2} + \cos\theta \frac{dG}{d\theta} + \lambda \sin\theta = 0$$

$$0 \le \rho \le b \qquad\qquad\qquad\qquad 0 \le \theta \le \pi$$

Recall the equation (2.69) and note the equation for $G = G(\theta)$ is a singular Sturm-Liouville problem with eigenvalues $\lambda = \lambda_n = n(n+1)$ and eigenfunctions given by $G_n(\theta) = P_n(\cos\theta)$ for $n = 0, 1, 2, 3, \ldots$. The differential equation for F, with parameter $\lambda = \lambda_n = n(n+1)$, is a Cauchy-Euler equation with the solution $F = F_n(\theta) = c_1\rho^n + c_2\rho^{-(n+1)}$. In order to have bounded solutions in the region $0 \le \rho \le b$ we require the solution for F to be a bounded solution and continuous for $\rho = 0$. This requires we set $c_2 = 0$ in the solution for the F-equation. Note boundedness is always implied in our boundary value problems as unbounded solutions only occur in physical problems under special resonance situations. By superposition we construct the more general solution

$$u = u(\rho, \theta) = \sum_{n=0}^{\infty} A_n \rho^n P_n(\cos\theta), \qquad 0 < \theta < \pi \tag{4.78}$$

where A_n are constants which are determined from the boundary condition

$$u(b, \theta) = f(\theta) = \sum_{n=0}^{\infty} A_n b^n P_n(\cos\theta), \qquad 0 < \theta < \pi. \tag{4.79}$$

The boundary condition requires $f(\theta)$ be represented as a Fourier-Legendre series. We use the orthogonality property of the Legendre polynomials, (see equation (2.70)), and find $A_n b^n$ are the Fourier coefficients determined from an inner product divided by a norm squared or

$$A_m b^m = \frac{(f(\theta), P_m(\cos\theta))}{\|P_m(\cos\theta)\|^2} \quad \text{or} \quad A_m = \frac{2m+1}{2b^m} \int_0^\pi f(\theta) P_m(\cos\theta) \sin\theta \, d\theta \tag{4.80}$$

for $m = 0, 1, 2, \ldots$. By changing m to n we find the solution can be represented in the form

$$u = u(\rho, \theta) = \sum_{n=0}^{\infty} C_n \left(\frac{\rho}{b}\right)^n P_n(\cos\theta)$$

$$\text{where} \qquad C_n = \frac{2n+1}{2} \int_0^\pi f(\theta) P_n(\cos\theta) \sin\theta \, d\theta \tag{4.81}$$

■

Heat equation in higher dimensions

In this section we shall examine solutions to the 2-dimensional and 3-dimensional heat equation.

$$\frac{\partial u}{\partial t} = K\nabla^2 u + q. \tag{4.82}$$

We will use the method of separation of variables and assume that the temperature $u = u(x, y, t)$ is expressed in terms of two space variables and a time variable.

We begin by examining the boundary value problem associated with the source free version of the heat equation over a rectangular region.

PDE:
$$\frac{\partial u}{\partial t} = K\left(\frac{\partial^2 u}{\partial x^2} + \frac{\partial^2 u}{\partial y^2}\right), \quad 0 < x < a, \quad 0 < y < b$$

BC:
$$u(0, y, t) = 0, \quad u(a, y, t) = 0 \quad \text{Dirichlet conditions}$$
$$u(x, 0, t) = 0, \quad u(x, b, t) = 0$$

IC:
$$u(x, y, 0) = f(x, y)$$

(4.83)

which represents the temperature distribution in a rectangle with given side conditions and a given initial temperature distribution.

Solution method 1: We can separate the variables one at a time. If we assume a solution $u = \mathcal{F}(x, y)T(t)$ and substitute into the PDE and BC's we obtain

$$\mathcal{F}T' = K\left(\frac{\partial^2 \mathcal{F}}{\partial x^2} + \frac{\partial^2 \mathcal{F}}{\partial y^2}\right)T \tag{4.84}$$

with boundary conditions

$$\mathcal{F}(0, y)T(t) = 0, \qquad \mathcal{F}(a, y)T(t) = 0$$
$$\mathcal{F}(x, 0)T(t) = 0, \qquad \mathcal{F}(x, b)T(t) = 0.$$

We assume x, y and t are independent variables so we can separate the variables to obtain

$$\frac{T'}{KT} = \frac{1}{\mathcal{F}}\left(\frac{\partial^2 \mathcal{F}}{\partial x^2} + \frac{\partial^2 \mathcal{F}}{\partial y^2}\right) = \lambda_1 \tag{4.85}$$

where λ_1 is a separation constant. For a nonzero solution we require

$$\mathcal{F}(0, y) = 0, \qquad \mathcal{F}(a, y) = 0$$
$$\mathcal{F}(x, 0) = 0, \qquad \mathcal{F}(x, b) = 0. \tag{4.86}$$

From the equation (4.85) we obtain the two equations

$$\frac{dT}{dt} - \lambda_1 KT = 0, \qquad \frac{\partial^2 \mathcal{F}}{\partial x^2} + \frac{\partial^2 \mathcal{F}}{\partial y^2} - \lambda_1 \mathcal{F} = 0. \tag{4.87}$$

We now separate variables in the equation for $\mathcal{F}$ by assuming a solution $\mathcal{F} = F(x)G(y)$ and substitute into the PDE and BC's to obtain

$$F''G + FG'' - \lambda_1 FG = 0 \tag{4.88}$$

and

$$\mathcal{F}(0,y) = F(0)G(y) = 0, \qquad \mathcal{F}(a,y) = F(a)G(y) = 0$$
$$\mathcal{F}(x,0) = F(x)G(0) = 0, \qquad \mathcal{F}(x,b) = F(x)G(b) = 0. \tag{4.89}$$

We separate the variables in equation (4.88) and obtain

$$\frac{F''}{F} = \lambda_1 - \frac{G''}{G} = \lambda_2 \tag{4.90}$$

where λ_2 is a second separation constant. The equations (4.89) and (4.90) produce the Sturm-Liouville problems

$$F'' - \lambda_2 F = 0, \quad 0 < x < a \qquad G'' + (\lambda_2 - \lambda_1)G = 0, \quad 0 < y < b$$
$$F(0) = 0 \qquad\qquad\qquad G(0) = 0 \tag{4.91}$$
$$F(a) = 0 \qquad\qquad\qquad G(b) = 0$$

The separation constants λ_1 and λ_2 are just constants and so we can put them in any form we desire. For simplicity of the boundary value problems we select $\lambda_2 = -\omega^2$ and $\lambda_2 - \lambda_1 = \mu^2$. The equations (4.91) and (4.87) then become

$$F'' + \omega^2 F = 0 \qquad G'' + \mu^2 G = 0$$
$$F(0) = 0 \qquad\qquad G(0) = 0$$
$$\qquad\qquad\qquad\qquad\qquad\qquad T' + (\omega^2 + \mu^2)T = 0 \tag{4.92}$$
$$F(L) = 0 \qquad\qquad G(b) = 0$$
$$0 < x < a \qquad\qquad 0 < y < b$$

which represents a Sturm-Liouville problem in the x-direction, another Sturm-Liouville problem in the y-direction and a first order differential equation for $T = T(t)$. There is not just one solution for $F = F(x)$ but a whole family of orthogonal solutions with eigenvalues

$$\omega = \omega_n = \frac{n\pi}{a}, \qquad n = 1, 2, 3, \ldots$$

and eigenfunctions

$$F = F_n(x) = \sin\frac{n\pi x}{a}, \qquad n = 1, 2, 3, \ldots$$

Similarly in the y-direction we have the eigenvalues

$$\mu = \mu_m = \frac{m\pi}{b}, \qquad m = 1, 2, 3, \ldots$$

and eigenfunctions

$$G = G_m(y) = \sin\frac{m\pi y}{b}, \qquad m = 1, 2, 3, \ldots$$

This implies there is not one-solution for $T = T(t)$ but a double family of solutions

$$T = T_{nm}(t) = \exp\left[-\pi^2 Kt\left(\frac{n^2}{a^2} + \frac{m^2}{b^2}\right)\right]. \qquad (4.93)$$

Therefore, there are solutions to equation (4.83) of the form

$$u_{nm} = u_{nm}(x, y, t) = F_n(x)G_m(y)T_{nm}(t).$$

By superposition we write the general solution as

$$u(x, y, t) = \sum_{n=1}^{\infty}\sum_{m=1}^{\infty} b_{nm}\sin\frac{n\pi x}{a}\sin\frac{m\pi y}{b}\exp\left[-\pi^2 Kt\left(\frac{n^2}{a^2} + \frac{m^2}{b^2}\right)\right] \qquad (4.94)$$

where b_{nm} are constants selected such that the IC is satisfied. This requires

$$u(x, y, 0) = f(x, y) = \sum_{n=1}^{\infty}\sum_{m=1}^{\infty} b_{nm}\sin\frac{n\pi x}{a}\sin\frac{m\pi y}{b} \qquad (4.95)$$

or $f(x, y)$ is to be represented by a double Fourier sine series. We use the inner product divided by a norm squared approach twice which gives the Fourier coefficients

$$b_{nm} = \frac{((f, F_n), G_m)}{\|F_n\|^2\,\|G_m\|^2} = \frac{4}{ab}\int_0^a\int_0^b f(\xi, \eta)\sin\frac{n\pi\xi}{a}\sin\frac{m\pi\eta}{b}\,d\xi d\eta. \qquad (4.96)$$

The equations (4.94) and (4.96) constitute the double series form of the solution to equation (4.83).

Solution method 2: Instead of separating the variables one at a time we can start by assuming a solution of the form

$$u(x, y, t) = F(x)G(y)T(t) \qquad (4.97)$$

and substitute into the PDE and BC's to obtain

$$FGT' = K(F''GT + FG''T). \qquad (4.98)$$

We separate the variables and obtain

$$\frac{T'}{KT} = \frac{F''}{F} + \frac{G''}{G} = \lambda_1 \qquad (4.99)$$

or

$$T' - \lambda_1 KT = 0, \qquad \frac{F''}{F} + \frac{G''}{G} = \lambda_1. \qquad (4.100)$$

We separate the variables a second time and obtain

$$T' - \lambda_1 KT = 0, \qquad \frac{F''}{F} = \lambda_1 - \frac{G''}{G} = \lambda_2. \qquad (4.101)$$

Substituting the assumed solution into the BC's give the same BC's for F,G as in solution method 1. The final results from the equations (4.101) are

$$T' - \lambda_1 KT = 0, \qquad
\begin{array}{ll}
F'' - \lambda_2 F = 0, & G'' + (\lambda_2 - \lambda_1)G = 0 \\
F(0) = 0, & G(0) = 0 \\
F(a) = 0, & G(b) = 0 \\
0 < x < a, & 0 < y < b
\end{array}
\qquad (4.102)$$

which are the same equations obtain by the previous solution method. This solution method now continues as discussed in the earlier solution method 1.

These separation of variable concepts can be applied to higher dimensions. Consider the homogeneous heat equation in 3-dimensions for the temperature $u = u(x, y, z, t)$ throughout a parallelepiped having sides of lengths a,b,c. The temperature distribution is determined from the heat equation boundary value problem

PDE: $\qquad \dfrac{\partial u}{\partial t} = k\left(\dfrac{\partial^2 u}{\partial x^2} + \dfrac{\partial^2 u}{\partial y^2} + \dfrac{\partial^2 u}{\partial z^2} \right), \quad 0 < x < a, \quad 0 < y < b, \quad 0 < z < c$

BC:
$\qquad u(0, y, z, t) = 0, \quad u(a, y, z, t) = 0 \qquad$ Dirichlet conditions

$\qquad u(x, 0, z, t) = 0, \quad u(x, b, z, t) = 0$

$\qquad u(x, y, 0, t) = 0, \quad u(x, y, c, t) = 0$

IC: $\qquad u(x, y, z, 0) = f(x, y, z)$

$$(4.103)$$

which represents the temperature in a parallelepiped where the sides are held at zero temperature after the parallelepiped has been given an initial temperature distribution. We assume a product solution

$$u = u(x, y, z, t) = F(x)G(y)H(z)T(t) \qquad (4.104)$$

and substitute this assumed solution into the PDE and BC's to obtain

$$FGHT' = K(F''GHT + FG''HT + FGH''T) \qquad (4.105)$$

and

$$F(0)G(y)H(z)T(t) = 0, \quad F(x)G(0)H(z)T(t) = 0, \quad F(x)G(y)H(0)T(t) = 0$$
$$F(a)G(y)H(z)T(t) = 0, \quad F(x)G(b)H(z)T(t) = 0, \quad F(x)G(y)H(c)T(t) = 0 \qquad (4.106)$$

We assume x, y, z, t are independent variables and separate the variables in equation (4.105) by dividing each term by $KFGHT$ and find

$$\frac{T'}{KT} = \frac{F''}{F} + \frac{G''}{G} + \frac{H''}{H} = \lambda_1. \qquad (4.107)$$

This gives two equations

$$T' - \lambda_1 KT = 0 \quad \text{and} \quad \frac{F''}{F} + \frac{G''}{G} + \frac{H''}{H} = \lambda_1. \qquad (4.108)$$

We separate the variables in the second equation and find

$$\frac{F''}{F} + \frac{G''}{G} = \lambda_1 - \frac{H''}{H} = \lambda_2 \qquad (4.109)$$

or

$$\frac{F''}{F} + \frac{G''}{G} = \lambda_2, \qquad H'' + (\lambda_2 - \lambda_1)H = 0. \qquad (4.110)$$

A final separation of variables for the F and G equation gives

$$\frac{F''}{F} = \lambda_2 - \frac{G''}{G} = \lambda_3$$

or

$$F'' - \lambda_3 F = 0, \qquad G'' + (\lambda_3 - \lambda_2)G = 0 \qquad (4.111)$$

where λ_1, λ_2 and λ_3 are separation constants and can be selected to have any convenient form. The boundary conditions are obtained from the equations (4.106) where we impose the condition that the assumed solution, given by equation (4.104), is to be a nonzero solution. The equations (4.108), (4.109), (4.111) and implied boundary conditions from equations (4.106) produce the Sturm-Liouville problems

$$\begin{array}{lll}
F'' - \lambda_3 F = 0, & G'' + (\lambda_3 - \lambda_2)G = 0, & H'' + (\lambda_2 - \lambda_1)H = 0 \\
F(0) = 0, & G(0) = 0, & H(0) = 0 \\
F(a) = 0, & G(b) = 0, & H(c) = 0 \\
0 < x < a, & 0 < y < b, & 0 < z < c
\end{array} \qquad (4.112)$$

together with the time varying function

$$T' - \lambda_1 KT = 0. \tag{4.113}$$

We select the separation constants in such a way that

$$\lambda_3 = -\omega^2, \qquad \lambda_3 - \lambda_2 = \mu^2, \qquad \lambda_2 - \lambda_1 = \nu^2 \tag{4.114}$$

which gives the eigenvalues and eigenfunction solutions

$$\omega = \omega_n = \frac{n\pi}{a}, \qquad \mu = \mu_m = \frac{m\pi}{b}, \qquad \nu = \nu_\ell = \frac{\ell\pi}{c}$$
$$F = F_n(x) = \sin\frac{n\pi x}{a}, \qquad G = G_m(y) = \sin\frac{m\pi y}{b}, \qquad H = H_\ell(z) = \sin\frac{\ell\pi z}{c} \tag{4.115}$$

for $n = 1, 2, 3, \ldots$, $m = 1, 2, 3, \ldots$ and $\ell = 1, 2, 3, \ldots$. This gives multiple combinations of solutions for $T = T(t) = T_{\ell mn}(t)$ where

$$\frac{dT_{\ell mn}}{dt} + \pi^2 K \left(\frac{n^2}{a^2} + \frac{m^2}{b^2} + \frac{\ell^2}{c^2}\right) T_{\ell mn} = 0 \tag{4.116}$$

or

$$T_{\ell mn} = T_{\ell mn}(t) = \exp\left[-\pi^2 Kt \left(\frac{n^2}{a^2} + \frac{m^2}{b^2} + \frac{\ell^2}{c^2}\right)\right]. \tag{4.117}$$

Using superposition of the previous results we find the 3-dimensional homogeneous heat equation has the general solution

$$u = u(x, y, z, t) = \sum_{n=1}^{\infty} \sum_{m=1}^{\infty} \sum_{\ell=1}^{\infty} b_{\ell mn} F_n(x) G_m(y) H_\ell(z) T_{\ell mn}(t) \tag{4.118}$$

with coefficients $b_{\ell mn}$ selected to satisfy the IC

$$u(x, y, z, 0) = f(x, y, z) = \sum_{n=1}^{\infty} \sum_{m=1}^{\infty} \sum_{\ell=1}^{\infty} b_{\ell mn} F_n(x) G_m(y) H_\ell(z) \tag{4.119}$$

This is nothing more than a triple Fourier sine series representation of the function $f(x, y, z)$ over the region $0 < x < a$, $0 < y < b$, $0 < z < c$. We find the Fourier coefficients by taking inner products divided by a norm squared for each of the directions x, y and z and find

$$b_{\ell mn} = \frac{(((f, F_n), G_m), H_\ell)}{\|F_n\|^2 \|G_m\|^2 \|H_\ell\|^2}. \tag{4.120}$$

This gives the solution

$$u = \sum_{n=1}^{\infty}\sum_{m=1}^{\infty}\sum_{\ell=1}^{\infty} b_{\ell mn} \sin\frac{n\pi x}{a}\sin\frac{m\pi y}{b}\sin\frac{\ell\pi z}{c}\exp\left[-\pi^2 Kt\left(\frac{n^2}{a^2}+\frac{m^2}{b^2}+\frac{\ell^2}{c^2}\right)\right] \quad (4.121)$$

where

$$b_{\ell mn} = \frac{8}{abc}\int_0^a\int_0^b\int_0^c f(\xi,\eta,\zeta)\sin\frac{n\pi\xi}{a}\sin\frac{m\pi\eta}{b}\sin\frac{\ell\pi\zeta}{c}\,d\xi d\eta d\zeta. \quad (4.122)$$

Continuous spectrum

The principle of superposition states that if $u_1, u_2, \ldots$ are solutions of a linear homogeneous partial differential equation, then the linear combination

$$u = \sum_{i=1}^{\infty} c_i u_i = c_1 u_1 + c_2 u_2 + \cdots,$$

for arbitrary constants $c_1, c_2, \ldots$, is also a solution. A generalization of this concept is that if $u_1(x,t;\omega), u_2(x,t;\omega), \ldots$ are solutions of a linear homogeneous partial differential equation which contain a parameter ω, then

$$u = u(x,t) = \int_0^{\infty} [c_1(\omega)u_1(x,t;\omega) + c_2(\omega)u_2(x,t;\omega) + \cdots]\,d\omega,$$

for arbitrary functions $c_1(\omega), c_2(\omega), \ldots$, is also a solution. Here we have used a continuous sum over all values $\omega \geq 0$. This is a more general principle of superposition which can be utilized for eigenvalue problems having a continuous spectrum of eigenvalues.

In contrast to separation of variable methods leading to Sturm-Liouville problems and discrete spectrum of eigenvalues and superposition, we consider the following example which leads to a parameter in the solution giving rise to a continuous spectrum of eigenvalues and use of the more general superposition principle involving integration for the representation of the solution.

Example 4-9. (Heat equation infinite domain.)
Solve the boundary value problem for the temperature $u = u(x,t)$ in an infinite thin bar.

PDE: $\quad \frac{\partial u}{\partial t} = K\frac{\partial^2 u}{\partial x^2}, \quad -\infty < x < \infty, \quad t > 0$

IC: $\quad u(x,0) = f(x)$ $\quad (4.123)$

Solution: For problems with infinite domains there is always some type of implied boundary condition. For this problem we assume bounded solutions are required. This condition is expressed $|u(x,t)| < M$ for all x, where M is some constant. We assume a solution

$$u = u(x,t) = F(x)T(t) \tag{4.124}$$

and substitute into the PDE. We obtain

$$FT' = KF''T$$

and then separate the variables to get

$$\frac{F''}{F} = \frac{T'}{KT} = \lambda.$$

This produces two ordinary differential equations

$$F''(x) - \lambda F(x) = 0, \quad -\infty < x < \infty, \qquad T'(t) - \lambda KT(t) = 0 \tag{4.125}$$

To obtain bounded solutions we let $\lambda = -\omega^2$ with $\omega \geq 0$ and obtain the solutions

$$F = F(x) = A\cos\omega x + B\sin\omega x, \qquad T = T(t) = e^{-\omega^2 Kt}. \tag{4.126}$$

Here there are solutions for all values of $\omega > 0$ or a continuous spectrum of values for ω exist in contrast to the discrete spectrum of values considered earlier. Therefore, there exists solutions of the form

$$u = u(x,t;\omega) = F(x)T(t) = [A\cos\omega x + B\sin\omega x]\, e^{-\omega^2 Kt} \tag{4.127}$$

involving a parameter ω. Using superposition of solutions involving a parameter we can let the coefficients A and B be arbitrary functions of ω and obtain the more general solution

$$u = u(x,t) = \int_0^\infty [A(\omega)\cos\omega x + B(\omega)\sin\omega x]\, e^{-\omega^2 Kt}\, d\omega. \tag{4.128}$$

The initial condition requires

$$u(x,0) = f(x) = \int_0^\infty [A(\omega)\cos\omega x + B(\omega)\sin\omega x]\, d\omega \tag{4.129}$$

which requires $f(x)$ be represented as a Fourier integral. The coefficients are therefore given by

$$A(\omega) = \frac{1}{\pi} \int_{-\infty}^{\infty} f(\xi) \cos \omega x \, d\xi$$

$$B(\omega) = \frac{1}{\pi} \int_{-\infty}^{\infty} f(\xi) \sin \omega \xi \, d\xi. \tag{4.130}$$

The solution is then represented in the form

$$u = u(x,t) = \frac{1}{\pi} \int_{0}^{\infty} \left[\int_{-\infty}^{\infty} f(\xi) \left(\cos \omega \xi \cos \omega x + \sin \omega \xi \sin \omega x \right) d\xi \right] e^{-\omega^2 K t} d\omega.$$

We simplify the above integral and interchange the order of integration and write the solution in the form

$$u = u(x,t) = \frac{1}{\pi} \int_{-\infty}^{\infty} f(\xi) \left[\int_{0}^{\infty} \cos \omega (\xi - x) e^{-\omega^2 K t} \, d\omega \right] d\xi. \tag{4.131}$$

From the table of integrals in Appendix D we make use of the result

$$\int_{0}^{\infty} \cos \omega (\xi - x) e^{-\omega^2 K t} \, d\omega = \sqrt{\frac{\pi}{4Kt}} e^{-\frac{(\xi-x)^2}{4Kt}}, \qquad t > 0. \tag{4.132}$$

The solution given by equation (4.131) can therefore be written in the form

$$u = u(x,t) = \frac{1}{\sqrt{4K\pi t}} \int_{-\infty}^{\infty} f(\xi) e^{-\frac{(\xi-x)^2}{4Kt}} \, d\xi \tag{4.133}$$

■

Example 4-10. **(Heat equation infinite domain.)**
Solve for the temperature distribution $u = u(x,t)$ in a long thin rod where

$$\begin{aligned}
\text{PDE:} \quad & \frac{\partial u}{\partial t} = \alpha^2 \frac{\partial^2 u}{\partial x^2}, \quad 0 < x < \infty, \quad t > 0 \\
\text{BC:} \quad & u(0,t) = 0, \quad t > 0 \\
\text{IC:} \quad & u(x,0) = f(x), \quad 0 < x < \infty
\end{aligned} \tag{4.134}$$

Solution: Here we have replaced the diffusivity K by α^2 because it simplifies the form for the final solution. We use separation of variables and assume a solution of the form $u(x,t) = F(x)G(t)$ as in the previous problem. This leads to the ordinary differential equations

$$F''(x) - \lambda F(x) = 0, \quad 0 < x < \infty \qquad \text{and} \qquad T'(t) - \lambda \alpha^2 T(t) = 0.$$

For $\lambda = -\omega^2$ we find the solution $F(x) = A\cos\omega x + B\sin\omega x$ must satisfy the end condition $F(0) = A = 0$. This gives the solution

$$u = u(x,t;\omega) = B\sin(\omega x)\, e^{-\omega^2\alpha^2 t} \tag{4.135}$$

which contains a parameter $\omega > 0$. By superposition of solutions with a parameter we can assume B is also a function of ω and write the general solution in the form

$$u = u(x,t) = \int_0^\infty B(\omega)\sin(\omega x)\, e^{-\omega^2\alpha^2 t}\, d\omega. \tag{4.136}$$

The initial condition requires

$$u(x,0) = f(x) = \int_0^\infty B(\omega)\sin(\omega x)\, d\omega \tag{4.137}$$

which is a Fourier sine integral representation of the initial condition. We find the coefficient

$$B(\omega) = \frac{2}{\pi}\int_0^\infty f(\xi)\sin\omega\xi d\xi. \tag{4.138}$$

One should note that if there is a question concerning the convergence of the above integral it is sometimes convenient to define the integral by the limiting process

$$B(\omega) = \lim_{\beta\to 0}\frac{2}{\pi}\int_0^\infty e^{-\beta\xi} f(\xi)\sin\omega\xi\, d\xi.$$

Substituting the coefficient from equation (4.138) into the equation (4.136) gives the solution

$$u(x,t) = \frac{2}{\pi}\int_0^\infty \int_0^\infty f(\xi)\sin(\omega\xi)\sin(\omega x)\, e^{-\omega^2\alpha^2 t}\, d\xi d\omega. \tag{4.139}$$

Using the trigonometric identity

$$\sin\omega\xi\sin\omega x = \frac{1}{2}\cos\omega(\xi - x) - \frac{1}{2}\cos\omega(\xi + x)$$

and interchanging the order of integration, the solution can be expressed

$$u = u(x,t) = \frac{1}{\pi}\int_0^\infty \int_0^\infty f(\xi)\left[\cos\omega(\xi - x) - \cos\omega(\xi + x)\right] e^{-\omega^2\alpha^2 t} d\omega d\xi.$$

Using the integral relationship given in equation (4.132) we simplify the above integral to the form

$$u = u(x,t) = \frac{1}{\pi}\int_0^\infty f(\xi)\left[\frac{\sqrt{\pi}}{2\alpha\sqrt{t}}e^{-\frac{(\xi - x)^2}{4\alpha^2 t}} - \frac{\sqrt{\pi}}{2\alpha\sqrt{t}}e^{-\frac{(\xi + x)^2}{4\alpha^2 t}}\right] d\xi.$$

In the first integral make the substitution $\dfrac{(\xi - x)^2}{4\alpha^2 t} = \nu^2$ and in the second integral make the substitution $\dfrac{(\xi + x)^2}{4\alpha^2 t} = \nu^2$ and write the solution for u as

$$u = u(x,t) = \frac{1}{\sqrt{\pi}}\left[\int_{\frac{-x}{2\alpha\sqrt{t}}}^{\infty} f(x + 2\nu\alpha\sqrt{t})e^{-\nu^2}\,d\nu - \int_{\frac{x}{2\alpha\sqrt{t}}}^{\infty} f(-x + 2\nu\alpha\sqrt{t})e^{-\nu^2}\,d\nu\right]. \quad (4.140)$$

The error function is defined[‡]

$$\text{erf}(x) = \frac{2}{\sqrt{\pi}}\int_0^x e^{-\xi^2}\,d\xi \quad (4.141)$$

and the complementary error function is defined

$$\text{erfc}(x) = 1 - \text{erf}(x) = \frac{2}{\sqrt{\pi}}\int_x^{\infty} e^{-\xi^2}\,d\xi \quad (4.142)$$

and represent those areas under the probability curve $\dfrac{2}{\sqrt{\pi}}e^{-x^2}$ which are illustrated in the figure 4-4.

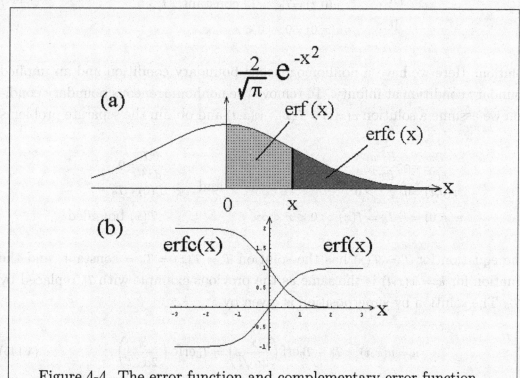

Figure 4-4. The error function and complementary error function.
(a) As areas under the probability curve.
(b) As functions of x.

[‡] The error function can be defined in different ways.

In the special case $f(x) = T_0$ is a constant, the solution given by equation (4.140) can be expressed in terms of the error function. One can show that the solution given by equation (4.140) simplifies to the form

$$u = u(x, t) = \frac{2T_0}{\sqrt{\pi}} \int_0^{x/2\alpha\sqrt{t}} e^{-\nu^2} \, d\nu \quad \text{or} \quad u = u(x, t) = T_0 \, \text{erf}\left(\frac{x}{2\alpha\sqrt{t}}\right). \tag{4.143}$$

∎

Example 4-11. (Heat equation infinite domain.)

Solve for the temperature distribution $u = u(x, t)$ in a long thin rod where

$$
\begin{aligned}
\text{PDE:} \quad & \frac{\partial u}{\partial t} = \alpha^2 \frac{\partial^2 u}{\partial x^2}, \quad 0 < x < \infty, \quad t > 0 \\
\text{BC:} \quad & u(0, t) = T_0, \quad T_0 \text{ constant} \quad t > 0 \\
\text{IC:} \quad & u(x, 0) = 0, \quad 0 < x < \infty
\end{aligned} \tag{4.144}
$$

Solution: Here we have a nonhomogeneous boundary condition and an implied boundary condition at infinity. To remove the nonhomogeneous boundary condition we assume a solution $u(x, t) = T(x) + w(x, t)$ and obtain the separate problems

$$
\begin{aligned}
\frac{\partial w}{\partial t} &= \alpha^2 \frac{\partial^2 w}{\partial x^2} \\
w(0, t) &= 0, \quad t > 0 \\
w(x, 0) &= -T_0 = f(x), \quad 0 < x < \infty
\end{aligned}
\quad \text{and} \quad
\begin{aligned}
\frac{d^2 T}{dx^2} &= 0 \\
T(0) &= T_0 \\
T(x) & \text{ bounded}
\end{aligned}
$$

The equation for $T = T(x)$ has the solution $T = T(x) = T_0 = $ constant, and the equation for $w = w(x, t)$ is the same as the previous example with T_0 replaced by $-T_0$. The solution by superposition is given by

$$u = u(x, t) = T_0 - T_0 \text{erf}\left(\frac{x}{2\alpha\sqrt{t}}\right) = T_0 \text{erfc}\left(\frac{x}{2\alpha\sqrt{t}}\right). \tag{4.145}$$

A graph of this complementary error function solution is illustrated in the figure 4-5 for various values of the time t. The solution indicates that as you move in the x-direction, the temperature increases with time.

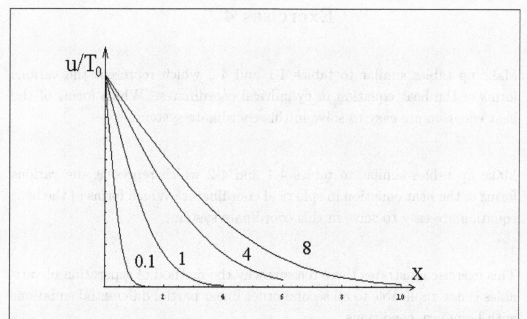

Figure 4-5. The complementary error function solution $u/T_0 = \text{erfc}(x/2\alpha\sqrt{t})$.

Example 4-12. (The error function and complementary error function.)

(a) $\dfrac{2}{\sqrt{\pi}} \displaystyle\int_x^y e^{-\nu^2}\, d\nu = \dfrac{2}{\sqrt{\pi}} \int_x^0 e^{-\nu^2}\, d\nu + \dfrac{2}{\sqrt{\pi}} \int_0^y e^{-\nu^2}\, d\nu = \text{erf}(y) - \text{erf}(x)$

(b) $\dfrac{2}{\sqrt{\pi}} \displaystyle\int_{-x}^x e^{-\nu^2}\, d\nu = \dfrac{4}{\sqrt{\pi}} \int_0^x e^{-\nu^2}\, d\nu = 2\,\text{erf}(x)$

(c) From statistics we have $\Phi(x) = \dfrac{1}{\sqrt{2\pi}} \displaystyle\int_{-\infty}^x e^{-\xi^2/2}\, d\xi$

as the area under the standard normal curve from $-\infty$ to x.

$\Phi(x) = \dfrac{1}{\sqrt{2\pi}} \displaystyle\int_{-\infty}^0 e^{-\xi^2/2}\, d\xi + \dfrac{1}{\sqrt{2\pi}} \int_0^x e^{-\xi^2/2}\, d\xi = \dfrac{1}{2} + \dfrac{1}{2}\text{erf}\left(x/\sqrt{2}\right)$

or $\text{erf}(x) = 2\Phi(\sqrt{2}\,x) - 1$

(d) $\text{erf}(-x) = -\text{erf}(x), \qquad \text{erf}(0) = 0, \qquad \text{erf}(\infty) = 1$

$\text{erfc}(0) = 1, \qquad \text{erfc}(\infty) = 0, \qquad \text{erfc}(x) = 1 - \text{erf}(x) = \dfrac{2}{\sqrt{\pi}} \displaystyle\int_x^{\infty} e^{-\xi^2}\, d\xi$

Exercises 4

▶ **1.**

Make up tables similar to tables 4-1 and 4-2 which represent the various forms of the heat equation in cylindrical coordinates. Which forms of the heat equation are easy to solve in this coordinate system?

▶ **2.**

Make up tables similar to tables 4-1 and 4-2 which represent the various forms of the heat equation in spherical coordinates. Which forms of the heat equation are easy to solve in this coordinate system?

▶ **3.**

This exercise illustrates three reasons why the method of separation of variables is not applicable to all second order linear partial differential equations with boundary conditions.

(a) Assume a solution $u = u(x, y) = F(x)G(y)$ to the linear second order partial differential equation

$$L(u) = \alpha_1 \frac{\partial^2 u}{\partial x^2} + \alpha_2 \frac{\partial^2 u}{\partial x \partial y} + \alpha_3 \frac{\partial^2 u}{\partial y^2} + \alpha_4 \frac{\partial u}{\partial x} + \alpha_5 \frac{\partial u}{\partial y} + \alpha_6 u = 0 \quad (x, y) \in R$$

where the region R is some simple and easy to handle geometry. Show that in order for the method of separation of variables to be applicable

(i) α_2 must equal zero.

(ii) α_1 and α_4 must be constants or functions of x.

(iii) α_3 and α_5 must be constants or functions of y.

(iv) α_6 can be a constant or a sum of a function of x and a function of y
   (i.e. $\alpha_6 = f(x) + g(y)$)

(v) Comment upon how the region R effects the resulting ordinary differential equations.

(b) Show $u = x^2 - y^2$ is a solution of $L(u) = \dfrac{\partial^2 u}{\partial x^2} + \dfrac{\partial^2 u}{\partial y^2} = 0$. This simple example illustrates there can be some solutions to a partial differential equation which are not of the form $u = F(x)G(y)$.

(c) The region R or the boundary conditions may be so complicated that an assumed solution $u = F(x)G(y)$ leads to ordinary differential equations that cannot be solved.

▶ **4.**

(a) Show the heat equation for a thin rod or wire which considers lateral heat loss is given by

$$\frac{\partial u}{\partial t} = K\frac{\partial^2 u}{\partial x^2} - \beta u, \qquad \beta \text{ constant.}$$

Show this equation can be solved by assuming a solution of the form

$$u = u(x,t) = e^{-\beta t}v(x,t).$$

Find the resulting equation for $v = v(x,t)$

(b) Solve the boundary value problem

PDE: $\quad \dfrac{\partial u}{\partial t} = K\dfrac{\partial^2 u}{\partial x^2} - \beta u, \qquad 0 < x < L,\ t > 0$

BC: $\quad u(0,t) = 0, \qquad u(L,t) = 0$

IC: $\quad u(x,0) = f(x)$

▶ **5.**

(a) Show the heat equation which considers convection

$$\frac{\partial u}{\partial t} = K\frac{\partial^2 u}{\partial x^2} - V\frac{\partial u}{\partial x}, \qquad K, V \text{ constants}$$

can be solved by assuming a solution of the form

$$u = u(x,t) = \exp\left(\frac{2Vx - V^2 t}{4K}\right)v(x,t).$$

Find the resulting equation for $v = v(x,t)$.

(b) Solve the boundary value problem

PDE: $\quad \dfrac{\partial u}{\partial t} = K\dfrac{\partial^2 u}{\partial x^2} - V\dfrac{\partial u}{\partial x}, \qquad 0 < x < L,\ t > 0$

BC: $\quad u(0,t) = 0, \qquad u(L,t) = 0$

IC: $\quad u(x,0) = f(x)$

▶ **6.**

Solve for the steady state temperature distribution $u = u(x)$ in a thin wire $0 < x < L$ which is laterally insulated. Assume boundary conditions are $u(0) = T_0$ and $u(L) = T_1$ where T_0 and T_1 are constants.

► **7.**

Solve for the steady state temperature distribution in polar coordinates where the temperature is given by $u = u(r)$ in a thin circular disk $0 < r < b$ which is insulated on its faces. Assume boundary condition $u(b) = T_0$.

► **8.**

Solve for the steady state temperature distribution in polar coordinates where the temperature is given by $u = u(r)$ in the annular region $a < r < b$ subject to the boundary conditions $u(a) = T_a$ and $u(b) = T_b$ where T_a and T_b are constants.

► **9.**

Determine the steady state temperature $u = u(r, z)$ in a solid cylinder bounded by the surfaces $r = 1, z = 0$ and $z = 1$ when the bottom face is insulated, the top face is held at a fixed temperature T_0 and the side $r = 1$ is held at a zero temperature.

► **10.**

Find the steady state temperature $u = u(r, z)$ in the circular cylinder $0 \leq r \leq b$, $0 \leq z \leq h$ subject to the top and bottom boundary conditions $u(r, h) = 0$, $u(r, 0) = 0$ and side condition $u(b, z) = f(z)$.

► **11.**

Solve the boundary value problem
$$\frac{\partial u}{\partial t} = K \left(\frac{\partial^2 u}{\partial r^2} + \frac{1}{r} \frac{\partial u}{\partial r} \right), \quad 0 < r < b, \quad t > 0$$

subject to the boundary conditions $\frac{\partial u(b,t)}{\partial r} + \beta u(b, t) = 0$ and initial condition $u(r, 0) = f(r)$. Hint: Multiply the BC by b and define $b\beta = h > 0$ as a new constant. Note that β represents a ratio of surface conductance divided by the thermal conductivity of the cylinder material and the BC is a form of Newton's law of cooling.

► **12.**

Solve the steady state heat equation
$$\nabla^2 u = \frac{\partial^2 u}{\partial x^2} + \frac{\partial^2 u}{\partial y^2} = 0, \quad x, y \in R = \{x, y \, | \, 0 < x < \infty, \; 0 < y < b\}$$

subject to the boundary conditions $u(x, 0) = 0$, $u(x, b) = 0$ and $u(0, y) = f(y)$ where $u(x, y)$ is to be bounded for all $x, y \in R$.

▶ **13.**

(a) Find the temperature distribution in an infinite bar where the temperature $u = u(x, t)$ satisfies

$$\frac{\partial u}{\partial t} = \alpha^2 \frac{\partial^2 u}{\partial x^2}, \qquad -\infty < x < \infty, \quad t > 0$$

subject to the initial condition $u(x, 0) = f(x)$.

(b) Evaluate your answer in part (a) when $f(x) = \begin{cases} T_0, & |x| < 1 \\ 0, & |x| > 1 \end{cases}$ where T_0 is a constant. Express the answer in terms of the error function.

▶ **14.**

Solve the boundary value problem $\dfrac{\partial^2 u}{\partial x^2} + \dfrac{\partial^2 u}{\partial y^2} = 0$, $\quad 0 < x < L$, $\quad y > 0$ subject to the boundary conditions $u(0, y) = 0$, $u(L, y) = 0$, $0 \le y < \infty$ and $u(x, 0) = f(x)$, $0 < x < L$.

▶ **15.**

Solve the nonhomogeneous heat equation $\dfrac{\partial u}{\partial t} = K \dfrac{\partial^2 u}{\partial x^2} + g(x)$, $\quad 0 < x < \pi$, $t > 0$ where $g(x) = \begin{cases} x, & 0 \le x \le \pi/2 \\ \pi - x, & \pi/2 \le x \le \pi \end{cases}$ subject to the boundary conditions $u(0, t) = 0$ and $u(\pi, t) = 0$, and initial condition $u(x, 0) = x(\pi - x)$.

▶ **16.**

In a uniform slab $0 \le x \le 1$ the thermal conductivity is given by $K = \frac{\kappa}{\varrho c} = 1 \left[\frac{\text{cm}^2}{\text{sec}}\right]$ and heat is generated at the constant rate where the ratio $\frac{H}{\varrho c} = Q = 2 \left(\frac{°C}{\text{sec}}\right)$. Assume the ends $x = 0$ and $x = 1$ are held at temperatures of 0 and 1 respectively. Find the temperature in the slab as a function of time t if the initial temperature distribution is given by x^2.

▶ **17.**

Solve for the temperature $u = u(r, t)$ in a circular disk which is a solution of the boundary value problem

$$\frac{\partial u}{\partial t} = \alpha^2 \left(\frac{\partial^2 u}{\partial r^2} + \frac{1}{r}\frac{\partial u}{\partial r}\right), \qquad 0 < r < 2, \ t > 0$$
$$u(2, t) = 0, \quad u(r, 0) = f(r), \quad |u(r, t)| < M.$$

► **18.**

Solve for the steady state temperature distribution in spherical coordinates where $u = u(\rho)$ for a sphere $0 < \rho < \rho_0$ with boundary condition $u(\rho_0) = T_0$.

► **19.**

Solve for the steady state temperature distribution in spherical coordinates where $u = u(\rho)$ for a hollow sphere $\rho_0 < \rho < \rho_1$ subject to the boundary conditions $u(\rho_0) = T_0$ and $u(\rho_1) = T_1$ where T_0 and T_1 are constants.

► **20.**

Give a physical interpretation to the given boundary value problem and solve.

$$\text{PDE:} \quad \frac{\partial u}{\partial t} = K \frac{\partial^2 u}{\partial x^2} + q(x,t), \qquad 0 < x < L,\ t > 0$$
$$\text{BC:} \quad u(0,t) = 0, \qquad u(L,t) = 0$$
$$\text{IC:} \quad u(x,0) = f(x)$$

► **21.**

Solve the given boundary value problem and express your answer in terms of the error function.

$$\text{PDE:} \quad \frac{\partial u}{\partial t} = \alpha^2 \frac{\partial^2 u}{\partial x^2}, \qquad 0 < x < \infty,\ t > 0$$
$$\text{BC:} \quad u(0,t) = 0$$
$$\text{IC:} \quad u(x,0) = 1$$

Hint: See example 4-10 and consider a limiting case. See also Appendix D.

► **22.**

Give a physical interpretation to the given boundary value problem and solve.

$$\text{PDE:} \quad \frac{\partial u}{\partial t} = K \frac{\partial^2 u}{\partial x^2}, \qquad 0 < x < L,\ t > 0$$
$$\text{BC:} \quad \frac{\partial u(0,t)}{\partial x} = 0, \qquad \frac{\partial u(L,t)}{\partial x} = 0$$
$$\text{IC:} \quad u(x,0) = f(x)$$

▶ **23.**

Give a physical interpretation to the given boundary value problem and solve.

$$\text{PDE:} \qquad \frac{\partial u}{\partial t} = K \frac{\partial^2 u}{\partial x^2}, \qquad 0 < x < L, \ t > 0$$

$$\text{BC:} \qquad u(0,t) = \alpha(t), \qquad u(L,t) = 0$$

$$\text{IC:} \qquad u(x,0) = f(x)$$

▶ **24.**

Give a physical interpretation to the given boundary value problem and solve.

$$\text{PDE:} \qquad \frac{\partial u}{\partial t} = K \frac{\partial^2 u}{\partial x^2}, \qquad 0 < x < L, \ t > 0$$

$$\text{BC:} \qquad u(0,t) = 0, \qquad u(L,t) = \beta(t)$$

$$\text{IC:} \qquad u(x,0) = f(x)$$

▶ **25.**

Solve the Dirichlet problem for the steady state temperature in a cylinder $0 < r < a$ and $0 < z < h$ if

$$\text{PDE:} \qquad \nabla^2 u = \frac{\partial^2 u}{\partial r^2} + \frac{1}{r}\frac{\partial u}{\partial r} + \frac{\partial^2 u}{\partial z^2} = 0, \quad 0 < z < h, \ 0 < r < a, \ t > 0$$

$$\text{BC:} \qquad u(r,0) = 0, \qquad u(r,h) = 0, \qquad u(a,z) = f(z)$$

▶ **26.**

Give a physical interpretation to the given boundary value problem and solve.

$$\text{PDE:} \qquad \frac{\partial u}{\partial t} = K \frac{\partial^2 u}{\partial x^2}, \quad 0 < x < L$$

$$\text{BC:} \qquad u(0,t) = 0, \qquad \frac{\partial u(L,t)}{\partial x} = 0$$

$$\text{IC:} \qquad u(x,0) = f(x)$$

▶ **27.**

Give a physical interpretation to the given boundary value problem and solve.

$$\text{PDE:} \qquad \frac{\partial u}{\partial t} = K \frac{\partial^2 u}{\partial x^2}, \qquad 0 < x < \pi, \ t > 0$$

$$\text{BC:} \qquad \frac{\partial u(0,t)}{\partial x} = 0, \qquad \frac{\partial u(\pi,t)}{\partial x} = 0$$

$$\text{IC:} \qquad u(x,0) = f(x)$$

▶ **28.**

Give a physical interpretation to the given boundary value problem and solve.

PDE: $\dfrac{\partial u}{\partial t} = K\dfrac{\partial^2 u}{\partial x^2} + q_0$, q_0 constant, $0 < x < L,\ t > 0$

BC: $\dfrac{\partial u(0,t)}{\partial x} = 0$, $\dfrac{\partial u(L,t)}{\partial x} + hu(L,t) = 0$, $h > 0$ and constant.

IC: $u(x,0) = f(x)$

▶ **29.**

Give a physical interpretation to the given boundary value problem and solve.

PDE: $\dfrac{\partial u}{\partial t} = K\dfrac{\partial^2 u}{\partial x^2} + q(t)$, $0 < x < L,\ t > 0$, K constant.

BC: $\dfrac{\partial u(0,t)}{\partial x} = 0$, $\dfrac{\partial u(L,t)}{\partial x} + hu(L,t) = 0$, $h > 0$ and constant.

IC: $u(x,0) = f(x)$

▶ **30.**

Consider the Dirichlet problem for the heat equation which describes the temperature $u = u(x,t)$ in a long thin rod with sides that are insulated and end conditions are specified.

PDE: $\dfrac{\partial u}{\partial t} = K\dfrac{\partial^2 u}{\partial x^2}$, $0 < x < L,\ t > 0$, K constant

BC: $u(0,t) = 0$, $u(L,t) = 0$

IC: $u(x,0) = \sin\dfrac{n\pi x}{L}$

▶ **31.**

Consider the Dirichlet problem for the heat equation which describes the temperature $u = u(x,t)$ in a long thin rod where the sides are not insulated and end conditions are specified.

PDE: $\dfrac{\partial u}{\partial t} = K\dfrac{\partial^2 u}{\partial x^2} - \beta u$, $0 < x < L,\ t > 0$, K, β constant

BC: $u(0,t) = 0$, $u(L,t) = 0$

IC: $u(x,0) = \sin\dfrac{n\pi x}{L}$

▶ **32.**

Give a physical interpretation to the given boundary value problem and solve.

PDE: $\qquad \dfrac{\partial u}{\partial t} = K \dfrac{\partial^2 u}{\partial x^2} + q(x,t), \qquad 0 < x < \pi, \ t > 0$

BC: $\qquad u(0,t) = 0, \qquad u(\pi, t) = 0$

IC: $\qquad u(x,0) = f(x)$

▶ **33.**

(a) Show the axially symmetric heat equation

$$\frac{\partial u}{\partial t} = K \left(\frac{\partial^2 u}{\partial r^2} + \frac{1}{r} \frac{\partial u}{\partial r} \right), \qquad 0 \le r \le r_0$$

can be normalized to the form

$$\frac{\partial u}{\partial \tau} = \frac{\partial^2 u}{\partial R^2} + \frac{1}{R} \frac{\partial u}{\partial R}, \qquad 0 \le R \le 1$$

by introducing scaled variables $\tau = t/t_0$, $R = r/r_0$ for appropriate constant t_0.

(b) Solve the normalized axially symmetric heat equation for $u = u(R, \tau)$ where

$$\frac{\partial u}{\partial \tau} = \frac{\partial^2 u}{\partial R^2} + \frac{1}{R} \frac{\partial u}{\partial R}, \qquad 0 \le R \le 1, \quad \tau > 0$$

subject to the insulated boundary condition $\dfrac{\partial u}{\partial R}\Big|_{R=1} = 0$ and satisfying the initial condition $u(R, 0) = f(R)$.

(c) Find the steady state solution.

▶ **34.**

(a) Use the method of separation of variables to solve

$$\frac{\partial u}{\partial t} = \alpha^2 \frac{\partial^2 u}{\partial x^2}, \qquad -\infty < x < \infty, \quad t > 0 \qquad\qquad (34a)$$

subject to the initial condition $u(x,0) = f(x)$, $-\infty < x < \infty$ and show that the solution can be written in the form

$$u = u(x,t) = \frac{1}{\pi} \int_{-\infty}^{\infty} f(\xi)\, d\xi \int_{0}^{\infty} e^{-\alpha^2 \omega^2 t} \cos \omega (\xi - x)\, d\omega.$$

(b) Use the integral relation $\displaystyle\int_{0}^{\infty} e^{-\alpha^2 \omega^2 t} \cos \omega(\xi - x)\, d\omega = \frac{1}{2\alpha}\sqrt{\frac{\pi}{t}} e^{-(x-\xi)^2/4\alpha^2 t}$ to show the solution can also be represented as

$$u = u(x,t) = \frac{1}{2\alpha} \int_{-\infty}^{\infty} \frac{f(\xi)}{\sqrt{\pi t}} e^{-(x-\xi)^2/4\alpha^2 t}\, d\xi.$$

(c) Verify that the function $g(x,t) = \dfrac{1}{2\alpha\sqrt{\pi t}} e^{-(x-\xi)^2/4\alpha^2 t}$ is a solution to the equation (34a) for all values of x and t for which $x \ne \xi$ and $t > 0$.

212

▶ **35.**

Solve the Robin problem for the heat equation which describes the temperature $u = u(x,t)$ in a long thin rod with sides that are insulated and the ends obey the Newton law of cooling.

PDE: $\quad \dfrac{\partial u}{\partial t} - K\dfrac{\partial^2 u}{\partial x^2} = 0, \quad 0 < x < L,\ t > 0$

BC: $\quad -\dfrac{\partial u(0,t)}{\partial x} = -h(u(0,t) - T_0), \quad h > 0$ and $K > 0$ are constants.

$\qquad \dfrac{\partial u(L,t)}{\partial x} = -h(u(L,t) - T_0), \qquad T_0$ constant

IC: $\quad u(x,0) = f(x)$

Hint: Let $U(x,t) = u(x,t) - T_0$ and observe that if one assumes a solution $U(x,t) = F(x)G(t)$ one obtains equations of the form

$$B\omega - Ah = 0$$

$$B[\omega\cos\omega L + h\sin\omega L] + A[h\cos\omega L - \omega\sin\omega L] = 0$$

which must be satisfied for nonzero values for A and B. This requires that the determinant of the coefficients be zero, which implies that $2\cot\omega L = \dfrac{\omega L}{hL} - \dfrac{hL}{\omega L}$. One can now plot graphs of $y_1 = 2\cot\omega L$ and $y_2 = \dfrac{\omega L}{hL} - \dfrac{hL}{\omega L}$ vs ωL to obtain the graph illustrated. Hence, one can write $2\cot\lambda_n = \dfrac{\lambda_n}{hL} - \dfrac{hL}{\lambda_n}$ for $n = 1, 2, 3, \ldots$, where $\lambda_n = \omega_n L$.

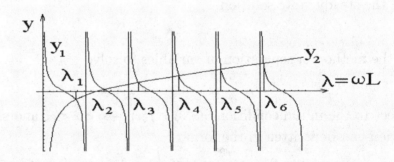

▶ **36.**

Solve the Dirichlet problem for the temperature $u = u(\rho,t)$ in spherical coordinates $0 < \rho < b$

PDE: $\quad \dfrac{\partial u}{\partial t} = K\left(\dfrac{\partial^2 u}{\partial\rho^2} + \dfrac{2}{\rho}\dfrac{\partial u}{\partial\rho}\right), \qquad 0 < \rho < b,\ t > 0$

BC: $\quad u(b,t) = 0$

IC: $\quad u(\rho,0) = f(\rho)$

Hint: Let $v(\rho,t) = \rho u(\rho,t)$ and note that $v(0,t) = 0$.

▶ **37.**

Interpret the boundary value problem and solve

PDE: $\dfrac{\partial u}{\partial t} = K\left(\dfrac{\partial^2 u}{\partial \rho^2} + \dfrac{2}{\rho}\dfrac{\partial u}{\partial \rho}\right), \quad 0 < \rho < b$

BC: $\dfrac{\partial u(b,t)}{\partial \rho} + hu(b,t) = 0, \qquad h$ constant such that $hb - 1 = c > 0$

IC: $u(\rho,0) = f(\rho)$

Hint: Let $v(\rho,t) = \rho u(\rho,t)$ and note $v(0,t) = 0$.

▶ **38.**

Assume $u_1 = u_1(x,t)$ is a solution of the partial differential equation

$$\frac{\partial u_1}{\partial t} = K\frac{\partial^2 u_1}{\partial x^2}, \quad 0 < x < a, \ t > 0$$

and that $u_2 = u_2(y,t)$ is a solution of the partial differential equation

$$\frac{\partial u_2}{\partial t} = K\frac{\partial^2 u_2}{\partial y^2}, \quad 0 < y < b, \ t > 0.$$

Show the product $u = u(x,y,t) = u_1(x,t)u_2(y,t)$ satisfies the two-dimensional heat equation

$$\frac{\partial u}{\partial t} = K\left(\frac{\partial^2 u}{\partial x^2} + \frac{\partial^2 u}{\partial y^2}\right), \qquad 0 < x < a, \ 0 < y < b, \ t > 0.$$

▶ **39.**

Give a physical interpretation to the given boundary value problem and solve.

PDE: $\dfrac{\partial u}{\partial t} = K\left(\dfrac{\partial^2 u}{\partial r^2} + \dfrac{1}{r}\dfrac{\partial u}{\partial r}\right) + q(r,t), \qquad 0 < r < b, \ t > 0$

BC: $u(b,t) = 0$

IC: $u(r,0) = f(r)$

▶ **40.**

Solve the Dirichlet boundary value problem for the temperature $u = r(r,t)$ in the annular region $a < r < b$

PDE: $\dfrac{\partial u}{\partial t} = K\left(\dfrac{\partial^2 u}{\partial r^2} + \dfrac{1}{r}\dfrac{\partial u}{\partial r}\right), \quad a < r < b, \ t > 0$

BC: $u(a,t) = 0, \qquad u(b,t) = 0$

IC: $u(r,0) = f(r)$

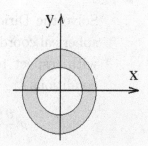

214

▶ 41.

Solve the Dirichlet boundary value problem for the annular region $a < r < b$

PDE: $\qquad \dfrac{\partial u}{\partial t} = K\left(\dfrac{\partial^2 u}{\partial r^2} + \dfrac{1}{r}\dfrac{\partial u}{\partial r}\right) + q(r,t), \qquad a < r < b, \ t > 0$

BC: $\qquad u(a,t) = 0, \qquad u(b,t) = 0$

IC: $\qquad u(r,0) = f(r)$

▶ 42.

Solve the Dirichlet boundary value problem for temperature $u = r(r,t)$ in the annular region $a < r < b$ where the boundaries are held at the constant temperatures T_0 and T_1.

PDE: $\qquad \dfrac{\partial u}{\partial t} = K\left(\dfrac{\partial^2 u}{\partial r^2} + \dfrac{1}{r}\dfrac{\partial u}{\partial r}\right), \qquad a < r < b, \ t > 0$

BC: $\qquad u(a,t) = T_0, \qquad u(b,t) = T_1$

IC: $\qquad u(r,0) = f(r)$

▶ 43.

Solve the boundary value problem for the annular region $a < r < b$ where the boundaries are held at the constant temperatures T_0 and T_1.

PDE: $\qquad \dfrac{\partial u}{\partial t} = K\left(\dfrac{\partial^2 u}{\partial r^2} + \dfrac{1}{r}\dfrac{\partial u}{\partial r}\right) + q(r,t), \qquad a < r < b, \ t > 0$

BC: $\qquad u(a,t) = T_0, \qquad u(b,t) = T_1$

IC: $\qquad u(r,0) = f(r)$

▶ 44.

Solve the Dirichlet problem for the steady state temperature $u = u(\rho, \theta)$ in spherical coordinates (ρ, θ) over the region specified, where there is symmetry with respect to ϕ, and the solution is subject to the specified boundary conditions .

$$\rho^2 \frac{\partial^2 u}{\partial \rho^2} + 2\rho \frac{\partial u}{\partial \rho} + \frac{\partial^2 u}{\partial \theta^2} + \cot\theta \frac{\partial u}{\partial \theta} = 0, \quad b < \rho < \infty, \quad 0 < \theta < \pi$$

$$u(b, \theta) = f(\theta)$$

▶ **45.**

(a) Show the solution to the steady state heat equation $\nabla^2 u = \dfrac{\partial^2 u}{\partial x^2} + \dfrac{\partial^2 u}{\partial y^2} = 0$ over the rectangular region $R = \{(x,y)| \ 0 < x < a, \ 0 < y < b\}$ subject to the Dirichlet boundary conditions

$$u(0,y) = f_1(y), \qquad u(a,y) = f_2(y)$$

$$u(x,0) = g_4(x), \qquad u(x,b) = g_3(x)$$

can be broken up into four separate problems.

Problem 1: $\nabla^2 u_1 = 0$ for $(x,y) \in R$ subject to the boundary conditions

$$u_1(0,y) = 0, \qquad u_1(a,y) = f_1(y)$$

$$u_1(x,0) = 0, \qquad u_1(x,b) = 0$$

Problem 2: $\nabla^2 u_2 = 0$ for $(x,y) \in R$ subject to the boundary conditions

$$u_2(0,y) = f_2(y), \qquad u_2(a,y) = 0$$

$$u_2(x,0) = 0, \qquad u_2(x,b) = 0$$

Problem 3: $\nabla^2 u_3 = 0$ for $(x,y) \in R$ subject to the boundary conditions

$$u_3(0,y) = 0, \qquad u_3(a,y) = 0$$

$$u_3(x,0) = 0, \qquad u_3(x,b) = g_3(x)$$

Problem 4: $\nabla^2 u_4 = 0$ for $(x,y) \in R$ subject to the boundary conditions

$$u_4(0,y) = 0, \qquad u_4(a,y) = 0$$

$$u_4(x,0) = g_4(x), \qquad u_4(x,b) = 0$$

(b) Use separation of variables to solve each of the above problems.

Hint: Assume a solution $u = u(x,y) = u_1(x,y) + u_2(x,y) + u_3(x,y) + u_4(x,y)$.

▶ **46.**

(a) In the heat equation $\dfrac{\partial u}{\partial t} = K \dfrac{\partial^2 u}{\partial x^2}$ make the change of variables

$$U = \frac{u}{u_0}, \qquad \tau = \frac{t}{t_0}, \qquad \xi = \frac{x}{x_0}$$

where u_0, t_0, x_0 are scaling constants. Show that if t_0 and x_0 are selected such that $Kt_0 = x_0^2$, then there results the scaled equation $\dfrac{\partial U}{\partial \tau} = \dfrac{\partial^2 U}{\partial \xi^2}$.

(b) Solve the scaled partial differential equation

$$\frac{\partial U}{\partial \tau} = \frac{\partial^2 U}{\partial \xi^2}, \qquad 0 < \xi < \pi, \ \tau > 0$$

$$U(0,\tau) = 0, \qquad U(\pi,\tau) = 0$$

$$U(\xi,0) = 4\sin 3\xi$$

▶ **47.**

Solve the nonhomogeneous heat equation

$$\frac{\partial u}{\partial t} = K\frac{\partial^2 u}{\partial x^2} + g(x), \qquad 0 < x < \pi, \ t > 0$$

subject to the boundary conditions $u(0,t) = 0$, $u(\pi,t) = 0$ and initial condition $u(x,0) = x(\pi - x)$.

Determine the solution to the above heat equation in the special case where

$$g(x) = \begin{cases} x, & 0 \le x \le \pi/2 \\ \pi - x, & \pi/2 \le x \le \pi \end{cases}$$

Hints:

$$x(\pi - x) = \frac{4}{\pi}\sum_{n=1}^{\infty}\frac{(1 - (-1)^n)}{n^3}\sin nx, \qquad 0 < x < \pi$$

$$g(x) = \begin{cases} x, & 0 \le x \le \pi/2 \\ \pi - x, & \pi/2 \le x \le \pi \end{cases} = \frac{4}{\pi}\sum_{n=1}^{\infty}\frac{\sin\left(\frac{n}{2}\right)}{n^2}\sin nx, \qquad 0 < x < \pi$$

▶ **48.**

(Duhamel's† principle for the heat equation) Show that if $v = v(x,t;\tau)$ is a solution of the boundary-initial value problem

$$\begin{aligned} \text{PDE:} && \frac{\partial v}{\partial t} &= K\frac{\partial^2 v}{\partial x^2}, & 0 < x < L, \ t > \tau \\ \text{BC:} && v(0,t;\tau) &= 0, & t > \tau, \qquad v(L,t;\tau) = 0 \\ \text{IC:} && v(x,\tau;\tau) &= f(x,\tau), & 0 < x < L \end{aligned}$$

where $\tau \ge 0$, then the solution of the related nonhomogeneous heat equation

$$\begin{aligned} \text{PDE:} && \frac{\partial u}{\partial t} &= K\frac{\partial^2 u}{\partial x^2} + f(x,t), & 0 < x < L, \ t > 0 \\ \text{BC:} && u(0,t) &= 0, & t > 0, \qquad u(L,t) = 0 \\ \text{IC:} && u(x,0) &= 0, & 0 < x < L \end{aligned}$$

is given by

$$u = u(x,t) = \int_0^t v(x,t;\tau)\,d\tau$$

Hint: Use Leibnitz's† rule for differentiating an integral

$$\frac{d}{dt}\int_{\alpha(t)}^{\beta(t)} f(t,\tau)\,d\tau = \int_{\alpha(t)}^{\beta(t)}\frac{\partial f(t,\tau)}{\partial t}\,d\tau + f[t,\beta(t)]\frac{d\beta}{dt} - f[t,\alpha(t)]\frac{d\alpha}{dt}.$$

---

† See Appendix C

▶ **49.**

(**Duhamel's**[†] **principle for the heat equation**) The solution $u = u(x,t)$ of the nonhomogeneous heat equation

$$\frac{\partial u}{\partial t} - \alpha^2 \frac{\partial^2 u}{\partial x^2} = f(x), \qquad -\infty < x < \infty \tag{49a}$$

with initial conditions $u(x,0) = 0$, is related to the homogeneous initial value problem

$$\frac{\partial v}{\partial t} - \alpha^2 \frac{\partial^2 v}{\partial x^2} = 0, \qquad -\infty < x < \infty, \quad t > \tau \tag{49b}$$

which has an initial value at $t = \tau$ given by $v(x,\tau;\tau) = f(x)$. Here τ is a parameter representing the time the initial condition is applied and so the problem (49b) is sometimes referred to as a 'pulsed' initial value problem. The relation between the two problems is given by

$$u(x,t) = \int_0^t v(x,t;\tau)\, d\tau \tag{49c}$$

From equation (49c) one can see that the initial condition $u(x,0) = 0$ is satisfied. Differentiating the equation (49c) with respect to t gives

$$\frac{\partial u}{\partial t} = v(x,t;t) + \int_0^t \frac{\partial v}{\partial t}\, d\tau \tag{49d}$$

Interchanging the order of differentiation and integration and using the equation (49b) one can write

$$\frac{\partial u}{\partial t} = f(x) + \int_0^t \alpha^2 \frac{\partial^2 v(x,t;\tau)}{\partial xx^2}\, d\tau$$

$$= f(x) + \alpha^2 \frac{\partial^2}{\partial x^2} \int_0^t v(x,t;\tau)\, d\tau$$

$$= f(x) + \alpha^2 \frac{\partial^2 u}{\partial x^2}$$

(a) Use separation of variables to solve the equation (49b) and show that

$$v = v(x,t;\tau) = \frac{1}{\pi} \int_0^\infty \int_{-\infty}^\infty \exp\left[-\alpha^2\omega^2(t - \tau)\right] \cos[\omega(x - \xi)]f(\xi)\, d\xi d\omega \tag{49e}$$

(b) Use the integral relation

$$\int_0^\infty e^{-a^2 r^2} \cos(br)\, dr = \frac{\sqrt{\pi}}{2a} e^{-b^2/4a^2} \tag{49f}$$

[†] See Appendix C

to show equation (49e) can be reduced to the form

$$v(x, t; \tau) = \frac{1}{2\alpha\sqrt{\pi(t-\tau)}} \int_{-\infty}^{\infty} \exp\left[-\frac{(x-\xi)^2}{4\alpha^2(t-\tau)}\right] f(\xi)\, d\xi \tag{49g}$$

(c) Show that

$$u = u(x, t) = \int_0^t \frac{1}{2\alpha\sqrt{\pi(t-\tau)}} \int_{-\infty}^{\infty} \exp\left[-\frac{(x-\xi)^2}{4\alpha^2(t-\tau)}\right] f(\xi)\, d\xi. \tag{49h}$$

▶ **50.**

Source free heat conduction in a semi-infinite solid $x > 0$ is described by the boundary value problem

PDE: $\quad \dfrac{\partial u}{\partial t} - K\dfrac{\partial^2 u}{\partial x^2} = 0, \qquad 0 < x < \infty,\ t > 0,\ K = \text{constant}$

BC: $\qquad u(0, t) = 0, \quad |u(x, t)| < M = \text{constant},\ t > 0$

IC: $\qquad u(x, 0) = g(x)$

Solve the above boundary value problem for the following conditions.

(a) $g(x) = T_0 = \text{constant}$

(b) $g(x) = \begin{cases} 1, & 0 < x < L \\ 0, & x > L \end{cases}$

▶ **51.**

Source free heat conduction in an infinite solid $-\infty < x < \infty$ is described by the boundary value problem

PDE: $\quad \dfrac{\partial u}{\partial t} - K\dfrac{\partial^2 u}{\partial x^2} = 0, \qquad -\infty < x < \infty,\ t > 0,\ K = \text{constant}$

IC: $\qquad u(x, 0) = f(x)$

Here a boundedness condition, $|u(x, t)| < M = \text{constant}$, is assumed. Solve the above boundary value problem for the following conditions.

(a) $\quad f(x) = \begin{cases} 0, & x < 0 \\ 1, & x > 0 \end{cases}$

(b) $\quad f(x) = \begin{cases} 1, & -L < x < L \\ 0, & \text{otherwise} \end{cases}$

Chapter 5
Hyperbolic Equations

The wave equation is an important hyperbolic equation of mathematical physics which occurs in studying such disciplines as vibration of strings and membranes, sound waves, tidal waves, elastic waves, electric and magnetic waves. Wave motion is the movement of a disturbance from some source. In the process energy and momentum are transferred from the source. The wave motion can be longitudinal, transverse or torsional.

Waves are classified as either mechanical or electromagnetic. Mechanical waves require some kind of elastic material medium for their propagation and so they are sometimes called elastic waves. An elastic medium is characterized by a continuum of points where a displacement of one point (source) immediately experiences reaction forces from neighboring points. The reaction forces at the neighboring points produce forces on their neighboring points and by continuing this process the initial displacement gets propagated into the elastic medium by way of neighboring points being affected as a function of time. An electromagnetic wave differs from a mechanical wave in that it can propagate through a vacuum.

The figure 5-1(a) illustrates a long string which is given an initial displacement and then suddenly released. The elastic properties of the string cause forces which try to pull the initial displacement of the string back to its equilibrium position. The restoring forces affect points immediately to the right of the displacement. These neighboring points are pulled downward and so the displacement moves to the right at a definite speed which depends upon the material properties of the string. This is an example of a transverse wave where the displacements of the elastic medium are perpendicular to the direction of wave propagation. In figure 5-1(b) there is illustrated masses connected together by springs to form a chain. The masses might represent atoms and the springs might represent the forces between the atoms. When one mass is displaced longitudinally there is created spring elongations and compressions which produce forces which act on the neighboring masses. If the left most mass is given an initial displacement to the right, then a compression disturbance is created which in turn propagates to the right along the chain. This is an example of a longitudinal or compression wave.

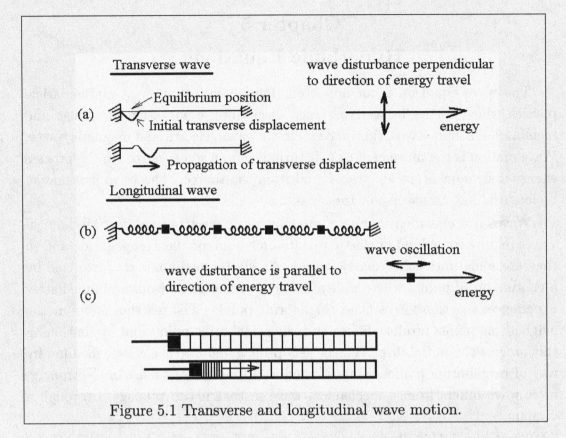

Figure 5.1 Transverse and longitudinal wave motion.

The figure 5-1(c) illustrates a gas with density ρ inside a circular pipe with cross-sectional area A. Imagine a piston given an impulsive force F which causes the piston to move with velocity c over a time interval Δt. This sudden motion compresses the gas inside the pipe and generates a compression wave moving to the right. This is also an example of a longitudinal wave.

Sound waves are examples of longitudinal waves while a vibrating string is an example of a transverse wave motion. Longitudinal waves move in the direction of energy propagation while transverse waves move perpendicular to the direction of energy propagation. A wave front is a moving surface in three-dimensions, which reduces to a moving curve in two-dimensions and a moving point in one-dimension. The wave front is characterized by the wave speed being the same at all points on the wave front. For example, from a point source there can be a spherical wave moving out in all directions. This would be represented by a circle moving in two-dimensions and a point moving in one direction. If the source is a line, then the wave front would be a moving cylinder in three-dimensions and a moving line in two-dimensions. If the source is a plane, then the wave front would be a plane wave.

We first examine the mathematical representation of a moving wave and then find the partial differential equation satisfied by this wave. This is followed by some selected areas of application where the wave equation occurs.

Wave Motion

Let us begin by developing the partial differential equation for a general wave motion. Consider an arbitrary but continuous curve $y = f(x)$ which represents some kind of wave shape or wave profile taken at some instant of time. A nominal wave shape is given by the curve illustrated in the figure 5-2(a). We shall assume the wave profile retains its shape as it moves. This is not the case if there is any distortion caused by the medium in which the wave is moving.

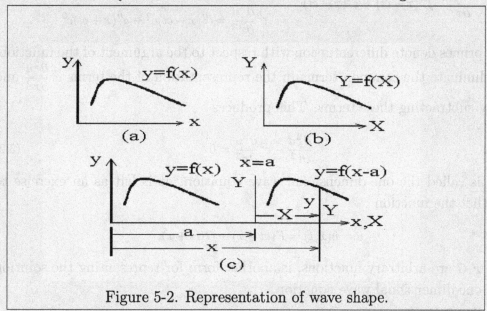

Figure 5-2. Representation of wave shape.

Now change all the symbols on this graph by replacing x by X and y by Y everywhere to obtain the figure 5-2(b). Now place the X, Y graph upon the x, y graph so that the Y-axis lies on the line $x = a$ as illustrated in the figure 5-2(c). Now represent the curve $Y = f(X)$ in terms of x and y. The figure 5-2(c) gives the transformation equations $x = a + X$ and $y = Y$ so that the curve $Y = f(X)$, with respect to the X, Y axes, becomes $y = f(x - a)$ with respect to the x, y axes. Let c denote a constant velocity and t denote time. If the distance a is represented

$a = ct$, then the curve $y = f(x - ct)$ represents a wave motion to the right along the x-axis. In a similar manner one can show for an arbitrary wave shape $g(x)$ the function $y = g(x + ct)$ represents a wave motion to the left along the x-axis.

Consider the general wave motion

$$u = u(x, t) = f(x - ct) + g(x + ct) \tag{5.1}$$

where one wave shape moves right and another wave shape moves left, where c is the constant velocity of wave propagation and f, g represent arbitrary wave shapes. We form the derivatives

$$\frac{\partial u}{\partial x} = f'(x - ct) + g'(x + ct)$$
$$\frac{\partial^2 u}{\partial x^2} = f''(x - ct) + g''(x + ct)$$

$$\frac{\partial u}{\partial t} = f'(x - ct)(-c) + g'(x + ct)(c)$$
$$\frac{\partial^2 u}{\partial t^2} = f''(x - ct)c^2 + g''(x + ct)c^2,$$
$$c^2\frac{\partial^2 u}{\partial x^2} = f''(x - ct)c^2 + g''(x + ct)c^2$$

where primes denote differentiation with respect to the argument of the function. Now eliminate the common terms in the representation of the terms $c^2\dfrac{\partial^2 u}{\partial x^2}$ and $\dfrac{\partial^2 u}{\partial t^2}$ by subtracting these terms. This produces

$$\frac{\partial^2 u}{\partial t^2} - c^2\frac{\partial^2 u}{\partial x^2} = 0 \tag{5.2}$$

which is called the one-dimensional wave equation. It is left as an exercise to show that the function

$$u = u(x, t) = F(ct - x) + G(ct + x),$$

where F, G are arbitrary functions, is another form for representing the solution of the one-dimensional wave equation.

In summary we have:

The solution to the one-dimensional wave equation $\dfrac{\partial^2 u}{\partial t^2} = c^2\dfrac{\partial^2 u}{\partial x^2}$ can be represented in either of the forms

$$u = u(x, t) = f(x - ct) + g(x + ct)$$
$$\text{or} \quad u = u(x, t) = F(ct - x) + G(ct + x)$$

where f, g, F, G represent arbitrary wave shapes.

In Cartesian coordinates the two-dimensional and three-dimensional forms for the wave equation are

$$\frac{\partial^2 u}{\partial t^2} = c^2 \left(\frac{\partial^2 u}{\partial x^2} + \frac{\partial^2 u}{\partial y^2} \right) = c^2 \nabla^2 u$$

$$\frac{\partial^2 u}{\partial t^2} = c^2 \left(\frac{\partial^2 u}{\partial x^2} + \frac{\partial^2 u}{\partial y^2} + \frac{\partial^2 u}{\partial z^2} \right) = c^2 \nabla^2 u \tag{5.3}$$

where c represents a constant velocity and ∇^2 is the Laplacian operator in the appropriate dimension. To the above equations one can add source terms to make them more general. We could also add damping terms to change the wave shapes. A damping term is usually assumed proportional to the velocity of the wave motion and so would have the form $-\beta \frac{\partial u}{\partial t}$ added to the right-hand side of equation (5.3). The wave equation can be represented in other coordinate systems by replacing the Laplacian operator $\nabla^2 u$ in the appropriate form.

The following are a rectangular, cylindrical and spherical representation of the three-dimensional wave equation

$$\frac{\partial^2 u}{\partial t^2} = c^2 \nabla^2 u + q(x, y, z, t),$$

where q represents some general source term. In Cartesian coordinates (x, y, z) the wave equation is written

$$\frac{\partial^2 u}{\partial t^2} = c^2 \left(\frac{\partial^2 u}{\partial x^2} + \frac{\partial^2 u}{\partial y^2} + \frac{\partial^2 u}{\partial z^2} \right) + q(x, y, z, t) = c^2 \nabla^2 u + q(x, y, z, t).$$

In cylindrical coordinates (r, θ, z) the above form is written

$$\frac{\partial^2 u}{\partial t^2} = c^2 \left[\frac{1}{r} \frac{\partial}{\partial r} \left(r \frac{\partial u}{\partial r} \right) + \frac{1}{r^2} \frac{\partial^2 u}{\partial \theta^2} + \frac{\partial^2 u}{\partial z^2} \right] + q(r, \theta, z, t)$$

In spherical coordinates (ρ, θ, ϕ) one obtains the equation

$$\frac{\partial^2 u}{\partial t^2} = c^2 \left[\frac{1}{\rho^2} \frac{\partial}{\partial \rho} \left(\rho^2 \frac{\partial u}{\partial \rho} \right) + \frac{1}{\rho^2 \sin \theta} \frac{\partial}{\partial \theta} \left(\sin \theta \frac{\partial u}{\partial \theta} \right) + \frac{1}{\rho^2 \sin^2 \theta} \frac{\partial^2 u}{\partial \phi^2} \right] + q(\rho, \theta, \phi, t)$$

It is left as an exercise to write out the various special cases of the above forms and construct tables similar to the tables 4.1 and 4.2 of chapter 4.

The wave equation occurs in many physical problems. We consider the motion of a vibrating string, wave motion of sound waves in a gas or liquid medium, the motion of electric and magnetic waves, wave motion in a solid and wave motion associated with a vibrating membrane as being selected examples for illustrating how the wave equation arises in applied problems.

Vibrating string.

Consider the motion of a string which is stretched along all or part of the x-axis. We shall assume the string is allowed to perform transverse vibrations over an interval $0 \leq x \leq L$ in a viscous medium under the action of external forces. We introduce the notations

$u = u(x,t)$ =Transverse displacement of string,   (cm)

$\varrho = \varrho(x)$ =lineal density of string,   (gm/cm)

$T = T(x)$ =Tension in string at point x,   (dynes)

$w = w(x,t)$ =Transverse external force per unit length ,   (dynes/cm)

$u_t = \dfrac{\partial u}{\partial t}$ =Transverse velocity,   (cm/sec)

$u_{tt} = \dfrac{\partial^2 u}{\partial t^2}$ =Transverse acceleration ,   (cm/sec^2)

β =Linear velocity damping force per unit length ,   (dynes-sec/cm^2)

We assume a damping force which is proportional to the velocity of the string. Consider an element of string with density ϱ per unit length located between x and $x + \Delta x$ as illustrated in the figure 5-3. The figure 5-3 we find the derivatives $\frac{\partial u(x,t)}{\partial x}$ and $\frac{\partial u(x+\Delta x,t)}{\partial x}$ represent the slopes of the tangent lines at the end points of the Δx section. In terms of the angles θ_1 and θ_2 these slopes are represented

$$\tan \theta_2 = \frac{\partial u(x + \Delta x, t)}{\partial x} \qquad \tan \theta_1 = \frac{\partial u(x,t)}{\partial x}.$$

The external forces acting on this Δx section of the string are approximated by

$w\Delta x =$ External force (dynes)   and $\beta u_t \Delta x =$ Damping force (dynes)

where it is assumed the damping force is proportional to the string velocity and the string weight has been neglected. If the motion is purely vertical, then the horizontal forces must be in equilibrium and so we impose a force balance

$$T(x + \Delta x) \cos \theta_2 = T(x) \cos \theta_1 = T_0.$$

In the vertical direction we apply Newton's[†] second law of motion that mass times acceleration must equal the sum of forces acting on the string. We sum

---

[†] See Appendix C

forces at the center point of the string element to obtain

$$\varrho(x + \tfrac{\Delta x}{2})\Delta x \frac{\partial^2 u(x + \tfrac{\Delta x}{2}, t)}{\partial t^2} = T(x + \Delta x)\sin\theta_2 - T(x)\sin\theta_1$$

$$+ w(x + \tfrac{\Delta x}{2})\Delta x - \beta \Delta x \frac{\partial u(x + \tfrac{\Delta x}{2}, t)}{\partial t}$$

$$= T_0(\tan\theta_2 - \tan\theta_1) + w(x + \tfrac{\Delta x}{2})\Delta x - \beta\Delta x \frac{\partial u(x + \tfrac{\Delta x}{2}, t)}{\partial t}$$

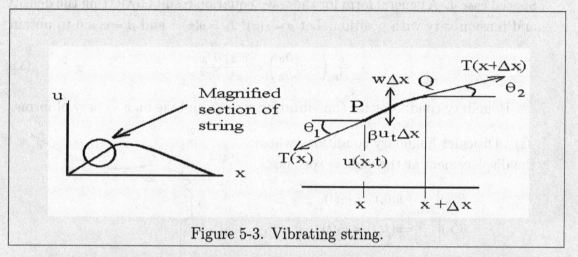

Figure 5-3. Vibrating string.

Now divide this equation by Δx and take the limit as Δx approaches zero. Note the second derivative is defined

$$\lim_{\Delta x \to 0} \left[\frac{\frac{\partial[T_0 u(x+\Delta x, t)]}{\partial x} - \frac{\partial[T_0 u(x,t)]}{\partial x}}{\Delta x} \right] = \frac{\partial}{\partial x}\left(T_0 \frac{\partial u(x,t)}{\partial x} \right)$$

and consequently the equation of motion of the string can be represented.

$$\varrho(x)\frac{\partial^2 u}{\partial t^2} = \frac{\partial}{\partial x}\left(T_0 \frac{\partial u(x,t)}{\partial x} \right) + w(x,t) - \beta\frac{\partial u}{\partial t} \tag{5.4}$$

Various special cases of equation (5.4) are listed.

Special case 1. The density and tension ϱ, T_0 are constants with $\beta = 0$ and $w = 0$. Let $c^2 = T_0/\varrho$ to obtain the one-dimensional wave equation

$$c^2\frac{\partial^2 u}{\partial x^2} = \frac{\partial^2 u}{\partial t^2} \tag{5.5}$$

Special case 2. The density and tension ϱ, T_0 are constants, $w = 0$. Let $c^2 = T_0/\varrho$ and $k = \beta/\varrho$ to obtain the telegraphy equation

$$c^2\frac{\partial^2 u}{\partial x^2} = \frac{\partial^2 u}{\partial t^2} + k\frac{\partial u}{\partial t} \tag{5.6}$$

Special case 3. The density and tension ϱ, T_0 are constants, $\beta = 0$ and $w = -\varrho g$ where g is the acceleration of gravity. Here the weight of the string is not neglected so there results the wave equation with damping plus a forcing term

$$c^2 \frac{\partial^2 u}{\partial x^2} = \frac{\partial^2 u}{\partial t^2} + k\frac{\partial u}{\partial t} + g \tag{5.7}$$

Special case 4. A general form for the wave equation results by letting the density and tension vary with position. Let $\varrho = r(x)$, $T_0 = p(x)c^2$ and $\beta = w = 0$ to obtain

$$\frac{\partial}{\partial x}\left(p(x)\frac{\partial u}{\partial x}\right) = \frac{r(x)}{c^2}\frac{\partial^2 u}{\partial t^2}. \tag{5.8}$$

Boundary conditions for the vibrating string can take on a variety of forms.

(i) Dirichlet boundary conditions where the displacement at the ends is specified.

$$B(u)\Big|_{x=0} = u(0,t) = g_1(t)$$

$$B(u)\Big|_{x=L} = u(L,t) = g_2(t)$$

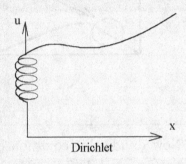
Dirichlet

where $g_1(t)$, $g_2(t)$ are given functions of t.

(ii) Neumann boundary conditions are where the derivatives at the ends are specified

$$B(u)\Big|_{x=0} = \frac{\partial u(0,t)}{\partial x} = g_3(t)$$

$$B(u)\Big|_{x=L} = \frac{\partial u(L,t)}{\partial x} = g_4(t)$$

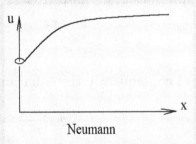
Neumann

where $g_3(t), g_4(t)$ are specified functions.

(iii) Robin type boundary conditions are where some linear combination of the displacement and slope are specified at a boundary.

$$B(u) = \left(\frac{\partial u}{\partial n} + hu\right)\Big|_{x=0} = g_5(t)$$

$$B(u) = \left(\frac{\partial u}{\partial n} + hu\right)\Big|_{x=L} = g_6(t).$$

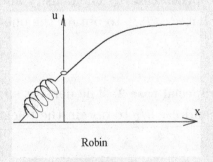
Robin

where $g_5(t), g_6(t)$ are specified functions and $\frac{\partial u}{\partial n}$ denotes a normal derivative given by $\frac{\partial u}{\partial n} = \operatorname{grad} u \cdot \hat{n}$ where $\hat{n}$ is a unit normal to the boundary. These types of boundary conditions result when certain types of elastic springs are attached to the ends of the string. Under such circumstances the boundary conditions are referred to as elastic boundary conditions.

Initial conditions for the vibrating string consist of

$$u(x,0) = f(x) \quad \text{Initial shape of string } f(x) \text{ specified.}$$

$$\frac{\partial u(x,0)}{\partial t} = g(x) \quad \text{Initial velocity of string } g(x) \text{ specified.}$$

Consider the small element of string denoted by PQ in the figure 5-3. This element has an arc length $ds = \sqrt{1 + \left(\frac{\partial u}{\partial x}\right)^2}\, dx \approx [1 + \frac{1}{2}\left(\frac{\partial u}{\partial x}\right)^2]\, dx$. The work done in moving a length of string dx on $y = 0$ to the position PQ of length ds is the potential energy of the element PQ. The work done in stretching the string from dx to ds is represented by a force times a distance. Assume there is a constant tension T_0 in the string, then the potential energy (PE) can be written as

$$PE = T_0[ds - dx] = \frac{T_0}{2}\left(\frac{\partial u}{\partial x}\right)^2 dx \quad \text{(dyne-cm).} \tag{5.9}$$

The kinetic energy (KE) of the element PQ is defined as one-half the mass times the velocity squared and can be represented

$$KE = \frac{1}{2}(\varrho\, dx)\left(\frac{\partial u}{\partial t}\right)^2 \quad \text{(dyne-cm).} \tag{5.10}$$

The total energy (E) associated with the string is the sum of the kinetic energy plus potential energy and is obtained by summing over all elements along the string. This produces the total energy

$$E = KE + PE = \frac{\varrho}{2}\int_0^L \left[\left(\frac{\partial u}{\partial t}\right)^2 + c^2\left(\frac{\partial u}{\partial x}\right)^2\right] dx, \tag{5.11}$$

where $c^2 = T_0/\varrho$ is constant with units of $(\text{cm/sec})^2$.

Consider the boundary-initial value problem for a vibrating string with forcing

$$\text{PDE:} \quad \frac{\partial^2 u}{\partial t^2} = c^2 \frac{\partial^2 u}{\partial x^2} + w(x,t) \quad 0 \le x \le L, \quad t > 0$$

$$\text{BC:} \quad u(0,t) = 0, \qquad u(L,t) = 0$$

$$\text{IC:} \quad u(x,0) = f(x), \qquad \frac{\partial u(x,0)}{\partial t} = g(x)$$

where $f(x)$ represents the initial wave shape and $g(x)$ represents the initial wave velocity. We assume that the functions f, g and w are well behaved continuous functions. We now show that when we obtain a solution of this boundary-initial value problem then the solution will be unique. To show this we assume there exists two different solutions, say $u_1(x, t)$ and $u_2(x, t)$ which both satisfy the partial differential equation as well as the boundary and initial conditions. We form the difference $v(x, t) = u_1(x, t) - u_2(x, t)$ and find $v(x, t)$ satisfies

$$
\begin{aligned}
\text{PDE:} \qquad & \frac{\partial^2 v}{\partial t^2} - c^2 \frac{\partial^2 v}{\partial x^2} = 0, \quad 0 \le x \le L, \quad t > 0 \\
\text{BC:} \qquad & v(0, t) = 0, \qquad v(L, t) = 0 \\
\text{IC:} \qquad & v(x, 0) = 0, \qquad \frac{\partial v(x, 0)}{\partial t} = 0.
\end{aligned}
$$

That is, $v(x, t)$ satisfies a homogeneous PDE, homogeneous BC, and homogeneous IC. The physical problem described by this type of partial differential equation is a string released from rest while it is in its equilibrium position. There are no external forces present and so we would expect that the string doesn't move. This is indeed the case and can be verified by an examination of the energy equation (5.11) for $v(x, t)$ which satisfies

$$
E = E(t) = \frac{\varrho}{2} \int_0^L \left[\left(\frac{\partial v}{\partial t} \right)^2 + c^2 \left(\frac{\partial v}{\partial x} \right)^2 \right] dx
$$

and

$$
\frac{dE}{dt} = \varrho \int_0^L \left(\frac{\partial v}{\partial t} \frac{\partial^2 v}{\partial t^2} + c^2 \frac{\partial v}{\partial x} \frac{\partial^2 v}{\partial x \partial t} \right) dx
$$

$$
\frac{dE}{dt} = \varrho \int_0^L \frac{\partial v}{\partial t} \left(\frac{\partial^2 v}{\partial t^2} - c^2 \frac{\partial^2 v}{\partial x^2} \right) dx + \varrho c^2 \int_0^L \frac{\partial}{\partial x} \left(\frac{\partial v}{\partial t} \frac{\partial v}{\partial x} \right) dx
$$

$$
\frac{dE}{dt} = \varrho c^2 \left[\frac{\partial v(L, t)}{\partial t} \frac{\partial v(L, t)}{\partial x} - \frac{\partial v(0, t)}{\partial t} \frac{\partial v(0, t)}{\partial x} \right] = 0.
$$

Hence, E is a constant. But $E(0) = 0$ so that E is zero for all time. The integrand for the energy is assumed continuous and it is nonnegative. If E is always zero, then $v(x, t)$ is a constant. However, $v(x, 0) = 0$ so that $v(x, t) = u_1(x, t) - u_2(x, t) = 0$ for all t, with $0 \le x \le L$. Therefore, we must have $u_1(x, t) = u_2(x, t)$ and so the solution is unique.

Wave equation with boundaries

The figure 5-4 illustrates a wave moving toward a fixed boundary along with a reflected negative virtual image of the wave. The moving wave travels along a string and produces an upward or downward transverse force depending upon the wave shape. When the transverse wave hits a fixed boundary an equal and opposite force is produced on the string by the wall (Newton's third law). This causes a reflected wave to come off from the wall with its shape reversed. This can be viewed as a superposition of the incoming wave and a negative virtual wave as they move past the boundary. Note that longitudinal waves just get reflected from the fixed boundary. Longitudinal waves in solids can reflect off free ends.

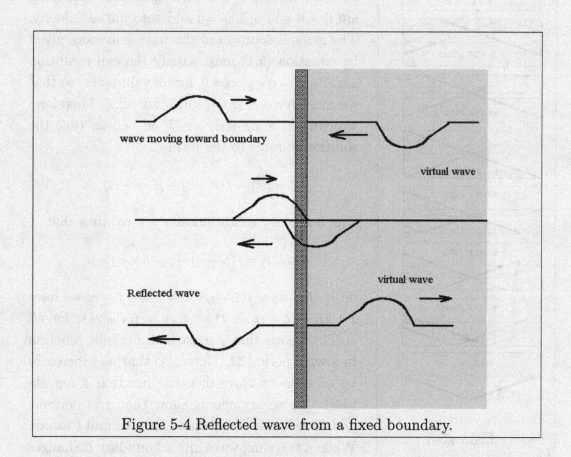

Figure 5-4 Reflected wave from a fixed boundary.

We consider wave motion with boundaries for the examples of a vibrating string and membrane. These physical problems will serve to illustrate the application of the method of separation of variables in one and two-dimensions.

Example 5-1.   (Vibrating string with fixed ends.)

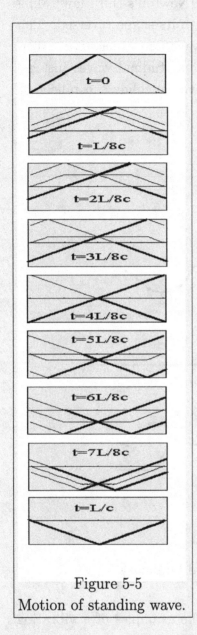

Figure 5-5
Motion of standing wave.

Examine equation (5.1) which represents the general solution of the wave equation. If we assume that $f(x)$ and $g(x)$ are only defined on the interval $0 \leq x \leq L$, then the general solution is only valid for the region $ct \leq x \leq L - ct$ as illustrated in the figure 5-5. Consider a vibrating string over the interval $0 < x < L$ with fixed ends satisfying $u(0,t) = 0$ and $u(L,t) = 0$ and zero initial velocity. The general solution of the wave equation, given by equation (5.1) must satisfy the end condition $u(0,t) = f(-ct) + g(ct) = 0$, for all values of t, so that we must have $f(-x) = -g(x)$, for all x. Therefore, we can write $g(x + ct) = -f(-x - ct)$ so that the solution reduces to the form

$$u(x,t) = f(x - ct) - f(-x - ct).$$

The boundary condition at $x = L$ requires that

$$u(L,t) = f(L - ct) - f(-L - ct) = 0$$

or $f(-L - ct) = f(L - ct)$. For $x = -L - ct$ we have $x + 2L = L - ct$ so that $f(x) = f(x + 2L)$, for all x. This shows that f must be a periodic function in x with period $2L$. Note also that as t increases by an amount $2L/c$, then the function f repeats itself. These arguments show that f is periodic and therefore is well defined for all x and t values. When a traveling wave hits a boundary it changes sign and gets reflected back. The figure 5-5 illustrates a string over the region $0 < x < L$ with fixed ends which has been given the initial shape

$$f(x) = \begin{cases} x, & 0 < x < \frac{L}{2} \\ L - x, & \frac{L}{2} < x < L \end{cases}$$

and initial velocity zero. In terms of traveling waves, this initial wave shape moves both right and left as time increases. The average value of the left moving wave shape and right moving wave shape is the middle curve in figure 5-5. Note also the reflected waves from the right and left moving wave shapes. Thus as time increases the resulting motion of the right and left moving wave is a standing vibrating wave which moves up and down. This is illustrated by the motion of the center curve in each figure.

∎

Example 5-2. (Vibrating string with no external force.)

Solve the boundary value problem

$$\text{PDE:} \qquad L(u) = \frac{\partial^2 u}{\partial t^2} - c^2 \frac{\partial^2 u}{\partial x^2} = 0, \quad 0 < x < L, \quad t > 0$$

$$\text{BC:} \qquad u(0,t) = 0, \quad u(L,t) = 0, \quad t > 0 \quad \text{fixed ends}$$

$$\text{IC:} \qquad u(x,0) = f(x), \quad \frac{\partial u(x,0)}{\partial t} = g(x)$$

Solution: Assume a solution $u(x,t) = F(x)G(t)$ and obtain the Sturm-Liouville system

$$F''(x) + \lambda F(x) = 0, \qquad F(0) = 0, \quad F(L) = 0$$

with eigenvalues $\lambda = \lambda_n = \left(\frac{n\pi}{L}\right)^2$ for $n = 1,2,3,\ldots$ and eigenfunctions given by $F(x) = F_n(x) = \sin \frac{n\pi x}{L}$ for $n = 1,2,3,\ldots$. The resulting differential equation for $G = G(t)$ is given by

$$G''(t) + c^2 \lambda G(t) = 0$$

which must be solved for each eigenvalue. The solutions are easily found to be

$$G = G_n(t) = A_n \cos\left(\frac{n\pi ct}{L}\right) + B_n \sin\left(\frac{n\pi ct}{L}\right)$$

and hence separable solutions have the form

$$u = u_n(x,t) = F_n(x)G_n(t) = \left[A_n \cos\left(\frac{n\pi ct}{L}\right) + B_n \sin\left(\frac{n\pi ct}{L}\right)\right] \sin \frac{n\pi x}{L}$$

for $n = 1,2,3,\ldots$. The functions $u_n(x,t)$ are called *normal modes* of vibration. The numbers $\omega_n = \frac{n\pi c}{L}$, for $n = 1,2,3,\ldots$ are called the *normal* frequencies. (Sometimes called *characteristic* frequencies.) The lowest frequency is called the fundamental frequency. When higher frequencies are integral multiples of the fundamental frequency the vibrations are referred to as harmonics. The fundamental frequency is called the first harmonic, and values of $n = 2, 3, \ldots$ are referred to as second harmonics, third harmonics, etc.

Sometimes $u_n(x, t)$ is written in the form $u_n(x, t) = C_n \sin\left(\frac{n\pi x}{L}\right) \cos(\omega_n t - \gamma_n)$ where $C_n = (A_n^2 + B_n^2)^{1/2}$ represents the amplitude of the vibration and the phase shift is denoted by $\gamma_n = \arctan(B_n/A_n)$. Note that each normal mode is a standing wave. See for example the figure 5-6. The function $u_n(x, t)$ is called a harmonic vibration with maximum displacement given by $\pm C_n \sin \frac{n\pi x}{L}$. The period of the vibration is $P = 2\pi/\omega_n$. In terms of sound vibrations, the frequency is the reciprocal of the period $f = 1/P = \omega_n/2\pi$ and determines the pitch of the sound, while the quantity C_n^2 determines the intensity of the sound.

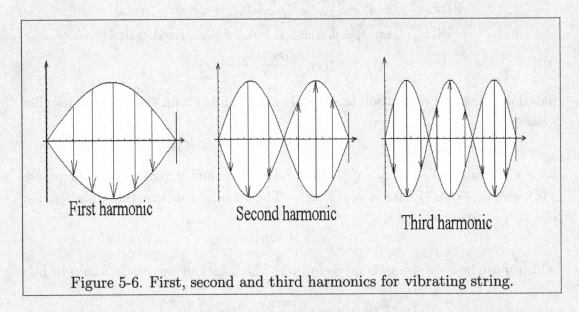

First harmonic   Second harmonic   Third harmonic

Figure 5-6. First, second and third harmonics for vibrating string.

By superposition the general solution can be expressed in the form

$$u = u(x, t) = \sum_{n=1}^{\infty} u_n(x, t)$$

$$u = u(x, t) = \sum_{n=1}^{\infty} [A_n \cos \omega_n t + B_n \sin \omega_n t] \sin\left(\frac{n\pi x}{L}\right). \tag{5.12}$$

This solution has the derivative

$$\frac{\partial u(x, t)}{\partial t} = \sum_{n=1}^{\infty} [-\omega_n A_n \sin \omega_n t + \omega_n B_n \cos \omega_n t] \sin\left(\frac{n\pi x}{L}\right).$$

The initial conditions require that

$$u(x,0) = \sum_{n=1}^{\infty} A_n \sin \frac{n\pi x}{L} = f(x)$$

and $\quad \dfrac{\partial u(x,0)}{\partial t} = \displaystyle\sum_{n=1}^{\infty} \omega_n B_n \sin \frac{n\pi x}{L} = g(x), \quad 0 < x < L.$

One can calculate the Fourier coefficients for these series by taking an inner product divided by a norm squared to obtain

$$A_n = \frac{(f, F_n)}{\|F_n\|^2} = \frac{2}{L} \int_0^L f(x) \sin \frac{n\pi x}{L}\, dx$$

and $\quad B_n = \dfrac{1}{\omega_n} \dfrac{(g, F_n)}{\|F_n\|^2} = \dfrac{2}{\omega_n L} \displaystyle\int_0^L g(x) \sin \frac{n\pi x}{L}\, dx.$

(5.13)

The equations (5.12) and (5.13) produce the solution to the original problem.

∎

Example 5-3. (**Vibrating string with external force.**)
Consider the boundary value problem

PDE: $\quad L(u) = \dfrac{\partial^2 u}{\partial t^2} - c^2 \dfrac{\partial^2 u}{\partial x^2} = h(x,t), \quad 0 < x < L, \quad t > 0$

BC: $\quad u(0,t) = 0, \qquad u(L,t) = 0, \quad t > 0 \quad$ fixed ends

IC: $\quad u(x,0) = 0, \qquad \dfrac{\partial u(x,0)}{\partial t} = 0$

Solution: This problem has homogeneous BC and IC's. Here an external driving force is applied to the string which is released from rest in its equilibrium position. We use the results from the previous example because the calculated eigenfunctions $F_n(x) = \sin \dfrac{n\pi x}{L}$ satisfy the given boundary conditions. We therefore assume a solution of the form

$$u = u(x,t) = \sum_{n=1}^{\infty} A_n(t) \sin \frac{n\pi x}{L} = \sum_{n=1}^{\infty} A_n(t) F_n(x) \tag{5.14}$$

where $A_n(t)$ are functions to be determined such that equation (5.14) satisfies the given initial conditions. This requires that $A_n(t)$ satisfy

$$A_n(0) = 0 \quad \text{and} \quad A_n'(0) = 0 \quad \text{for} \quad n = 1, 2, 3, \ldots \tag{5.15}$$

We substitute the assumed solution (5.14) into the PDE and obtain

$$\sum_{n=1}^{\infty} (A_n''(t) + \omega_n^2 A_n(t)) F_n(x) = h(x,t) \tag{5.16}$$

where $\omega_n = \dfrac{cn\pi}{L}$ for $n = 1, 2, 3, \ldots$. Now multiply both sides of equation (5.16) by $F_m(x)\,dx$ and then integrate from 0 to L to make use of the orthogonality properties of the set $\{F_n(x)\}$ over the interval $0 < x < L$, to obtain

$$\sum_{n=1}^{\infty} (A_n''(t) + \omega_n^2 A_n(t))(F_n, F_m) = \int_0^L h(x,t) F_m(x)\,dx. \tag{5.17}$$

The orthogonality of the set $\{F_n(x)\}$ makes all inner products on the left-hand side of equation (5.17) zero except in the case where the summation index n equals m. When n equals m the equation (5.17) simplifies to

$$(A_m''(t) + \omega_m^2 A_m(t))\|F_m\|^2 = \int_0^L h(x,t) F_m(x)\,dx \tag{5.18}$$

which gives the ordinary nonhomogeneous differential equation

$$A_m''(t) + \omega_m^2 A_m(t) = H_m(t) \tag{5.19}$$

where we have defined

$$H_m(t) = \frac{1}{\|F_m\|^2} \int_0^L h(x,t) F_m(x)\,dx = \frac{2}{L} \int_0^L h(x,t) \sin\frac{m\pi x}{L}\,dx. \tag{5.20}$$

The solution to equation (5.19) which satisfies the initial conditions given by equation (5.15) is obtained using the method of variation of parameters. That is, the homogeneous equation $A_m''(t) + \omega_m^2 A_m(t) = 0$ has the complementary solution

$$A_m(t) = c_1 \cos \omega_m t + c_2 \sin \omega_m t.$$

which suggests the form to assume for the nonhomogeneous equation. To solve the nonhomogeneous differential equation we assume a solution of the form

$$A_m(t) = \alpha(t) \cos \omega_m t + \beta(t) \sin \omega_m t \tag{5.21}$$

where $\alpha(t)$ and $\beta(t)$ are selected to satisfy the conditions

$$\alpha'(t) \cos \omega_m t + \beta'(t) \sin \omega_m t = 0$$
$$\alpha'(t)(-\omega_m \sin \omega_m t) + \beta'(t)(\omega_m \cos \omega_m t) = H_m(t). \tag{5.22}$$

This requires that

$$\alpha'(t) = \frac{-1}{\omega_m} H_m(t) \sin \omega_m t \quad \text{and} \quad \beta'(t) = \frac{1}{\omega_m} H_m(t) \cos \omega_m t \tag{5.23}$$

An integration gives the values

$$\alpha = \alpha(t) = -\int_0^t \frac{1}{\omega_m} H_m(\xi) \sin \omega_m \xi \, d\xi$$

$$\text{and} \quad \beta = \beta(t) = \int_0^t \frac{1}{\omega_m} H_m(\xi) \cos \omega_m \xi \, d\xi. \tag{5.24}$$

Substituting these values for α and β into the equation (5.21) we obtain the solution

$$A_m = A_m(t) = \frac{1}{\omega_m} \int_0^t H_m(\xi)[\cos \omega_m \xi \sin \omega_m t - \cos \omega_m t \sin \omega_m \xi] \, d\xi$$

$$A_m(t) = \frac{1}{\omega_m} \int_0^t H_m(\xi) \sin \omega_m(t - \xi) \, d\xi \tag{5.25}$$

We can now change m back to n and write the solution to the original PDE in the form

$$u = u(x,t) = \sum_{n=1}^{\infty} \left\{ \frac{1}{\omega_n} \int_0^t H_n(\xi) \sin \omega_n(t - \xi) \, d\xi \right\} \sin \frac{n\pi x}{L} \tag{5.26}$$

where the functions $H_n(t)$ are defined by equation (5.20).

∎

Example 5-4. (Superposition of previous results.)
Note that $L(u) = \dfrac{\partial^2 u}{\partial t^2} - c^2 \dfrac{\partial^2 u}{\partial x^2}$ is a linear operator so that we can write the solution to the boundary value problem

$$
\begin{aligned}
\text{PDE:} \quad & L(u) = \frac{\partial^2 u}{\partial t^2} - c^2 \frac{\partial^2 u}{\partial x^2} = h(x,t) \quad 0 < x < L \quad t > 0 \\
\text{BC:} \quad & u(0,t) = 0, \quad u(L,t) = 0, \quad t > 0 \quad \text{fixed ends} \\
\text{IC:} \quad & u(x,0) = f(x), \quad \frac{\partial u(x,0)}{\partial t} = g(x)
\end{aligned}
$$

in the form $u(x,t) = v(x,t) + w(x,t)$ where we select v to satisfy

$$
\begin{aligned}
\text{PDE:} \quad & L(v) = \frac{\partial^2 v}{\partial t^2} - c^2 \frac{\partial^2 v}{\partial x^2} = h(x,t), \quad 0 < x < L \quad t > 0 \\
\text{BC:} \quad & v(0,t) = 0, \quad v(L,t) = 0 \\
\text{IC:} \quad & v(x,0) = 0, \quad \frac{\partial v(x,0)}{\partial t} = 0
\end{aligned}
$$

(see example 5-3) and we select w to satisfy (see example 5-2)

$$\text{PDE:} \quad L(w) = \frac{\partial^2 w}{\partial t^2} - c^2 \frac{\partial^2 w}{\partial x^2} = 0, \quad 0 < x < L \quad t > 0$$

$$\text{BC:} \quad w(0,t) = 0, \quad w(L,t) = 0$$

$$\text{IC:} \quad w(x,0) = f(x), \quad \frac{\partial w(x,0)}{\partial t} = g(x).$$

To show that superposition of previous solutions satisfies the given problem, consider the following explanation. The operator L is a linear operator and so

$$\text{PDE:} \quad L(u) = L(v+w) = L(v) + L(w) = h(x,t) + 0 = h(x,t)$$

$$\text{BC:} \quad u(0,t) = v(0,t) + w(0,t) = 0 + 0 = 0$$

$$u(L,t) = v(L,t) + w(L,t) = 0 + 0 = 0$$

$$\text{IC:} \quad u(x,0) = v(x,0) + w(x,0) = 0 + f(x) = f(x)$$

$$\frac{\partial u(x,0)}{\partial t} = \frac{\partial v(x,0)}{\partial t} + \frac{\partial w(x,0)}{\partial t} = 0 + g(x) = g(x).$$

This gives the solution

$$u(x,t) = \sum_{n=1}^{\infty} (A_n \cos \omega_n t + B_n \sin \omega_n t) \sin \frac{n\pi x}{L} + \sum_{n=1}^{\infty} \left\{ \frac{1}{\omega_n} \int_0^t H_n(\xi) \sin \omega_n (t-\xi) \, d\xi \right\} \sin \frac{n\pi x}{L}$$

where A_n, B_n and H_n are given by the equations (5.13) and (5.20).

■

Sound waves in a gas or liquid

Sound waves are longitudinal elastic waves which are created by such things as vibrating strings, air columns, and vibrating surfaces. The word sound is used by most people to mean the vibrational wave motion that is capable of producing the sensation of hearing. The figure 5-7 illustrates a sound spectrum in terms of frequency f (number of oscillations performed in a second) which is measured in units called Hertz[†] (Hz). The frequency f of a simple harmonic motion is related to the period P of the oscillation by the reciprocal relation f (Hz) $= 1/P$ (sec^{-1}) = cycles per second. The physicist and mathematician are interested in all disturbances which produce some type of wave motion and do not restrict themselves to the study of sound in the region audible to humans. Some examples illustrating the importance of knowing and understanding the complete sound spectrum are:

---

† See Appendix C

(A) Subsonic waves: (i) Are produced by earthquakes. (ii) They are used in echo soundings and nondestructive testing of materials (finding flaws within the material). (iii) Locating tumors in the human body.

(B)) Ultrasonic waves: (i) Are used to remove kidney stones. (ii) Used to clean materials. (iii) Used to destroy living cells. (iv) Are used to perform many different types of medical diagnostic tests.

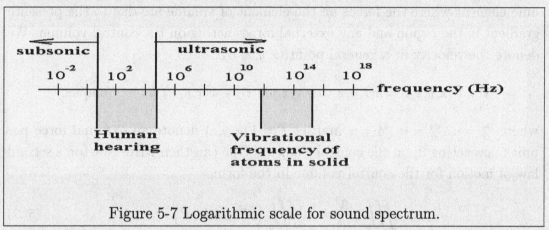

Figure 5-7 Logarithmic scale for sound spectrum.

We investigate sound waves that result from the compressibility of a fluid or gas having a density ϱ (gm/cm³). Imagine a surface S enclosing a volume V within the fluid. We examine an element of volume $d\tau = dxdydz$ (cm³), positioned at a point (x, y, z) inside the volume. This imagined volume is called a control volume. We will examine all the forces acting on this control volume to determine how it moves. Let $P = P(x, y, z)$ (dyne/cm²), denote a normal pressure (force per unit area) acting in the direction of the inward normal at each point of the surface. We sum all the pressure forces over the surface of the control volume to obtain a resultant force. This resultant force is given by

$$-\iint_S P(x, y, z)\hat{n}\, d\sigma \qquad \text{(dyne)}$$

where $\hat{n}$ is a unit exterior normal to the control surface and $d\sigma$ (cm²), is an element of surface area. The Gauss divergence theorem allows us to represent this resultant force in the form

$$-\iint_S P(x, y, z)\hat{n}\, d\sigma = -\iiint_V \text{grad}\, P(x, y, x)\, d\tau$$

where $d\tau$ is an element of volume. Let

$$\vec{F} = \vec{F}(x, y, z) = F_1(x, y, z)\hat{i} + F_2(x, y, z)\hat{j} + F_3(x, y, z)\hat{k} \qquad \text{(dyne/gm)}$$

denote an external force per unit mass acting at each point of the control volume. The resultant force is obtained by summing the external forces over all points within the control volume to obtain

$$\iiint_V \rho \vec{F} \, d\tau \qquad \text{(dyne)}$$

where $d\tau$ is an element of volume. We apply Newton's second law to the volume element where the forces on the element of volume are due to the pressure gradient in the region and any external forces acting on the control volume. We denote the velocity at a general point (x, y, z) by

$$\vec{V} = \vec{V}(x, y, z, t) = u(x, y, z, t)\hat{i} + v(x, y, z, t)\hat{j} + w(x, y, z, t)\hat{k}, \qquad \text{(cm/sec)}$$

where $\frac{dx}{dt} = u$, $\frac{dy}{dt} = v$, $\frac{dz}{dt} = w$ and let $\vec{F} = \vec{F}(x, y, z)$ denote an external force per unit mass acting upon the control element. One can then write Newton's second law of motion for the control volume in the form

$$\iiint_V \varrho \frac{d\vec{V}}{dt} \, d\tau = \iiint_V \left(\varrho \vec{F} - \operatorname{grad} P \right) d\tau. \qquad (5.27)$$

This implies that for an arbitrary volume V

$$\varrho \frac{d\vec{V}}{dt} = \varrho \vec{F} - \operatorname{grad} P. \qquad (5.28)$$

The equation (5.28) can also be represented by the scalar equations

$$
\begin{aligned}
\frac{\partial u}{\partial t} + \frac{\partial u}{\partial x} u + \frac{\partial u}{\partial y} v + \frac{\partial u}{\partial z} w &= F_1(x, y, z) - \frac{1}{\varrho} \frac{\partial P}{\partial x} \\
\frac{\partial v}{\partial t} + \frac{\partial v}{\partial x} u + \frac{\partial v}{\partial y} v + \frac{\partial v}{\partial z} w &= F_2(x, y, z) - \frac{1}{\varrho} \frac{\partial P}{\partial y} \\
\frac{\partial w}{\partial t} + \frac{\partial w}{\partial x} u + \frac{\partial w}{\partial y} v + \frac{\partial w}{\partial z} w &= F_3(x, y, z) - \frac{1}{\varrho} \frac{\partial P}{\partial z}
\end{aligned}
\qquad (5.29)
$$

which are called the Euler equations of motion for an ideal fluid or gas.

In addition the amount of fluid or gas entering the control volume must equal the amount of fluid or gas leaving the control volume. The mass of material inside the control volume is $m = \iiint_V \varrho \, d\tau$ and the change in mass with time is

$$\frac{\partial m}{\partial t} = \iiint_V \frac{\partial \varrho}{\partial t} \, d\tau. \qquad (5.30)$$

The flux of material crossing the surface of the control volume is

$$\iint_S -\varrho \vec{V} \cdot \hat{n}\, d\sigma = \iiint_V -\nabla(\varrho \vec{V})\, d\tau \tag{5.31}$$

where we have used the Gauss divergence theorem in order to express the surface integral in terms of a volume integral. Equating the equations (5.30) and (5.31) we obtain

$$\iiint_V \left(\frac{\partial \varrho}{\partial t} + \nabla(\varrho \vec{V}) \right) d\tau = 0.$$

The above equation represents conservation of mass and requires for an arbitrary control volume that the density ϱ and velocity V be related by the continuity equation

$$\frac{\partial \varrho}{\partial t} + \operatorname{div}(\varrho \vec{V}) = 0. \tag{5.32}$$

The equations (5.29) and (5.32) represent four equations in the five unknowns u, v, w, ϱ, P and so we need an additional equation to work with. One relation between pressure and density which is used quite often is

$$P = P_0 \left(\frac{\varrho}{\varrho_0} \right)^\gamma, \qquad \gamma = \frac{C_p}{C_v} \tag{5.33}$$

where P_0 is an initial pressure, ϱ_0 is an initial density, C_p is the gas specific heat at constant pressure, C_v is the gas specific heat at constant volume and γ is their ratio. Here P_0 and ϱ_0 represent the constant pressure and density of an undisturbed system. The equation (5.33) results from thermodynamic considerations using the assumption that the entropy of the system is constant (isentropic process).

In summary, the motion of sound waves can be described by the system of equations representing the compressibility of a fluid or gas.

$$\varrho \left(\frac{\partial \vec{V}}{\partial t} + \vec{V} \cdot \nabla \vec{V} \right) = \vec{F} - \operatorname{grad} P \tag{5.34}$$

$$\frac{\partial \varrho}{\partial t} + \nabla \left(\varrho \vec{V} \right) = 0 \tag{5.35}$$

$$P = P_0 \left(\frac{\varrho}{\varrho_0} \right)^\gamma, \quad \gamma = \frac{C_p}{C_v} \tag{5.36}$$

where P denotes the pressure, ϱ is the density of the gas or liquid and $\vec{V}$ has the velocity components u, v, w. The above system represents five equations in five unknowns.

The system of equations (5.34), (5.35), (5.36) are nonlinear equations and can be difficult to solve. By making suitable assumptions the system of nonlinear equations can be linearized to a form where they are more tractable. Toward this goal we make the following definition and assumptions:

1. Define the condensation

$$s = s(x, y, z, t) = \frac{\varrho - \varrho_0}{\varrho_0} \tag{5.37}$$

as representing the relative change of the gas or fluid density.

2. Assume the velocity components, density and condensation are small so that product terms like $\vec{V} \cdot \nabla \vec{V}$, $s\nabla \vec{V}$, $\vec{V} \cdot \nabla \varrho$ and s^2 are small and can be neglected.

3. There are no external forces.

The above assumptions reduce the equations (5.34) to the form

$$\frac{\partial u}{\partial t} = -\frac{1}{\varrho}\frac{\partial P}{\partial x}, \qquad \frac{\partial v}{\partial t} = -\frac{1}{\varrho}\frac{\partial P}{\partial y}, \qquad \frac{\partial w}{\partial t} = -\frac{1}{\varrho}\frac{\partial P}{\partial z} \tag{5.38}$$

$$\text{or} \qquad \frac{\partial \vec{V}}{\partial t} = -\nabla P.$$

The equation (5.37) implies that

$$\varrho = \varrho_0(1 + s) \tag{5.39}$$

and so the equation (5.36) can be represented as a power series in s having the form

$$P = P_0(1 + s)^\gamma = P_0 \left(1 + \gamma s + \gamma(\gamma - 1)\frac{s^2}{2!} + \cdots \right).$$

The linear form of the above equation is obtained by taking the first two terms of this series and neglecting the other terms. This gives the linear relation

$$P = P_0(1 + \gamma s). \tag{5.40}$$

We substitute the equations (5.40) and (5.39) into the continuity equation (5.35) and obtain

$$\varrho_0 \frac{\partial s}{\partial t} + \varrho_0 \nabla \vec{V} + \varrho s \nabla \vec{V} + \nabla \varrho \cdot \vec{V} = 0.$$

Neglecting small product terms produces the linearized form of the continuity equation

$$\frac{\partial s}{\partial t} + \nabla \vec{V} = 0. \tag{5.41}$$

Substituting the equations (5.40) and (5.39) into the equations (5.38) and using the approximation

$$\frac{P_0\gamma}{\varrho_0(1+s)} \approx \frac{P_0\gamma}{\varrho_0} = a^2 = \text{a constant,}$$

then the Euler equations reduce to the linearized form

$$\frac{\partial u}{\partial t} = -a^2\frac{\partial s}{\partial x}, \qquad \frac{\partial v}{\partial t} = -a^2\frac{\partial s}{\partial y}, \qquad \frac{\partial w}{\partial t} = -a^2\frac{\partial s}{\partial z}$$

$$\text{or} \qquad \frac{\partial \vec{V}}{\partial t} = -a^2\nabla s. \tag{5.42}$$

In summary, the linearized form of the equations (5.34), (5.35), (5.36) are given by

$$\frac{\partial \vec{V}}{\partial t} = -a^2\nabla s \qquad \text{where} \qquad a^2 = \frac{P_0\gamma}{\varrho_0} \tag{5.43}$$

$$\frac{\partial s}{\partial t} + \nabla \vec{V} = 0 \tag{5.44}$$

$$P = P_0(1+\gamma s) \tag{5.45}$$

where $\gamma = C_p/C_v$ is the ratio of specific heats and $s = (\varrho - \varrho_0)/\varrho_0$ is the condensation. These represent five equations in the five unknowns u, v, w, P, s.

Now differentiate the equation (5.41) with respect to time t and use equations (5.42) to obtain

$$\frac{\partial^2 s}{\partial t^2} = -\frac{\partial}{\partial t}\nabla \vec{V} = -\nabla\left(\frac{\partial \vec{V}}{\partial t}\right) = a^2\nabla^2 s = a^2\left(\frac{\partial^2 s}{\partial x^2} + \frac{\partial^2 s}{\partial y^2} + \frac{\partial^2 s}{\partial z^2}\right). \tag{5.46}$$

We find the condensation s is a solution of the three-dimensional wave equation. Note also that one can differentiate the equation (5.40) to obtain

$$\frac{\partial^2 P}{\partial t^2} = P_0\gamma\frac{\partial^2 s}{\partial t^2} \quad \text{and} \quad \nabla^2 P = P_0\gamma\nabla^2 s \quad \text{so that} \quad \frac{\partial^2 P}{\partial t^2} = a^2\nabla^2 P$$

which shows the pressure $P = P(x, y, z, t)$ also satisfies the three-dimensional wave equation. Similarly, it can be shown that the velocity $\vec{V}$ satisfies the wave equation since

$$\frac{\partial^2 \vec{V}}{\partial t^2} = -a^2\nabla\left(\frac{\partial s}{\partial t}\right) = -a^2\nabla\left(-\nabla \vec{V}\right) = a^2\nabla^2 \vec{V}.$$

If the velocity $\vec{V}(x,y,z,t)$ is derivable from a potential function $U = U(x,y,z,t)$ such that

$$\vec{V} = \vec{V}(x,y,z,t) = -\text{grad}\,U(x,y,z,t) = -\nabla U \qquad (5.47)$$

then from equation (5.42) one can write

$$\vec{V} = -a^2 \int_0^t \text{grad}\,s\,dt + \vec{C} = -a^2\text{grad}\left(\int_0^t s\,dt\right) + \vec{C} \qquad (5.48)$$

where $\vec{C}$ is a constant of integration. We assign a value for the constant $\vec{C}$ by setting $t = 0$. This gives $\vec{V}(x,y,z,0) = \vec{C} = -\text{grad}\,U(x,y,z,0)$ and so we can write

$$\vec{V} = -\text{grad}\left(U(x,y,z,0) + a^2 \int_0^t s\,dt\right) = -\text{grad}\,U(x,y,z,t) = -\nabla U \qquad (5.49)$$

or

$$U = U(x,y,z,t) = U(x,y,z,0) + a^2 \int_0^t s\,dt. \qquad (5.50)$$

Differentiate the equation (5.50) and use equation (5.41) to show

$$\begin{aligned}
\frac{\partial U}{\partial t} &= a^2 s \\
\frac{\partial^2 U}{\partial t^2} &= a^2 \frac{\partial s}{\partial t} = -a^2 \nabla \vec{V} = -a^2 \nabla(-\nabla U) = a^2 \nabla^2 U
\end{aligned} \qquad (5.51)$$

which shows the potential function U also satisfies the three-dimensional wave equation.

The figure 5-8 illustrates sound from a source which moves through some kind of gas medium. If the source is fixed the sound speed remains constant and depends upon the physical properties of the gas. The frequency or the wavelength of the sound waves is constant when the source of the sound is fixed. Subsonic conditions are said to exist if the source moves slower than the speed of sound. No matter what the speed of the source, the sound waves move out from it at the speed of sound.[‡] As the source moves faster the waves upstream from the direction of motion tend to bunch up and the wavelength shortens while downstream the waves spread out and the wavelength increases. This produces a doppler effect. Sonic or transonic sound conditions are said to exist when the speed of the source approaches the speed of sound. As the source approaches the speed of sound the upstream wavelength approaches zero and the waves bunch

---

[‡] $\approx 344\,\text{m/s}$ in air at $20°C$ and $\approx 1482\,\text{m/s}$ in water at $20°C$

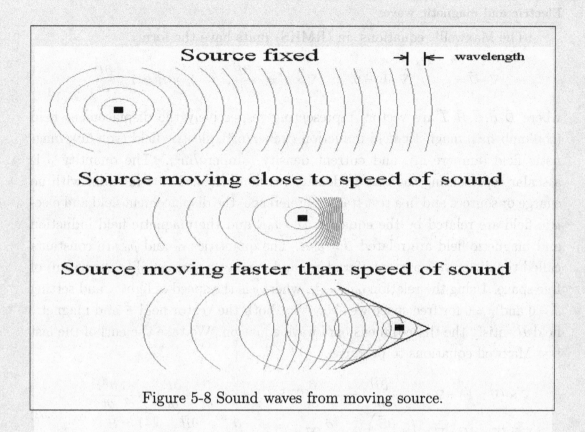

Source fixed   →| |← wavelength

Source moving close to speed of sound

Source moving faster than speed of sound

Figure 5-8 Sound waves from moving source.

up or collect into a single wave called a shock wave. The flow conditions change dramatically across a shock wave and special considerations must be given to a moving shock wave. Conditions where the source moves faster than the speed of sound are said to be supersonic. The wave fronts upstream produce a conical shock wave. The angle of the cone depends upon the relation between the speed of the source and the speed of sound. The cone angle gets smaller and smaller as the source speed increases above the speed of sound which remains constant. The ratio defined as the speed of the source divided by the speed of sound is called the Mach[†] number. For Mach numbers less than one, subsonic conditions exist, while for Mach numbers near one, transonic conditions exist and for Mach numbers greater than one, supersonic conditions occur.

[†] See Appendix C

Electric and magnetic waves

The Maxwell[†] equations in (RMKS) units have the form

$$\nabla \cdot \vec{D} = \varrho, \qquad \nabla \cdot \vec{B} = 0, \qquad \nabla \times \vec{E} = -\frac{\partial \vec{B}}{\partial t}, \qquad \nabla \times \vec{H} = \vec{J} + \frac{\partial \vec{D}}{\partial t}$$

where $\vec{D}, \vec{B}, \vec{E}, \vec{H}, \vec{J}$ are vectors representing respectively the displacement field (coulomb/m²), magnetic field induction (weber/m²), electric field (volt/m), magnetic field (ampere/m), and current density (ampere/m²). The quantity ϱ is a scalar representing the charge density (coulomb/m³). For a medium with no charge or sources and in a rest frame of reference, the displacement field and electric field are related by the equation $\vec{D} = \epsilon_0 \vec{E}$ and the magnetic field induction and magnetic field are related $\vec{B} = \mu_0 \vec{H}$. The quantities ϵ_0 and μ_0 are constants called the dielectric constant (farad/m) and magnetic permeability (henry/m) of free space. Using the relation $\epsilon_0 \mu_0 = \frac{1}{c^2}$, where c is the speed of light[†] , and setting $\vec{J} = 0$ and $\varrho = 0$ for free space we show that both the vector field $\vec{E}$ and magnetic field $\vec{H}$ satisfy the three-dimensional wave equation. We take the curl of the last two Maxwell equations to produce

$$\nabla \times (\nabla \times \vec{E}) = \nabla \times (-\mu_0 \frac{\partial \vec{H}}{\partial t}) = -\mu_0 \frac{\partial}{\partial t}(\nabla \times \vec{H}) = -\mu_0 \frac{\partial}{\partial t}(\epsilon_0 \frac{\partial \vec{E}}{\partial t}) = \frac{-1}{c^2}\frac{\partial^2 \vec{E}}{\partial t^2}$$

$$\nabla \times (\nabla \times \vec{H}) = \nabla \times \left(\epsilon_0 \frac{\partial \vec{E}}{\partial t}\right) = \epsilon_0 \frac{\partial}{\partial t}(\nabla \times \vec{E}) = \epsilon_0 \frac{\partial}{\partial t}(-\mu_0 \frac{\partial \vec{H}}{\partial t}) = \frac{-1}{c^2}\frac{\partial^2 \vec{H}}{\partial t^2}$$

(5.52)

Using the vector identity $\nabla \times (\nabla \times \vec{A}) = -\nabla^2 \vec{A} + \nabla(\nabla \cdot \vec{A})$ we can express the equations (5.52) in the form of the wave equations

$$\frac{\partial^2 \vec{E}}{\partial t^2} = c^2 \nabla^2 \vec{E} \qquad \text{and} \qquad \frac{\partial^2 \vec{H}}{\partial t^2} = c^2 \nabla^2 \vec{H}. \qquad (5.53)$$

That is, each component (E_1, E_2, E_3) of $\vec{E}$ and each component (H_1, H_2, H_3) of $\vec{H}$ must satisfy the wave equation.

Example 5-5. (Polarization and Maxwell's equations).

Let us examine a vibrational wave motion along a semi-infinite string. Recall that a transverse wave occurs when the wave displacement is perpendicular to the wave direction of energy propagation. Consider the special case where the

---

[†] Speed of light in vacuum is $c = 2.997925(10)^8$ m/s.

string displacements are restricted to the x–y plane. In this special case the wave motion is said to be plane polarized or linearly polarized and the plane is called the plane of polarization. Now consider the same wave motion but this time assume displacements are restricted to lie in the x-z plane as illustrated in the figure 5-9.

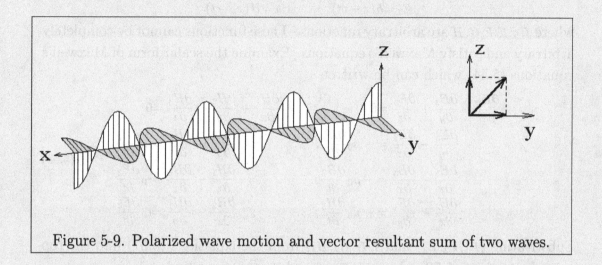

Figure 5-9. Polarized wave motion and vector resultant sum of two waves.

If a string transmits two or more waves all having different planes of polarization, then by superposition one can determine the resultant wave motion. In the case of polarized plane waves the superposition must be a vector sum rather than an algebraic sum of the wave displacements.

We have shown that for free space where $\rho = 0$ and $\vec{J} = 0$, the electric field $\vec{E}$ and magnetic field $\vec{H}$ must satisfy the Maxwell equations

$$\nabla \cdot \vec{E} = 0, \qquad \nabla \cdot \vec{H} = 0, \qquad \nabla \times \vec{E} = -\mu_0 \frac{\partial \vec{H}}{\partial t}, \qquad \nabla \times \vec{H} = \epsilon_0 \frac{\partial \vec{E}}{\partial t} \qquad (5.54)$$

where both $\vec{E}$ and $\vec{H}$ satisfy the wave equations

$$\frac{\partial^2 \vec{E}}{\partial t^2} = c^2 \nabla^2 \vec{E} \quad \text{and} \quad \frac{\partial^2 \vec{H}}{\partial t^2} = c^2 \nabla^2 \vec{H}. \qquad (5.55)$$

where $c^2 = \frac{1}{\mu_0 \epsilon_0}$. We write $\vec{E}$ and $\vec{H}$ in the component form

$$\vec{E} = E_1 \hat{i} + E_2 \hat{j} + E_3 \hat{k} \qquad \text{and} \qquad \vec{H} = H_1 \hat{i} + H_2 \hat{j} + H_3 \hat{k}, \qquad (5.56)$$

then each component $\{E_1, E_2, E_3\}$ of $\vec{E}$ and each component $\{H_1, H_2, H_3\}$ of $\vec{H}$ must be solutions of the wave equation. The solutions of the wave equations for $\vec{E}$ and

$\vec{H}$ cannot be arbitrary waves as Maxwell's equations puts certain restrictions on the wave forms. Consider for example a polarized wave in the x-y plane moving in the positive x-direction. We assume solutions to the equations (5.55) of the form

$$E_1 = f(x - ct), \qquad H_1 = F(x - ct)$$
$$E_2 = g(x - xt), \qquad H_2 = G(x - ct) \qquad (5.57)$$
$$E_3 = h(x - ct), \qquad H_3 = H(x - ct)$$

where f, g, h, F, G, H are arbitrary functions. These functions cannot be completely arbitrary and satisfy Maxwell's equations. Examine the scalar form of Maxwell's equations (5.54) which can be written

$$\frac{\partial E_1}{\partial x} + \frac{\partial E_2}{\partial y} + \frac{\partial E_3}{\partial z} = 0, \qquad \frac{\partial H_1}{\partial x} + \frac{\partial H_2}{\partial y} + \frac{\partial H_3}{\partial z} = 0$$
$$\frac{\partial E_3}{\partial y} - \frac{\partial E_2}{\partial z} = -\mu_0 \frac{\partial H_1}{\partial t}, \qquad \frac{\partial H_3}{\partial y} - \frac{\partial H_2}{\partial z} = \epsilon_0 \frac{\partial E_1}{\partial t}$$
$$\frac{\partial E_1}{\partial z} - \frac{\partial E_3}{\partial x} = -\mu_0 \frac{\partial H_2}{\partial t}, \qquad \frac{\partial H_1}{\partial z} - \frac{\partial H_3}{\partial x} = \epsilon_0 \frac{\partial E_2}{\partial t} \qquad (5.58)$$
$$\frac{\partial E_2}{\partial x} - \frac{\partial E_1}{\partial y} = -\mu_0 \frac{\partial H_3}{\partial t}, \qquad \frac{\partial H_2}{\partial x} - \frac{\partial H_1}{\partial y} = \epsilon_0 \frac{\partial E_3}{\partial t}$$

Substituting $\{E_1, E_2, E_3\}$ and $\{H_1, H_2, H_3\}$ from the equations (5.57) into the top row of equations (5.58) where $\nabla \cdot \vec{E} = 0$ and $\nabla \cdot \vec{H} = 0$, we find

$$\frac{\partial f}{\partial x} = 0 \qquad \text{and} \qquad \frac{\partial F}{\partial x} = 0$$

which tells us that f and F must not depend upon x. This also tells us that Maxwell's equations do not permit longitudinal waves (i.e. E_1 and H_1 displacements in the direction of the wave motion.) If we set $f = F = 0$, then only E_2, E_3 and H_2, H_3 components remain and these give transverse waves which are perpendicular to the wave motion. This can be demonstrated by representing the $\vec{E}$ wave motion for the remaining E_2, E_3 components as a polarized wave moving in the x-y plane. Let $H_1 = 0$ with H_2, H_3 to be determined and set

$$E_1 = 0, \qquad E_2 = g(x - ct), \qquad E_3 = 0, \qquad (5.59)$$

then the last three rows of Maxwell's equations (5.58) reduce to

$$0 - 0 = 0, \qquad\qquad 0 - 0 = 0$$
$$0 - 0 = -\mu_0 \frac{\partial H_2}{\partial t}, \qquad 0 - \frac{\partial H_3}{\partial x} = \epsilon_0 \left(-c \frac{dg}{d\xi} \right) \qquad (5.60)$$
$$\frac{dg}{d\xi} - 0 = -\mu_0 \frac{\partial H_3}{\partial t}, \qquad \frac{\partial H_2}{\partial x} - 0 = 0$$

where $\xi = x - ct$. The equations (5.60) require that H_2 be a constant which we take as zero. The equations (5.60) which involve H_3 can be integrated. We write

$$\frac{\partial H_3}{\partial t} = \frac{-1}{\mu_0}\frac{dg}{d\xi}, \qquad \frac{\partial H_3}{\partial x} = c\epsilon_0 \frac{dg}{d\xi}$$

$$\frac{dH_3}{d\xi}(-c) = \frac{-1}{\mu_0}\frac{dg}{d\xi}, \qquad \frac{dH_3}{d\xi} = c\epsilon_0 \frac{dg}{d\xi}$$

which gives

$$H_3 = \frac{1}{c\mu_0}g = c\epsilon_0 g = \sqrt{\frac{\epsilon_0}{\mu_0}}g$$

because $c^2 = 1/\mu_0\epsilon_0$ and hence $c\epsilon_0 = 1/c\mu_0 = \sqrt{\epsilon_0/\mu_0}$. This gives the special waves

$$
\begin{aligned}
E_1 &= 0, & H_1 &= 0 \\
E_2 &= g(x - ct), & H_2 &= 0 \\
E_3 &= 0, & H_3 &= \frac{\epsilon_0}{\mu_0}g(x - ct).
\end{aligned}
\tag{5.61}
$$

Hence, Maxwell's equations require that both the electric and magnetic waves must be transverse waves of the same form and the plane of polarization for the electric wave must be perpendicular to the plane of polarization for the magnetic wave. ∎

If periodic solutions to the wave equations given by equations (5.55) are desired, it is customary to substitute into the equations (5.55) the following change of variables:

$$\vec{E}(x,y,z,t) = \vec{\mathcal{E}}(x,y,z)e^{-i\omega t} \quad \text{and} \quad \vec{H}(x,y,x,t) = \vec{\mathcal{H}}(x,y,z)e^{-i\omega t}. \tag{5.62}$$

This produces the new equations

$$\nabla^2\vec{\mathcal{E}} = -c^2\omega^2\vec{\mathcal{E}} \quad \text{and} \quad \nabla^2\vec{\mathcal{H}} = -c^2\omega^2\vec{\mathcal{H}} \tag{5.63}$$

which implies that the components $\mathcal{E}_i$ of $\vec{\mathcal{E}}$ and $\mathcal{H}_i$ of $\vec{\mathcal{H}}$, for $i = 1, 2, 3$ must satisfy the Helmholtz equation

$$\frac{\partial^2 u}{\partial x^2} + \frac{\partial^2 u}{\partial y^2} + \frac{\partial^2 u}{\partial z^2} + k^2 u = 0 \tag{5.64}$$

with $k^2 = c^2\omega^2$. The corresponding electric and magnetic fields are the real parts of the quantities $\vec{\mathcal{E}}e^{-i\omega t}$ and $\vec{\mathcal{H}}e^{-i\omega t}$. The components of the electric vector $\vec{\mathcal{E}}$ and

248

the magnetic vector $\vec{\mathcal{H}}$ determine the electromagnetic field of a wave. These components are solutions of the Helmholtz equation (5.64). Boundary conditions associated with the Helmholtz equation require that at a boundary, the tangential component of $\vec{\mathcal{E}}$ and the normal component of $\vec{\mathcal{H}}$ must vanish. These conditions enable us to construct one of the following electromagnetic fields:

$(0, \mathcal{E}_2, 0); (\mathcal{H}_1, 0, \mathcal{H}_2)$ TE wave (Transverse electric).

$(\mathcal{E}_1, 0, \mathcal{E}_3) : (0, \mathcal{H}_2, 0)$ TM wave (Transverse magnetic).   (5.65)

$(\mathcal{E}_1, \mathcal{E}_2, 0); (\mathcal{H}_1, \mathcal{H}_2, 0)$ TEM wave (Transverse electromagnetic).

In the electromagnetic fields given by the equations (5.65) the z-axis or $\hat{k}$ direction is assumed to be the direction of wave propagation. The TE wave has an electric vector perpendicular to the direction of the wave motion and the magnetic vector is different from zero along the direction of wave motion and is perpendicular to the $\mathcal{E}$ field. The TM wave is opposite that of the TE wave. The TEM wave has both electric and magnetic components perpendicular to the direction of motion. Electric fields are usually described as some combination of the above three fields.

Wave motion in a solid

Hooke's[†] experiment considered the extension of a wire, of uniform cross-section A, by a tensile force W as illustrated in the figure 5-10. The stress and strain are defined

$$\text{Strain} = \epsilon = \frac{\Delta L}{L} = \frac{\text{Change in length}}{\text{Original length}}$$

$$\text{Stress} = \sigma = \frac{W}{A} = \frac{\text{Force}}{\text{Area}}$$

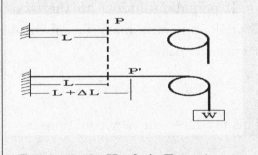

Figure 5-10. Hooke's Experiment.

Hooke's law states that stress is proportional to strain $\sigma = E\epsilon$ where E is the proportionality constant called the Young's[†] modulus of elasticity. The value of E is constant for uniform and homogeneous materials, but for nonuniform nonhomogeneous material $E = E(x)$ is a function of position.

[†]   See Appendix C

Pochammer's[†] experiment is to consider a bar of variable cross section denoted by $A = A(x)$, Young's modulus of elasticity $E = E(x)$ and having variable density $\varrho = \varrho(x)$ such as the bar illustrated in the figure 5-11. Let the bar undergo longitudinal vibrations so that at time t the cross-sectional area at P, at point x, is given a longitudinal displacement $u(x,t)$ so that P moves to the point P'. A neighboring cross-sectional area at Q, at point $x + \Delta x$, will undergo the displacement $u(x+\Delta x, t)$ and move to the point Q'. The strain in the element PQ at time t is calculated as

$$\text{strain} = \epsilon = \frac{P'Q' - PQ}{PQ} = \frac{[x + \Delta x + u(x + \Delta x, t) - x - u(x,t)] - \Delta x}{\Delta x}$$

$$= \frac{u(x + \Delta x, t) - u(x,t)}{\Delta x}$$

Now let $\Delta x \to 0$, and find the strain at time t is given by

$$\epsilon = \epsilon(x) = \frac{\partial u(x,t)}{\partial x}.$$

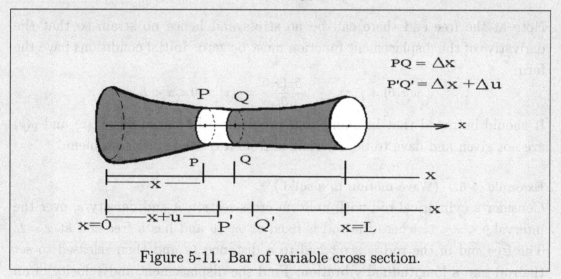

$$PQ = \Delta x$$
$$P'Q' = \Delta x + \Delta u$$

Figure 5-11. Bar of variable cross section.

Therefore, the stress at position x and time t is given by

$$\sigma(x,t) = E(x)\epsilon(x,t) = E(x)\frac{\partial u(x,t)}{\partial x}.$$

Applying Newton's second law of motion to the element PQ we find

$$\varrho(x + \frac{\Delta x}{2})\Delta x \frac{\partial^2 u(x + \frac{\Delta x}{2}, t)}{\partial t^2} = A(x + \Delta x)\sigma(x + \Delta x, t) - A(x)\sigma(x,t)$$

$$= A(x + \Delta x)E(x + \Delta x)\frac{\partial u(x + \Delta x, t)}{\partial x} - A(x)E(x)\frac{\partial u(x,t)}{\partial x}$$

---

†   See Appendix C

Upon dividing by Δx and taking the limit as $\Delta x \to 0$, we obtain the equation of motion

$$\frac{\partial}{\partial x}\left(A(x)E(x)\frac{\partial u}{\partial x}\right) = \varrho(x)\frac{\partial^2 u}{\partial t^2}$$

known as Pochammer's equation In the special case A, E, ϱ are constants, this equation reduces to the one-dimensional wave equation

$$\frac{\partial^2 u}{\partial t^2} = \alpha^2 \frac{\partial^2 u}{\partial x^2} \qquad \text{where} \qquad \alpha^2 = \frac{AE}{\varrho}.$$

Typical boundary conditions are:

$$\text{Dirichlet (fixed ends)} \qquad u(0,t) = 0, \quad u(L,t) = 0 \quad t > 0$$

$$\begin{array}{l}\text{Mixed (One end fixed and}\\ \text{the other end stress free.)}\end{array} \qquad u(0,t) = 0, \quad \frac{\partial u(L,t)}{\partial x} = 0$$

Note at the free end there can be no stress and hence no strain so that the derivative of the displacement function must be zero. Initial conditions have the form

$$u(x,0) = f(x), \qquad \frac{\partial u(x,0)}{\partial t} = g(x), \quad 0 < x < L$$

It should be noted that in many "bar problems" the initial data $f(x)$ and $g(x)$ are not given and have to be determined from a related statics problem.

Example 5-6.   (Wave motion in a solid.)

Consider a cylindrical rod with uniform cross-section A and density ρ, over the interval $0 \le x \le L$, where the rod is fixed at $x = 0$ and has a free end at $x = L$. The free end of the rod is stretched to a distance L_1 and then released to set the rod into a longitudinal vibration. Find the displacement and velocity of an arbitrary cross-section.

Solution: We solve the wave equation

$$\frac{\partial^2 u}{\partial t^2} = \alpha^2 \frac{\partial^2 u}{\partial x^2}, \qquad \alpha^2 = \frac{AE}{\rho} \tag{5.66}$$

subject to the boundary conditions

$$u(0,t) = 0 \qquad \text{and} \qquad \frac{\partial u(L,t)}{\partial x} = 0 \tag{5.67}$$

where $u = u(x,t)$ denotes the displacement of a cross-section at position x at the time t. Initially the strain in the rod is given by

$$\text{strain} = \frac{\text{change in length}}{\text{original length}} = \frac{L_1 - L}{L} = \epsilon = \frac{du}{dx} \tag{5.68}$$

which gives the initial displacement

$$u(x,0) = \epsilon x, \qquad 0 < x < L \tag{5.69}$$

An arbitrary cross-section is initially at rest and so we write the initially velocity as

$$\frac{\partial u(x,0)}{\partial t} = 0, \qquad 0 < x < L \tag{5.70}$$

We write the general solution to the wave equation (5.66) in the form

$$u = u(x,t) = F(\alpha t - x) + G(\alpha t + x) \tag{5.71}$$

where F, G are arbitrary functions to be determined from the boundary and initial conditions. The boundary condition $u(0,t) = 0$ requires that

$$u(0,t) = F(\alpha t) + G(\alpha t) = 0. \tag{5.72}$$

This condition implies

$$G(z) = -F(z), \qquad z = \alpha t \tag{5.73}$$

and allows one to write the solution given by equation (5.71) in the form

$$u(x,t) = F(\alpha t - x) - F(\alpha t + x). \tag{5.74}$$

The initial condition given by equation (5.70) requires

$$\frac{\partial u(x,0)}{\partial t} = F'(-x)\alpha - F'(x)\alpha = 0 \qquad \text{or} \qquad F'(-x) = F'(x) \tag{5.75}$$

Now differentiate the equation (5.74) with respect to x and show

$$\frac{\partial u}{\partial x} = F'(\alpha t - x)(-1) - F'(\alpha t + x)(1) \tag{5.76}$$

where primes $'$ denote differentiation with respect to the argument of the function. The boundary condition at $x = L$, from equation (5.67), requires

$$-F'(\alpha t - L) - F'(\alpha t + L) = 0. \tag{5.77}$$

Let $\alpha t + L = x$ with $\alpha t - L = x - 2L$ and show the equation (5.77) implies

$$F'(x) = -F'(x - 2L) \tag{5.78}$$

Note also the initial condition from equation (5.69) requires

$$u(x,0) = \epsilon x = F(-x) - F(x), \qquad 0 < x < L \tag{5.79}$$

and by differentiating with respect to x we find

$$\epsilon = F'(-x)(-1) - F'(x), \qquad 0 < x < L. \tag{5.80}$$

We solve the equations (5.75) and (5.80) simultaneously and show

$$F'(x) = -\frac{\epsilon}{2}, \qquad -L < x < L \tag{5.81}$$

In the boundary condition equation (5.78) we observe that when $x = L^-$ we have $F'(L^-) = -F'(-L^+) = \epsilon/2$ and when $x = 3L^-$, we have $F'(3L^-) = F'(L^-) = \epsilon/2$. This requires that

$$F'(x) = \frac{\epsilon}{2}, \qquad L < x < 3L \tag{5.82}$$

The above equations imply that $F(x)$ should be periodic with period $4L$. The requirement $F(x) = F(x + 4L)$ defines $F(x)$ for all values of x. The solution is therefore

$$u = u(x,t) = F(\alpha t - x) - F(\alpha t + x) \tag{5.83}$$

where

$$F(x) = \begin{cases} -\epsilon x/2, & -L < x < L \\ \epsilon x/2, & L < x < 3L \end{cases} \tag{5.84}$$

with $F(x + 4L) = F(x)$ for all x values. Note that the solution given by equation (5.83) represents a forward moving wave and a backward moving wave. When these waves hit a boundary (free or fixed) they are reflected back in the direction from where they came. Consider the backward moving wave starting at the end $x = L$. The time interval for the backward moving wave to travel to a point x in the bar is $\Delta t = (L - x)/\alpha$. The cross-section at x then begins to vibrate. The backward moving wave continues to $x = 0$ where it gets reflected and travels back to the point x in the time interval $\Delta t = 2x/\alpha$. Similarly the time interval for the forward moving wave, starting at x and traveling a distance $L - x$ to the end where it gets reflected back to the point x, is given by $\Delta t = 2(L-x)/\alpha$. The velocity of the moving cross-section at point x is given by

$$\frac{\partial u(x,t)}{\partial t} = v(x,t) = \alpha F'(\alpha t - x) - \alpha F'(\alpha t + x). \tag{5.85}$$

We can now determine the velocity of the cross-section at the point x by analyzing the following table.

Analysis of velocity terms.				
Δt	t varies	$\alpha t - x$ range	$\alpha t + x$ range	velocity
$(L-x)/\alpha$	$0 < t < (L-x)/\alpha$	$(-L, L)$	$(0, L)$	0
$2x/\alpha$	$(L-x)/\alpha < t < (L+x)/\alpha$	$(-L, L)$	$(L, 3L)$	$-\epsilon\alpha$
$2(L-x)/\alpha$	$(L+x)/\alpha < t < (3L-x)/\alpha$	$(L, 3L)$	$(L, 3L)$	0
$2x/\alpha$	$(3L-x)/\alpha < t < (3L+x)/\alpha$	$(L, 3L)$	$(3L, 5L)$	$\epsilon\alpha$
$2(L-x)/\alpha$	$(3L+x)/\alpha < t < (5L-x)/\alpha$	$(3L, 5L)$	$(3L, 5L)$	0

Since the function $F'(x)$ is periodic with period $4L$ the propagation of the above wave motion is repeated. ∎

Vibrating membrane

Consider the motion of a membrane which is stretched out over a frame in the x-y plane and is made to perform transverse vibrations in a viscous medium under the action of an external force. We imagine the membrane as illustrated in the figure 5-12.

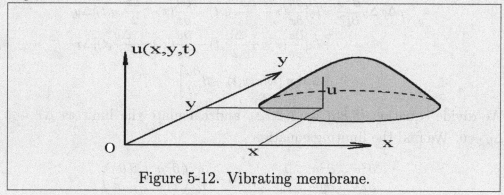

Figure 5-12. Vibrating membrane.

We introduce the following notation and units of measurement

$u = u(x,y,t)$ Transverse displacement of membrane,   (cm)

ρ Constant area density of membrane,   (gm/cm^2)

T   Constant surface tension per unit length,   (dynes/cm)

$w = w(x,y,t)$ Transverse external force per unit area,   (dynes/cm^2)

$F_e = w(x,y,t)\Delta x\Delta y$ Transverse external force,   (dynes)

$F_d = \beta\Delta x\Delta y\dfrac{\partial u}{\partial t}$ Damping force,   (dynes)

β Linear velocity damping coefficient,   (dynes/cm^2 cm/sec)

t time,   (sec)

Now consider a small section of the membrane at a point (x,y) as illustrated in the figure 5-13.

We consider small displacements where the angles $\alpha_1, \alpha_2, \beta_1, \beta_2$ are small and sum forces acting upon the element in the x and y-directions. For small displacements we assume equilibrium of forces in the x and y direction to obtain

$$T\cos\alpha_2 = T\cos\alpha_1 = T_0, \qquad\qquad T\cos\beta_2 = T\cos\beta_1 = T_0.$$

Then by Newton's second law of motion we have summing forces in the vertical direction

$$
\begin{aligned}
\rho\Delta x\Delta y\frac{\partial^2 u}{\partial t^2} =& T\Delta y[\sin\alpha_2 - \sin\alpha_1]\\
&+T\Delta x[\sin\beta_2 - \sin\beta_1]\\
&+w(x,y,t)\Delta x\Delta y - \beta\Delta x\Delta y\frac{\partial u}{\partial t}\\
\rho\Delta x\Delta y\frac{\partial^2 u}{\partial t^2} =& T_0[\frac{\partial u}{\partial x}(x+\frac{\Delta x}{2},y,t) - \frac{\partial u}{\partial x}(x-\frac{\Delta x}{2},y,t)]\Delta y\\
&+T_0[\frac{\partial u}{\partial y}(x,y+\frac{\Delta y}{2},t) - \frac{\partial u}{\partial y}(x,y-\frac{\Delta y}{2},t)]\Delta x\\
&+\Delta x\Delta y\left[w(x,y,t) - \beta\frac{\partial u}{\partial t}\right]
\end{aligned}
\tag{5.86}
$$

We divide equation (5.86) by $T_0\Delta x\Delta y$ and calculate the limit as $\Delta x \to 0$ and $\Delta y \to 0$. We find the limiting equation

$$\frac{\partial^2 u}{\partial x^2} + \frac{\partial^2 u}{\partial y^2} + \frac{1}{T_0}w(x,y,t) = \frac{\rho}{T_0}\left(\frac{\partial^2 u}{\partial t^2} + \frac{\beta}{\rho}\frac{\partial u}{\partial t}\right) \tag{5.87}$$

which is written in the form

$$\frac{\partial^2 u}{\partial x^2} + \frac{\partial^2 u}{\partial y^2} + \frac{1}{T_0} w(x, y, t) = \frac{1}{c^2}\left[\frac{\partial^2 u}{\partial t^2} + k\frac{\partial u}{\partial t}\right] \tag{5.88}$$

where $c^2 = \frac{T_0}{\rho}$ and $k = \frac{\beta}{\rho}$. The equation (5.88) is called the vibrating membrane equation.

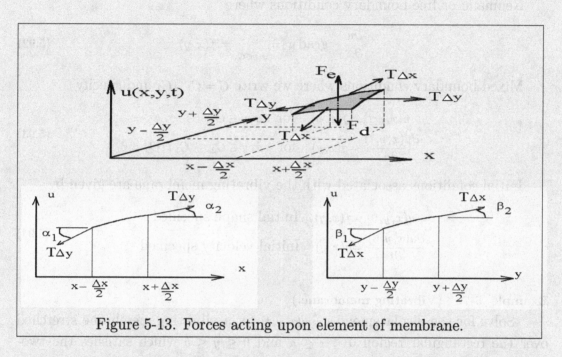

Figure 5-13. Forces acting upon element of membrane.

We have the following special cases associated with the vibrating membrane equation.

Case 1: For $\beta = 0$ and $w = 0$ there results the two-dimensional wave equation

$$\frac{\partial^2 u}{\partial x^2} + \frac{\partial^2 u}{\partial y^2} = \frac{1}{c^2}\frac{\partial^2 u}{\partial t^2} \tag{5.89}$$

Case 2: For static deflection of the membrane there results the Poisson[†] equation

$$\frac{\partial^2 u}{\partial x^2} + \frac{\partial^2 u}{\partial y^2} = \frac{-1}{T_0} w(x, y). \tag{5.90}$$

Let C denote a simple closed curve which represent the frame defining the boundary of the membrane which lies in the x, y plane and let $\hat{n}$ denote the

[†] See Appendix C

outward unit normal to this boundary curve . Typical boundary conditions associated with the vibrating membrane equation are:

Dirichlet or fixed boundary conditions where

$$u(x,y,t)\Big|_{x,y \in C} = f(x,y) \tag{5.91}$$

Neumann or free boundary conditions where

$$\frac{\partial u}{\partial n} = \operatorname{grad} u \cdot \hat{n}\Big|_{x,y \in C} = f(x,y) \tag{5.92}$$

Mixed boundary conditions where we write $C = C_1 \cup C_2$ and specify

$$\begin{aligned} u(x,y,t) &= f(x,y) \quad \text{for} \quad x,y \in C_1 \\ \frac{\partial u(x,y,t)}{\partial n} &= g(x,y) \quad \text{for} \quad x,y \in C_2, \quad C_1 \cap C_2 = \emptyset \end{aligned} \tag{5.93}$$

Initial conditions associated with the vibrating membrane are given by

$$\begin{aligned} u(x,y,0) &= f(x,y) \quad \text{Initial shape specified} \\ \frac{\partial u(x,y,0)}{\partial t} &= g(x,y) \quad \text{Initial velocity specified} \end{aligned} \tag{5.94}$$

Example 5-7.   (Vibrating membrane.)

Solve for the displacement $u = u(x,y,t)$ for a vibrating membrane stretched over the rectangular region $0 \le x \le a$ and $0 \le y \le b$ which satisfies the two-dimensional wave equation

$$\begin{aligned} \text{PDE:} \quad & \frac{\partial^2 u}{\partial t^2} = c^2\left(\frac{\partial^2 u}{\partial x^2} + \frac{\partial^2 u}{\partial y^2}\right), \quad 0 \le x \le a, \quad 0 \le y \le b, \quad t > 0 \\ \text{BC:} \quad & u(0,y,t) = 0, \quad u(a,y,t) = 0, \quad u(x,0,t) = 0, \quad u(x,b,t) = 0, \quad t \ge 0 \\ \text{IC:} \quad & u(x,y,0) = f(x,y), \quad \frac{\partial u(x,y,0)}{\partial t} = g(x,y) \end{aligned}$$

Solution We use the method of separation of variables and assume a product solution of the form

$$u = u(x,y,t) = F(x)G(y)H(t) \tag{5.95}$$

with derivatives

$$\frac{\partial^2 u}{\partial x^2} = F''(x)G(y)H(t), \quad \frac{\partial^2 u}{\partial y^2} = F(x)G''(y)H(t), \quad \frac{\partial^2 u}{\partial t^2} = F(x)G(y)H''(t). \tag{5.96}$$

We substitute the assumed solution into the given partial differential equation to obtain

$$F(x)G(y)H''(t) = c^2 \left(F''(x)G(y)H(t) + F(x)G''(y)H(t) \right)$$

Now divide each term of this equation by $c^2 F(x)G(y)H(t)$ to separate the t-variable and obtain

$$\frac{H''(t)}{c^2 H(t)} = \frac{F''(x)}{F(x)} + \frac{G''(y)}{G(y)} = \lambda_1$$

where λ_1 is a separation constant. This produces the two equations

$$H''(t) - c^2 \lambda_1 H(t) = 0, \qquad \frac{F''(x)}{F(x)} + \frac{G''(y)}{G(y)} = \lambda_1. \tag{5.97}$$

We separate the variables in the second equation to obtain

$$\frac{F''(x)}{F(x)} = \lambda_1 - \frac{G''(y)}{G(y)} = \lambda_2 \tag{5.98}$$

where λ_1 and λ_2 are separation constants that we can select such that nonzero solutions are produced. The given boundary conditions require that

$$u(0,y,t) = F(0)G(y)H(t) = 0, \qquad u(x,0,t) = F(x)G(0)H(t) = 0$$
$$u(a,y,t) = F(a)G(y)H(t) = 0, \qquad u(x,b,t) = F(x)G(b)H(t) = 0$$

which in turn produces the differential equations

$$F''(x) - \lambda_2 F(x) = 0, \quad G''(y) + (\lambda_2 - \lambda_1)G(y) = 0,$$
$$F(0) = 0, \quad F(a) = 0, \qquad G(0) = 0, \quad G(b) = 0, \quad H''(t) - c^2 \lambda_1 H(t) = 0. \tag{5.99}$$
$$0 < x < a \qquad\qquad\qquad 0 < y < b$$

The first two equations are Sturm-Liouville problems with parameters λ_2 and $\lambda_2 - \lambda_1$. From our previous experience we select these parameters as $\lambda_2 = -\omega_1^2$ and $\lambda_2 - \lambda_1 = \omega_2^2$ as this produces the eigenvalues and eigenfunctions

$$\omega_1^2 = \left(\frac{n\pi}{a}\right)^2, \qquad\qquad\qquad \omega_2^2 = \left(\frac{m\pi}{b}\right)^2$$
$$F(x) = F_n(x) = \sin\frac{n\pi x}{a}, \qquad\qquad G(y) = G_m(y) = \sin\frac{m\pi y}{b} \tag{5.100}$$

for $n = 1, 2, 3, \ldots$ and $m = 1, 2, 3, \ldots$. These solutions give us orthogonal sets of functions over the x and y interval of the membrane. The eigenvalues determined can now be used to define the separation constant in the $H(t)$ equation. We find

$$\lambda_1 = \lambda_2 - \omega_2^2 = -\omega_1^2 - \omega_2^2 = -\left(\frac{n\pi}{a}\right)^2 - \left(\frac{m\pi}{b}\right)^2$$

is a function of both integers n and m and so we define

$$\lambda_{nm}^2 = \left(\frac{n\pi}{a}\right)^2 + \left(\frac{m\pi}{b}\right)^2 \qquad (5.101)$$

and write the equation for $H(t)$ in the form

$$H''(t) + c^2\lambda_{nm}^2 H(t) = 0. \qquad (5.102)$$

The equation (5.102) is easily solved to obtain the solution combinations

$$H(t) = H_{mn}(t) = A_{mn}\cos c\lambda_{nm}t + B_{mn}\sin c\lambda_{nm}t \qquad (5.103)$$

where A_{mn} and B_{mn} are arbitrary constants. The method of separation of variables has produced three ordinary differential equations to solve and produces solutions

$$
\begin{aligned}
u = u_{mn}(x,y,t) &= H_{mn}(t)F_n(x)G_m(y) \\
u = u_{mn}(x,y,t) &= (A_{mn}\cos c\lambda_{nm}t + B_{mn}\sin c\lambda_{nm}t)\sin\frac{n\pi x}{a}\sin\frac{m\pi y}{b}
\end{aligned}
\qquad (5.104)
$$

By superposition we obtain the more general solution

$$
\begin{aligned}
u = u(x,y,t) &= \sum_{m=1}^{\infty}\sum_{n=1}^{\infty} u_{mn}(x,y,t) = \sum_{m=1}^{\infty}\sum_{n=1}^{\infty} H_{mn}(t)F_n(x)G_m(y) \\
u = u(x,y,t) &= \sum_{m=1}^{\infty}\sum_{n=1}^{\infty} (A_{mn}\cos c\lambda_{nm}t + B_{mn}\sin c\lambda_{nm}t)\sin\frac{n\pi x}{a}\sin\frac{m\pi y}{b}
\end{aligned}
\qquad (5.105)
$$

The initial conditions require that the constants A_{mn} and B_{mn} be selected to satisfy the conditions

$$
\begin{aligned}
u(x,y,0) &= f(x,y) = \sum_{m=1}^{\infty}\sum_{n=1}^{\infty} A_{mn}F_n(x)G_m(y) \\
\frac{\partial u(x,y,0)}{\partial t} &= g(x,y) = \sum_{m=1}^{\infty}\sum_{n=1}^{\infty} c\lambda_{nm}B_{nm}F_n(x)G_m(y)
\end{aligned}
\qquad (5.106)
$$

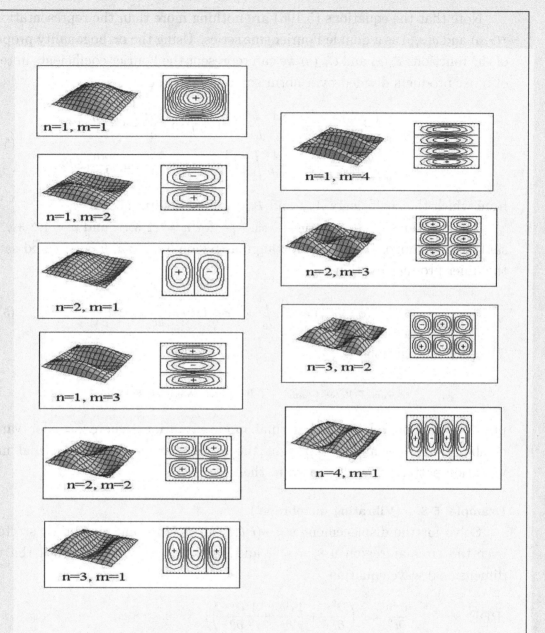

Figure 5-14. Surface and contour plots associated with normal
modes $\phi_{nm}(x,y) = \sin \dfrac{n\pi x}{a} \sin \dfrac{m\pi y}{b}$

Note that the equations (5.106) are nothing more than the representation of $f(x,y)$ and $g(x,y)$ as a double Fourier sine series. Using the orthogonality property of the functions $F_n(x)$ and $G_m(y)$ we can represent the Fourier coefficients in terms of inner products divided by a norm squared. We find

$$A_{mn} = \frac{((f,F_n),G_m)}{\| F_n \|^2 \| G_m \|^2} = \frac{4}{ab} \int_0^a \int_0^b f(\xi,\eta) \sin\frac{n\pi\xi}{a} \sin\frac{m\pi\eta}{b} \, d\xi d\eta$$

$$c\lambda_{mn} B_{nm} = \frac{((g,F_n),G_m)}{\| F_n \|^2 \| G_m \|^2} = \frac{4}{ab} \int_0^a \int_0^b g(\xi,\eta) \sin\frac{n\pi\xi}{a} \sin\frac{m\pi\eta}{b} \, d\xi d\eta$$

(5.107)

from which the coefficients A_{mn} and B_{mn} can be determined.

The functions $\phi_{nm}(x,y) = \sin\frac{n\pi x}{a} \sin\frac{m\pi y}{b}$ for $n = 1,2,3,\ldots$ and $m = 1,2,3\ldots$ are said to be orthogonal over the rectangular region $0 < x < a$, $0 < y < b$ and satisfy the inner product relation

$$(\phi_{nm},\phi_{ij}) = \int_0^a \int_0^b \phi_{nm}\phi_{ij} \, dxdy = \frac{4}{ab}\delta_{ni}\delta_{mj}$$

(5.108)

The product solutions

$$\phi_{nm}(x,y)\cos c\lambda_{nm}t \qquad \text{and} \qquad \phi_{nm}(x,y)\sin c\lambda_{nm}t$$

produce vibrations known as normal mode vibrations where ϕ_{nm}, for various combinations of n and m represent the mode shapes. Various normal mode vibration patterns are illustrated in the figure 5-14.

Example 5-8. (Vibrating membrane.)

Solve for the displacement $u = u(r,\theta,t)$ for a vibrating membrane stretched over the circular region $0 \le r \le a$ and $-\pi \le \theta \le \pi$ which satisfies the two-dimensional wave equation

PDE: $\qquad \dfrac{\partial^2 u}{\partial t^2} = c^2\left(\dfrac{\partial^2 u}{\partial r^2} + \dfrac{1}{r}\dfrac{\partial u}{\partial r} + \dfrac{1}{r^2}\dfrac{\partial^2 u}{\partial \theta^2}\right), \qquad 0 \le r \le a, \quad -\pi \le \theta \le \pi, \quad t > 0$

BC: $\qquad u(a,\theta,t) = 0, \quad -\pi \le \theta \le \pi, \quad t \ge 0$

IC: $\qquad u(r,\theta,0) = f(r,\theta), \qquad \dfrac{\partial u(r,\theta,0)}{\partial t} = 0$

Solution The geometry of the problem and requirement for continuity of the solution implies periodic boundary conditions in the variable θ. We also require realistic solutions to be bounded over the interval $0 \le r \le a$ in the radial direction.

We use the method of separation of variables and assume a product solution of the form $u = u(r, \theta, t) = F(r)G(\theta)H(t)$. We substitute the assumed solution into the differential equation to obtain

$$FGH'' = c^2 \left(F''GH + \frac{1}{r}F'GH + \frac{1}{r^2}FG''H \right).$$

We divide each term of this equation by $c^2 FGH$ and separate out the time variable to obtain

$$\frac{H''}{c^2 H} = \frac{F''}{F} + \frac{1}{r}\frac{F'}{F} + \frac{1}{r^2}\frac{G''}{G} = \lambda_1 \qquad (5.109)$$

where λ_1 is a separation constant. This gives the two equations

$$H'' - \lambda_1 c^2 H = 0, \qquad \frac{F''}{F} + \frac{1}{r}\frac{F'}{F} + \frac{1}{r^2}\frac{G''}{G} = \lambda_1. \qquad (5.110)$$

One can separate the variables in the second equation above to obtain

$$r^2\frac{F''}{F} + r\frac{F'}{F} - \lambda_1 r^2 = -\frac{G''}{G} = \lambda_2 \qquad (5.111)$$

where λ_2 is a second separation constant. We now have three ordinary differential equations with boundary conditions on two of the equations. These equations can be written

$$H'' - \lambda_1 c^2 H = 0, \qquad \begin{array}{l} r^2 F'' + rF' - (\lambda_2 + \lambda_1 r^2)F = 0, \\ 0 < x < a, \quad F \text{ bounded} \end{array} \qquad \begin{array}{l} G'' + \lambda_2 G = 0, \\ -\pi < \theta < \pi, \\ G(-\pi) = G(\pi), \ G'(-\pi) = G'(\pi) \end{array}$$

where we are now free to select λ_1 and λ_2 to be constants of any convenient form. We recognize the G-equation as a Sturm-Liouville problem with periodic boundary conditions. We select $\lambda_2 = \omega^2$ and obtain the eigenvalues $\{0, n^2\}$ and eigenfunctions $\{1, \sin n\theta, \cos n\theta\}$ for the G-equation where n has the integer values, $n = 1, 2, 3, \ldots$. These produce non-trivial solutions of the form

$$G = G_n(\theta) = A_n \cos n\theta + B_n \sin n\theta \quad n = 0, 1, 2, 3, \ldots \qquad (5.112)$$

Observe that both the H-equation and F-equation will be easy to solve if we select $\lambda_1 = -\lambda^2$. This choice of a separation constant produces the singular Sturm-Liouville problem for F involving the Bessel differential equation

$$r^2 F'' + rF' + (\lambda^2 r^2 - n^2)F = 0 \quad \text{for } n = 0, 1, 2, \ldots \qquad (5.113)$$

with solutions

$$F = F(r) = c_1 J_n(\lambda r) + c_2 Y_n(\lambda r) \qquad 0 \le r \le a.$$

For bounded solutions over the interval $0 \le r \le a$ we require the constant $c_2 = 0$. This gives the solution $F = F(r) = J_n(\lambda r)$ where we have selected $c_1 = 1$ for convenience. The remaining boundary condition requires that $F(a) = J_n(\lambda a) = 0$ and so we obtain the eigenvalues $\lambda = \lambda_{mn} = \frac{\xi_{nm}}{a}$ where for each fixed value of n, ξ_{nm} $m = 1, 2, 3, \ldots$ represent the mth zero of the nth order Bessel function of the first kind. This gives, for each fixed value of n, the eigenfunctions $F_m^{(n)}(r) = J_n(\lambda_{nm}r)$ for $m = 1, 2, 3, \ldots$. The equation for H can now be written

$$H'' + \lambda_{nm}^2 c^2 H = 0$$

with solutions

$$H = H_{nm}(t) = A_{nm} \cos \lambda_{nm} ct + B_{nm} \sin \lambda_{nm} ct.$$

We have found that there are many solutions of the assumed form $u = FGH$ which satisfy the given boundary conditions. These many solutions can be represented

$$\begin{aligned} u = u_{mn}(r, \theta, t) = &[(A_{mn} \cos \lambda_{nm} ct + B_{nm} \sin \lambda_{nm} ct) \cos n\theta \\ &+ (C_{nm} \cos \lambda_{nm} ct + D_{nm} \sin \lambda_{nm} ct) \sin n\theta] J_n(\lambda_{nm} r) \end{aligned} \qquad (5.114)$$

where $A_{nm}, B_{nm}, C_{nm}, D_{nm}$ are constants. By the principle of superposition we obtain the general solution

$$\begin{aligned} u = u(r, \theta, t) = \sum_{n=0}^{\infty} \sum_{m=1}^{\infty} &[(A_{mn} \cos \lambda_{nm} ct + B_{nm} \sin \lambda_{nm} ct) \cos n\theta \\ &+ (C_{nm} \cos \lambda_{nm} ct + D_{nm} \sin \lambda_{nm} ct) \sin n\theta] J_n(\lambda_{nm} r) \end{aligned} \qquad (5.115)$$

The initial condition $\dfrac{\partial u(r, \theta, 0)}{\partial t} = 0$ requires that the coefficients B_{0m}, B_{nm}, D_{nm} be zero. The initial condition $u(r, \theta, 0)$ requires that

$$u(r, \theta, 0) = f(r, \theta) = \sum_{m=1}^{\infty} A_{0m} J_0(\lambda_{0m}r) + \sum_{n=1}^{\infty} \sum_{m=1}^{\infty} [A_{nm} \cos n\theta + C_{nm} \sin n\theta] J_n(\lambda_{nm}r).$$

We recognize the above series as a double Fourier series involving sine and cosine functions together with Bessel functions. We can make use of the orthogonality of these functions to solve for the Fourier coefficients A_{0m}, A_{nm}, C_{nm} which we will represent in terms of inner products and norms squared. Note that we have separated out the case $n = 0$ to treat it separately. We calculate the remaining nonzero coefficients from the equations

$$A_{0m} = \frac{((f,1), F_m^{(0)})}{||1||^2 \, ||F_m^{(0)}||^2} = \frac{1}{\pi a^2 J_1^2(\lambda_{0m}a)} \int_0^a \int_{-\pi}^{\pi} f(r,\theta) J_0(\lambda_{0m}r) \, d\theta r dr$$

$$A_{nm} = \frac{((f,\cos n\theta), F_m^{(n)})}{||\cos n\theta||^2 \, ||F_m^{(n)}||^2} = \frac{2}{\pi a^2 J_{n+1}^2(\lambda_{nm}a)} \int_0^a \int_{-\pi}^{\pi} f(r,\theta) J_n(\lambda_{mn}r) \cos n\theta \, d\theta r dr \quad (5.116)$$

$$C_{nm} = \frac{((f,\sin n\theta), F_m^{(n)})}{||\sin n\theta||^2 \, ||F_m^{(n)}||^2} = \frac{2}{\pi a^2 J_{n+1}^2(\lambda_{mn}a)} \int_0^a \int_{-\pi}^{\pi} f(r,\theta) J_n(\lambda_{mn}r) \sin n\theta \, d\theta r dr$$

for $m = 1, 2, 3, \ldots$ and $n = 1, 2, 3, \ldots$.

The set of functions

$$\phi_{nm}^{(e)}(r,\theta) = J_n(\xi_{nm}\frac{r}{a}) \cos n\theta$$

$$\phi_{nm}^{(o)}(r,\theta) = J_n(\xi_{nm}\frac{r}{a}) \sin n\theta$$

$$(5.117)$$

are associated with the vibrations

$$u_{nm}(r,\theta,t) = \left[A_{nm}\phi_{nm}^{(e)}(r,\theta) + C_{nm}\phi_{nm}^{(o)}(r,\theta) \right] \cos \lambda_{nm}ct$$
$$+ \left[B_{nm}\phi_{nm}^{(e)}(r,\theta) + D_{nm}\phi_{nm}^{(o)}(r,\theta) \right] \sin \lambda_{nm}ct$$

called normal mode vibrations. The functions $\sin \lambda_{nm}ct, \cos \lambda_{nm}ct$ oscillate between $+1$ and -1 and produce the movement of the wave shapes $\phi_{nm}^{(e)}(r,\dot{\theta})$ and $\phi_{nm}^{(o)}(r,\theta)$ called standing waves. For example, consider the special initial conditions $u(r,\theta,0) = \phi_{nm}^{(e)}(r,\theta)$ and $\frac{\partial u(r,\theta,0)}{\partial t} = 0$ for fixed values of m and n. These initial conditions produce the special solution

$$u(r,\theta,t) = \phi_{nm}^{(e)}(r,\theta) \cos \lambda_{nm}ct$$

where m and n have fixed values. The solution vibrates with a frequency of $\lambda_{nm}c$ with period $T = 2\pi/\lambda_{nm}c$. The figure 5-15 illustrates some of these special wave shapes.

∎

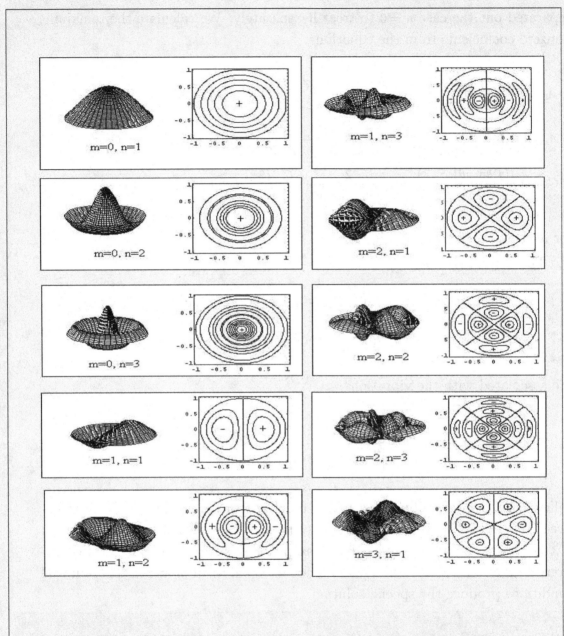

Figure 5-15. Surface and contour plot associated with special
normal mode vibrational wave shapes $\phi_{nm}^{(e)}(r,\theta) = J_n\left(\xi_{nm}\dfrac{r}{a}\right)\cos n\theta$

Example 5-9. (Wave equation canonical form.)

For the one-dimensional wave equation $\dfrac{\partial^2 u}{\partial t^2} = c^2 \dfrac{\partial^2 u}{\partial x^2}$, $-\infty < x < \infty$, for $t > 0$, with $u = u(x,t)$, we make the change of variables $\xi = x + ct$ and $\eta = x - ct$. We treat $u = u(\xi, \eta)$ as a function of ξ, η and use chain rule differentiation to calculate the derivatives

$$\frac{\partial u}{\partial x} = \frac{\partial u}{\partial \xi}\frac{\partial \xi}{\partial x} + \frac{\partial u}{\partial \eta}\frac{\partial \eta}{\partial x} = \frac{\partial u}{\partial \xi} + \frac{\partial u}{\partial \eta}$$

$$\frac{\partial^2 u}{\partial x^2} = \frac{\partial^2 u}{\partial \xi^2}\frac{\partial \xi}{\partial x} + \frac{\partial^2 u}{\partial \xi \partial \eta}\frac{\partial \eta}{\partial x} + \frac{\partial^2 u}{\partial \eta \partial \xi}\frac{\partial \xi}{\partial x} + \frac{\partial^2 u}{\partial \eta^2}\frac{\partial \eta}{\partial x} = \frac{\partial^2 u}{\partial \xi^2} + 2\frac{\partial^2 u}{\partial \xi \partial \eta} + \frac{\partial^2 u}{\partial \eta^2}$$

$$\frac{\partial u}{\partial t} = \frac{\partial u}{\partial \xi}\frac{\partial \xi}{\partial t} + \frac{\partial u}{\partial \eta}\frac{\partial \eta}{\partial t} = \frac{\partial u}{\partial \xi}c + \frac{\partial u}{\partial \eta}(-c)$$

$$\frac{\partial^2 u}{\partial t^2} = c\left(\frac{\partial^2 u}{\partial \xi^2}\frac{\partial \xi}{\partial t} + \frac{\partial^2 u}{\partial \xi \partial \eta}\frac{\partial \eta}{\partial t}\right) - c\left(\frac{\partial^2 u}{\partial \eta \partial \xi}\frac{\partial \xi}{\partial t} + \frac{\partial^2 u}{\partial \eta^2}\frac{\partial \eta}{\partial t}\right) = c^2\frac{\partial^2 u}{\partial \xi^2} - 2c^2\frac{\partial^2 u}{\partial \xi \partial \eta} + c^2\frac{\partial^2 u}{\partial \eta^2}$$

Substituting these derivatives into the above wave equation we find that if one makes the change of variables $\xi = x + ct$ and $\eta = x - ct$, then the wave equation is transformed from the form $\frac{\partial^2 u}{\partial t^2} = c^2\frac{\partial^2 u}{\partial x^2}$ to the canonical form $\frac{\partial^2 u}{\partial \xi \partial \eta} = 0$. This canonical form is easily integrated. We integrate with respect to η and obtain $\frac{\partial u}{\partial \xi} = G(\xi)$ where $G(\xi)$ is an arbitrary function of ξ. We then integrate with respect to ξ and obtain the solution

$$u = u(\xi, \eta) = f(\eta) + g(\xi) \tag{5.118}$$

where $g(\xi) = \int G(\xi)\,d\xi$ is some new arbitrary function of ξ and $f(\eta)$ is an arbitrary function of η. The solution of the original wave equation can then be represented

$$u = u(x,t) = f(x - ct) + g(x + ct) \tag{5.119}$$

where f and g are arbitrary functions representing wave shapes. This solution is the superposition of two wave motions. One wave moving to the right and the other wave moving to the left.

■

Example 5-10. (D'Alembert solution.)

Solve the wave equation initial value problem for an infinite string

PDE: $\quad \text{L}(u) = \frac{\partial^2 u}{\partial t^2} - c^2\frac{\partial^2 u}{\partial x^2} = 0, \qquad -\infty < x < \infty$

IC: $\quad u(x,0) = f(x), \qquad \frac{\partial u(x,0)}{\partial t} = g(x)$

Solution: From the previous example we know the wave equation has the general solution $u = u(x,t) = H(x - ct) + G(x + ct)$ where H, G are arbitrary functions. The solution to the initial value problem requires H, G be selected to satisfy the conditions

$$u(x,0) = H(x) + G(x) = f(x)$$

$$\text{and} \quad \frac{\partial u(x,0)}{\partial t} = H'(x)(-c) + G'(x)(c) = g(x) \tag{5.120}$$

where the prime $'$ denotes differentiation with respect to the argument of the function. We integrate the second equation in (5.120) to obtain

$$-H(x) + G(x) = \frac{1}{c} \int_{x_0}^{x} g(x)\, dx + K \tag{5.121}$$

where x_0 is an arbitrary point and K is some constant of integration. We can now add the first equation in (5.120) to the equation (5.121) giving

$$2G(x) = f(x) + \frac{1}{c} \int_{x_0}^{x} g(\xi)\, d\xi + K. \tag{5.122}$$

By subtracting these same equations we find

$$2H(x) = f(x) - \frac{1}{c} \int_{x_0}^{x} g(\xi)\, d\xi - K. \tag{5.123}$$

From equations (5.122) and (5.123) we solve for H, G and find

$$H(x) = \frac{f(x)}{2} - \frac{1}{2c} \int_{x_0}^{x} g(\xi)\, d\xi - \frac{K}{2}$$

$$\text{and} \quad G(x) = \frac{f(x)}{2} + \frac{1}{2c} \int_{x_0}^{x} g(\xi)\, d\xi + \frac{K}{2} \tag{5.124}$$

from which we find

$$H(x - ct) = \frac{1}{2} f(x - ct) - \frac{1}{2c} \int_{x_0}^{x-ct} g(\xi)\, d\xi - \frac{K}{2}$$

$$G(x + ct) = \frac{1}{2} f(x + ct) + \frac{1}{2c} \int_{x_0}^{x+ct} g(\xi)\, d\xi + \frac{K}{2}. \tag{5.125}$$

This gives the solution

$$u = u(x,t) = H(x - ct) + G(x + ct)$$

$$u = u(x,t) = \frac{1}{2} [f(x - ct) + f(x + ct)] + \frac{1}{2} \int_{x-ct}^{x+ct} g(\xi)\, d\xi \tag{5.126}$$

which is known as the D'Alembert[†] solution of the wave equation. The D'Alembert solution can be interpreted in terms of the characteristic curves given by the family of lines $x - ct = constant$ and $x + ct = constant$. By selecting a point (x_0, t_0) in the x, t-plane we obtain the two characteristic curves $x + ct = x_0 + ct_0$ and $x - ct = x_0 - ct_0$ which are lines through the point (x_0, t_0) as illustrated in the figure 5-16.

[†] See Appendix C

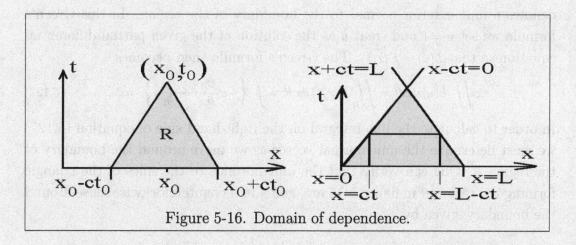
Figure 5-16. Domain of dependence.

The solution at time $t = t_0$ and position $x = x_0$ depends only upon the initial data between the points $x_0 - ct_0$ and $x_0 + ct_0$. The first term in the D'Alembert solution is the average value of the initial wave shape moving to the right and the initial wave shape moving to the left when evaluated at (x_0, t_0). The second term is an integration of g between the points $x_0 - ct_0$ and $x_0 + ct_0$. The domain $x_0 - ct_0 < x < x_0 + ct_0$ is called the domain of dependence.

■

Example 5-11. (D'Alembert solution nonhomogeneous wave equation.)
Solve the one-dimensional nonhomogeneous wave equation for $u = u(x, t)$ if

$$L(u) = \frac{\partial^2 u}{\partial t^2} - c^2 \frac{\partial^2 u}{\partial x^2} = F(x, t), \qquad -\infty < x < \infty, \quad t > 0$$

subject to the initial conditions

$$u(x, 0) = f(x), \qquad \frac{\partial u(x, 0)}{\partial t} = g(x), \quad -\infty < x < \infty$$

Solution: This is a Cauchy problem where the initial curve is the line $t = 0$ or the x-axis. In the figure 5-16 let (x_0, t_0) denote a fixed point and consider the Green's formula for the wave operator $L(u) = \frac{\partial^2 u}{\partial t^2} - c^2 \frac{\partial^2 u}{\partial x^2}$ involving the region R and its boundary. From exercise 36 of chapter one we obtain the Green's formula for the wave operator

$$\iint_R (vL(u) - uL^*(v)) \, dx dt = \oint_C \left\{ -c^2 \left(v \frac{\partial u}{\partial x} - u \frac{\partial v}{\partial x} \right) \hat{i} - \left(u \frac{\partial v}{\partial t} - v \frac{\partial u}{\partial t} \right) \hat{j} \right\} \cdot \hat{n} \, ds$$

where u and v are arbitrary functions, C is the boundary curve around R and $\hat{n}$ denotes a unit exterior normal to the boundary of the region. In this Green's formula we set $v = 1$ and treat u as the solution of the given partial differential equation so that $L(u) = F(x, t)$. The Green's formula then becomes

$$\iint_R L(u)\, dx dt = \iint_R F(x, t)\, dx dt = \oint_C \left\{ -c^2 \frac{\partial u}{\partial x} \hat{i} + \frac{\partial u}{\partial t} \hat{j} \right\} \cdot \hat{n}\, ds. \tag{5.127}$$

In order to calculate the line integral on the right-hand side of equation (5.127) we must determine the unit normal vector as we move around the boundary of the region R. One can verify that the unit normals to the sides of the triangle forming the region R in figure 5-16 are, going in a counterclockwise sense around the boundary, given by

$$\hat{n} = -\hat{j}, \qquad \hat{n} = \frac{\hat{i} + c\hat{j}}{\sqrt{1 + c^2}}, \qquad \hat{n} = \frac{-\hat{i} + c\hat{j}}{\sqrt{1 + c^2}}.$$

The line integral on the right-hand side of equation (5.127) can therefore be broken up into three line integrals. On the bottom portion of the triangle we have $ds = dx$ and $\hat{n} = -\hat{j}$ and so the line integral along this portion of the boundary, where t is zero, becomes

$$I_1 = \int_{x_0 - ct_0}^{x_0 + ct_0} -\frac{\partial u(x, 0)}{\partial t}\, dx = -\int_{x_0 - ct_0}^{x_0 + ct_0} g(x)\, dx. \tag{5.128}$$

because the boundary conditions are specified along the side $t = 0$. On the right-hand side of the triangle in figure 5-16 the unit normal is $\hat{n} = \frac{\hat{i} + c\hat{j}}{\sqrt{1 + c^2}}$ and so along this portion of the boundary where $x - x_0 + c(t - t_0) = 0$ we obtain the line integral

$$I_2 = \int_{(x_0 + ct_0, 0)}^{(x_0, t_0)} -c^2 \frac{\partial u}{\partial x}\, dt - \frac{\partial u}{\partial t}\, dx. \tag{5.129}$$

Along the line $x - x_0 + c(t - t_0) = 0$ there is a definite relationship between dx and dt. From the slope of the line we find $dt = -\frac{1}{c} dx$ or $dx = -c\, dt$. Substituting these values into the line integral given by equation (5.129) we obtain

$$\begin{aligned}
I_2 &= \int_{(x_0 + ct_0, 0)}^{(x_0, t_0)} c \left(\frac{\partial u}{\partial x}\, dx + \frac{\partial u}{\partial t}\, dt \right) \\
&= c \int_{(x_0 + ct_0, 0)}^{(x_0, t_0)} du \\
&= cu(x, t) \Big|_{(x_0 + ct_0, 0)}^{(x_0, t_0)} = c[u(x_0, t_0) - u(x_0 + ct_0, 0)] \\
&= c[u(x_0, t_0) - f(x_0 + ct_0)]
\end{aligned}$$

where again we have made use of the given initial conditions. Along the left side of the triangle there results the line integral

$$I_3 = \int_{(x_0,t_0)}^{(x_0-ct_0,0)} -c^2 \frac{\partial u}{\partial x}\, dt - \frac{\partial u}{\partial t}\, dx. \tag{5.130}$$

Along the line $x - x_0 - c(t - t_0) = 0$ defining the left-hand side of the triangle in figure 5-16, there is a definite relationship between dx and dt. From the slope of this line we find $dt = \frac{1}{c}dx$ or $dx = cdt$. Substituting these values into the line integral given by equation (5.130) we find

$$\begin{aligned}
I_3 &= \int_{(x_0,t_0)}^{(x_0-ct_0,0)} -c\left(\frac{\partial u}{\partial x}\, dx + \frac{\partial u}{\partial t}\, dt\right) \\
&= \int_{(x_0,t_0)}^{(x_0-ct_0,0)} -cdu = -cu(x,t)\Big|_{(x_0,t_0)}^{(x_0-ct_0,0)} \\
&= -c[u(x_0 - ct_0,0) - u(x_0,t_0)] \\
&= c[u(x_0,t_0) - f(x_0 - ct_0)].
\end{aligned}$$

Adding the results from I_1, I_2, I_3 the Green's formula (5.127) can be written in the form

$$\int_R F(x,t)\, dxdt = c[2u(x_0,t_0) - f(x_0 + ct_0) - f(x_0 - ct_0)] - \int_{x_0-ct_0}^{x_0+ct_0} g(x)\, dx$$

Now let (x_0,t_0) vary by dropping the subscripts to obtain the solution

$$u(x,t) = \frac{1}{2}\left[f(x + ct) + f(x - ct)\right] + \frac{1}{2c}\int_{x-ct}^{x+ct} g(\xi)\, d\xi + \frac{1}{2c}\int_0^t \int_{x-c(t-\tau)}^{x+c(t-\tau)} F(\xi,\tau)\, d\xi d\tau \tag{5.131}$$

which is a more general form of the D'Alembert[†] solution.

∎

Wave equation without boundaries

We shall now develop solutions of the wave equation without boundaries. We consider three-dimensional spherical waves, two-dimensional cylindrical waves and one-dimensional plane waves. We develop spherical waves for the three-dimensional wave equation and then use the method of descent to construct cylindrical wave solutions to the two-dimensional wave equation. The D'Alembert solution to the one-dimensional wave equation is obtained from the method of descent from two-dimensions to one-dimension.

[†] See Appendix C

Plane waves

The example 1-17 shows that the solution to the one-dimensional wave equation $\frac{\partial^2 u}{\partial t^2} = c^2 \frac{\partial^2 u}{\partial x^2}$ given by equation (5.119) can also be written in the form

$$u = u(x,t) = f[N_1(x - ct)] + g[N_1(x + ct)] \tag{5.132}$$

where N_1 is some real nonzero constant. This latter form is more in line with the representation of wave motion in higher dimensions. Define the quantity $\omega = N_1 c$, with dimensions of [sec^{-1}], as the angular frequency and write the solution (5.132) of the one-dimensional wave equation in the form

$$u = u(x,t) = f(N_1 x - \omega t) + g(N_1 x + \omega t). \tag{5.133}$$

Plane harmonic waves result when the functions f, g have the exponential form $Ae^{i(N_1 x \pm \omega t)}$. This produces sine and cosine wave forms

$$A\sin(N_1 x \pm \omega t) \quad \text{and} \quad A\cos(N_1 x \pm \omega t)$$

where $\omega = N_1 c$. These represent simple harmonic motions in both space and time. Some terminology associated with these wave forms are as follows. The constant $N_1 = \omega/c$ [cm^{-1}] is called a space frequency and is sometime referred to as a wave number, c [cm/sec] is the wave velocity, A is the amplitude of the wave, $\lambda = 2\pi/N_1$ [cm] is called the wave length, $1/\lambda$ [cm^{-1}] is called the wave number, $p = \lambda/c$ [sec] is the wave period and $f = 1/p$ [sec^{-1}] is the wave frequency. For example, if $u(x,t) = A\sin(N_1 x - \omega t)$ represents a wave moving to the right along the x-axis with speed c, then $u(x,0) = A\sin N_1 x$ is the wave shape, $\lambda = 2\pi/N_1$ [cm] is the wavelength, $p = 2\pi/N_1 c = 2\pi/\omega$ [sec] is the wave period or interval of time it takes for the wave shape to move past a fixed point. Alternatively, $f = 1/p = \omega/2\pi$ [sec^{-1}] represents the number of waves which move past a fixed point in one second.

Standing waves result from the superposition of certain moving wave forms. For example, the function

$$u = u(x,t) = \alpha \cos(N_1 x - \omega t) + \alpha \cos(N_1 x + \omega t)$$

represents the superposition of cosine waves, one moving to the right and the other moving to the left. Basic trigonometric identities allow us to write this wave as

$$u = u(x,t) = (\cos \omega t)\, 2\alpha \cos N_1 x.$$

This shows the wave shape $u(x,0) = 2\alpha \cos N_1 x$ is changed by the amplitude factor $\cos \omega t$ as time changes. This produces a standing wave motion like a vibrating string.

The solution form given by equation (5.133) is representative of the form of plane wave solutions to the two-dimensional wave equation

$$\frac{\partial^2 u}{\partial t^2} = c^2 \left(\frac{\partial^2 u}{\partial x^2} + \frac{\partial^2 u}{\partial y^2} \right) \tag{5.134}$$

and three-dimensional wave equation

$$\frac{\partial^2 u}{\partial t^2} = c^2 \left(\frac{\partial^2 u}{\partial x^2} + \frac{\partial^2 u}{\partial y^2} + \frac{\partial^2 u}{\partial z^2} \right) \tag{5.135}$$

which we shall now investigate. Recall that for the equation (5.135) to have a solution in terms of an arbitrary function f, then there must exist a functionally invariant pair (ξ, η) such that the solution can be represented in the form

$$u = u(x,y,z,t) = \eta(x,y,z,t) f(\xi(x,y,z,t)) \tag{5.136}$$

where ξ and η must satisfy certain conditions. These conditions are obtained by differentiating the equation (5.136) and substituting into the equation (5.135). We find the derivatives

$$u_x = \eta f'(\xi)\xi_x + \eta_x f(\xi)$$
$$u_{xx} = \eta f'(\xi)\xi_{xx} + \eta f''(\xi)\xi_x^2 + 2\eta_x f'(\xi)\xi_x + \eta_{xx} f(\xi)$$

with similar results for the other derivatives. Here subscripts denote partial differentiation. For example, $\xi_x = \frac{\partial \xi}{\partial x}$, $\xi_{xx} = \frac{\partial^2 \xi}{\partial x^2}$, etc. Substituting these derivatives into the equation (5.135) and rearranging terms produces

$$f''(\xi) \left[\eta \left(\xi_x^2 + \xi_y^2 + \xi_z^2 - \frac{1}{c^2}\xi_t^2 \right) \right] +$$
$$f'(\xi) \left[\eta \left(\xi_{xx} + \xi_{yy} + \xi_{zz} - \frac{1}{c^2}\xi_{tt} \right) + 2 \left(\eta_x \xi_x + \eta_y \xi_y + \eta_z \xi_z - \frac{1}{c^2}\eta_t \xi_t \right) \right] + \tag{5.137}$$
$$f(\xi) \left[\eta_{xx} + \eta_{yy} + \eta_{zz} - \frac{1}{c^2}\eta_{tt} \right] = 0.$$

This illustrates that if ξ and η are to be a functionally invariant pair, then for arbitrary f we require the terms inside the brackets of equation (5.137) equal zero. This gives the three equations

$$\xi_x^2 + \xi_y^2 + \xi_z^2 - \frac{1}{c^2}\xi_t^2 = 0$$

$$\eta\left(\xi_{xx} + \xi_{yy} + \xi_{zz} - \frac{1}{c^2}\xi_{tt}\right) + 2\left(\eta_x\xi_x + \eta_y\xi_y + \eta_z\xi_z - \frac{1}{c^2}\eta_t\xi_t\right) = 0 \qquad (5.138)$$

$$\eta_{xx} + \eta_{yy} + \eta_{zz} - \frac{1}{c^2}\eta_{tt} = 0$$

One possible solution to the equations (5.138) are functions ξ and η of the form

$$\eta = 1, \qquad \xi = \hat{n}\cdot\vec{r} \pm ct = n_1 x + n_2 y + n_3 z \pm ct$$

where $n_1^2 + n_2^2 + n_3^2 = 1$. Here $\hat{n}$ or $-\hat{n}$ represents a unit normal to the plane wave moving with velocity c. The wave fronts are moving planes which are perpendicular to $\hat{n}$. This can be shown by considering two points $\vec{r_1}$ and $\vec{r_2}$ lying in the plane $\hat{n}\cdot\vec{r} + ct_0 = k = constant$, where t_0 is held constant. We have

$$\hat{n}\cdot\vec{r_1} + ct_0 = k, \qquad \hat{n}\cdot\vec{r_2} + ct_0 = k$$

so that $\hat{n}\cdot(\vec{r_1} - \vec{r_2}) = 0$ which shows $\hat{n}$ is a unit vector perpendicular to this wave front. Here $\vec{r_1} - \vec{r_2}$ is a vector lying in the planar wave front, which moves with velocity c. In three-dimensions, waves of the form

$$u = f(\hat{n}\cdot\vec{r} + ct) + g(\hat{n}\cdot\vec{r} - ct), \qquad |\hat{n}| = 1$$

or $\quad u = f(\vec{N}\cdot\vec{r} + \omega t) + g(\vec{N}\cdot\vec{r} - \omega t), \qquad \omega = c|\vec{N}| = ck,$

where $\vec{N} = k\hat{n}$ is a nonzero vector, represent plane waves normal to $\vec{N}$. Here $f(\vec{N}\cdot\vec{r} + \omega t)$ represents a plane wave moving with velocity c in the direction $-\vec{N}$ and $g(\vec{N}\cdot\vec{r} - \omega t)$ represents a plane wave moving with velocity c in the positive direction of the vector $\vec{N}$. Note the form of the plane wave solutions

$$u = f(N_1 x + N_2 y + N_3 z - \omega t) \qquad \text{3-dimensions}$$
$$u = f(N_1 x + N_2 y - \omega t) \qquad \text{2-dimensions}$$
$$u = f(N_1 x - \omega t) \qquad \text{1-dimension}$$

where $\omega = c|\vec{N}|$. These type of solutions are known as plane waves.

Poisson's Solution for spherical and cylindrical waves

We shall now develop a more general solution which satisfies the three-dimensional wave equation

$$\frac{\partial^2 u}{\partial t^2} = c^2 \left(\frac{\partial^2 u}{\partial x^2} + \frac{\partial^2 u}{\partial y^2} + \frac{\partial^2 u}{\partial z^2} \right) \tag{5.139}$$

and from this solution we can use the method of descent to obtain solutions to the two-dimensional and one-dimensional forms of the wave equation.

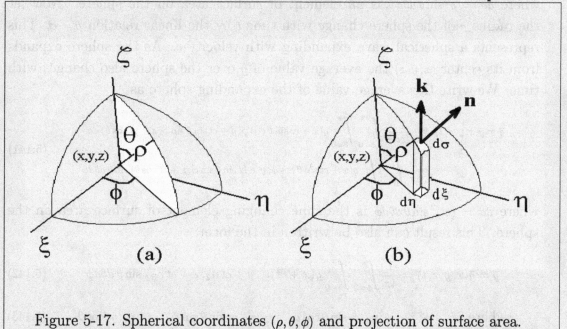

Figure 5-17. Spherical coordinates (ρ, θ, ϕ) and projection of surface area.

Consider a sphere of radius ρ centered about the point (x, y, z) which can be represented in Cartesian coordinates

$$(\xi - x)^2 + (\eta - y)^2 + (\zeta - z)^2 = \rho^2$$

or we can represent a point (ξ, η, ζ) on the surface of the sphere using the parametric equations

$$\xi = x + \rho \sin\theta \cos\phi, \qquad \eta = y + \rho \sin\theta \sin\phi, \qquad \zeta = z + \rho \cos\theta$$

obtained from the figure 5-17 where $0 < \theta < \pi$ and $0 < \phi < 2\pi$.

Consider the mean value of a function $g = g(\xi, \eta, \zeta)$ taken over the surface of this sphere. To calculate the mean value of g over the spherical surface we sum the values of g over the surface of the sphere and then divide by the surface area of the sphere. The mean value is denoted by $\bar{g}$ and can be calculated using the double integral

$$\bar{g} = \bar{g}(x, y, z) = \frac{1}{4\pi\rho^2} \int_{\phi=0}^{2\pi} \int_{\theta=0}^{\pi} g(x + \rho\sin\theta\cos\phi, y + \rho\sin\theta\sin\phi, z + \rho\cos\theta) \, d\sigma \quad (5.140)$$

where $d\sigma = \rho^2 \sin\theta \, d\theta d\phi$ is an element of surface area on the sphere. Now let the radius ρ of the sphere change with time t by the linear relation $\rho = ct$. This represents a spherical wave expanding with velocity c. As the sphere expands from its center (x, y, z) the average value of g over the sphere also changes with time. We write this average value of the expanding sphere as

$$\bar{g} = \bar{g}(x, y, z, t) = \frac{1}{4\pi(ct)^2} \int_{\phi=0}^{2\pi} \int_{\theta=0}^{\pi} g(x + ct\sin\theta\cos\phi, y + ct\sin\theta\sin\phi, z + ct\cos\theta) \, d\sigma$$
$$= \frac{1}{4\pi} \int_0^{2\pi} \int_0^{\pi} g(x + ct\sin\theta\cos\phi, y + ct\sin\theta\sin\phi, z + ct\cos\theta) \sin\theta \, d\theta d\phi \quad (5.141)$$

where $d\sigma = (ct)^2 \sin\theta \, d\theta d\phi$ is the time changing element of surface area on the sphere. This result can also be written in the form

$$\bar{g} = \bar{g}(x, y, z, t) = \frac{1}{4\pi} \int_{\phi=0}^{2\pi} \int_{\theta=0}^{\pi} g(x + ct\, n_1, y + ct\, n_2, z + ct\, n_3) \sin\theta \, d\theta d\phi \quad (5.142)$$

where $\qquad\qquad n_1 = \sin\theta\cos\phi, \qquad n_2 = \sin\theta\sin\phi, \qquad n_3 = \cos\theta \quad (5.143)$

are the direction cosines of a unit exterior normal to the expanding sphere.

The following three theorems resulted from work started by Poisson[†] sometime around 1818.

Theorem: The solution $u = u(x, y, z, t)$ to the three-dimensional wave equation

$$\frac{\partial^2 u}{\partial t^2} = c^2 \left(\frac{\partial^2 u}{\partial x^2} + \frac{\partial^2 u}{\partial y^2} + \frac{\partial^2 u}{\partial z^2} \right), \qquad -\infty < x, y, z < \infty \quad (5.144)$$

satisfying the initial conditions

$$u(x, y, z, 0) = 0, \qquad \frac{\partial u(x, y, x, 0)}{\partial t} = g(x, y, z), \quad (5.145)$$

---

† See Appendix C

where g has continuous derivatives through the second order, is given by

$$u = \frac{t}{4\pi} \int_{\phi=0}^{2\pi} \int_{\theta=0}^{\pi} g(x + ct \sin\theta \cos\phi, y + ct \sin\theta \sin\phi, z + ct \cos\theta) \sin\theta \, d\theta d\phi \qquad (5.146)$$

$$u = t \, \overline{g}$$

Note that the solution $u = u(x, y, z, t)$, at time t, depends upon the integration of the initial data g over the surface of a sphere of radius $\rho = ct$ about the point (x, y, z). Consider the special case where the initial data g is defined

$$g = g(X, Y, Z) = \begin{cases} 1, & X, Y, Z \in R \\ 0, & X, Y, Z \notin R \end{cases}$$

where R is the small region illustrated in the figure 5-18 and the point (x, y, z), where the solution $u = u(x, y, z, t)$ is desired, is outside of this region.

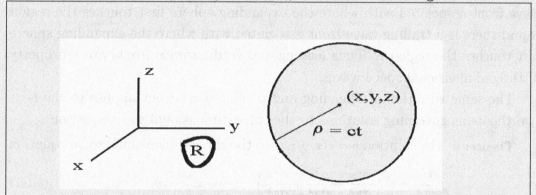

Figure 5-18. Solution dependent upon position of spherical surface $\rho = ct$.

When the integrand in equation (5.146) is zero, then the solution $u = u(x, y, z, t)$ is also zero. Hence, the solution is zero over a time interval $0 \le t \le t_1$, where t_1 is the time it takes for the expanding sphere, centered at (x, y, z), to reach the boundary of the region R. It is not until this time t_1 is reach before the point (x, y, z) "experiences" the effects of the region R and the integrand in the solution for $u = u(x, y, z, t)$ starts taking on nonzero values from the surface of the sphere $\rho = ct$. that is, if there exists a nonzero solution of the three-dimensional wave equation at a fixed time τ, then the integrand g, evaluated on the surface of the fixed sphere $\rho = c\tau$ about the point (x, y, z), must start taking on nonzero values. The first nonzero value for the integrand, in the surface integral defining the solution, occurs at a distance $d_{min} = ct_1$ from the point (x, y, z). At some time $t_2 > t_1$ the surface of the expanding sphere passes beyond the region R so that the region R becomes interior to the expanding sphere and so the function g no longer has a value on the surface of the sphere and therefore the solution u is again zero. This last point where the surface of the sphere passes beyond the region R occurs at some distance $d_{max} = ct_2$ from the point (x, y, z). In other words, with respect to the region R, the point (x, y, z) views the region R as a source of an initial disturbance where each point of R transmits as a wave front to other points in space. This produces a leading wave front and a trailing wave front moving from the region R and "experienced" by the point (x, y, z). Relative to the surface of the initial disturbance, one can construct line segments of distance d_{min} on all outward normals to the region R. These outward directed distances form an envelope of the leading wave front. Similarly, if one constructs line segments of distance d_{max} on all inward normals to the region R, then the envelope of these distances represents the trailing wave front. There is a leading wave front associated with where the expanding sphere first touches the region R and there is a trailing wave front associated with where the expanding sphere last touches the region R. Thus leading and trailing wave fronts are a property of three-dimensional space waves.

The same situation of a leading and trailing wave front applies to the next two theorems governing solutions to the three-dimensional wave equation.

Theorem: The solution $v = v(x, y, z, t)$ to the three-dimensional wave equation

$$\frac{\partial^2 v}{\partial t^2} = c^2 \left(\frac{\partial^2 v}{\partial x^2} + \frac{\partial^2 v}{\partial y^2} + \frac{\partial^2 v}{\partial z^2} \right), \qquad -\infty < x, y, z < \infty \qquad (5.147)$$

which satisfies the initial conditions

$$v(x, y, z, 0) = f(x, y, z), \qquad \frac{\partial v(x, y, x, 0)}{\partial t} = 0, \qquad (5.148)$$

where f has continuous derivatives through the third order, is given by

$$v = \frac{1}{4\pi} \frac{\partial}{\partial t} \int_{\phi=0}^{2\pi} \int_{\theta=0}^{\pi} t\, f(x + ct \sin\theta \cos\phi, y + ct \sin\theta \sin\phi, z + ct \cos\theta) \sin\theta\, d\theta d\phi$$

$$(5.149)$$

$$v = \frac{\partial}{\partial t} [t\, \overline{f}]$$

Observe that the solutions given in these theorems depend upon the values of the integrand on the surface of a sphere $\rho = ct$ which is expanding from the center point (x, y, z). It is only those nonzero values of the integrand evaluated on the surface of the sphere which produce the leading and trailing wave fronts.

Theorem: The solution $w = w(x, y, z, t)$ to the three-dimensional wave equation

$$\frac{\partial^2 w}{\partial t^2} = c^2 \left(\frac{\partial^2 w}{\partial x^2} + \frac{\partial^2 w}{\partial y^2} + \frac{\partial^2 w}{\partial z^2} \right), \qquad -\infty < x, y, z < \infty \qquad (5.150)$$

which satisfies the initial conditions

$$w(x, y, z, 0) = f(x, y, z), \qquad \frac{\partial w(x, y, z, 0)}{\partial t} = g(x, y, z) \qquad (5.151)$$

where $g = g(x, y, z)$ has continuous derivatives through the second order, and $f = f(x, y, z)$ has continuous derivatives through the third order, is given by

$$w = w(x, y, z, t) = \frac{t}{4\pi} \int_{\phi=0}^{2\pi} \int_{\theta=0}^{\pi} g(x + ct \sin\theta \cos\phi, y + ct \sin\theta \sin\phi, z + ct \cos\theta) \sin\theta \, d\theta d\phi$$

$$+ \frac{1}{4\pi} \frac{\partial}{\partial t} \left[\int_{\phi=0}^{2\pi} \int_{\theta=0}^{\pi} t f(x + ct \sin\theta \cos\phi, y + ct \sin\theta \sin\phi, z + ct \cos\theta) \sin\theta \, d\theta d\phi \right] \qquad (5.152)$$

$$w = w(x, y, x, t) = t \overline{g} + \frac{\partial}{\partial t} [t \overline{f}]$$

The equation (5.146) is known as Kirchhoff's[†] formula while the equation (5.152) is known as Poisson's formula.

To show that the equation (5.146) is a solution of the wave equation (5.144) which satisfies the initial conditions (5.145) we write the solution in the form $u = t \overline{g}$ and calculate the derivatives

$$\frac{\partial u}{\partial t} = t \frac{\partial \overline{g}}{\partial t} + \overline{g}, \qquad \frac{\partial^2 u}{\partial t^2} = t \frac{\partial^2 \overline{g}}{\partial t^2} + 2 \frac{\partial \overline{g}}{\partial t} \qquad (5.153)$$

One can also verify the derivatives

$$\frac{\partial u}{\partial x} = t \frac{\partial \overline{g}}{\partial x}, \qquad \frac{\partial u}{\partial y} = t \frac{\partial \overline{g}}{\partial y}, \qquad \frac{\partial u}{\partial z} = t \frac{\partial \overline{g}}{\partial z}$$

$$\frac{\partial^2 u}{\partial x^2} = t \frac{\partial^2 \overline{g}}{\partial x^2}, \qquad \frac{\partial^2 u}{\partial y^2} = t \frac{\partial^2 \overline{g}}{\partial y^2}, \qquad \frac{\partial^2 u}{\partial z^2} = t \frac{\partial^2 \overline{g}}{\partial z^2}.$$

We find that if $g = g(\xi, \eta, \zeta)$ and its first two derivatives are continuous, then

$$c^2 \left(\frac{\partial^2 u}{\partial x^2} + \frac{\partial^2 u}{\partial y^2} + \frac{\partial^2 u}{\partial z^2} \right) = \frac{c^2 t}{4\pi} \int_0^{2\pi} \int_0^{\pi} \left(\frac{\partial^2 g}{\partial \xi^2} + \frac{\partial^2 g}{\partial \eta^2} + \frac{\partial^2 g}{\partial \zeta^2} \right) \Big|_{\rho = ct} \sin\theta \, d\theta d\phi \qquad (5.154)$$

One can also calculate the derivative

$$\frac{\partial \overline{g}}{\partial t} = \frac{c}{4\pi} \int_0^{2\pi} \int_0^{\pi} \operatorname{grad} g \cdot \hat{n} \sin\theta \, d\theta d\phi = \frac{1}{4\pi c t^2} \int_0^{2\pi} \int_0^{\pi} \operatorname{grad} g \cdot \hat{n} \, d\sigma \qquad (5.155)$$

[†] See Appendix C

where $d\sigma = (ct)^2 \sin\theta\, d\theta d\phi$ is an element of surface area on the sphere of radius $\rho = ct$. Using the divergence theorem of Gauss, the equation (5.155) becomes

$$\frac{\partial \overline{g}}{\partial t} = \frac{1}{4\pi ct^2} \int_0^{ct} \int_0^{2\pi} \int_0^{\pi} \left(\frac{\partial^2 g}{\partial \xi^2} + \frac{\partial^2 g}{\partial \eta^2} + \frac{\partial^2 g}{\partial \zeta^2} \right) \rho^2 \sin\theta d\rho d\theta d\phi \qquad (5.156)$$

so that

$$\frac{\partial^2 \overline{g}}{\partial t^2} = \frac{c^2}{4\pi} \left[\int_0^{2\pi} \int_0^{\pi} \left(\frac{\partial^2 g}{\partial \xi^2} + \frac{\partial^2 g}{\partial \eta^2} + \frac{\partial^2 g}{\partial \zeta^2} \right) \Big|_{\rho=ct} \sin\theta d\theta d\phi \right]$$
$$- \frac{1}{2\pi ct^3} \int_0^{ct} \int_0^{2\pi} \int_0^{\pi} \left(\frac{\partial^2 g}{\partial \xi^2} + \frac{\partial^2 g}{\partial \eta^2} + \frac{\partial^2 g}{\partial \zeta^2} \right) \rho^2 \sin\theta d\rho d\theta d\phi \qquad (5.157)$$

Substituting the equations (5.157) and (5.156) into the equation (5.153) we find that

$$\frac{\partial^2 u}{\partial t^2} = \frac{c^2 t}{4\pi} \int_0^{2\pi} \int_0^{\pi} \left(\frac{\partial^2 g}{\partial \xi^2} + \frac{\partial^2 g}{\partial \eta^2} + \frac{\partial^2 g}{\partial \zeta^2} \right) \Big|_{\rho=ct} \sin\theta\, d\theta d\phi \qquad (5.158)$$

The derivatives from equations (5.158) and (5.154) show that the wave equation is identically satisfied by the solution (5.146). It is readily verify that the solution given by equation (5.146) also satisfies the initial conditions specified by the equations (5.145).

To show the equation (5.149) is a solution of the wave equation (5.147) and also satisfies the initial conditions given by equation (5.148), we write the solution in the form

$$v = v(x,y,z,t) = \frac{\partial}{\partial t}[t\overline{f}] = t\frac{\partial \overline{f}}{\partial t} + \overline{f}$$

then the initial condition given by equation (5.148) is satisfied because in the limit $\lim_{t\to 0} v(x,y,z,t) = \overline{f}(x,y,z,0) = f(x,y,z)$. Note that if we set $u = t\overline{f}$, then $v = \frac{\partial u}{\partial t}$ and $\frac{\partial v}{\partial t} = \frac{\partial^2 u}{\partial t^2}$ is given by equation (5.158) with g replace by f. The equation (5.158) shows that $\lim_{t\to 0} \frac{\partial v}{\partial t} = 0$ and so the solution $v = v(x,y,z,t)$ given by equation (5.149) satisfies the initial conditions. Also note that if u satisfies the wave equation

$$\frac{\partial^2 u}{\partial t^2} = c^2 \left(\frac{\partial^2 u}{\partial x^2} + \frac{\partial^2 u}{\partial y^2} + \frac{\partial^2 u}{\partial z^2} \right), \qquad -\infty < x,y,z < \infty$$

then differentiation with respect to time t gives the result

$$\frac{\partial^2}{\partial t^2}\left[\frac{\partial u}{\partial t} \right] = c^2 \left(\frac{\partial^2}{\partial x^2}\left[\frac{\partial u}{\partial t} \right] + \frac{\partial^2}{\partial y^2}\left[\frac{\partial u}{\partial t} \right] + \frac{\partial^2}{\partial z^2}\left[\frac{\partial u}{\partial t} \right] \right)$$

and so $v = \frac{\partial u}{\partial t}$ is also a solution of the wave equation.

The solution w given by equation (5.152) of the initial value problem defined by equations (5.150) and (5.151) results from a superposition of the solutions from the previous problems for u and v and $w = u + v$.

Method of descent

Observe that the solutions given by equations (5.146), (5.149), (5.152) can be expressed in terms of the surface element $d\sigma = (ct)^2 \sin\theta d\theta d\phi$ by multiplying and dividing by the quantity $(ct)^2$. For example, the solution equation (5.152) can be expressed in the form

$$
\begin{aligned}
w = w(x, y, z, t) = &\frac{1}{4\pi c^2 t} \int_{\phi=0}^{2\pi} \int_{\theta=0}^{\pi} g(x + ct\sin\theta\cos\phi, y + ct\sin\theta\sin\phi, z + ct\cos\theta)\, d\sigma \\
&+ \frac{1}{4\pi c^2} \frac{\partial}{\partial t} \left[\int_{\phi=0}^{2\pi} \int_{\theta=0}^{\pi} \frac{1}{t} f(x + ct\sin\theta\cos\phi, y + ct\sin\theta\sin\phi, z + ct\cos\theta)\, d\sigma \right]
\end{aligned}
\tag{5.159}
$$

An examination of the figure 5-17(b) shows the surface element $d\sigma$ can be projected onto the area element $d\xi d\eta$ of the ξ, η-plane. Recall the projection formula relating these area elements is given by $d\sigma = \dfrac{d\xi d\eta}{|\hat{n} \cdot \hat{k}|}$ where $\hat{n}$ is a unit normal to the surface of the sphere

$$
\phi(\xi, \eta, \zeta) = (\xi - x)^2 + (\eta - y)^2 + (\zeta - z)^2 - \rho^2 = 0
$$

where $\rho = ct$. If the unit normal is calculated

$$
\hat{n} = \frac{\operatorname{grad}\phi}{|\operatorname{grad}\phi|} = \frac{(\xi - x)\hat{i} + (\eta - y)\hat{j} + (\zeta - z)\hat{k}}{\sqrt{(\xi - x)^2 + (\eta - y)^2 + (\zeta - z)^2}},
$$

then one finds that $\hat{n} \cdot \hat{k} = \dfrac{|\zeta - z|}{ct}$ so that the equation (5.159) can be expressed in the form

$$
\begin{aligned}
w = w(x, y, z, t) = &\frac{1}{2\pi c} \int_{\xi=x-ct}^{\xi=x+ct} \int_{\eta=y-\sqrt{c^2t^2-(\xi-x)^2}}^{\eta=y+\sqrt{c^2t^2-(\xi-x)^2}} \frac{g(\xi, \eta, \zeta)}{\sqrt{c^2t^2 - (\xi-x)^2 - (\eta-y)^2}}\, d\xi d\eta \\
&+ \frac{1}{2\pi c} \frac{\partial}{\partial t} \left[\int_{\xi=x-ct}^{\xi=x+ct} \int_{\eta=y-\sqrt{c^2t^2-(\xi-x)^2}}^{\eta=y+\sqrt{c^2t^2-(\xi-x)^2}} \frac{f(\xi, \eta, \zeta)}{\sqrt{c^2t^2 - (\xi-x)^2 - (\eta-y)^2}}\, d\xi d\eta \right]
\end{aligned}
\tag{5.160}
$$

where now the integration is over the circle where the sphere intersects the ξ, η-plane. Note that a factor of two has been used to account for the upper and lower surface of the sphere.

The method of descent uses the result given by equation (5.160) in the special cases where f and g are independent of the variable z.

Theorem: The solution $w = w(x, y, t)$ to the two-dimensional wave equation

$$\frac{\partial^2 w}{\partial t^2} = c^2 \left(\frac{\partial^2 w}{\partial x^2} + \frac{\partial^2 w}{\partial y^2} \right), \qquad -\infty < x, y < \infty \tag{5.161}$$

which satisfies the initial conditions

$$w(x, y, 0) = f(x, y), \qquad \frac{\partial w(x, y, 0)}{\partial t} = g(x, y) \tag{5.162}$$

is given by

$$\begin{aligned}
w = w(x, y, t) = {} & \frac{1}{2\pi c} \int_{\xi=x-ct}^{\xi=x+ct} \int_{\eta=y-\sqrt{c^2 t^2 - (\xi-x)^2}}^{\eta=y+\sqrt{c^2 t^2 - (\xi-x)^2}} \frac{g(\xi, \eta)}{\sqrt{c^2 t^2 - (\xi-x)^2 - (\eta-y)^2}} \, d\xi \, d\eta \\
& + \frac{1}{2\pi c} \frac{\partial}{\partial t} \left[\int_{\xi=x-ct}^{\xi=x+ct} \int_{\eta=y-\sqrt{c^2 t^2 - (\xi-x)^2}}^{\eta=y+\sqrt{c^2 t^2 - (\xi-x)^2}} \frac{f(\xi, \eta)}{\sqrt{c^2 t^2 - (\xi-x)^2 - (\eta-y)^2}} \, d\xi \, d\eta \right]
\end{aligned} \tag{5.163}$$

which can also be represented as

$$\begin{aligned}
w = w(x, y, t) = {} & \frac{1}{2\pi c} \int_0^{2\pi} \int_0^{ct} \frac{g(x + r\cos\theta, y + r\sin\theta)}{\sqrt{c^2 t^2 - r^2}} \, r \, dr \, d\theta \\
& + \frac{1}{2\pi c} \frac{\partial}{\partial t} \int_0^{2\pi} \int_0^{ct} \frac{f(x + r\cos\theta, y + r\sin\theta)}{\sqrt{c^2 t^2 - r^2}} \, r \, dr \, d\theta
\end{aligned}$$

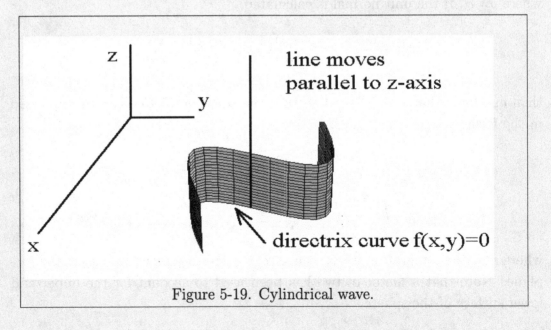

Figure 5-19. Cylindrical wave.

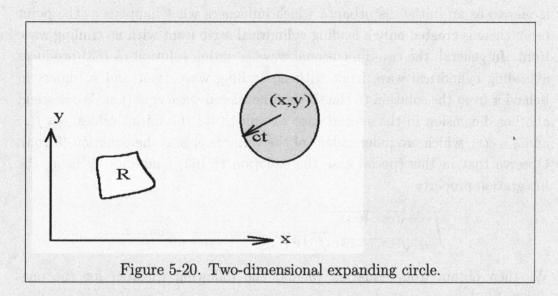

Figure 5-20. Two-dimensional expanding circle.

Solutions of the wave equation which are independent of the variable z are called cylindrical waves. Recall that a cylinder is the locus of points on a line which moves in such a way that it is always parallel to a fixed straight line while at the same time intersecting a fixed plane curve called the directrix curve of the cylinder. The figure 5-19 illustrates a directrix curve $f(x,y) = 0$ along with a line which moves parallel to the z-axis.

The wave motion in two-dimensions is unlike the wave motion in three-dimensions. To illustrate this difference we consider the special disturbance

$$w(x,y,0) = f(x,y), \qquad \frac{\partial w(x,y,0)}{\partial t} = 0 \tag{5.164}$$

where

$$f(X,Y) = \begin{cases} 1, & X,Y \in R \\ 0, & X,Y \notin R \end{cases}$$

where R is the region illustrated in the figure 5-20. Notice that in this special case the solution (5.163) is an area integral dependent upon the value of the integrand f inside the expanding circle of radius ct and so the solution remains zero while the expanding circle does not contain the region R, or any part of R, in its interior. The solution remains zero up to some time t_1 when the expanding circle first touches the region R. For all times $t > t_1$ we will have some part or all of the region R inside the expanding circle and therefore the value of the area integral will contain a nonzero integrand. With respect to the region R, which is viewed as an initial disturbance which influences what happens at the point (x, y), there is created only a leading cylindrical wave front with no trailing wave front. In general, the two-dimensional wave equation solution (5.163) produces a leading cylindrical wave front with no trailing wave front and so differs in behavior from the solution to the three-dimensional wave equation. We descend another dimension in the special case we substitute the initial values $f = f(x)$ and $g = g(x)$, which are independent of the variable y, into the equation (5.163). Observe that in this special case the equation (5.163) simplifies by using the integration property

$$\int_{\eta=y-\sqrt{c^2t^2-(\xi-x)^2}}^{\eta=y+\sqrt{c^2t^2-(\xi-x)^2}} \frac{1}{\sqrt{(ct)^2 - (\xi - x)^2 - (\eta - y)^2}} \, d\eta = \pi.$$

We then obtain from equation (5.163) the following theorem for the one-dimensional wave equation.

Theorem: The solution $w = w(x)$ to the one-dimensional wave equation

$$\frac{\partial^2 w}{\partial t^2} = c^2 \frac{\partial^2 w}{\partial x^2}, \qquad -\infty < x < \infty \tag{5.165}$$

which satisfies the initial conditions

$$w(x, 0) = f(x), \qquad \frac{\partial w(x, 0)}{\partial t} = g(x) \tag{5.166}$$

is given by

$$
\begin{aligned}
w = w(x, t) &= \frac{1}{2c} \int_{\xi=x-ct}^{\xi=x+ct} g(\xi) \, d\xi + \frac{1}{2c} \frac{\partial}{\partial t} \left[\int_{\xi=x-ct}^{\xi=x+ct} f(\xi) \, d\xi \right] \\
w = w(x, t) &= \frac{1}{2c} \int_{\xi=x-ct}^{\xi=x+ct} g(\xi) \, d\xi + \frac{1}{2} \left[f(x + ct) + f(x - ct) \right].
\end{aligned}
\tag{5.167}
$$

This result agrees with our previous D'Alembert solution of the one-dimensional wave-equation.

Exercises 5

▶ **1.**

Use separation of variables to solve for the displacement $y = y(x, t)$ of a vibrating string if

$$\frac{\partial^2 y}{\partial t^2} - c^2 \frac{\partial^2 y}{\partial x^2} = 0, \quad 0 < x < L, \quad t > 0$$

subject to the boundary conditions $y(0, t) = 0$, $y(L, t) = 0$ and initial conditions $y(x, 0) = f(x)$ and $\frac{\partial y(x, 0)}{\partial t} = 0$.

▶ **2.**

Use separation of variables to solve for the displacement $y = y(x, t)$ of a vibrating string if

$$\frac{\partial^2 y}{\partial t^2} = c^2 \frac{\partial^2 y}{\partial x^2}, \quad 0 < x < L, \quad t > 0$$

subject to the boundary conditions $y(0, t) = 0$, $y(L, t) = 0$ and initial conditions $y(x, 0) = f(x)$ and $\frac{\partial y(x, 0)}{\partial t} = g(x)$.

▶ **3.**

Use separation of variables to solve for the displacement $y = y(x, t)$ of a vibrating string if

$$\frac{\partial^2 y}{\partial t^2} = c^2 \frac{\partial^2 y}{\partial x^2}, \quad 0 < x < L, \quad t > 0$$

subject to the boundary conditions $\frac{\partial y(0, t)}{\partial x} = 0$, $\frac{\partial y(L, t)}{\partial x} = 0$ and initial conditions $y(x, 0) = f(x)$ and $\frac{\partial y(x, 0)}{\partial t} = 0$.

▶ **4.**

The vibrations of a circular membrane with fixed edge conditions satisfies the partial differential equation

$$\frac{\partial^2 u}{\partial t^2} = c^2 \left(\frac{\partial^2 u}{\partial r^2} + \frac{1}{r} \frac{\partial u}{\partial r} \right), \quad 0 < r < b, \quad t > 0$$

subject to the boundary condition $u(b, t) = 0$ and initial conditions $u(r, 0) = f(r)$ and $\frac{\partial u(r, 0)}{\partial t} = 0$. Use separation of variables to find the displacement $u = u(r, t)$.

▶ 5.

Let $L_{xt}(V) = \dfrac{\partial^2 V}{\partial t^2} - \alpha^2 \dfrac{\partial^2 V}{\partial x^2}$ and consider the wave equation with forcing function having an initial shape and velocity specified.

$$
\begin{aligned}
&\text{PDE:} && L_{xt}(V) = G(x,t), && 0 < x < L, && t > 0 \\
&\text{BC:} && V(0,t) = 0, && V(L,t) = 0 && \text{fixed ends} \\
&\text{IC:} && V(x,0) = f(x), && \dfrac{\partial V(x,0)}{\partial t} = g(x)
\end{aligned}
\qquad 5(a)
$$

(a) Assume a solution $V = U + W$ and show that equation 5(a) can be replaced by the following two problems

$$
\begin{aligned}
&\text{PDE:} && L_{xt}(U) = 0, && 0 < x < L, && t > 0 \\
&\text{BC:} && U(0,t) = 0, && U(L,t) = 0 \\
&\text{IC:} && U(x,0) = f(x), && \dfrac{\partial U(x,0)}{\partial t} = g(x)
\end{aligned}
\qquad 5(b)
$$

and

$$
\begin{aligned}
&\text{PDE:} && L_{xt}(W) = G(x,t), && 0 < x < L, && t > 0 \\
&\text{BC:} && W(0,t) = 0, && W(L,t) = 0 \\
&\text{IC:} && W(x,0) = 0, && \dfrac{\partial W(x,0)}{\partial t} = 0
\end{aligned}
\qquad 5(c)
$$

(b) Assume a solution to the equation 5(c) of the form

$$
W = W(x,t) = \sum_{n=1}^{\infty} w_n(t) \sin \frac{n\pi x}{L}
$$

and show that

$$
\frac{d^2 w_n}{dt^2} + a_n^2 w_n = g_n(t), \quad w_n(0) = 0, \quad \frac{dw_n(0)}{dt} = 0
\qquad 5(d)
$$

where $a_n = \alpha \frac{n\pi}{L}$ and $g_n(t) = \frac{2}{L} \int_0^L G(x,t) \sin \frac{n\pi x}{L}\, dx$.

(c) Solve the ordinary differential equation given by equation 5(d).

(d) Write out the complete solution to the original partial differential equation given by equation 5(a).

▶ 6.

Consider the wave equation with forcing term and time varying nonhomogeneous boundary conditions given by

$$\text{PDE:} \quad L_{xt}(U) = \frac{\partial^2 U}{\partial t^2} - \alpha^2 \frac{\partial^2 U}{\partial x^2} = F(x,t), \quad 0 < x < L, \quad t > 0$$

$$\text{BC:} \quad U(0,t) = h_1(t), \quad U(L,t) = h_2(t) \qquad\qquad 6(a)$$

$$\text{IC:} \quad U(x,0) = f_1(x), \quad \frac{\partial U(x,0)}{\partial t} = f_2(x)$$

(a) Find a function $W = W(x,t)$ which satisfies $\frac{\partial^2 W}{\partial x^2} = 0$ and

$$W(0,t) = h_1(t), \qquad W(L,t) = h_2(t)$$

(b) Assume a solution to the equation 6(a) of the form $U(x,t) = W(x,t) + V(x,t)$ and show that $V = V(x,t)$ must satisfy a partial differential equation having the general form

$$\text{PDE:} \quad L_{xt}(V) = G(x,t), \quad 0 < x < L, \quad t > 0$$

$$\text{BC:} \quad V(0,t) = 0, \quad V(L,t) = 0 \qquad\qquad 6(b)$$

$$\text{IC:} \quad V(x,0) = f(x), \quad \frac{\partial V(x,0)}{\partial t} = g(x)$$

Find appropriate values for the functions $G(x,t)$, $f(x)$ and $g(x)$ such that $U(x,t) = V(x,t) + W(x,t)$ is a solution of the equation 6(a).
(c) Solve the partial differential equation 6(a).

▶ 7.

The ends $x = 0$ and $x = 1$ of a uniform string are fixed. Discuss the motion which results if the string is given an initial displacement $x(1-x)$ and then released from rest.

▶ 8.

Solve for the displacement $u = u(r,t)$ of a vibrating membrane which satisfies

$$\frac{\partial^2 u}{\partial t^2} = c^2 \left(\frac{\partial^2 u}{\partial r^2} + \frac{1}{r} \frac{\partial u}{\partial r} \right), \qquad 0 < r < b, \ t > 0$$

subject to the boundary condition $u(b,t) = 0$ and initial conditions $u(r,0) = 0$ and $\frac{\partial u(r,0)}{\partial t} = g(r)$

► 9.

Assume that u_1 is a solution of the boundary value problem

$$\text{PDE:} \quad \frac{\partial^2 u}{\partial t^2} - c^2 \frac{\partial^2 u}{\partial x^2} = H_1(x,t), \quad 0 < x < L, \, t > 0$$

$$\text{BC:} \quad u(0,t) = f_1(t), \quad u(L,t) = g_1(t)$$

$$\text{IC:} \quad u(x,0) = F_1(x), \quad \frac{\partial u(x,0)}{\partial t} = G_1(x)$$

and that u_2 is a solution of the boundary value problem

$$\text{PDE:} \quad \frac{\partial^2 u}{\partial t^2} - c^2 \frac{\partial^2 u}{\partial x^2} = H_2(x,t), \quad 0 < x < L, \, t > 0$$

$$\text{BC:} \quad u(0,t) = f_2(t), \quad u(L,t) = g_2(t)$$

$$\text{IC:} \quad u(x,0) = F_2(x), \quad \frac{\partial u(x,0)}{\partial t} = G_2(x).$$

Does the principle of superposition apply? What partial differential equation and boundary conditions are satisfied by $u(x,t) = u_1(x,t) + u_2(x,t)$?

► 10.

Solve the boundary value problem

$$\text{PDE:} \quad \frac{\partial^2 u}{\partial t^2} - 4 \frac{\partial^2 u}{\partial x^2} = 0, \quad 0 < x < 5, \, t > 0$$

$$\text{BC:} \quad u(0,t) = 0, \quad u(5,t) = 0$$

$$\text{IC:} \quad u(x,0) = 0, \quad \frac{\partial u(x,0)}{\partial t} = \sin \frac{2\pi x}{5}$$

► 11.

A membrane $0 \le x \le \pi$, $0 \le y \le \pi$ has all its edges fixed. Find the displacement as a function of position and time if the membrane is given an initial displacement $u(x,y,0) = f(x,y)$ and released from rest.

► 12.

Solve the boundary value problem

$$\text{PDE:} \quad \frac{\partial^2 u}{\partial t^2} = c^2 \frac{\partial^2 u}{\partial x^2} + c^4 \cos(cx), \quad 0 < x < L, \, t > 0, \, u = u(x,t)$$

$$\text{BC:} \quad u(0,t) = \alpha, \quad u(L,t) = \beta \quad \alpha, \beta \text{ are constants}$$

$$\text{IC:} \quad u(x,0) = f(x), \quad \frac{\partial u(x,0)}{\partial t} = g(x)$$

where c is constant.

► **13.**

Solve the boundary value problem

$$\text{PDE:} \quad \frac{\partial^2 u}{\partial t^2} - c^2 \frac{\partial^2 u}{\partial x^2} = 0, \quad 0 < x < L, \ t > 0$$

$$\text{BC:} \quad u(0,t) = 0, \quad \frac{\partial u(L,t)}{\partial x} = 0$$

$$\text{IC:} \quad u(x,0) = f(x), \quad \frac{\partial u(x,0)}{\partial t} = g(x)$$

► **14.**

Solve the boundary value problem

$$\text{PDE:} \quad \frac{\partial^2 u}{\partial t^2} - c^2 \frac{\partial^2 u}{\partial x^2} = 0, \quad 0 < x < L, \ t > 0$$

$$\text{BC:} \quad u(0,t) = 0, \quad \frac{\partial u(L,t)}{\partial x} + u(L,t) = 0$$

$$\text{IC:} \quad u(x,0) = f(x), \quad \frac{\partial u(x,0)}{\partial t} = g(x)$$

► **15.**

Solve the vibrating string with damping

$$L_{xt}(u) = \frac{\partial^2 u}{\partial t^2} + \beta \frac{\partial u}{\partial t} - \alpha^2 \frac{\partial^2 u}{\partial x^2} = 0, \qquad 0 < x < L, \ t > 0$$

subject to the boundary conditions $u(0,t) = 0$, $u(L,t) = 0$ and initial conditions $u(x,0) = f(x)$, $\frac{\partial u(x,0)}{\partial t} = g(x)$.

(a) Assume that $\beta < \frac{2\pi\alpha}{L}$ and show that the method of separation of variables leads to solutions of the form

$$u_n(x,t) = e^{-\beta t/2} \left(A_n \cos \gamma_n t + B_n \sin \gamma_n t \right) \sin \frac{n\pi x}{L}$$

where $\gamma_n^2 = \frac{n^2 \pi^2 \alpha^2}{L^2} - \frac{\beta^2}{4}$, for $n = 1, 2, 3, \ldots$.

(b) Show the solution satisfying the given initial conditions is

$$u(x,t) = e^{-\beta t/2} \sum_{n=1}^{\infty} \left(f_n \cos \gamma_n t + \frac{1}{\gamma_n} \left(g_n + \frac{\beta}{2} f_n \right) \sin \gamma_n t \right) \sin \frac{n\pi x}{L}$$

$$\text{where} \quad f_n = \frac{2}{L} \int_0^L f(\xi) \sin \frac{n\pi \xi}{L} \, d\xi$$

$$g_n = \frac{2}{L} \int_0^L g(\xi) \sin \frac{n\pi \xi}{L} \, d\xi$$

(c) Assume a solution $u(x,t) = \sum_{n=1}^{\infty} u_n(t) \sin \frac{n\pi x}{L}$ which satisfies the boundary conditions $u(0,t) = 0$ and $u(L,t) = 0$.

(i) Show that $u_n(0) = f_n$ and $\frac{du_n(0)}{dt} = g_n$.

(ii) Show that

$$\frac{d^2 u_n}{dt^2} + \beta \frac{du_n}{dt} + (\gamma_n^2 + \frac{\beta^2}{4}) u_n = 0, \quad u_n(0) = f_n, \quad \frac{du_n(0)}{dt} = g_n \qquad (15a)$$

and then solve equation (15a) for u_n and derive the previous solution given in part (b). What happens if $\beta > 2\pi\alpha/L$? What happens if the damping coefficient satisfies $\beta = 2\pi\alpha/L$?

▶ **16.**

Solve the Dirichlet boundary value problem

$$\text{PDE:} \qquad \frac{\partial^2 u}{\partial t^2} - c^2 \frac{\partial^2 u}{\partial x^2} = 0, \quad 0 < x < L, \ t > 0$$

$$\text{BC:} \qquad u(0,t) = F(t), \qquad u(L,t) = G(t)$$

$$\text{IC:} \qquad u(x,0) = f(x), \qquad \frac{\partial u(x,0)}{\partial t} = g(x)$$

▶ **17.** (Vibrating membrane)

Use separation of variables to solve for the displacement $u = u(r,\theta,t)$ which satisfies the wave equation

$$\frac{\partial^2 u}{\partial t^2} = \alpha^2 \left(\frac{\partial^2 u}{\partial r^2} + \frac{1}{r}\frac{\partial u}{\partial r} + \frac{1}{r^2}\frac{\partial^2 u}{\partial \theta^2} \right), \quad 0 < r < b, \ -\pi < \theta < \pi, \ t > 0 \qquad (17a)$$

and is subject to the Dirichlet boundary condition $u(b,\theta,t) = 0$ together with the initial conditions $u(r,\theta,0) = f(r,\theta)$ and $\frac{\partial u(r,\theta,0)}{\partial t} = 0$.

(a) Assume a solution $u = F(r)G(\theta)T(t)$ and show that there results the differential equations

$$r^2 \frac{d^2 F}{dr^2} + r \frac{dF}{dr} + (\lambda_1 r^2 - \lambda_2) F = 0, \qquad \begin{array}{l} \frac{d^2 G}{d\theta^2} + \lambda_2 G = 0, \\ G(-\pi) = G(\pi), \\ \frac{dG(-\pi)}{d\theta} = \frac{dG(\pi)}{d\theta}, \end{array} \qquad \begin{array}{l} \frac{d^2 T}{dt^2} + \lambda_1 \alpha^2 T = 0 \\ \frac{dT(0)}{dt} = 0 \end{array}$$

$$F(b) = 0$$

(b) Show that for $\lambda_2 = n^2$, with $n = 0, 1, 2, 3, \ldots$ and $\lambda_1 = \lambda^2$ the general solution of equation (17a) can be represented

$$u = u(r, \theta, t) = \sum_{m=1}^{\infty} A_{0m} J_0(\lambda_{0m} r) \cos \lambda_{0m} \alpha t$$

$$+ \sum_{n=1}^{\infty} \sum_{m=1}^{\infty} J_n(\lambda_{nm} r) \left(A_{nm} \cos n\theta + B_{nm} \sin n\theta \right) \cos \lambda_{nm} \alpha t$$

where $J_n(\lambda b) = 0$ for $m = 0, 1, 2, 3, \ldots$ gives $\lambda = \lambda_{nm} = \xi_{nm}/b$ where ξ_{nm} denotes the mth zero of the nth order Bessel function.

(c) Show the coefficients which satisfy the given initial conditions are given by

$$A_{0m} = \frac{1}{\pi b^2 J_1^2(\xi_{0m})} \int_0^b \int_{-\pi}^{\pi} r f(r, \theta) J_0(\lambda_{0m} r) \, d\theta dr$$

$$A_{nm} = \frac{2}{\pi b^2 J_{n+1}^2(\xi_{nm})} \int_0^b \int_{-\pi}^{\pi} r f(r, \theta) J_n(\lambda_{nm} r) \cos n\theta \, d\theta dr$$

$$B_{nm} = \frac{2}{\pi b^2 J_{n+1}^2(\xi_{nm})} \int_0^b \int_{-\pi}^{\pi} r f(r, \theta) J_n(\lambda_{nm} r) \sin n\theta \, d\theta dr$$

for $m = 1, 2, 3, \ldots$ and $n = 1, 2, 3, \ldots$.

▶ **18.**

Solve the Neumann problem for the one-dimensional wave equation $0 < x < L$ which is describe by the boundary value problem

PDE:
$$\frac{\partial^2 u}{\partial t^2} - c^2 \frac{\partial^2 u}{\partial x^2} = F(x, t), \quad 0 < x < L, \ t > 0$$

BC:
$$\frac{\partial u(0, t)}{\partial x} = 0, \qquad \frac{\partial u(L, t)}{\partial x} = 0$$

IC:
$$u(x, 0) = f(x), \qquad \frac{\partial u(x, 0)}{\partial t} = g(x)$$

which represents a vibrating string where the ends of the string are required to have a zero slope.

▶ **19.**

Determine conditions that must be placed upon the constants $\{n_1, n_2, n_3\}$ and k such that $u = \exp\left[\pm\imath(n_1 x + n_2 y + n_3 z + kct)\right]$ is a solution of the wave equation

$$\frac{\partial^2 u}{\partial t^2} = c^2 \nabla^2 u.$$

▶ **20.**

(a) Verify for arbitrary f that each of the functions:

$$(i) \quad u = f(N_1 x + N_2 y + N_3 z - \omega t)$$

$$(ii) \quad u = f(N_1 x + N_2 y - \omega t)$$

$$(iii) \quad u = f(N_1 x - \omega t)$$

are solutions of the wave equation

$$\frac{\partial^2 u}{\partial t^2} = c^2 \left(\frac{\partial^2 u}{\partial x^2} + \frac{\partial^2 u}{\partial y^2} + \frac{\partial^2 u}{\partial z^2} \right).$$

In each case find the relation existing between
the quantities $\vec{N} = (N_1, N_2, N_3), \omega$ and c.

(b) Give a physical interpretation to the wave motion for the cases:

$(i) \quad \vec{N} = (1,1,1) \qquad (ii) \quad \vec{N} = (1,1,0) \qquad (iii) \quad \vec{N} = (1,0,0) \qquad (iv) \quad \vec{N} = (0,1,0)$

▶ **21.**

(a) For $\vec{F} = 0$, assume perturbational solutions to the nonlinear equations (5.34),
(5.35), and (5.36) having the form

$$P = P_0 + p \qquad \varrho = \varrho_0 + r \qquad \vec{V} = \vec{v}$$

where P_0 and ϱ_0 are constants. Show that if product terms of the perturbational variables $p, r, \vec{v}$ are considered small and negligible, then the equations
(5.34), (5.35), and (5.36) reduce to the form

$$\varrho_0 \frac{\partial \vec{v}}{\partial t} = -\nabla p, \qquad \frac{\partial r}{\partial t} + \varrho_0 \nabla \vec{v} = 0, \qquad p = a^2 r$$

where $a^2 = P_0 \gamma / \varrho_0$.

(b) Show the pressure p satisfies the wave equation $\dfrac{\partial^2 p}{\partial t^2} = a^2 \nabla^2 p$.

▶ **22.**

Solve the boundary value problem

$$\text{PDE:} \qquad \frac{\partial^2 u}{\partial t^2} = c^2 \frac{\partial^2 u}{\partial x^2} - g, \qquad 0 < x < L, \ t > 0$$

$$\text{BC:} \qquad u(0,t) = 0, \qquad u(L,t) = 0$$

$$\text{IC:} \qquad u(x,0) = 0, \qquad \frac{\partial u(L,t)}{\partial t} = 0$$

where g is the acceleration of gravity.

▶ 23.

Solve boundary value problem for the one-dimensional wave equation associated with a semi-infinite string which is given by

PDE: $\dfrac{\partial^2 u}{\partial t^2} - c^2 \dfrac{\partial^2 u}{\partial x^2} = 0, \quad 0 < x < \infty, \, t > 0$

BC: $u(0, t) = 0, \qquad |u(x, t)| < M, \quad \text{as } x \to \infty$

IC: $u(x, 0) = f(x), \qquad \dfrac{\partial u(x, 0)}{\partial t} = 0$

▶ 24.

Solve boundary value problem for the one-dimensional wave equation associated with a semi-infinite string which is given by

PDE: $\dfrac{\partial^2 u}{\partial t^2} - c^2 \dfrac{\partial^2 u}{\partial x^2} = 0, \quad 0 < x < \infty, \, t > 0$

BC: $u(0, t) = 0, \qquad |u(x, t)| < M \text{ as } x \to \infty$

IC: $u(x, 0) = 0, \qquad \dfrac{\partial u(x, 0)}{\partial t} = g(x)$

▶ 25.

Solve the initial value problem for the one-dimensional wave equation associated with a infinite string which is given by

PDE: $\dfrac{\partial^2 u}{\partial t^2} - c^2 \dfrac{\partial^2 u}{\partial x^2} = 0, \quad -\infty < x < \infty, \, t > 0$

IC: $u(x, 0) = \sin x, \qquad \dfrac{\partial u(x, 0)}{\partial t} = 0$

▶ 26.

Solve the initial value problem

$$\dfrac{\partial^2 u}{\partial t^2} = c^2 \dfrac{\partial^2 u}{\partial x^2}, \quad -\infty < x < \infty, \, t > 0$$

subject to the initial displacement and initial velocity given by the equations $u(x, 0) = 0$ and $\dfrac{\partial u(x, 0)}{\partial t} = \sin x$.

▶ **27.**

The three-dimensional wave equation in spherical coordinates (ρ, θ, ϕ) independent of the angular variables θ, ϕ is given by

$$\frac{\partial^2 u}{\partial t^2} = c^2 \left(\frac{\partial^2 u}{\partial \rho^2} + \frac{2}{\rho} \frac{\partial u}{\partial \rho} \right) \tag{27a}$$

(a) Show the transformation $w = \rho u$ reduces the equation (27a) to the form

$$\frac{\partial^2 w}{\partial t^2} = c^2 \frac{\partial^2 w}{\partial \rho^2} \tag{27b}$$

which has the solutions

$$w = w(\rho, t) = f(\rho + ct) + g(\rho - ct)$$

where f, g are arbitrary wave shapes.

(b) Show the solution to equation (27a) can be written in the form

$$u = u(\rho, t) = \frac{f(\rho + ct)}{\rho} + \frac{g(\rho - ct)}{\rho}$$

which represents spherical waves propagated in the radial direction ρ with velocity c from the region where the initial disturbance is specified. The solution $\frac{f(\rho + ct)}{\rho}$ represents a spherical wave moving away from the origin in the radial direction and is called and expanding or outgoing wave. The solution $\frac{g(\rho - ct)}{\rho}$ represents a spherical wave moving toward the origin in the radial direction and is called a contracting or in going wave.

▶ **28.**

Solve the damped wave equation

$$\text{PDE:} \quad \frac{\partial^2 u}{\partial t^2} = \frac{\partial^2 u}{\partial x^2} - 2\beta \frac{\partial u}{\partial t}, \quad 0 < x < \pi \quad 0 < \beta < 1$$

$$\text{BC:} \quad u(0, t) = 0, \quad u(\pi, t) = 0$$

$$\text{IC:} \quad u(x, 0) = f(x), \quad \frac{\partial u(x, 0)}{\partial t} = g(x)$$

▶ **29.**

Solve the given boundary value problem

$$\text{PDE:} \quad \frac{\partial^2 H}{\partial t^2} = \alpha^2 \frac{\partial^2 H}{\partial x^2} + G(x, t), \quad 0 < x < L, \quad t > 0$$

$$\text{BC:} \quad H(0, t) = 0, \quad H(L, t) = 0$$

$$\text{IC:} \quad H(x, 0) = 0, \quad \frac{\partial H(x, 0)}{\partial t} = 0$$

▶ **30.**

Solve for the displacement $u = u(x,t)$ of a vibrating string governed by the partial differential equation

$$\frac{\partial^2 u}{\partial t^2} = \frac{\partial^2 u}{\partial x^2}, \quad 0 < x < \pi, \ t > 0$$

and subject to the boundary conditions $u(0,t) = 0$ and $u(\pi,t) = 0$ and initial conditions $u(x,0) = \begin{cases} x, & 0 \le x \le \pi/2 \\ \pi - x, & \pi/2 \le x \le \pi \end{cases}$ and $\dfrac{\partial u(x,0)}{\partial t} = x(\pi - x)$.

▶ **31.**

(a) Show that the solution to the wave equation

$$\frac{\partial^2 u}{\partial t^2} - \alpha^2 \frac{\partial^2 u}{\partial x^2} = 0, \quad -\infty < x < \infty, \ t > 0$$

which satisfies $u(x,0) = f(x)$ and $\dfrac{\partial u(x,0)}{\partial t} = 0$ is given by

$$u = u(x,t) = \frac{1}{2}\left[f(x + \alpha t) + f(x - \alpha t)\right]$$

(b) Show that the solution to the wave equation

$$\frac{\partial^2 u}{\partial t^2} - \alpha^2 \frac{\partial^2 u}{\partial x^2} = 0, \quad -\infty < x < \infty, \ t > 0$$

which satisfies $u(x,0) = 0$ and $\dfrac{\partial u(x,0)}{\partial t} = g(x)$ is given by

$$u = u(x,t) = \frac{1}{2\alpha} \int_{x-\alpha t}^{x+\alpha t} g(\xi)\, d\xi$$

(c) Solve the wave equation

$$\frac{\partial^2 u}{\partial t^2} - \alpha^2 \frac{\partial^2 u}{\partial x^2} = 0, \quad -\infty < x < \infty, \ t > 0$$

which satisfies $u(x,0) = f(x)$ and $\dfrac{\partial u(x,0)}{\partial t} = g(x)$.

▶ **32.** (Wave guides)

Consider a wave moving in the z-direction in a dielectric medium between the planes $x = 0$ and $x = a$. The wave motion is assumed periodic and the components of $\vec{\mathcal{E}}$ and $\vec{\mathcal{H}}$ must satisfy the two-dimensional form of the Helmholtz equation given by equation (5.64).

(a) Write out the Maxwell equations in component form when the wave travels in the z- direction and is independent of the y-direction.

Hint: Use $\vec{E} = \vec{\mathcal{E}}(x, z)e^{-i\omega t}$ and $\vec{H} = \vec{\mathcal{H}}(x, z)e^{-i\omega t}$.

(b) (TE waves $(0, \mathcal{E}_2, 0); (\mathcal{H}_1, 0, \mathcal{H}_3)$) If $(0, \mathcal{E}_2, 0)$ is known, how would you use the Maxwell equations to determine the $(\mathcal{H}_1, 0, \mathcal{H}_2)$ components? Find the form of the solutions to

$$\frac{\partial^2 \mathcal{E}_2}{\partial x^2} + \frac{\partial^2 \mathcal{E}_2}{\partial z^2} + k^2 \mathcal{E}_2 = 0, \quad 0 < x < a, \ z > 0$$

satisfying

$$\mathcal{E}_2(0, z) = 0, \quad \mathcal{E}_2(a, z) = 0, \quad |\mathcal{E}_2| < M, \quad \text{with} \quad \frac{\pi}{a} > k.$$

and then show how to calculate the form of $\mathcal{H}_1$ and $\mathcal{H}_3$.

(c) (TM waves $(\mathcal{E}_1, 0, \mathcal{E}_3); (0, \mathcal{H}_2, 0)$) If $(0, \mathcal{H}_2, 0)$ is known, how would you use the Maxwell equations to determine the $(\mathcal{E}_1, 0, \mathcal{E}_3)$ components? Find the form of the solution to

$$\frac{\partial^2 \mathcal{H}_2}{\partial x^2} + \frac{\partial^2 \mathcal{H}_2}{\partial z^2} + k^2 \mathcal{H}_2 = 0, \quad 0 < x < a, \ z > 0$$

satisfying the boundary conditions

$$\frac{\partial \mathcal{H}_2(0, z)}{\partial x} = 0, \quad \frac{\partial \mathcal{H}_2(a, z)}{\partial x} = 0, \quad |\mathcal{H}_2| < M, \quad \text{with} \quad \frac{\pi}{a} > k$$

and then show how to calculate the form of $\mathcal{E}_1$ and $\mathcal{E}_3$.

▶ **33.**

Solve the boundary value problem

$$\text{PDE:} \quad \frac{\partial^2 u}{\partial t^2} + 2\beta \frac{\partial u}{\partial t} = c^2 \frac{\partial^2 u}{\partial x^2} - g, \quad 0 < x < L, \ t > 0, \ c\pi > \beta L$$

$$\text{BC:} \quad u(0, t) = 0, \quad u(L, t) = 0$$

$$\text{IC:} \quad u(x, 0) = 0, \quad \frac{\partial u(L, t)}{\partial t} = 0$$

where g is the acceleration of gravity and β is constant.

▶ **34.**

(Duhamel's principle for the wave equation) Show that if $v = v(x, t; \tau)$ is a solution of the initial value problem

$$\text{PDE:} \qquad \frac{\partial^2 v}{\partial t^2} - \alpha^2 \frac{\partial^2 v}{\partial x^2} = 0, \qquad -\infty < x < \infty,$$

$$\text{IC:} \qquad v(x, \tau; \tau) = 0$$

$$\frac{\partial v(x, \tau; \tau)}{\partial t} = f(x, \tau), \qquad -\infty < x < \infty$$

where $\tau \geq 0$, then the solution of the related nonhomogeneous wave equation initial value problem

$$\text{PDE:} \qquad \frac{\partial^2 u}{\partial t^2} - \alpha^2 \frac{\partial^2 u}{\partial x^2} = f(x, t), \qquad -\infty < x < \infty, \; t > 0$$

$$\text{IC:} \qquad u(x, 0) = 0$$

$$\frac{\partial u(x, 0)}{\partial t} = 0, \qquad -\infty < x < \infty$$

is given by

$$u = u(x, t) = \int_0^t v(x, t; \tau) \, d\tau$$

Hint: Use Leibnitz's† rule for differentiating an integral

$$\frac{d}{dt} \int_{\alpha(t)}^{\beta(t)} f(t, \tau) \, d\tau = \int_{\alpha(t)}^{\beta(t)} \frac{\partial f(t, \tau)}{\partial t} \, d\tau + f[t, \beta(t)] \frac{d\beta}{dt} - f[t, \alpha(t)] \frac{d\alpha}{dt}.$$

▶ **35.**

Use Duhamel's principle to solve the following initial value problem

$$\text{PDE:} \qquad \frac{\partial^2 u}{\partial t^2} - \alpha^2 \frac{\partial^2 u}{\partial x^2} = f(x), \qquad -\infty < x < \infty, t > 0$$

$$\text{IC:} \qquad u(x, 0) = 0, \qquad \frac{\partial u(x, 0)}{\partial t} = 0, \qquad -\infty < x < \infty$$

---

† See Appendix C

▶ **36.**

(Duhamel's principle for the wave equation)

(a) Show that if $v = v(x, y, z, t; \tau)$ is a solution of the initial value problem

PDE: $\quad \dfrac{\partial^2 v}{\partial t^2} - \alpha^2 \left(\dfrac{\partial^2 v}{\partial x^2} + \dfrac{\partial^2 v}{\partial y^2} + \dfrac{\partial^2 v}{\partial z^2} \right) = 0, \qquad -\infty < x, y, z < \infty,$

IC: $\qquad\qquad\qquad\qquad v(x, y, z, \tau; \tau) = 0 \hspace{3cm}$ (36a)

$\qquad\qquad\qquad\qquad \dfrac{\partial v(x, y, z, \tau; \tau)}{\partial t} = f(x, y, z, \tau), \quad -\infty < x < \infty$

then $u = u(x, y, z, t) = \displaystyle\int_0^t v(x, y, z, t; \tau)\, d\tau$ is a solution of the initial value problem

PDE: $\quad \dfrac{\partial^2 u}{\partial t^2} - \alpha^2 \left(\dfrac{\partial^2 u}{\partial x^2} + \dfrac{\partial^2 u}{\partial y^2} + \dfrac{\partial^2 u}{\partial z^2} \right) = f(x, y, z, t), \qquad -\infty < x, y, z < \infty,$

IC: $\qquad\qquad\qquad\qquad u(x, y, z, 0) = 0$

$\qquad\qquad\qquad\qquad \dfrac{\partial u(x, y, z, 0)}{\partial t} = 0, \quad -\infty < x, y, z < \infty$

(b) Make a change of variable in equation (5.146) and let the initial conditions start at time $t = \tau$ rather than at time $t = 0$. This can be accomplished by replacing t by $t - \tau$. Consequently, show that for n_1, n_2, n_3 given by equations (5.143) one can write

$v = v(x, y, z, t; \tau) = (t - \tau)\overline{f}$

$= \dfrac{t - \tau}{4\pi} \displaystyle\int_0^{2\pi} \int_0^{\pi} f(x + n_1 c(t - \tau), y + n_2 c(t - \tau), z + n_3 c(t - \tau), \tau) \sin\theta \, d\theta d\phi$

as a solution to equation (36a).

▶ **37.**

Show how one can use superposition to solve the partial differential equation

PDE: $\quad \dfrac{\partial^2 u}{\partial t^2} - \alpha^2 \left(\dfrac{\partial^2 u}{\partial x^2} + \dfrac{\partial^2 u}{\partial y^2} + \dfrac{\partial^2 u}{\partial z^2} \right) = F(x, y, z, t), \qquad -\infty < x, y, z < \infty,$

IC: $\qquad\qquad\qquad\qquad u(x, y, z, 0) = f(x, y, z)$

$\qquad\qquad\qquad\qquad \dfrac{\partial u(x, y, z, 0)}{\partial t} = g(x, y, z), \quad -\infty < x, y, z < \infty$

Hint: See problem 36.

▶ **38.**

Consider the vibrating string problem

$$\text{PDE:} \quad \frac{\partial^2 u}{\partial t^2} = c^2 \frac{\partial^2 u}{\partial x^2}, \quad u = u(x,t), \quad 0 < x < L, \quad t > 0$$

$$\text{BC:} \quad u(0,t) = 0, \quad u(L,t) = 0$$

$$\text{IC:} \quad u(x,0) = f(x), \quad \frac{\partial u(x,0)}{\partial t} = g(x), \quad 0 < x < L$$

which is solved by the method of separation of variables.

(a) Assume that $f(x)$ and $g(x)$ have Fourier sine series expansions of the form

$$f(x) = \sum_{n=1}^{\infty} f_n \sin \frac{n\pi x}{L}, \qquad g(x) = \sum_{n=1}^{\infty} g_n \sin \frac{n\pi x}{L}.$$

Express the solution to the vibrating string problem in terms of the coefficients f_n and g_n.

(b) Use the trigonometric identities

$$\sin A \sin B = \frac{1}{2} \left[\cos(A - B) - \cos(A + B) \right] = \frac{1}{2} \int_{A-B}^{A+B} \sin \xi \, d\xi$$

$$\sin A \cos B = \frac{1}{2} \left[\sin(A + B) + \sin(A - B) \right]$$

and show the solution for the vibrating string problem can be written in the form

$$u = u(x,t) = \frac{1}{2} \left[\widetilde{F}(x + ct) + \widetilde{F}(x - ct) \right] + \frac{1}{2c} \int_{x-ct}^{x+ct} \widetilde{G}(\xi) \, d\xi$$

where $\widetilde{F}(x)$, $\widetilde{G}(x)$ are the periodic extensions of $f(x)$, $g(x)$ on the interval $0 < x < L$.

▶ **39.**

Solve the equation for a vibrating membrane

$$\frac{\partial^2 u}{\partial t^2} = c^2 \left(\frac{\partial^2 u}{\partial r^2} + \frac{1}{r} \frac{\partial u}{\partial r} + \frac{1}{r^2} \frac{\partial^2 u}{\partial \theta^2} \right), \quad -\pi < \theta < \pi, \ 0 < r < r_0, \ t > 0$$

subject to the boundary condition $u(r_0, \theta, t) = 0$ and initial conditions

$$u(r, \theta, 0) = 0, \qquad \frac{\partial u(r, \theta, 0)}{\partial t} = g(r, \theta)$$

(a) Show that by assuming a solution $u = F(r)G(\theta)H(t)$ there results the equations

$$\frac{d^2H}{dt^2} + \omega^2 c^2 H = 0, \qquad r^2\frac{d^2F}{dr^2} + r\frac{dF}{dr} + (\omega^2 r^2 - n^2)F = 0, \qquad \frac{d^2G}{d\theta^2} + n^2 G = 0$$

$$H(0) = 0, \qquad\qquad\qquad F(r_0) = 0, \qquad\qquad\qquad G(-\pi) - G(\pi) = 0$$

$$G'(-\pi) - G'(\pi) = 0$$

where n is an integer and ω is a constant .

(b) Show the solution can be represented

$$u = u(r, \theta, t) = \sum_{m=1}^{\infty} C_{0m} J_0(\xi_{0m}\frac{r}{r_0}) \sin\frac{\xi_{0m}}{r_0} ct$$

$$+ \sum_{n=1}^{\infty}\sum_{m=1}^{\infty} J_n(\xi_{nm}\frac{r}{r_0}) [C_{nm}\cos n\theta + D_{nm}\sin n\theta] \sin\frac{\xi_{nm}}{r_0} ct$$

where

$$C_{0m} = \frac{1}{\pi c^2 \xi_{0m}^2 J_1^2(\xi_{0m})} \int_0^{r_0}\int_{-\pi}^{\pi} r J_0(\xi_{0m}\frac{r}{r_0}) g(r, \theta)\, d\theta dr$$

$$C_{nm} = \frac{2}{\pi c^2 \xi_{nm}^2 J_{n+1}^2(\xi_{nm})} \int_0^{r_0}\int_{-\pi}^{\pi} r J_n(\xi_{nm}\frac{r}{r_0}) g(r, \theta) \cos n\theta\, d\theta dr$$

$$D_{nm} = \frac{2}{\pi c^2 \xi_{nm}^2 J_{n+1}^2(\xi_{nm})} \int_0^{r_0}\int_{-\pi}^{\pi} r J_n(\xi_{nm}\frac{r}{r_0}) g(r, \theta) \sin n\theta\, d\theta dr$$

where $J_n(\xi_{nm}) = 0$ for $m = 1, 2, 3, \ldots$.

▶ **40.**

Solve the equation for a vibrating membrane

$$\frac{\partial^2 u}{\partial t^2} = c^2\left(\frac{\partial^2 u}{\partial r^2} + \frac{1}{r}\frac{\partial u}{\partial r} + \frac{1}{r^2}\frac{\partial^2 u}{\partial\theta^2}\right), \qquad -\pi < \theta < \pi,\ 0 < r < r_0,\ t > 0$$

subject to the boundary condition $u(r_0, \theta, t) = 0$ and initial conditions

$$u(r, \theta, 0) = f(r, \theta), \qquad \frac{\partial u(r, \theta, 0)}{\partial t} = g(r, \theta).$$

Chapter 6
Elliptic Equations

The homogeneous Laplace equation

$$\nabla^2 u = 0 \qquad x, y, z \in \mathcal{R} \tag{6.1}$$

and its nonhomogeneous form

$$\nabla^2 u = f(x, y, z) \qquad x, y, z \in \mathcal{R} \tag{6.2}$$

known as Poisson's[†] equation, together with the Helmholtz[†] equation

$$\nabla^2 u + \lambda u = 0 \qquad x, y, z \in \mathcal{R} \quad \lambda \text{ constant,} \tag{6.3}$$

are important elliptic partial differential equations which occur quite frequently in a variety of science and engineering disciplines. These equations are classified as elliptic equations. The Helmholtz equation is sometimes referred to as an elliptic eigenvalue problem where one seeks values of λ for which nonzero solutions exist. The above equations can take on various forms depending upon how the Laplacian operator $\nabla^2 u$ is represented. Let us investigate various forms for the above equations and illustrated how these equations arise in some selected application areas from science and engineering.

The Laplacian Operator

The Laplace operator $\nabla^2 u$ can be represented in many different forms. The general form for an orthogonal coordinate system is given by equation (4.13), from which one can obtain the Cartesian, cylindrical and spherical coordinate forms used most often. Various one-dimensional, two-dimensional and three-dimensional forms for the Laplace, Poisson and Helmholtz equations are achieved by making appropriate simplifications of the above Laplacian operators. Some examples of the different coordinate forms for the Laplacian are:

Cartesian (x, y, z)

$$\begin{aligned}
u &= u(x, y, z), & \nabla^2 u &= \frac{\partial^2 u}{\partial x^2} + \frac{\partial^2 u}{\partial y^2} + \frac{\partial^2 u}{\partial z^2} \\
u &= u(x, y), & \nabla^2 u &= \frac{\partial^2 u}{\partial x^2} + \frac{\partial^2 u}{\partial y^2} \\
u &= u(x), & \nabla^2 u &= \frac{d^2 u}{dx^2}
\end{aligned} \tag{6.4}$$

Cylindrical (r, θ, z)

$$\begin{aligned}
u &= u(r, \theta, z), & \nabla^2 u &= \frac{\partial^2 u}{\partial r^2} + \frac{1}{r}\frac{\partial u}{\partial r} + \frac{1}{r^2}\frac{\partial^2 u}{\partial \theta^2} + \frac{\partial^2 u}{\partial z^2} \\
u &= u(r, \theta), & \nabla^2 u &= \frac{\partial^2 u}{\partial r^2} + \frac{1}{r}\frac{\partial u}{\partial r} + \frac{1}{r^2}\frac{\partial^2 u}{\partial \theta^2} \\
u &= u(r, z), & \nabla^2 u &= \frac{\partial^2 u}{\partial r^2} + \frac{1}{r}\frac{\partial u}{\partial r} + \frac{\partial^2 u}{\partial z^2} \\
u &= u(r), & \nabla^2 u &= \frac{d^2 u}{dr^2} + \frac{1}{r}\frac{du}{dr}
\end{aligned} \tag{6.5}$$

[†]   See Appendix C

where $\hat{n}$ is the outward normal to S and $\hat{n}_\epsilon$ is the inward normal on the spherical surface S_ϵ. (i.e. The surface normals in Green's identity are always exterior normals.) Note that the normal $\hat{n}_\epsilon = -\hat{e}_\rho$ is in the opposite direction of a unit normal $\hat{e}_\rho$ from the origin of the sphere in the radial direction. It is easily verified using the gradient information from equation (6.11) that

$$\frac{\partial}{\partial n_\epsilon}\left(\frac{1}{\rho}\right) = \text{grad}\left(\frac{1}{\rho}\right) \cdot (-\hat{e}_\rho) = \frac{1}{\rho^2}$$

$$\frac{\partial \psi}{\partial n_\epsilon} = \text{grad}\,\psi \cdot (-\hat{e}_\rho) = \nabla\psi \cdot (-\hat{e}_\rho) = -\frac{\partial \psi}{\partial \rho}$$

so that the last integral in equation (6.40) can be represented

$$\iint_{S_\epsilon}\left(\psi\frac{1}{\rho^2} + \frac{1}{\rho}\frac{\partial\psi}{\partial\rho}\right)\Big|_{\rho=\epsilon} d\sigma = \iint_{S_\epsilon}\left(\psi\frac{1}{\rho^2} + \frac{1}{\rho}\frac{\partial\psi}{\partial\rho}\right)\Big|_{\rho=\epsilon} \epsilon^2 \sin\theta d\theta d\phi$$

$$= \int_0^{2\pi}\int_0^\pi \psi\Big|_{\rho=\epsilon}\sin\theta d\theta d\phi + \epsilon\int_0^{2\pi}\int_0^\pi \frac{\partial\psi}{\partial\rho}\Big|_{\rho=\epsilon}\sin\theta d\theta d\phi$$

$$= 4\pi\overline{\psi} + \epsilon\int_0^{2\pi}\int_0^\pi \frac{\partial\psi}{\partial\rho}\Big|_{\rho=\epsilon}\sin\theta d\theta d\phi$$

where $\overline{\psi}$ is the average value of ψ over the surface of the sphere S_ϵ. Note that the element of volume is given by $d\tau = \rho^2 \sin\theta d\rho d\theta d\phi$ so that the volume integral $\iiint_{V-V_\epsilon}\frac{1}{\rho}\nabla^2\psi\, d\tau$ remains bounded as $\epsilon \to 0$. Therefore, in the limit as $\epsilon \to 0$ in equation (6.40) one obtains the Green's third identity

$$\psi(x_0, y_0, z_0) = \frac{-1}{4\pi}\iint_V \frac{1}{\rho}\nabla^2\psi\, d\tau + \frac{1}{4\pi}\iint_S \frac{1}{\rho}\frac{\partial\psi}{\partial n}d\sigma - \frac{1}{4\pi}\iint_S \psi\frac{\partial}{\partial n}\left(\frac{1}{\rho}\right)d\sigma \qquad (6.41)$$

which is referred to as a representation theorem for three-dimensional space. In the special case where S is a sphere of radius ρ_0 and ψ is a harmonic function, the volume integral term in equation (6.41) is zero. The second term vanishes by property 1 from table 6-1. In this special case the Green's third identity reduces to

$$\psi(x_0, y_0, z_0) = \frac{1}{4\pi\rho_0^2}\iint_S \psi\, d\sigma \qquad (6.42)$$

which gives the mean value property listed as item 5 in table 6-1.

The two-dimensional form of the above representation theorem is derived from the two-dimensional form of the Gauss divergence theorem given by equation (6.31). If we substitute $\vec{F} = v\nabla u$ into equation (6.31) we obtain

$$\iint_R v\nabla^2 u\, d\sigma + \iint_R\left(\frac{\partial v}{\partial x}\frac{\partial u}{\partial x} + \frac{\partial v}{\partial y}\frac{\partial u}{\partial y}\right) = \int_C v\nabla u \cdot \hat{n}\, ds = \int_c v\frac{\partial u}{\partial n}ds. \qquad (6.43)$$

In this equation we interchange the roles of u and v to obtain

$$\iint_R u\nabla^2 v \, d\sigma + \iint_R \left(\frac{\partial u}{\partial x}\frac{\partial v}{\partial x} + \frac{\partial u}{\partial y}\frac{\partial v}{\partial y}\right) = \int_C u\nabla v \cdot \hat{n}\, ds = \int_c u\frac{\partial v}{\partial n}\, ds. \tag{6.44}$$

In equations (6.43) and (6.44) we assume that the functions u and v have continuous second derivatives and so when we subtract the above two equations we obtain the two-dimensional form of Green's second identity

$$\iint_R (v\nabla^2 u - u\nabla^2 v)\, d\sigma = \int_C \left(v\frac{\partial u}{\partial n} - u\frac{\partial v}{\partial n}\right)\, ds. \tag{6.45}$$

where R is understood to represent a bounded region lying within the boundary curve C. Note the special case where $u = 1$ we obtain

$$\iint_R \nabla^2 v \, d\sigma = \int_C \frac{\partial v}{\partial n}\, ds. \tag{6.46}$$

Observe that if v is a harmonic function everywhere within the region R, then

$$\int_C \frac{\partial u}{\partial n}\, ds = 0 \tag{6.47}$$

which is a special case of item 1 of table 6-1. Interpreted in terms of heat flow the condition given by equation (6.47) requires that for steady state conditions to exist the total heat flow across the boundary must equal zero otherwise the temperature would be changing with time. The condition given by equation (6.47) is sometimes referred to as a solvability condition for steady state solutions of the heat equation.

In equation (6.45) let

$$u = \log r = \log|\vec{r} - \vec{r}_0| = \log\sqrt{(x - x_0)^2 + (y - y_0)^2} \tag{6.48}$$

and modify the region of integration to avoid the singularity at (x_0, y_0) by placing a small circle of radius ϵ about the singular point (x_0, y_0). We denote the area of this small circle by R_ϵ and we denote the circumference of the circle by C_ϵ. We have $\nabla^2 u = 0$ everywhere interior to the region $R - R_\epsilon$ because the singular point is now avoided. The Green's second identity, equation (6.45) can then be expressed in the form

$$-\iint_{R-R_\epsilon}(\log r)\nabla^2 v \, d\sigma = \int_C \left(v\frac{\partial}{\partial n}\log r - \log r\frac{\partial v}{\partial n}\right)\, ds - \int_{C_\epsilon}\left(v\frac{1}{\epsilon} - (\log \epsilon)\frac{\partial v}{\partial n}\right)\Big|_{r=\epsilon}\epsilon\, d\theta \tag{6.49}$$

where ds is an element of arc length and all line integrals are to be taken in the positive sense. We divide by 2π and take the limit as $\epsilon \to 0$ to obtain

$$\lim_{\epsilon \to 0} \frac{1}{2\pi\epsilon} \int_{C_\epsilon} v\, \epsilon\, d\theta = \frac{1}{2\pi} \int_C \left(v \frac{\partial}{\partial n} \log r - \log r \frac{\partial v}{\partial n} \right) ds$$
$$+ \frac{1}{2\pi} \iint_R \log r \, \nabla^2 v \, d\sigma + \lim_{\epsilon \to 0} \frac{\log \epsilon}{2\pi} \int_{C_\epsilon} \frac{\partial v}{\partial n} \, ds. \tag{6.50}$$

Note that

$$\lim_{\epsilon \to 0} \frac{1}{2\pi\epsilon} \int_{C_\epsilon} v\, \epsilon\, d\theta = \lim_{\epsilon \to 0} \frac{1}{2\pi\epsilon} \int_0^{2\pi} v(x_0 + \epsilon \cos\theta, y_0 + \epsilon \sin\theta)\, \epsilon\, d\theta = v(x_0, y_0)$$

and using equation (6.46) we can write

$$\left| \frac{\log \epsilon}{2\pi} \int_{C_\epsilon} \frac{\partial v}{\partial n} \, ds \right| = \left| \frac{\log \epsilon}{2\pi} \iint_{R_\epsilon} \nabla^2 v \, d\sigma \right| \le \frac{1}{2\pi} |\log \epsilon| M (\pi \epsilon^2)$$

where M is the maximum value of $\nabla^2 v$ over the region R_ϵ. The above inequality can now be used to show that

$$\lim_{\epsilon \to 0} \left| \frac{\log \epsilon}{2\pi} \int_{C_\epsilon} \frac{\partial v}{\partial n} \, ds \right| = 0$$

and consequently the equation (6.50) reduces to the two-dimensional representation theorem

$$v(x_0, y_0) = \frac{1}{2\pi} \int_C \left(v \frac{\partial}{\partial n} \log r - \log r \frac{\partial v}{\partial n} \right) ds + \frac{1}{2\pi} \iint_R \log r \, \nabla^2 v \, d\sigma. \tag{6.51}$$

The two-dimensional and three-dimensional representation theorems are important in that they illustrate that the solution to Laplace's equation within a region R can be expressed in terms of known values of the function and normal derivative on the boundary of the given region R. This is the property 3 listed in the table 6-1.

The Laplacian operator ∇^2 operating on a function $u(x, y)$ and evaluated at a point (x_0, y_0) is given the physical interpretation of producing a value which is proportional to the difference between the average value $\bar{u}$ of u about the point (x_0, y_0) and the functional value at (x_0, y_0). This can be written

$$\nabla^2 u \Big|_{(x_0, y_0)} \approx \alpha \left[\bar{u} - u(x_0, y_0) \right] \tag{6.52}$$

where α is a proportionality constant. In two-dimensions note that the average value of $u(x, y)$ on a circle of radius ϵ about the point (x_0, y_0) is given by

$$\overline{u}(x_0, y_0) = \frac{1}{2\pi\epsilon} \int_0^{2\pi} u(x_0 + \epsilon \cos\theta, y_0 + \epsilon \sin\theta)\, \epsilon d\theta.$$

Expanding the integrand in a Taylor[†] series about the point $\epsilon = 0$ produces the result

$$\overline{u}(x_0, y_0) = \frac{1}{2\pi} \int_0^{2\pi} \left[u(x_0, y_0) + \left(\frac{\partial u}{\partial x} \cos\theta + \frac{\partial u}{\partial y} \sin\theta \right) \epsilon \right.$$
$$\left. + \left(\frac{\partial^2 u}{\partial x^2} \cos^2\theta + 2\frac{\partial^2 u}{\partial x \partial y} \sin\theta \cos\theta + \frac{\partial^2 u}{\partial y^2} \sin^2\theta \right) \frac{\epsilon^2}{2!} + \text{higher order terms} \right] d\theta$$

which when integrated simplifies to

$$\overline{u} = \frac{1}{2\pi} \left[u(x_0, y_0)(2\pi) + 0 + \frac{\pi}{2} \left(\frac{\partial^2 u}{\partial x^2} + \frac{\partial^2 u}{\partial y^2} \right) \epsilon^2 + \text{higher order terms} \right].$$

Neglecting the higher order terms in the expansion, we obtain the approximation

$$\nabla^2 u \approx \frac{4}{\epsilon^2} \left[\overline{u} - u_0 \right].$$

A similar approximation is obtained in 3-dimensions and is left for the exercises. This physical interpretation associated with the Laplacian operator can be used to help analyze partial differential equations which contain the Laplacian operator. For example the Laplace equation $\nabla^2 u = 0$ applied to equation (6.52) tells us the solution u at a point (x, y) represents the average value of its neighboring points. The equation (6.52) also tells us that if $\nabla^2 u > 0$, then the average value of u on a circle about a point (x, y) will be greater than the value of u at the center of the circle. Similarly, if $\nabla^2 u < 0$, the average value of u on a circle about a point (x, y) will be less than the value of u at the center of the circle.

We can apply this approximation to the heat equation and wave equation. The heat equation $\frac{\partial u}{\partial t} = \alpha^2 \nabla^2 u$ can be interpreted that the change in temperature with time is proportional to $\nabla^2 u$. If $\nabla^2 u > 0$, then the temperature about a point is increasing since its temperature is less than the average of the temperatures of the surrounding points. If $\nabla^2 u < 0$, then the temperature about a point is decreasing since the average temperatures of the surrounding points is less than

---

[†] See Appendix C

the temperature at the central point. If $\nabla^2 u = 0$, then steady state conditions exist and the temperature at a point is the average of the temperatures of the surrounding points.

The wave equation can also be analyzed in terms of the Laplacian operator. The wave equation $\frac{\partial^2 u}{\partial t^2} = \alpha^2 \nabla^2 u$ states that the acceleration of the displacement u, or driving force for the displacement u, is proportional to $\nabla^2 u$. In terms of a vibrating membrane, if $\nabla^2 u > 0$, then the membrane experiences an upward force at a point if the value of the displacement at that point is less than the average value of the displacements of the surrounding neighboring points.

A nonconstant harmonic function must attain its maximum or minimum values on the boundary of the region R. The proof is by contradiction. Assume that u has a maximum at an interior point $P \in R$. This maximum value must be the average of all points on any circle (two-dimensions) or sphere (three-dimensions) about the point P. It is impossible for the function u to have a maximum value which is an average of surrounding values unless u is a constant everywhere. That is, u is a maximum value at P only if u takes on its maximum value at all surrounding points on the circle or sphere. Hence u would be constant everywhere in the domain R. This is the only situation where u can take on a maximum (or minimum) value within the domain R. All other situations require that a harmonic function u take on its maximum or minimum values on the boundary of the region R. This is the property 7 listed in the table 6-1.

The property 8 listed in the table 6-1, states that the Dirichlet problem associated with the Laplace equation must be unique. Assume the Dirichlet problem to solve

$$\nabla^2 u = 0 \quad \text{for} \quad x, y, z \in R$$

subject to the boundary condition

$$u(x, y, z) = f(x, y, z) \quad \text{for} \quad x, y, z \in \partial R$$

has two distinct solutions, say u_1 and u_2. Then the difference $v = u_1 - u_2$ must satisfy the Dirichlet problem

$$\nabla^2 v = 0 \quad \text{for} \quad x, y, z \in R$$

subject to the homogeneous boundary condition that

$$v(x, y, z) = 0 \quad \text{for} \quad x, y, z \in \partial R.$$

Using the property 2 from table 6-1, we can say that if $v = 0$ on ∂R, then $v = 0$ everywhere in R and so $u_1 = u_2$ and the solution is unique.

The property 9, listed in the table 6-1, states that a harmonic function remains harmonic under certain types of variable changes. This can be verified by using chain rule differentiation and showing the harmonic function in the new variables also satisfies the Laplace equation. The following is a partial list of variable changes one can use on a harmonic function $\Phi = \Phi(x, y)$ for $x, y \in R$ to obtain a harmonic function $\Phi = \Phi(u, v)$ for u, v belonging to the image of R under the given change of variables.

1. **Translation of axes.**

$$u = x + \alpha, \quad v = y + \beta \quad \text{or} \quad x = u - \alpha, \quad y = v - \beta$$

where α and β are constants.

2. **Rotation of axes.**

$$u = \cos\theta\, x + \sin\theta\, y, \quad v = -\sin\theta\, x + \cos\theta\, y \quad \text{or} \quad x = \cos\theta\, u - \sin\theta\, v, \quad y = \sin\theta\, u + \cos\theta\, v$$

where θ is the angle of rotation.

3. **Reflection about a line.** The reflection about a line is accomplished by a translation of axes followed by a rotation of axes. Special reflections of interest are:

(a) Reflection about the y-axis.

$$u = -x, \quad v = y \qquad \text{or} \qquad x = -u, \quad y = v$$

(b) Reflection about the x-axis.

$$u = x, \quad v = -y \qquad \text{or} \qquad x = u, \quad y = -v$$

(c) Reflection about the line $y = x$.

$$u = y, \quad v = x \qquad \text{or} \qquad x = v, \quad y = u$$

(d) Reflection about the line $y = mx + b$ with $m \neq 0$.

For $\vec{r} = (x_0, y_0)$ the reflected image point is given by $\vec{r}^* = (2x_0^* - x_0, 2y_0^* - y_0)$ where
$x_0^* = \dfrac{y_0 - b + x_0/m}{m + 1/m}, \quad y_0^* = mx_0^* + b$

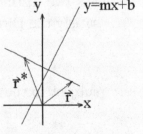

4. **Conformal mappings.** Let $z = x + iy$ denote a complex variable, where i is an imaginary unit with $i^2 = -1$, and let

$$w = f(z) = u(x,y) + iv(x,y) \qquad (6.53)$$

denote a function of the complex variable z. The function $w = f(z)$ is called an analytic function if the mapping given by equation (6.53) is a 1-to-1 mapping and the derivative of $f(z)$ satisfies the following equations

$$\frac{dw}{dz} = f'(z) = \frac{\partial u}{\partial x} + i\frac{\partial v}{\partial x} = \frac{\partial v}{\partial y} - i\frac{\partial u}{\partial y}.$$

If the derivative exists, then the Cauchy-Riemann[†] equations

$$\frac{\partial u}{\partial x} = \frac{\partial v}{\partial y}, \qquad \frac{\partial u}{\partial y} = -\frac{\partial v}{\partial x} \qquad (6.54)$$

must be satisfied. This implies that both the real part u and the imaginary part v of the complex function $w = f(z)$ are harmonic functions and satisfy

$$\nabla^2 u = \frac{\partial^2 u}{\partial x^2} + \frac{\partial^2 u}{\partial y^2} = 0 \qquad \text{and} \qquad \nabla^2 v = \frac{\partial^2 v}{\partial x^2} + \frac{\partial^2 v}{\partial y^2} = 0 \qquad (6.55)$$

as one can readily verify by differentiating the Cauchy-Riemann equations. Also note that from the Cauchy-Riemann equations one can show

$$\frac{\partial u}{\partial x}\frac{\partial v}{\partial x} + \frac{\partial u}{\partial y}\frac{\partial v}{\partial y} = 0 \qquad (6.56)$$

and

$$|f'(z)|^2 = \left|\frac{\partial u}{\partial x} + i\frac{\partial v}{\partial x}\right|^2 = \left(\frac{\partial v}{\partial x}\right)^2 + \left(\frac{\partial v}{\partial y}\right)^2 = \left(\frac{\partial u}{\partial x}\right)^2 + \left(\frac{\partial u}{\partial y}\right)^2. \qquad (6.57)$$

An analytic mapping is called conformal at points $z = x + iy$ where $|f'(z)| \neq 0$. If $\Phi = \Phi(x,y)$ is a harmonic function satisfying $\dfrac{\partial^2 \Phi}{\partial x^2} + \dfrac{\partial^2 \Phi}{\partial y^2} = 0$, then with a change of variable

$$u = u(x,y), \qquad v = v(x,y), \qquad (6.58)$$

where u, v are the real and imaginary parts of the conformal mapping given by $w = f(z) = u(x,y) + iv(x,y)$, we wish to show that $\Phi = \Phi(u,v)$ is also a harmonic function. We have, using the chain rule for differentiation, the first derivatives

$$\frac{\partial \Phi}{\partial x} = \frac{\partial \Phi}{\partial u}\frac{\partial u}{\partial x} + \frac{\partial \Phi}{\partial v}\frac{\partial v}{\partial x}, \qquad \frac{\partial \Phi}{\partial y} = \frac{\partial \Phi}{\partial u}\frac{\partial u}{\partial y} + \frac{\partial \Phi}{\partial v}\frac{\partial v}{\partial y} \qquad (6.59)$$

[†] See Appendix C

and second derivatives

$$\frac{\partial^2 \Phi}{\partial x^2} = \frac{\partial \Phi}{\partial u}\frac{\partial^2 u}{\partial x^2} + \frac{\partial u}{\partial x}\left[\frac{\partial^2 \Phi}{\partial u^2}\frac{\partial u}{\partial x} + \frac{\partial^2 \Phi}{\partial u \partial v}\frac{\partial v}{\partial x}\right]$$
$$+ \frac{\partial \Phi}{\partial v}\frac{\partial^2 v}{\partial x^2} + \frac{\partial v}{\partial x}\left[\frac{\partial^2 \Phi}{\partial v \partial u}\frac{\partial u}{\partial x} + \frac{\partial^2 \Phi}{\partial v^2}\frac{\partial v}{\partial x}\right]$$
$$\frac{\partial^2 \Phi}{\partial y^2} = \frac{\partial \Phi}{\partial u}\frac{\partial^2 u}{\partial y^2} + \frac{\partial u}{\partial y}\left[\frac{\partial^2 \Phi}{\partial u^2}\frac{\partial u}{\partial y} + \frac{\partial^2 \Phi}{\partial u \partial v}\frac{\partial v}{\partial y}\right]$$
$$+ \frac{\partial \Phi}{\partial v}\frac{\partial^2 v}{\partial y^2} + \frac{\partial v}{\partial y}\left[\frac{\partial^2 \Phi}{\partial v \partial u}\frac{\partial u}{\partial y} + \frac{\partial^2 \Phi}{\partial v^2}\frac{\partial v}{\partial y}\right] \tag{6.60}$$

We add the second derivatives and use equation (6.57) to simplify the resulting equation and obtain

$$\frac{\partial^2 \Phi}{\partial x^2} + \frac{\partial^2 \Phi}{\partial y^2} = \frac{\partial \Phi}{\partial u}\left(\frac{\partial^2 u}{\partial x^2} + \frac{\partial^2 u}{\partial y^2}\right) + \frac{\partial \Phi}{\partial v}\left(\frac{\partial^2 v}{\partial x^2} + \frac{\partial^2 v}{\partial y^2}\right)$$
$$+ 2\frac{\partial^2 \Phi}{\partial u \partial v}\left(\frac{\partial u}{\partial x}\frac{\partial v}{\partial x} + \frac{\partial u}{\partial y}\frac{\partial v}{\partial y}\right) + |f'(z)|^2\left(\frac{\partial^2 \Phi}{\partial u^2} + \frac{\partial^2 \Phi}{\partial v^2}\right). \tag{6.61}$$

The equations (6.55) and (6.56) further simplify the equation (6.61) and we find

$$\frac{\partial^2 \Phi}{\partial x^2} + \frac{\partial^2 \Phi}{\partial y^2} = |f'(z)|^2\left(\frac{\partial^2 \Phi}{\partial u^2} + \frac{\partial^2 \Phi}{\partial v^2}\right). \tag{6.62}$$

The equation (6.62) shows that if $|f'(z)| \neq 0$, then if Φ is harmonic in the x, y variables, then it is also harmonic in the u, v variables. One can say that a harmonic function remains harmonic under a 1-to-1 conformal mapping.

5. Inversion with respect to circle.

Let $C(0, \alpha)$ denote a circle centered at the origin with radius α. In polar coordinates we say two points (r, θ) and (r^*, θ^*) are inverses with respect to the circle $C(0, \alpha)$ if

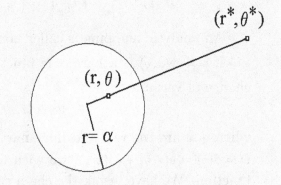

$$rr^* = \alpha^2, \quad \theta^* = \theta. \tag{6.63}$$

Thus, points that are inverses with respect to $C(0, \alpha)$ define a mapping

$$r^* = \frac{\alpha^2}{r}, \quad \theta^* = \theta \quad \text{or} \quad r = \frac{\alpha^2}{r^*}, \quad \theta = \theta^* \tag{6.64}$$

where (r, θ) gets mapped to the point (r^*, θ^*). If $\Phi(r, \theta)$ is a harmonic function in a region R, then a new harmonic function $\Phi(r^*, \theta^*)$ can be created using the above mapping. The new harmonic function $\Phi(r^*, \theta^*)$ is created by replacing r by α^2/r^* and θ by θ^*. The resulting function $\Phi(r^*, \theta^*)$ is harmonic in a region R^* which is the image of R under the above mapping.

6. **Inversion with respect to sphere.**

Let $S(0, \alpha)$ denote a sphere centered at the origin with radius α. In spherical coordinates two points (ρ, θ, ϕ) and $(\rho^*, \theta^*, \phi^*)$ are said to be inverses with respect to the sphere $S(0, \alpha)$ if

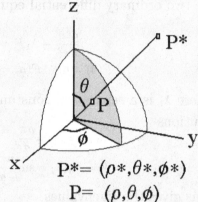

$$\rho\rho^* = \alpha^2, \quad \theta^* = \theta, \quad \phi^* = \phi$$

$$\text{or} \quad \rho = \alpha^2/\rho^*, \quad \theta = \theta^*, \quad \phi = \phi^* \tag{6.65}$$

$$P^* = (\rho*, \theta*, \phi*)$$
$$P = (\rho, \theta, \phi)$$

If $\Phi(\rho, \theta, \phi)$ is a harmonic function, then one can construct the new function

$$\Psi(\rho^*, \theta^*, \phi^*) = \frac{\alpha}{\rho^*} \Phi\left(\frac{\alpha^2}{\rho^*}, \theta^*, \phi^*\right) \tag{6.66}$$

which is also a harmonic function. The proof is left for the exercises.

Note that in two-dimensions if $\Phi(r, \theta)$ is harmonic in a region R, then $\Phi(r^*, \theta^*)$ is harmonic in a region R^* which is the image of R and calculated from the inversion mapping given by equation (6.64). Similarly, in three-dimensions, if $\Phi(\rho, \theta, \phi)$ is harmonic in a region R, then $\Psi(\rho^*, \theta^*, \phi^*)$ is harmonic in a region R^* which is the image of R and calculated from the inverse mapping given by equation (6.65).

Poisson equation for a rectangle

Consider the Poisson equation over a rectangle

$$\nabla^2 u = f(x, y), \qquad (x, y) \in R = \{x, y \mid 0 < x < a, \, 0 < y < b\} \tag{6.67}$$

subject to the boundary conditions

$$u(x, y)\Big|_{(x,y) \in \partial R} = g(x, y) \tag{6.68}$$

where $g(x, y)$ is a specified function. The solution to this problem can be expressed in terms of solutions to the two-dimensional eigenvalue problem

$$\nabla^2 \phi + \lambda \phi = 0, \qquad (x, y) \in R \tag{6.69}$$

subject to the homogeneous boundary condition

$$\phi\Big|_{(x,y)\in\partial R} = 0 \tag{6.70}$$

Assume a solution $\phi(x,y) = F(x)G(y)$ and show separation of variables produces the two ordinary differential equations

$$\frac{d^2F}{dx^2} - \lambda_1 F = 0, \qquad\qquad \frac{d^2G}{dy^2} + (\lambda + \lambda_1)G = 0 \tag{6.71}$$
$$F(0) = 0, \quad F(a) = 0 \qquad\qquad G(0) = 0, \quad G(b) = 0$$

where λ_1 is a separation constant. Let $\lambda_1 = -\omega^2$ and $\lambda + \lambda_1 = \mu^2$ and obtain the solutions

$$\omega = \omega_n = \frac{n\pi}{a}, \qquad\qquad \mu = \mu_m = \frac{m\pi}{b}$$
$$F = F_n(x) = \sin\frac{n\pi x}{a}, \qquad\qquad G = G_m(y) = \sin\frac{m\pi y}{b}$$

This gives the eigenvalues

$$\lambda = \lambda_{nm} = \left(\frac{n\pi}{a}\right)^2 + \left(\frac{m\pi}{b}\right)^2$$

and eigenfunctions

$$\phi = \phi_{nm}(x,y) = \sin\frac{n\pi x}{a}\sin\frac{m\pi y}{b}$$

where

$$\nabla^2\phi_{nm} = -\lambda_{nm}\phi_{nm} \tag{6.72}$$

and

$$\iint_R \phi_{nm}(x,y)\phi_{ij}(x,y)\,dxdy = \begin{cases} 0, & (i,j) \neq (n,m) \\ \|\phi_{nm}\|^2, & (i,j) = (n,m) \end{cases} \tag{6.73}$$

where

$$\|\phi_{nm}\|^2 = \frac{ab}{4}. \tag{6.74}$$

We now assume a solution to the Poisson equation (6.67) of the form

$$u = u(x,y) = \sum_{n=1}^{\infty}\sum_{m=1}^{\infty} A_{nm}\phi_{nm}(x,y). \tag{6.75}$$

We make use of the orthogonality of the functions ϕ_{nm} over the rectangle R and multiply the equation (6.75) by $\phi_{ij}(x,y)\,dxdy$ and integrate over the rectangular region R. This produces

$$\iint_R u\phi_{ij}\,dxdy = \sum_{n=1}^{\infty}\sum_{m=1}^{\infty} A_{nm}\iint_R \phi_{nm}\phi_{ij}\,dxdy \tag{6.76}$$

Note that all terms in the summations are zero except when the indices for n and m take on the values $n = i$ and $m = j$. The equation (6.76) therefore reduces to

$$\iint_R u\phi_{ij}\, dxdy = A_{ij} \parallel \phi_{ij} \parallel^2$$

or after changing i, j back to n, m one can write

$$A_{nm} = \frac{4}{ab} \iint_R u\phi_{nm}\, dxdy \qquad (6.77)$$

The equation (6.72) allows us to write $\phi_{nm} = \frac{-1}{\lambda_{nm}}\nabla^2\phi_{nm}$ so that equation (6.77) can be written as

$$A_{nm} = \frac{-4}{ab\lambda_{nm}} \iint_R u\nabla^2\phi_{nm}\, dxdy \qquad (6.78)$$

The two-dimensional Green's formula given by equation (6.45) can be used to evaluate the integral given by equation (6.78). Substitute $v = \phi_{nm}$ into equation (6.45) and write the Green's formula in the form

$$\iint_R \left(u\nabla^2\phi_{nm} - \phi_{nm}\nabla^2 u\right) dxdy = \oint (u\nabla\phi_{nm} - \phi_{nm}\nabla u) \cdot \hat{n}\, ds \qquad (6.79)$$

where $\hat{n}$ is the unit exterior normal to the boundary of the rectangle and $C = \partial R$ denotes the boundary of the rectangle. Now if u satisfies $\nabla^2 u = f(x, y)$ for $(x, y) \in R$ and $u(x, y)\big|_{(x,y)\in\partial R} = g(x, y)$ on the boundary, then equation (6.79) becomes

$$\iint_R u\nabla^2\phi_{nm}\, dxdy = \iint_R \phi_{nm}f(x, y)\, dxdy + \oint (g(x, y)\nabla\phi_{nm} - \phi_{nm}\nabla g) \cdot \hat{n}\, ds \qquad (6.80)$$

Note also that $\phi_{nm} = 0$ on $C = \partial R$ so the equation (6.80) further simplifies to

$$\iint_R u\nabla^2\phi_{nm}\, dxdy = \iint_R \phi_{nm}f(x, y)\, dxdy + \oint g(x, y)\nabla\phi_{nm} \cdot \hat{n}\, ds \qquad (6.81)$$

and so the coefficients A_{nm} can be evaluated in terms of the eigenfunction ϕ_{nm}, the eigenvalues λ_{nm}, the boundary values $g(x, y)$, and the right-hand side $f(x, y)$ of Poisson's equation.

One can apply a similar type of analysis to Poisson equation for (x, y) belonging to a rectangle with Neumann or Robins boundary conditions. Other regions which differ from the rectangle can also be considered. These other solutions are dependent upon the eigenvalues and eigenfunctions associated with the Helmholtz equation and boundary conditions for the new region.

Example 6-3. (Poisson's integral formula for circle.)

Solve for the steady state temperature $u = u(r, \theta)$ if

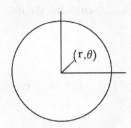

$$\nabla^2 u = \frac{\partial^2 u}{\partial r^2} + \frac{1}{r}\frac{\partial u}{\partial r} + \frac{1}{r^2}\frac{\partial^2 u}{\partial \theta^2} = 0$$
$$-\pi \le \theta \le \pi, \quad 0 \le r \le r_0$$

subject to the boundary condition $u(r_0, \theta) = h(\theta)$.

Solution: We use separation of variables and assume a solution of the form $u = u(r, \theta) = F(r)G(\theta)$. We substitute the assumed solution into the given partial differential equation and find F, G must be selected to satisfy

$$F''G + \frac{1}{r}F'G + \frac{1}{r^2}FG'' = 0.$$

Now multiply by $\frac{r^2}{FG}$ and separate the variables to obtain

$$r^2\frac{F''}{F} + r\frac{F'}{F} = -\frac{G''}{G} = \lambda$$

where λ is a separation constant. This gives two ordinary differential equations

$$r^2 F'' + r F' - \lambda F = 0, \qquad G'' - \lambda G = 0.$$

We recognize the F-equation as a Cauchy-Euler equation and the G-equation as a harmonic oscillator. The given boundary value problem has two implied boundary conditions: (i) We desire bounded solutions in the interior of the circle and (ii) The solutions must be periodic and continuous so that $u(r, -\pi) = u(r, \pi)$ and $\frac{\partial u(r, -\pi)}{\partial \theta} = \frac{\partial u(r, \pi)}{\partial \theta}$ which implies that G must satisfy the periodicity conditions $G(-\pi) = G(\pi)$ and $G'(-\pi) = G'(\pi)$. The equation for $G = G(\theta)$ is therefore treated as a Sturm-Liouville problem with periodic boundary conditions. We examine the cases $\lambda = -\omega^2, \lambda = 0$ and $\lambda = \omega^2$ and obtain the eigenvalues and eigenfunctions

$$\lambda = \lambda_0 = 0, \qquad G = G_0(\theta) = 1$$
$$\lambda = \lambda_n = n^2, \qquad G = G_n(\theta) = \alpha_0 \sin n\theta + \alpha_1 \cos n\theta$$

for $n = 1, 2, 3, \ldots$. The corresponding solutions for the Cauchy-Euler equation are

$$\text{For } \lambda_0 = 0, \qquad F = F_0(r) = c_0 + c_1 \ln r$$
$$\lambda_n = n^2, \qquad F = F_n(r) = \beta_0 r^n + \beta_1 r^{-n}$$

where $\alpha_0, \alpha_1, \beta_0, \beta_1$ are arbitrary constants. We therefore obtain solutions to the original equation of the form

$$u_0(r, \theta) = c_0 + c_1 \ln r, \qquad u_n(r, \theta) = (\beta_0 r^n + \beta_1 r^{-n})(\alpha_0 \sin n\theta + \alpha_1 \cos n\theta).$$

For bounded solutions over the region $0 \leq r \leq r_0$ we must require that $c_1 = 0$ and $\beta_1 = 0$. We use superposition and relabel the constants to obtain the general solution

$$u = u(r, \theta) = a_0 + \sum_{n=1}^{\infty} r^n \left(a_n \cos n\theta + b_n \sin n\theta\right), \qquad -\pi < \theta < \pi, \quad 0 < r < r_0.$$

The given boundary condition requires that

$$u(r_0, \theta) = h(\theta) = a_0 + \sum_{n=1}^{\infty} r_0^n \left(a_n \cos n\theta + b_n \sin n\theta\right).$$

This is just a Fourier series representation of the function $h(\theta)$. We make use of the orthogonality properties of the eigenfunctions to represent each Fourier coefficient in terms of an inner product divided by a norm squared. We find

$$a_0 = \frac{(1, h)}{\| 1 \|^2} = \frac{1}{2\pi} \int_{-\pi}^{\pi} h(\theta) \, d\theta$$

$$r_0^n a_n = \frac{(\cos n\theta, h)}{\| \cos n\theta \|^2} = \frac{1}{\pi} \int_{-\pi}^{\pi} h(\theta) \cos n\theta \, d\theta$$

$$r_0^n b_n = \frac{(\sin n\theta, h)}{\| \sin n\theta \|^2} = \frac{1}{\pi} \int_{-\pi}^{\pi} h(\theta) \sin n\theta \, d\theta.$$

This gives the solution

$$u = u(r, \theta) = \frac{1}{2\pi} \int_{-\pi}^{\pi} h(\xi) \, d\xi + \frac{1}{\pi} \sum_{n=1}^{\infty} \left(\frac{r}{r_0}\right)^n \int_{-\pi}^{\pi} h(\xi) \left[\cos n\xi \cos n\theta + \sin n\xi \sin n\theta\right] \, d\xi,$$

which can also be written in the form

$$u = u(r, \theta) = \frac{1}{2\pi} \int_{-\pi}^{\pi} h(\xi) \left\{1 + 2 \sum_{n=1}^{\infty} \left(\frac{r}{r_0}\right)^n \cos n(\theta - \xi)\right\} \, d\xi. \qquad (6.82)$$

Note that a series of the form

$$S = 1 + 2 \sum_{n=1}^{\infty} B^n \cos nx \qquad (6.83)$$

can be summed by writing it in terms of a geometric series

$$S = a + ar + ar^2 + ar^3 + \cdots = \frac{a}{1-r}, \qquad |r| < 1.$$

We use the Euler identity $e^{i\psi} = \cos\psi + i\sin\psi$ and write $\cos nx = \frac{e^{inx}+e^{-inx}}{2}$ so that the series (6.83) takes on the form

$$S = 1 + 2\sum_{n=1}^{\infty} B^n \cos nx = 1 + \sum_{n=1}^{\infty} B^n e^{inx} + \sum_{n=1}^{\infty} B^n e^{-inx}.$$

The last two summations are geometric series which can be summed to produce

$$S = 1 + 2\sum_{n=1}^{\infty} B^n \cos nx = 1 + \frac{Be^{ix}}{1 - Be^{ix}} + \frac{Be^{-ix}}{1 - Be^{-ix}}, \qquad |B| < 1$$

$$= \frac{1 - B^2}{1 - 2B\cos x + B^2}, \qquad |B| < 1.$$

Therefore, the equation (6.82) can be represented in the form

$$u = u(r,\theta) = \frac{1}{2\pi} \int_{-\pi}^{\pi} \frac{h(\xi)\left[1 - \left(\frac{r}{r_0}\right)^2\right] d\xi}{1 - 2\left(\frac{r}{r_0}\right)\cos(\theta - \xi) + \left(\frac{r}{r_0}\right)^2}, \qquad \left|\frac{r}{r_0}\right| < 1$$

or

$$u = u(r,\theta) = \frac{r_0^2 - r^2}{2\pi} \int_{-\pi}^{\pi} \frac{h(\xi)\,d\xi}{r_0^2 - 2rr_0\cos(\theta - \xi) + r^2}, \qquad r < r_0 \qquad (6.84)$$

which is called Poisson's integral formula for the circle. Note that this formula gives us the value of a harmonic function $u = u(r,\theta)$ inside the circle $0 \le r < r_0$ based upon the known values of $u(r_0,\theta) = h(\theta)$ on the boundary. Make note of the value at the center of the circle. The Poisson integral formula gives us

$$u(0,0) = a_0 = \frac{1}{2\pi} \int_{-\pi}^{\pi} h(\theta)\,d\theta \qquad (6.85)$$

which represents the average of the u-values on the circumference of the circle. This agrees with our previous physical interpretation associated with the Laplacian operator.

Example 6-4. (Poisson's integral formula for the upper half plane.)

Solve for the steady state temperature $u = u(x, y)$ if

$$\nabla^2 u = \frac{\partial^2 u}{\partial x^2} + \frac{\partial^2 u}{\partial y^2} = 0$$

$$-\infty < x < \infty, \quad y \geq 0$$

subject to the boundary condition $u(x, 0) = h(x)$.

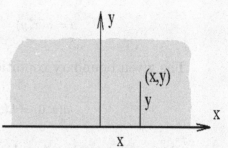

Solution: It is implied that $\lim_{|x| \to \infty} h(x) = 0$ and that bounded solutions are the only real solutions of interest. We use separation of variables and assume a product solution of the form $u = u(x, y) = F(x)G(y)$. We substitute the assumed solution into the partial differential equation and obtain

$$F''(x)G(y) + F(x)G''(y) = 0.$$

The variables are separated giving

$$\frac{F''(x)}{F(x)} = -\frac{G''(y)}{G(y)} = \lambda$$

where λ is a separation constant. This gives the two differential equations

$$F''(x) - \lambda F(x) = 0, \qquad G''(y) + \lambda G(y) = 0.$$

The given problem has the implied boundary condition of boundedness of solutions as $y \to \infty$. We therefore require that solutions satisfy $\lim_{y \to \infty} u(x, y) = 0$. We examine the cases of $\lambda = -\omega^2, \lambda = 0$, and $\lambda = \omega^2$ and find that only $\lambda = -\omega^2$, $\omega > 0$, gives bounded solutions for the y equation. We solve

$$F''(x) + \omega^2 F(x) = 0 \qquad \text{and} \qquad G''(y) - \omega^2 G(y) = 0$$

and obtain the solutions

$$F(x) = K_1 \cos \omega x + K_2 \sin \omega x, \qquad G(y) = C_1 e^{-\omega y} + C_2 e^{\omega y}.$$

We set $C_2 = 0$ for bounded solutions. We form the product solution $u = FG$ and relabel the constants to obtain solutions containing a parameter $\omega > 0$ of the form

$$u = u(x, y; \omega) = [A(\omega) \cos \omega x + B(\omega) \sin \omega x] e^{-\omega y} \qquad (6.86)$$

where $A(\omega)$ and $B(\omega)$ are arbitrary functions of the parameter ω. By superposition we form the more general solution

$$u = u(x, y) = \int_0^\infty [A(\omega) \cos \omega x + B(\omega) \sin \omega x] \, e^{-\omega y} \, d\omega. \tag{6.87}$$

The given boundary condition requires that

$$u(x, 0) = h(x) = \int_0^\infty [A(\omega) \cos \omega x + B(\omega) \sin \omega x] \, d\omega \tag{6.88}$$

which requires that $h(x)$ be represented in the form of a Fourier integral with coefficients

$$A(\omega) = \frac{1}{\pi} \int_{-\infty}^\infty h(\xi) \cos \omega \xi \, d\xi$$
$$B(\omega) = \frac{1}{\pi} \int_{-\infty}^\infty h(\xi) \sin \omega \xi \, d\xi. \tag{6.89}$$

This gives a solution which can be written in the form

$$u = u(x, y) = \frac{1}{\pi} \int_{-\infty}^\infty h(\xi) \left[\int_0^\infty (\cos \omega x \cos \omega \xi + \sin \omega x \sin \omega \xi) \, e^{-\omega y} \, d\omega \right] d\xi$$
$$u = u(x, y) = \frac{1}{\pi} \int_{-\infty}^\infty h(\xi) \left[\int_0^\infty \cos \omega(x - \xi) e^{-\omega y} \, d\omega \right] d\xi. \tag{6.90}$$

We find from the table of integrals in Appendix D the relation

$$\int_0^\infty e^{\omega \beta} \cos \omega(x - \xi) \, d\omega = \frac{\beta}{(x - \xi)^2 + \beta^2}, \tag{6.91}$$

which enables us to write equation (6.90) in the form

$$u = u(x, y) = \frac{y}{\pi} \int_{-\infty}^\infty \frac{h(\xi) \, d\xi}{(x - \xi)^2 + y^2} \tag{6.92}$$

which is known as Poisson's integral formula for the half-plane. The transformation $\xi = x + y \tan \theta$ enables one to express the Poisson integral formula for the plane in the alternate form

$$u(x, y) = \frac{1}{\pi} \int_{-\pi/2}^{\pi/2} h(x + y \tan \theta) \, d\theta. \tag{6.93}$$

■

Example 6-5. Laplace's equation for a rectangle.

Consider the problem to solve $\nabla^2 u = 0$ for $0 < x < a$, $0 < y < b$ subject to the boundary conditions

$$u(0,y) = f_\ell(y), \qquad u(x,0) = f_b(x)$$
$$u(a,y) = f_r(y), \qquad u(x,b) = f_t(x)$$

One can imagine the above problem as representing the equations satisfied by the steady state temperature $u = u(x,y)$ for (x,y) inside a rectangle $0 < x < a$, $0 < y < b$ which has nonhomogeneous boundary conditions on each side of the rectangle. The situation is illustrated in the figure 6-2.

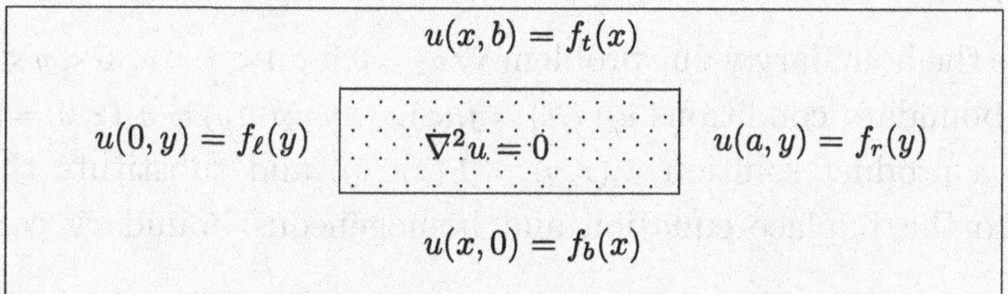

Figure 6-2. Laplace's equation for a rectangle.

Solution: In order to apply the method of separation of variables we must have homogeneous boundary conditions. Hence, we shall break the given problem up into four simpler problems which have the required homogeneous boundary conditions. We use superposition and assume a solution to the given problem of the form

$$u = u(x,y) = u_1(x,y) + u_2(x,y) + u_3(x,y) + u_4(x,y)$$

and substitute this solution into the Laplace equation and the given boundary conditions to obtain

$$\nabla^2 u = \nabla^2 u_1 + \nabla^2 u_2 + \nabla^2 u_3 + \nabla^2 u_4 = 0$$
$$u(x,0) = u_1(x,0) + u_2(x,0) + u_3(x,0) + u_4(x,0) = f_b(x)$$
$$u(x,b) = u_1(x,b) + u_2(x,b) + u_3(x,b) + u_4(x,b) = f_t(x)$$
$$u(0,y) = u_1(0,y) + u_2(0,y) + u_3(0,y) + u_4(0,y) = f_\ell(y)$$
$$u(a,y) = u_1(a,y) + u_2(a,y) + u_3(a,y) + u_4(a,y) = f_r(y).$$

Note that if we select $\nabla^2 u_1 = 0$ with u_1 satisfying $u_1(x,0) = f_b(x)$ and the other three sides satisfying $u_1(x,b) = u_1(0,y) = u_1(a,y) = 0$, then we obtain a problem for solving $\nabla^2 u_1 = 0$ for (x, y) inside a rectangle with homogeneous boundary conditions on three sides of the rectangle. We form three similar problems by letting $u_2(x, b) = f_t(x)$, $u_3(0, y) = f_\ell(y)$ and $u_4(a, y) = f_r(y)$, and require $\nabla^2 u_i = 0$ for $i = 2, 3, 4$. This gives three additional problems for solving $\nabla^2 u = 0$ inside a rectangle where each problem has homogeneous boundary conditions on three of the sides of the rectangle and a nonhomogeneous boundary condition on only one side of the rectangle. These four problems are illustrated in the figure 6-3. By superposition the sum of the solutions from each problem gives the solution to the original problem. Each of the problems formulated can now be solved by the method of separation of variables. We solve the problem 2 and leave the other problems for the exercises.

To solve the boundary value problem $\nabla^2 u_2 = 0$ for $0 < x < a$, $0 < y < b$ and subject to the boundary conditions $u_2(x, b) = f_t(x)$, $\qquad u_2(0, y) = u_2(x, 0) = u_2(a, y) = 0$, we assume a product solution $u_2(x, y) = F(x)G(y)$ and substitute the assumed solution into the Laplace equation and homogeneous boundary conditions to obtain

$$\frac{\partial^2 u_2}{\partial x^2} + \frac{\partial^2 u_2}{\partial y^2} = 0 \qquad\qquad F''(x)G(y) + F(x)G''(y) = 0$$
$$u_2(0, y) = 0, \quad u_2(a, y) = 0 \qquad F(0)G(y) = 0, \quad F(a)G(y) = 0$$
$$u_2(x, 0) = 0, \quad u_2(x, b) = f_t(x) \qquad\qquad F(x)G(0) = 0$$

and we leave the nonhomogeneous boundary condition until we reach the end of the problem. We separate the variables and obtain two ordinary differential equations

$$F''(x) - \lambda F(x) = 0, \quad 0 < x < a \qquad\qquad G''(y) + \lambda G(y) = 0, \quad 0 < y < b$$
$$F(0) = 0, \quad F(a) = 0 \qquad\qquad\qquad G(0) = 0$$

where λ is a separation constant.

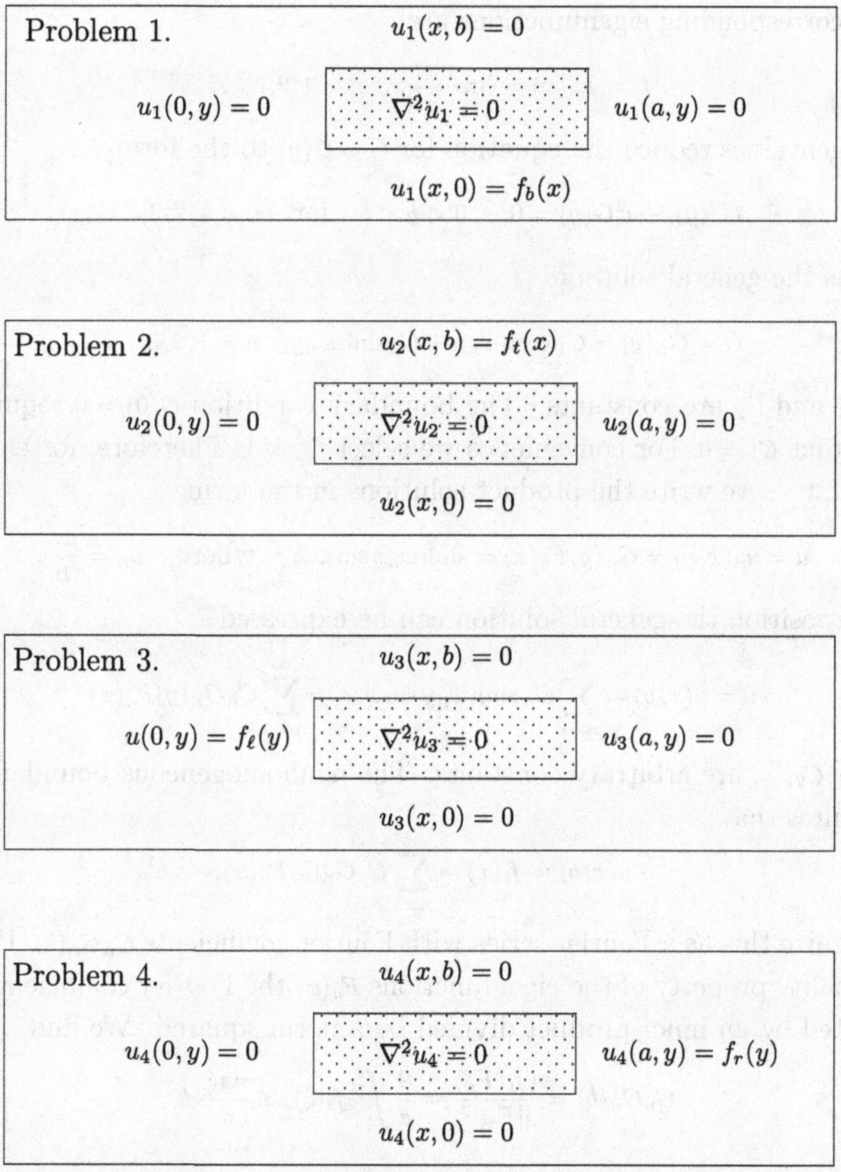

Figure 6-3. Laplace's equation for a rectangle and superposition.

The problem for $F = F(x)$ is a Sturm-Liouville problem which we have previously investigated. The eigenvalues are

$$\lambda = \lambda_n = -\omega_n^2 = -\left(\frac{n\pi}{L}\right)^2 \quad \text{for} \quad n = 1, 2, 3, \ldots$$

and the corresponding eigenfunctions are

$$F = F_n(x) = \sin\frac{n\pi x}{a}, \quad n = 1, 2, 3, \ldots.$$

These eigenvalues reduce the equation for $G = G(y)$ to the form

$$G''(y) - \omega_n^2 G(y) = 0, \quad 0 < y < b \quad \text{for} \quad n = 1, 2, 3, \ldots \tag{6.94}$$

which has the general solution

$$G = G_n(y) = C_1 \cosh\omega_n y + C_2 \sinh\omega_n y \quad n = 1, 2, 3, \ldots$$

where C_1 and C_2 are constants. The boundary condition $G(0) = 0$ requires that the constant $C_1 = 0$. For convenience we select $C_2 = 1$. Therefore, for each value of $n = 1, 2, 3, \ldots$ we write the product solutions in the form

$$u = u_n(x, y) = G_n(y)F_n(x) = \sinh\omega_n y \sin\omega_n x \quad \text{where} \quad \omega_n = \frac{n\pi}{a}.$$

By superposition the general solution can be expressed

$$u = u(x, y) = \sum_{n=1}^{\infty} C_n \sinh\omega_n y \sin\omega_n x = \sum_{n=1}^{\infty} C_n G_n(y) F_n(x) \tag{6.95}$$

where $C_1, C_2, \ldots$ are arbitrary constants. The nonhomogeneous boundary condition requires that

$$u(x, b) = f_t(x) = \sum_{n=1}^{\infty} C_n G_n(b) F_n(x).$$

We recognize this as a Fourier series with Fourier coefficients $C_n G_n(b)$. Using the orthogonality property of the eigenfunctions $F_n(x)$ the Fourier coefficients can be represented by an inner product divided by a norm squared. We find

$$C_n G_n(b) = \frac{(f_t, F_n)}{||F_n||^2} = \frac{2}{a} \int_0^a f_t(\xi) \sin\frac{n\pi\xi}{a} d\xi$$

or

$$C_n = \frac{2}{a \sinh(\frac{n\pi b}{a})} \int_0^a f_t(\xi) \sin\frac{n\pi\xi}{a} d\xi. \tag{6.96}$$

The solution of the problem 2 is given by the equations (6.95) and (6.96).

∎

Example 6-6. Laplace's equation for a cylinder.

Consider the problem to solve for the steady state temperature $u = u(r, \theta, z)$ inside a cylinder $0 < r < a$, $-\pi < \theta < \pi$, $0 < z < h$ which is subject to nonhomogeneous boundary conditions. The problem can be formulated to solve

$$\nabla^2 u = \frac{1}{r}\frac{\partial}{\partial r}\left(r\frac{\partial u}{\partial r}\right) + \frac{1}{r^2}\frac{\partial^2 u}{\partial \theta^2} + \frac{\partial^2 u}{\partial z^2} = 0$$

$$0 < r < a, \quad -\pi < \theta < \pi, \quad 0 < z < h$$

$$u(r, \theta, h) = f_t(r, \theta)$$

$$u(a, \theta, z) = f_s(\theta, z)$$

$$u(r, \theta, 0) = f_b(r, \theta)$$

$u(r, \theta, h) = f_t(r, \theta)$

$\nabla^2 u = 0$ $u(a, \theta, z) = f_s(\theta, z)$

$u(r, \theta, 0) = f_b(r, \theta)$

This is a Dirichlet problem to solve $\nabla^2 u = 0$ interior to the cylinder where f_t, f_s, f_b are specified boundary conditions on the top, side and bottom of the cylinder.

Solution: The nonhomogeneous boundary conditions can be approached in a manner similar to what has been done in the previous example. We use superposition and assume a solution

$$u = u(r, \theta, z) = u_1(r, \theta, z) + u_2(r, \theta, z) + u_3(r, \theta, z).$$

We substitute the assumed solution into the Laplace equation and boundary conditions to obtain

$$\nabla^2 u = \nabla^2 u_1 + \nabla^2 u_2 + \nabla^2 u_3 = 0$$

$$u(r, \theta, h) = u_1(r, \theta, h) + u_2(r, \theta, h) + u_3(r, \theta, h) = f_t(r, \theta)$$

$$u(a, \theta, z) = u_1(a, \theta, z) + u_2(a, \theta, z) + u_3(a, \theta, z) = f_s(\theta, z) \tag{6.97}$$

$$u(r, \theta, 0) = u_1(r, \theta, 0) + u_2(r, \theta, 0) + u_3(r, \theta, 0) = f_b(r, \theta).$$

We now can formulate three problems. For the first problem we require $\nabla^2 u_1 = 0$ subject to the boundary conditions

$$u_1(r, \theta, h) = f_t(r, \theta), \quad u_1(a, \theta, z) = 0, \quad u_1(r, \theta, 0) = 0.$$

For the second problem we require $\nabla^2 u_2 = 0$ subject to the boundary conditions

$$u_2(r, \theta, h) = 0, \quad u_2(a, \theta, z) = f_s(\theta, z), \quad u_2(r, \theta, 0) = 0.$$

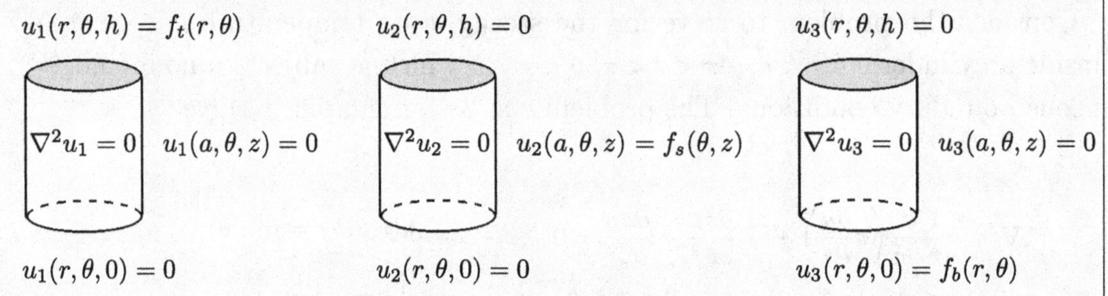

$u_1(r, \theta, h) = f_t(r, \theta)$ $u_2(r, \theta, h) = 0$ $u_3(r, \theta, h) = 0$

$\nabla^2 u_1 = 0$ $u_1(a, \theta, z) = 0$ $\nabla^2 u_2 = 0$ $u_2(a, \theta, z) = f_s(\theta, z)$ $\nabla^2 u_3 = 0$ $u_3(a, \theta, z) = 0$

$u_1(r, \theta, 0) = 0$ $u_2(r, \theta, 0) = 0$ $u_3(r, \theta, 0) = f_b(r, \theta)$

Figure 6-4. Superposition for Laplace's equation in a cylinder.

The first two problems simplify the equations (6.97) to give us a third problem to solve $\nabla^2 u_3 = 0$ subject to the boundary conditions

$$u_3(r, \theta, h) = 0, \quad u_3(a, \theta, z) = 0, \quad u_3(r, \theta, 0) = f_b(r, \theta).$$

These three problems are illustrated in the figure 6-4.

We shall solve the second problem where the temperature is specified on the side of the cylinder and leave the problems 1 and 3 for the exercises. To solve

$$\nabla^2 u_2 = \frac{1}{r}\frac{\partial}{\partial r}\left(r\frac{\partial u_2}{\partial r}\right) + \frac{1}{r^2}\frac{\partial^2 u_2}{\partial \theta^2} + \frac{\partial^2 u_2}{\partial z^2} = 0$$

$$0 < r < a, \quad -\pi < \theta < \pi, \quad 0 < z < h$$

$$u_2(r, \theta, h) = 0, \quad u_2(a, \theta, z) = f_s(\theta, z), \quad u_2(r, \theta, 0) = 0$$

we use the method of separation of variables and assume a product solution of the form

$$u_2(r, \theta, z) = u_2 = F(r)G(\theta)H(z).$$

We substitute the assumed solution into the Laplace equation and homogeneous boundary conditions to obtain

$$\frac{1}{r}\frac{\partial}{\partial r}\left(rF'GH\right) + \frac{1}{r^2}FG''H + FGH'' = 0$$

$$u_2(r, \theta, h) = F(r)G(\theta)H(h) = 0 \qquad (6.98)$$

$$u_2(r, \theta, 0) = F(r)G(\theta)H(0) = 0.$$

There is also implied, from the geometry of the problem, that for continuity of solutions the periodic boundary conditions $u_2(r, -\pi, z) = u_2(r, \pi, z)$ for all values of θ and $\dfrac{\partial u_2(r, -\pi, z)}{\partial \theta} = \dfrac{\partial u_2(r, \pi, z)}{\partial \theta}$ must be satisfied. Also implied is the desire for bounded solutions for all values of u in the interior of the cylinder. We separate the variables in equation (6.98) one at a time. We first separate the z-variable by dividing each term by FGH and write

$$\frac{1}{rF}\frac{d}{dr}(rF') + \frac{1}{r^2}\frac{G''}{G} = -\frac{H''}{H} = \lambda_1 \qquad (6.99)$$

where λ_1 is a separation constant. This produces the two equations

$$\frac{1}{rF}\frac{d}{dr}(rF') + \frac{1}{r^2}\frac{G''}{G} = \lambda_1, \qquad H'' + \lambda_1 H = 0. \qquad (6.100)$$

We next separate the r and θ variables so that the equations (6.100) reduce to

$$\frac{r}{F}\frac{d}{dr}(rF') - \lambda_1 r^2 = -\frac{G''}{G} = \lambda_2, \qquad H'' + \lambda_1 H = 0 \qquad (6.101)$$

where λ_2 is a second separation constant. Applying the boundary conditions we obtain the three ordinary differential equations

$$
\begin{array}{ccc}
r\dfrac{d}{dr}(rF') - (\lambda_2 + \lambda_1 r^2)F = 0 & \dfrac{d^2 G}{d\theta^2} + \lambda_2 G = 0 & \dfrac{d^2 H}{dz^2} + \lambda_1 H = 0 \\
0 < r < a & -\pi < \theta < \pi & 0 < z < h \\
F(r) \text{ is bounded} & G(-\pi) = G(\pi) & H(0) = 0, \quad H(h) = 0 \\
 & G'(-\pi) = G'(\pi) &
\end{array}
\qquad (6.102)
$$

where for convenience we can assign other forms for the separation constants λ_1 and λ_2. We observe that the equations for G and H are Sturm-Liouville problems. We examine the cases of $\lambda_2 = -\omega^2, \lambda_2 = 0, \lambda_2 = +\omega^2$ and find the eigenvalues

$$\lambda_2 = 0, \ (n = 0) \quad \text{and} \quad \lambda_2 = \omega_n^2 = n^2 \quad \text{for} \quad n = 1, 2, 3, \ldots$$

with corresponding eigenfunctions $\{1, \cos n\theta, \sin n\theta\}$ for $n = 1, 2, 3, \ldots$. The equation for H is also a Sturm-Liouville problem. We find the eigenvalues

$$\lambda_1 = \alpha_m^2 = \left(\frac{m\pi}{h}\right)^2 \quad \text{for} \quad m = 1, 2, 3, \ldots$$

and eigenfunctions

$$H = H_m(z) = \sin\frac{m\pi z}{h} \quad m = 1, 2, 3, \ldots.$$

The equation for F can now be written in the form

$$r^2\frac{d^2 F}{dr^2} + r\frac{dF}{dr} - (\alpha_m^2 r^2 + n^2)F = 0 \qquad (6.103)$$

which must be solved for all combinations of $n = 0, 1, 2, 3, \ldots$ and $m = 1, 2, 3, \ldots$. We recognize the equation (6.103) as a modified Bessel equation (See equation (1.45)), with general solution

$$F = F_{nm}(r) = c_1 I_n(\alpha_m r) + c_2 K_n(\alpha_m r). \tag{6.104}$$

For bounded solutions over the interval $0 \leq r \leq a$ we set $c_2 = 0$, because the modified Bessel function $K_n(\alpha_m r)$ is not bounded as $r \to 0$.

The method of separation of variables has given us solutions of the form

$$u_{nm}(r, \theta, z) = (A_{nm} \cos n\theta + B_{nm} \sin n\theta) F_{nm}(r) H_m(z)$$

for $n = 0, 1, 2, \ldots$ and $m = 1, 2, 3, \ldots$. Using superposition there results the more general solution

$$u_2(r, \theta, z) = \sum_{n=0}^{\infty} \sum_{m=1}^{\infty} A_{nm} \cos n\theta F_{nm}(r) H_m(z) + \sum_{n=1}^{\infty} \sum_{n=1}^{\infty} B_{nm} \sin n\theta F_{nm}(r) H_m(z) \tag{6.105}$$

where A_{nm} and B_{nm} are constants which can be determined from the remaining boundary condition. These constants must be selected such that the nonhomogeneous boundary condition $u_2(a, \theta, z) = f_s(\theta, z)$ is satisfied. This requires that

$$f_s(\theta, z) = \sum_{n=0}^{\infty} \sum_{m=1}^{\infty} A_{nm} \cos n\theta F_{nm}(a) H_m(z) + \sum_{n=1}^{\infty} \sum_{m=1}^{\infty} B_{nm} \sin n\theta F_{nm}(a) H_m(z) \tag{6.106}$$

The series given by equation (6.106) is a double Fourier series over the two dimensional region $R = \{-\pi \leq \theta \leq \pi, 0 \leq z \leq h\}$. Using the orthogonality properties of the functions $\{1, \cos n\theta, \sin n\theta\}$ and $\{H_m(z)\}$ we can solve for the Fourier coefficients A_{nm} and B_{nm}. For example, the inner product of 1 with the $\sin n\theta$ and $\cos n\theta$ functions gives zero and so if we multiply the Fourier series representation given by equation (6.106) by $(1)d\theta$ and then integrate over the interval $(-\pi, \pi)$ we obtain the inner products

$$(f_s, 1) = \sum_{m=1}^{\infty} A_{0m} F_{0m}(a) H_m(z) \|1\|^2. \tag{6.107}$$

Now multiply the equation (6.107) by $H_i(z) \, dz$ and integrate over the interval $(0, h)$ to obtain the inner products

$$((f_s, 1), H_i) = \sum_{m=1}^{\infty} A_{0m} F_{0m}(a) \|1\|^2 (H_m, H_i). \tag{6.108}$$

The only nonzero term on the right-hand side of equation (6.108) occurs when the index m equals i and so we obtain the result

$$((f_s, 1), H_i) = A_{0i} F_{0i}(a) ||1||^2 ||H_i||^2.$$

Changing i to m one can represent the coefficients A_{0m} in terms of inner products divided by norm squared terms or

$$A_{0m} = \frac{((f_s, 1), H_m)}{F_{0m}(a) ||1||^2 ||H_m||^2} = \frac{1}{\pi h F_{0m}(a)} \int_{-\pi}^{\pi} \int_0^h f_s(\theta, z) H_m(z) \, d\theta dz. \qquad (6.109)$$

In a similar manner we make use of the orthogonality property of the sets $\{\sin n\theta\}$, $\{H_m(z)\}$ and multiply the series (6.106) by $\sin j\theta H_i(z) \, d\theta dz$ and then integrate over the region R. This produces the results that

$$((f_s, \sin j\theta), H_i(z)) = B_{ji} ||\sin j\theta||^2 F_{ji}(a) ||H_i(z)||^2.$$

By changing j to n and i to m we obtain the results

$$B_{nm} = \frac{((f_s, \sin n\theta), H_m(z))}{F_{nm}(a) ||\sin n\theta||^2 ||H_m(z)||^2} = \frac{2}{\pi h F_{nm}(a)} \int_{-\pi}^{\pi} \int_0^h f_s(\theta, z) \sin n\theta H_m(z) \, d\theta dz \quad (6.110)$$

It is left as an exercise to verify the additional result

$$A_{nm} = \frac{((f_s, \cos n\theta), H_m(z))}{F_{nm}(a) ||\cos n\theta||^2 ||H_m(z)||^2} = \frac{2}{\pi h F_{nm}(a)} \int_{-\pi}^{\pi} \int_0^h f_s(\theta, z) \cos n\theta H_m(z) \, d\theta dz. \quad (6.111)$$

∎

Example 6-7. Laplace's equation for a sphere.

Consider the Dirichlet problem to solve for $u = u(\rho, \theta, \phi)$ which satisfies

$$\nabla^2 u = \frac{1}{\rho^2} \frac{\partial}{\partial \rho} \left(\rho^2 \frac{\partial u}{\partial \rho} \right) + \frac{1}{\rho^2 \sin \theta} \frac{\partial}{\partial \theta} \left(\sin \theta \frac{\partial u}{\partial \theta} \right) + \frac{1}{\rho^2 \sin^2 \theta} \frac{\partial^2 u}{\partial \phi^2} = 0$$

over the region $0 \le \rho \le a$, $0 \le \theta \le \pi$, $0 \le \phi \le 2\pi$ and subject to the boundary condition $u(a, \theta, \phi) = f(\theta, \phi)$. This problem corresponds to finding the electrostatic potential inside a sphere when the potential on the surface of the sphere is specified. Another physical problem represented by the above equation is to let u represent the steady state temperature inside a sphere where the temperature at the surface of the sphere is specified.

Solution: We use the method of separation of variables and assume a solution of the form $u = u(\rho, \theta, \phi) = F(\rho)Y(\theta, \phi)$ where the variables ρ, θ and ϕ are independent. We substitute the assumed solution into the Laplace equation and obtain

$$\frac{\partial}{\partial \rho}\left(\rho^2 \frac{dF}{d\rho}Y\right) + \frac{1}{\sin\theta}\frac{\partial}{\partial \theta}\left(\sin\theta F \frac{\partial Y}{\partial \theta}\right) + \frac{1}{\sin^2\theta}F\frac{\partial^2 Y}{\partial \phi^2} = 0. \qquad (6.112)$$

We separate the variables to obtain

$$\frac{1}{F}\frac{d}{d\rho}\left(\rho^2 \frac{dF}{d\rho}\right) = -\frac{1}{Y}\left[\frac{1}{\sin\theta}\frac{\partial}{\partial \theta}\left(\sin\theta \frac{\partial Y}{\partial \theta}\right) + \frac{1}{\sin^2\theta}\frac{\partial^2 Y}{\partial \phi^2}\right] = \lambda_1 \qquad (6.113)$$

where λ_1 is a separation constant. This produces the equations

$$\frac{1}{F}\frac{d}{d\rho}\left(\rho^2 \frac{dF}{d\rho}\right) = \lambda_1 \qquad \frac{1}{Y}\left[\frac{1}{\sin\theta}\frac{\partial}{\partial \theta}\left(\sin\theta \frac{\partial Y}{\partial \theta}\right) + \frac{1}{\sin^2\theta}\frac{\partial^2 Y}{\partial \phi^2}\right] = -\lambda_1 \qquad (6.114)$$

The equation for $F = F(\rho)$ is the Euler-Cauchy equation

$$\rho^2 \frac{d^2 F}{d\rho^2} + 2\rho \frac{dF}{d\rho} - \lambda_1 F = 0 \qquad (6.115)$$

and the equation for Y is the Helmholtz eigenvalue problem

$$\frac{1}{\sin\theta}\frac{\partial}{\partial \theta}\left(\sin\theta \frac{\partial Y}{\partial \theta}\right) + \frac{1}{\sin^2\theta}\frac{\partial^2 Y}{\partial \phi^2} + \lambda_1 Y = 0. \qquad (6.116)$$

We assume a solution $Y = Y(\theta, \phi) = P(\theta)H(\phi)$ and substitute into the equation (6.116) and then separate the variables to obtain

$$\frac{\sin\theta}{P}\frac{d}{d\theta}\left(\sin\theta \frac{dP}{d\theta}\right) + \lambda_1 \sin^2\theta = \frac{-1}{H}\frac{d^2 H}{d\phi^2} = \lambda_2 \qquad (6.117)$$

where λ_2 is a separation constant. This gives the two equations

$$\frac{d^2 H}{d\phi^2} + \lambda_2 H = 0, \qquad \sin\theta\frac{d}{d\theta}\left(\sin\theta \frac{dP}{d\theta}\right) + (\lambda_1 \sin^2\theta - \lambda_2)P = 0. \qquad (6.118)$$

The equation for H has the implied boundary conditions that H be periodic in the variable ϕ and so we treat the equation for H as a Sturm-Liouville problem with periodic boundary conditions to obtain the eigenvalues

$$\lambda_2 = 0, \ (m = 0) \quad \text{and} \quad \lambda_2 = \omega_m^2 = m^2 \quad \text{for} \quad m = 1, 2, 3, \ldots$$

This gives the set of eigenfunctions $\{1, \cos m\phi, \sin m\phi\}$ which are orthogonal over the interval $(0, 2\pi)$. The equation (6.118) for P then becomes the singular Sturm-Liouville problem

$$\frac{1}{\sin\theta}\frac{d}{d\theta}\left(\sin\theta \frac{dP}{d\theta}\right) + \left(\lambda_1 - \frac{m^2}{\sin^2\theta}\right)P = 0 \qquad 0 < \theta < \pi. \qquad (6.119)$$

We use the results from equation (2.74) and find the eigenvalues $\lambda_1 = n(n+1)$ and corresponding eigenfunctions $P_n^m(\cos\theta)$ which are orthogonal over the interval $(0,\pi)$ with respect to the weight function $\sin\theta$. Note that we have the additional requirement that $m \le n$ since $P_n^m(\cos\theta) = 0$ for $m > n$. The Cauchy-Euler equation (6.115) for F then becomes

$$\rho^2 \frac{d^2 F}{d\rho^2} + 2\rho \frac{dF}{d\rho} - n(n+1)F = 0 \qquad 0 \le \rho \le a \qquad (6.120)$$

with general solution

$$F = F(\rho) = c_1 \rho^n + c_2 \rho^{-(n+1)}.$$

For bounded solutions at $\rho = 0$ we require that $c_2 = 0$. This leaves the solutions $F = F_n(\rho) = \rho^n$ for $n = 0,1,2,\ldots$ where we have selected $c_1 = 1$.

The method of separation of variables has produced product solutions of the form

$$\rho^n P_n^m(\cos\theta), \quad \rho^n P_n^m(\cos\theta)\cos m\phi, \quad \rho^n P_n^m(\cos\theta)\sin m\phi.$$

We use the principle of superposition to write the general solution. Observe that $P_n^m(\cos\theta) = 0$ for $m > n$ so that a general solution can be written in either of the forms

$$u = u(\rho,\theta,\phi) = \sum_{m=0}^{\infty} \sum_{n=m}^{\infty} \rho^n \left(A_{nm}\cos m\phi + B_{nm}\sin m\phi\right) P_n^m(\cos\theta)$$

$$u = u(\rho,\theta,\phi) = \sum_{n=0}^{\infty} \sum_{m=0}^{n} \rho^n \left(A_{nm}\cos m\phi + B_{nm}\sin m\phi\right) P_n^m(\cos\theta)$$

(6.121)

where A_{nm}, B_{nm} are arbitrary constants. Note that the summation order can be reversed. An easy way to understand this is to examine the figure 6-5.

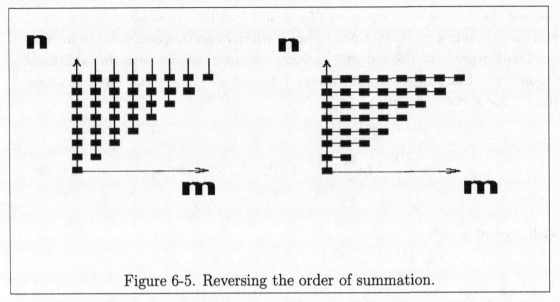

Figure 6-5. Reversing the order of summation.

The coefficients A_{nm} and B_{nm} are selected to satisfy the nonhomogeneous boundary condition $u(a, \theta, \phi) = f(\theta, \phi)$. This requires that

$$f(\theta, \phi) = \sum_{m=0}^{\infty} \sum_{n=m}^{\infty} a^n \left(A_{nm} \cos m\phi + B_{nm} \sin m\phi \right) P_n^m(\cos\theta) \qquad (6.122)$$

The orthogonality conditions enable us to calculate the above coefficients using inner products and norm squared terms. We find

$$\begin{aligned}
A_{nm} &= \frac{1}{a^n} \frac{((f, \cos m\phi), P_n^m(\cos\theta))}{\|\cos m\phi\|^2 \, \|P_n^m(\cos\theta)\|^2} \\
B_{nm} &= \frac{1}{a^n} \frac{((f, \sin m\phi), P_n^m(\cos\theta))}{\|\sin m\phi\|^2 \, \|P_n^m(\cos\theta)\|^2}.
\end{aligned} \qquad (6.123)$$

Example 6-8. Spherical harmonics.

The solution to the previous example can also be written in the form

$$u = u(\rho, \theta, \phi) = \sum_{n=0}^{\infty} \sum_{m=0}^{n} \rho^n \left[A_{nm} Y_{nm}^{(e)}(\theta, \phi) + B_{nm} Y_{nm}^{(o)}(\theta, \phi) \right]$$

where $Y_{nm}^{(e)}, Y_{nm}^{(o)}$ are the surface harmonics defined by equations (2.76). These surface harmonics are orthogonal over the surface of a unit sphere and satisfy the eigenvalue problem

$$\frac{1}{\sin\theta} \frac{\partial}{\partial\theta} \left(\sin\theta \frac{\partial Y}{\partial\theta} \right) + \frac{1}{\sin^2\theta} \frac{\partial^2 Y}{\partial\phi^2} + \lambda Y = 0$$

over the unit sphere $0 \leq \theta \leq \pi$, $0 \leq \phi \leq 2\pi$. The eigenvalues are $\lambda = \lambda_n = n(n+1)$ and the corresponding eigenfunctions can be any linear combination of the functions $P_n(\cos\theta), Y_{nm}^{(e)}, Y_{nm}^{(o)}$ and are called surface harmonics of degree n. These surface harmonics are expressed

$$Y_n(\theta, \phi) = A_{n0}P_n(\cos\theta) + \sum_{m=1}^{n} \left[A_{nm}Y_{nm}^{(e)}(\theta, \phi) + B_{nm}Y_{nm}^{(o)}(\theta, \phi) \right]$$

for constants A_{n0}, A_{nm}, B_{nm}.

The functions $P_n^0(\cos\theta) = P_n(\cos\theta)$ are called zonal harmonics because the solutions of the equation $P_n^0(\cos\theta) = 0$ are special values of θ which represent circles of constant latitude on the surface of a unit sphere. These circles divide the sphere into zones. The functions $P_m^m(\cos\theta)\cos m\phi$ are called sectorial harmonics because the values of ϕ which are solutions of the equation $P_m^m(\cos\theta)\cos m\phi = 0$ represent meridians which divide the surface of the unit sphere into sectors. The functions $P_n^m(\cos\theta)\sin m\phi$ and $P_n^m(\cos\theta)\cos m\phi$, with $m < n$, are called tesseral harmonics because the zeros of these functions divide the unit sphere up into a checkerboard or tesserae pattern. The figure 6-6 illustrates how these functions look on the surface of a unit sphere when positive and negative values are assigned different colors. The curves where the above functions are zero are called nodal lines and are represented on the surface of the unit sphere by the curves where the colors change.

Spherical harmonics are important functions used in many areas of physics, chemistry and engineering. They are used for the study of gravitational potentials in mechanics. They occur in the study of angular momentum in quantum mechanics. They are use to represent the structure of atoms and molecules in chemistry and physics. For example, wavefunctions associated with electron configurations in atoms and molecules are called orbitals and are related to probabilities of finding an electron in some particular region. Hydrogen orbitals have the form

$$\Psi_{n\ell m_\ell}(r, \theta, \phi) = R_{n\ell}(r)Y_{\ell m_\ell}(\theta, \phi)$$

where Y represents the spherical harmonics and the R function represents a radial dependence. Functions of this type are used to construct density distribution surfaces that represent 90 per cent capture probabilities for electrons.

■

346

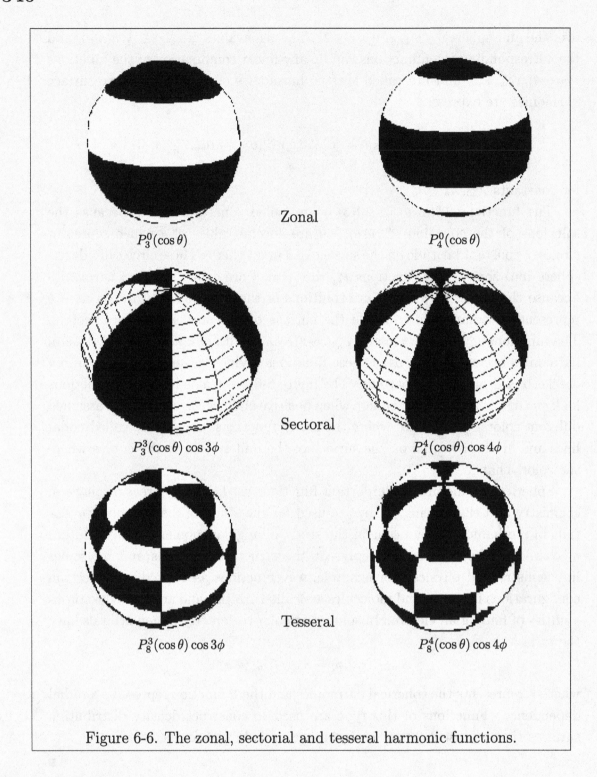

Zonal

$P_3^0(\cos\theta)$

$P_4^0(\cos\theta)$

Sectoral

$P_3^3(\cos\theta)\cos 3\phi$

$P_4^4(\cos\theta)\cos 4\phi$

Tesseral

$P_8^3(\cos\theta)\cos 3\phi$

$P_8^4(\cos\theta)\cos 4\phi$

Figure 6-6. The zonal, sectorial and tesseral harmonic functions.

Exercises 6

▶ **1.** Solve the given boundary value problems.

(a) Assume $u = u(x)$ for $a \le x \le b$ and solve

$$\nabla^2 u = 0, \quad a < x < b, \quad u(a) = T_a, \quad u(b) = T_b$$

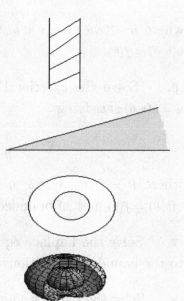

(b) Assume $u = u(\theta)$ for $0 \le \theta \le \theta_1$ and solve

$$\nabla^2 u = 0, \quad 0 < \theta < \theta_1, \quad u(0) = T_0, \quad u(\theta_1) = T_1$$

(c) Assume $u = u(r)$ for $r_0 \le r \le r_1$ and solve

$$\nabla^2 u = 0, \quad r_0 < r < r_1, \quad u(r_0) = T_0, \quad u(r_1) = T_1$$

(d) Assume $u = u(\rho)$ for $\rho_0 \le \rho \le \rho_1$ and solve

$$\nabla^2 u = 0, \quad \rho_0 < \rho < \rho_1, \quad u(\rho_0) = T_0, \quad u(\rho_1) = T_1$$

▶ **2.** Use separation of variables to solve the Dirichlet problem

$$\nabla^2 u = \frac{\partial^2 u}{\partial x^2} + \frac{\partial^2 u}{\partial y^2} = 0, \quad x, y \in R$$

where $R = \{x, y \mid 0 < x < a, 0 < y < b\}$ and subject to the boundary conditions $u(x,0) = 0$, $u(x,b) = 0$, $u(0,y) = 0$ and $u(a,y) = f(y)$.

▶ **3.** Use separation of variables to solve the Neumann problem

$$\nabla^2 u = \frac{\partial^2 u}{\partial x^2} + \frac{\partial^2 u}{\partial y^2} = 0, \quad x, y \in R$$

where $R = \{x, y \mid 0 < x < a, 0 < y < b\}$ and subject to the boundary conditions $u(x,0) = 0$, $u(x,b) = 0$, $\frac{\partial u(0,y)}{\partial x} = 0$ and $\frac{\partial u(a,y)}{\partial x} = f(y)$.

▶ **4.** Use separation of variables to solve the mixed boundary value problem

$$\nabla^2 u = \frac{\partial^2 u}{\partial x^2} + \frac{\partial^2 u}{\partial y^2} = 0, \quad x, y \in R$$

where $R = \{x, y \mid 0 < x < a, 0 < y < b\}$ and subject to the boundary conditions $u(x,0) = 0$, $u(x,b) = 0$, $u(0,y) = 0$ and $\frac{\partial u(a,y)}{\partial x} + hu(a,y) = f(y)$.

▶ **5.** Determine the temperature $u = u(r, \theta)$ associated with the interior Dirichlet problem satisfying

$$\nabla^2 u = \frac{\partial^2 u}{\partial r^2} + \frac{1}{r}\frac{\partial u}{\partial r} + \frac{1}{r^2}\frac{\partial^2 u}{\partial \theta^2} = 0, \qquad r, \theta \in R$$

where $R = \{r, \theta \mid 0 < r < b,\ 0 < \theta < 2\pi\}$ and subject to the boundary conditions $u(b, \theta) = f(\theta)$.

▶ **6.** Solve the exterior Dirichlet problem for the steady state temperature $u = u(r, \theta)$ satisfying

$$\nabla^2 u = \frac{\partial^2 u}{\partial r^2} + \frac{1}{r}\frac{\partial u}{\partial r} + \frac{1}{r^2}\frac{\partial^2 u}{\partial \theta^2} = 0, \qquad r, \theta \in R$$

where $R = \{r, \theta \mid r > b,\ 0 < \theta < 2\pi\}$ and subject to the boundary conditions $u(b, \theta) = f(\theta)$ and $|u|$ bounded as $r \to \infty$.

▶ **7.** Solve the Laplace equation over the annular region $a < r < b$ and subject to the boundary conditions $u(a, \theta) = f(\theta)$ and $u(b, \theta) = g(\theta)$.

▶ **8.** Solve the Laplace equation for the steady state temperature over the wedge shaped region $R = \{r, \theta \mid 0 < r < r_0,\ 0 < \theta < \theta_0\}$ subject to the boundary conditions $u(r, 0) = 0$, $u(r, \theta_0) = 0$ and $u(r_0, \theta) = f(\theta)$.

▶ **9.** Solve the Laplace equation for the steady state temperature inside the box $R = \{x, y, z \mid 0 < x < a,\ 0 < y < b,\ 0 < z < c\}$ which is subject to the boundary condition

$$u(0, y, z) = 0, \qquad u(x, 0, z) = 0 \qquad u(x, y, 0) = 0$$
$$u(a, y, z) = f(y, z), \qquad u(x, b, z) = 0 \qquad u(x, y, c) = 0$$

▶ **10.** Consider the boundary value problem to solve $\nabla^2 u = 0$ for $x, y \in R$ where $R = \{x, y \mid 0 < x < a,\ 0 < y < b\}$. Assume that u is subject to the boundary conditions

$$u(x, 0) = f_1(x), \qquad \frac{\partial u(0, y)}{\partial x} = f_2(y)$$
$$u(x, b) = 0, \qquad \frac{\partial u(a, y)}{\partial x} = f_3(y)$$

Set up appropriate equations so that one can use superposition of solutions to solve the above boundary value problem. Do not solve the equations you construct.

▶ **11.** Use separation of variables to solve the steady state heat equation

$$\nabla^2 u = \frac{\partial^2 u}{\partial x^2} + \frac{\partial^2 u}{\partial y^2} = 0, \quad 0 < x < L, \quad 0 < y < H$$

subject to the boundary conditions $u(0, y) = 0$, $u(x, 0) = 0$, $u(x, H) = 0$ and $u(L, y) = g(y)$.

▶ **12.** (Half-space $(z > 0)$ harmonic function)
(a) Show that for $\vec{r}_0 = (x_0, y_0, z_0)$ and $\vec{r} = (x, y, z)$ the function given by
$u = u(x, y, z) = \frac{-1}{4\pi|\vec{r} - \vec{r}_0|}$ is a solution of Laplace's equation over the upper half space where $z > 0$ and $(x, y, z) \neq (x_0, y_0, z_0)$. Show that this function satisfies the boundary condition $u(x, y, 0) = \frac{-1}{4\pi\sqrt{(x - x_0)^2 + (y - y_0)^2 + z_0^2}}$.
(b) Show that for $\vec{r}_0^* = (x_0, y_0, -z_0)$ and $\vec{r} = (x, y, z)$ the function given by
$u = u(x, y, z) = \frac{1}{4\pi|\vec{r} - \vec{r}_0^*|}$ is a solution of Laplace's equation over the upper half space where $z > 0$ and that this function satisfies the boundary condition
$u(x, y, 0) = \frac{1}{4\pi\sqrt{(x - x_0)^2 + (y - y_0)^2 + z_0^2}}$.
(c) Find a solution to $\nabla^2 u = 0$ for the upper half space $z > 0$ which satisfies the boundary condition $u(x, y, 0) = 0$.

▶ **13.** Solve for the steady state temperature $u = u(r, z)$ in the cylinder $0 < r < 1$, $0 < z < 1$ where

$$\nabla^2 u = \frac{\partial^2 u}{\partial r^2} + \frac{1}{r}\frac{\partial u}{\partial r} + \frac{\partial^2 u}{\partial z^2} = 0.$$

Assume that the top and bottom of the cylinder are held at a temperature of zero degrees and the curved surface $r = 1$ has the temperature distribution $f(z) = z(1 - z)$.

▶ **14.** Use the equation (6.32) to show the boundary value problem to solve

$$\nabla^2 \phi = f(x, y, z), \quad (x, y, z) \in V$$

subject to the boundary condition $\frac{\partial \phi}{\partial n} = g(x, y, z)$ for $(x, y, z) \in S$, where S is the boundary of V, has no solution unless the condition

$$\iiint_V f(x, y, z)\, d\tau = \iint_S g(x, y, z)\, d\sigma$$

is satisfied. Give a physical interpretation to this problem in terms of heat flow.

▶ **15.** Find the steady state temperature $u = u(r, \theta)$ in a long rod with semi-circular cross section given by $R = \{r, \theta \mid 0 < r < r_0,\ 0 < \theta < \pi\}$ subject to the boundary conditions $u(r_0, \theta) = f(\theta)$, $\dfrac{\partial u(r, 0)}{\partial \theta} = 0$, $\dfrac{\partial u(r, \pi)}{\partial \theta} = 0$. What is the solution in the special case $f(\theta) = T_0$, where T_0 is a constant temperature?

▶ **16.** Use separation of variables to solve for the potential $u = u(x, y)$ satisfying

$$\nabla^2 u = \frac{\partial^2 u}{\partial x^2} + \frac{\partial^2 u}{\partial y^2} = 0, \qquad x, y \in R$$

where $R = \{x, y \mid -\infty < x < \infty,\ y > 0\}$ and u is subject to the boundary conditions of $u(x, 0) = f(x)$. Assume that u is bounded far from the origin and determine the general solution.

(a) Write out the solution in the special case $u(x, 0) = f(x) = \begin{cases} -T_0, & -1 < x < 0 \\ T_0, & 0 < x < 1 \\ 0, & \text{otherwise} \end{cases}$

(b) Write out the solution in the special case $u(x, 0) = f(x) = \begin{cases} T_0, & -1 < x < 1 \\ 0, & \text{otherwise} \end{cases}$

▶ **17.** Use separation of variables to solve $\nabla^2 u = \dfrac{\partial^2 u}{\partial x^2} + \dfrac{\partial^2 u}{\partial y^2} = 0$, $x, y \in R$ where $R = \{x, y \mid 0 < x < \infty,\ y > 0\}$ and u is subject to the boundary conditions of $u(x, 0) = 0$, $u(0, y) = g(y)$ and u is bounded far from the origin.

▶ **18.** Use separation of variables to solve for the potential $u = u(x, y)$ satisfying $\nabla^2 u = \dfrac{\partial^2 u}{\partial x^2} + \dfrac{\partial^2 u}{\partial y^2} = 0$, $x, y \in R$ where $R = \{x, y \mid 0 < x < L,\ y > 0\}$ and u is subject to the boundary conditions of $u(x, 0) = 0$, $\dfrac{\partial u(0, y)}{\partial x} = 0$ and $u(L, y) = f(y)$.

▶ **19.** Solve the Poisson equation

$$\nabla^2 u = f(x, y), \qquad 0 < x < a, \quad 0 < y < b$$

subject to the boundary conditions

$$u(0, y) = 0, \quad u(a, y) = 0, \quad u(x, 0) = 0, \quad u(x, b) = 0.$$

▶ **20.** Use separation of variables to solve for the potential $u = u(x, y)$ satisfying

$$\nabla^2 u = \frac{\partial^2 u}{\partial x^2} + \frac{\partial^2 u}{\partial y^2} = 0, \qquad x, y \in R$$

where $R = \{x, y \mid 0 < x < L,\ -\infty < y < \infty\}$ and u is subject to the boundary conditions of $u(0, y) = f_1(y)$, $u(L, y) = f_2(y)$ with u remaining bounded as $|y| \to \infty$.

▶ **21.** Use separation of variables to solve for the potential $u = u(x, y)$ satisfying

$$\nabla^2 u = \frac{\partial^2 u}{\partial x^2} + \frac{\partial^2 u}{\partial y^2} = 0, \qquad x, y \in R$$

where $R = \{x, y \mid 0 < x < \infty, \ 0 < y < \infty\}$ and u is subject to the boundary conditions of $u(0, y) = 0$, $u(x, 0) = f(x)$ with u remaining bounded far from the origin.

(a) Show that for a general $f(x)$ the solution can be written

$$u = u(x, y) = \frac{2}{\pi} \int_0^\infty \int_0^\infty f(\xi) \sin \lambda \xi \sin \lambda x e^{-\lambda y} \, d\lambda dx$$

(b) Show using the identity $\sin \lambda \xi \sin \lambda x = \frac{1}{2}[\cos \lambda(\xi - x) - \cos \lambda(\xi + x)]$ that in the special case $f(x) = 1$, $0 < x < \infty$ the solution can be written in the form $u(x, y) = \frac{2}{\pi} \arctan(x/y)$.

▶ **22.** Consider the boundary value problem to solve

$$\nabla^2 u = \frac{\partial^2 u}{\partial x^2} + \frac{\partial^2 u}{\partial y^2} = 0, \qquad \text{for } x, y \in R$$

where $R = \{x, y \mid 0 < x < L, \ 0 < y < \infty\}$ subject to the boundary condition $u(0, y) = g_1(y)$, $u(L, y) = g_2(y)$ and $u(x, 0) = f(x)$.

(a) Show that by using superposition and assuming a solution $u = u_1 + u_2$ there results the two easier to solve problems

$$\nabla^2 u_1 = 0 \qquad\qquad\qquad\qquad \nabla^2 u_2 = 0$$

$$u_1(0, y) = g_1(y), \qquad u_1(L, y) = g_2(y) \qquad\qquad u_2(0, y) = 0, \qquad u_2(L, y) = 0$$

$$u_1(x, 0) = 0 \qquad\qquad\qquad\qquad\qquad u_2(x, 0) = f(x)$$

(b) Solve for u_2.
(c) Solve for u_1.

▶ **23.** Solve for the steady state temperature $u = u(r, \theta)$ satisfying

$$\nabla^2 u = \frac{\partial^2 u}{\partial r^2} + \frac{1}{r}\frac{\partial u}{\partial r} + \frac{1}{r^2}\frac{\partial^2 u}{\partial \theta^2} = 0, \qquad x, y \in R$$

where $R = \{r, \theta \mid 0 < r < b, \ 0 < \theta < \pi/2\}$ and subject to the boundary conditions $u(r, 0) = 0$, $u(r, \pi/2) = T_0$ and $u(b, \theta) = 0$, $0 < \theta < \pi/2$. Hint: Use superposition.

▶ **24.** Solve the boundary value problems in figure 6-3.

▶ **25.** Consider a sphere of radius ϵ and centered at the point (x_0, y_0, z_0) and having the parametric representation

$$x = x_0 + \epsilon \sin\theta \cos\phi, \quad y = y_0 + \epsilon \sin\theta \sin\phi, \quad z = z_0 + \epsilon, \cos\theta \quad 0 < \theta < \pi, \quad 0 < \phi < 2\pi.$$

The average value of a function $u = u(x, y, z)$ over the surface of this sphere is given by

$$\bar{u} = \frac{1}{4\pi\epsilon^2} \int_0^{2\pi} \int_0^\pi u(x_0 + \epsilon \sin\theta \cos\phi, y_0 + \epsilon \sin\theta \sin\phi, z_0 + \epsilon \cos\theta)\, \epsilon^2 \sin\theta d\theta d\phi.$$

Expand the integrand as a power series in ϵ and retain only the terms through order ϵ^2 and neglect higher powers of ϵ. Show that $\nabla^2 u$ in three dimensions is proportional to the difference $\bar{u} - u_0$ where $u_0 = u(x_0, y_0, z_0)$ is the value of u evaluated at the center of the sphere.

▶ **26.** Show the equations (6.123) have the expanded form

$$A_{nm} = \frac{2n+1}{2\pi a^n} \frac{(n-m)!}{(n+m)!} \int_0^{2\pi} \int_0^\pi f(\theta, \phi) P_n^m(\cos\theta) \cos m\phi \sin\theta \, d\theta d\phi$$

for $n = 1, 2, 3, \ldots$ and $m = 1, 2, \ldots, n$.

$$B_{nm} = \frac{2n+1}{2\pi a^n} \frac{(n-m)!}{(n+m)!} \int_0^{2\pi} \int_0^\pi f(\theta, \phi) P_n^m(\cos\theta) \sin m\phi \sin\theta \, d\theta d\phi$$

for $n = 1, 2, 3, \ldots$ and $m = 1, 2, \ldots, n$.

$$A_{n0} = \frac{2n+1}{4\pi a^n} \int_0^{2\pi} \int_0^\pi f(\theta, \phi) P_n(\cos\theta) \sin\theta \, d\theta d\phi$$

for $n = 0, 1, 2, 3, \ldots$.

▶ **27.** Find the steady state temperature distribution $u = u(r, z)$ in a cylinder of radius r_0 and length ℓ which satisfies

$$\nabla^2 u = \frac{1}{r} \frac{\partial}{\partial r}\left(r \frac{\partial u}{\partial r}\right) + \frac{\partial^2 u}{\partial z^2} = 0, \qquad 0 < r < r_0, \quad 0 < z < \ell$$

subject to the boundary conditions

$$u(r, 0) = 0, \qquad u(r, \ell) = 0, \qquad u(r_0, z) = T_0$$

where T_0 is a constant.

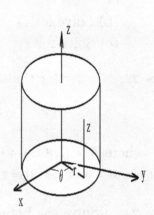

▶ **28.** Find the steady state temperature distribution $u = u(r, z)$ in a cylinder of radius r_0 and length ℓ which satisfies

$$\nabla^2 u = \frac{1}{r}\frac{\partial}{\partial r}\left(r\frac{\partial u}{\partial r}\right) + \frac{\partial^2 u}{\partial z^2} = 0, \qquad 0 < r < r_0, \quad 0 < z < \ell$$

subject to the boundary conditions

$$u(r, 0) = T_0, \qquad u(r, \ell) = 0, \qquad u(r_0, z) = 0$$

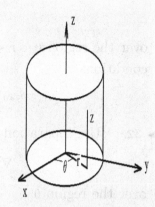

where T_0 is a constant.

▶ **29.** Find the steady state temperature distribution $u = u(r, z)$ in a cylinder of radius r_0 and length ℓ which satisfies

$$\nabla^2 u = \frac{1}{r}\frac{\partial}{\partial r}\left(r\frac{\partial u}{\partial r}\right) + \frac{\partial^2 u}{\partial z^2} = 0, \qquad 0 < r < r_0, \quad 0 < z < \ell$$

subject to the boundary conditions

$$u(r, 0) = 0, \qquad u(r, \ell) = T_0, \qquad u(r_0, z) = 0$$

where T_0 is a constant.

▶ **30.** Show how the Laplace equation representing the steady state temperature $u = u(r, \theta, z)$ in a cylinder can be broken up into three separate equations, with appropriate boundary conditions, which are easier to solve. The problem to be broken up is

$$\nabla^2 u = \frac{1}{r}\frac{\partial}{\partial r}\left(r\frac{\partial u}{\partial r}\right) + \frac{1}{r^2}\frac{\partial^2 u}{\partial \theta^2} + \frac{\partial^2 u}{\partial z^2} = 0$$

over the region $0 < r < r_0$, $-\pi < \theta < \pi$, $0 < z < z_0$ and subject to the boundary conditions

$$u(r, \theta, z_0) = f(r, \theta), \qquad u(r, \theta, 0) = g(r, \theta), \qquad u(r_0, \theta, z) = h(\theta, z)$$

Hint: Let $u = u_1 + u_2 + u_3$ and see the next three problems.

▶ **31.** Use separation of variables to solve the boundary value problem

$$\nabla^2 u_1 = \frac{1}{r}\frac{\partial}{\partial r}\left(r\frac{\partial u_1}{\partial r}\right) + \frac{1}{r^2}\frac{\partial^2 u_1}{\partial \theta^2} + \frac{\partial^2 u_1}{\partial z^2} = 0$$

over the region $0 < r < r_0,\ -\pi < \theta < \pi,\ 0 < z < z_0$ and subject to the boundary conditions

$$u_1(r,\theta,z_0) = 0, \qquad u_1(r,\theta,0) = 0, \qquad u_1(r_0,\theta,z) = h(\theta,z)$$

▶ **32.** Use separation of variables to solve the boundary value problem

$$\nabla^2 u_2 = \frac{1}{r}\frac{\partial}{\partial r}\left(r\frac{\partial u_2}{\partial r}\right) + \frac{1}{r^2}\frac{\partial^2 u_2}{\partial \theta^2} + \frac{\partial^2 u_2}{\partial z^2} = 0$$

over the region $0 < r < r_0,\ -\pi < \theta < \pi,\ 0 < z < z_0$ and subject to the boundary conditions

$$u_2(r,\theta,z_0) = f(r,\theta), \qquad u_2(r,\theta,0) = 0, \qquad u_2(r_0,\theta,z) = 0.$$

▶ **33.** Use separation of variables to solve the boundary value problem

$$\nabla^2 u_3 = \frac{1}{r}\frac{\partial}{\partial r}\left(r\frac{\partial u_3}{\partial r}\right) + \frac{1}{r^2}\frac{\partial^2 u_3}{\partial \theta^2} + \frac{\partial^2 u_3}{\partial z^2} = 0$$

over the region $0 < r < r_0,\ -\pi < \theta < \pi,\ 0 < z < z_0$ and subject to the boundary conditions

$$u_3(r,\theta,z_0) = 0, \qquad u_3(r,\theta,0) = g(r,\theta), \qquad u_3(r_0,\theta,z) = 0.$$

▶ **34.** Use separation of variables to solve for the electric potential $V = V(x,y)$ which satisfies

$$\nabla^2 V = \frac{\partial^2 V}{\partial x^2} + \frac{\partial^2 V}{\partial y^2} = 0, \qquad -a < x < a, \quad -b < y < b$$

subject to the boundary conditions

$$V(-a,y) = V_0, \quad V(a,y) = V_0, \quad V(x,-b) = 0, \quad V(x,b) = 0$$

where V_0 is a constant.

▶ **35.** Use separation of variables to solve for the electric potential $V = V(x,y)$ which satisfies

$$\nabla^2 V = \frac{\partial^2 V}{\partial x^2} + \frac{\partial^2 V}{\partial y^2} = 0, \quad 0 < x < \infty, \quad 0 < y < b$$

subject to the boundary conditions

$$V(x,b) = 0, \qquad V(x,0) = 0, \qquad V(0,y) = V_0$$

where V_0 is a constant.

Chapter 7
Transform Methods

In this chapter we will investigate the use of transform methods for solving linear partial differential equations. We begin with a review the Laplace transform and then introduce the Fourier exponential, Fourier sine and Fourier cosine transforms for solving boundary value problems associated with linear partial differential equations. These transform methods are applicable for certain types of boundary value problems with infinite or semi-infinite domains. We will then introduce some other types of transforms applicable for finite domains.

In our study of partial differential equations by transform methods it will be advantageous to introduce at this time the Heaviside[†] unit step function and the Dirac[†] delta function as these functions will be used in future sections for the modeling of physical problems and in the representation of certain solutions.

Heaviside Unit Step Function

The Heaviside unit step function is defined

$$H(\xi) = \begin{cases} 1 & \text{if} & \xi > 0 \\ 0 & \text{if} & \xi < 0. \end{cases} \tag{7.1}$$

The Heaviside step function has a value of unity when the argument of the function is positive and a value of zero when the argument of the function is negative. A graph of the Heaviside unit step function $H(x - x_0)$ is illustrated in the figure 7-1.

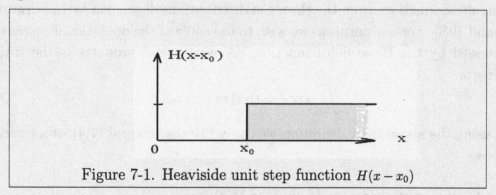

Figure 7-1. Heaviside unit step function $H(x - x_0)$

The Heaviside unit step function can be used to define the impulse function

$$\delta_\epsilon(x - x_0) = \frac{1}{\epsilon}\left[H(x - x_0) - H(x - (x_0 + \epsilon))\right] \tag{7.2}$$

[†]   See Appendix C

356

of height $1/\epsilon$ between x_0 and $x_0 + \epsilon$ and zero elsewhere. A graph of $\delta_\epsilon(x - x_0)$ is illustrated in the figure 7-2 along with the Heaviside unit step functions used to construct it. Note from the graph that the area under the impulse curve is always unity.

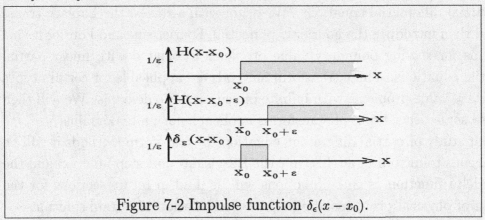

Figure 7-2 Impulse function $\delta_\epsilon(x - x_0)$.

The Dirac Delta Function

The Dirac delta function or unit impulse function $\delta(x - x_0)$ is defined

$$\delta(x - x_0) = \lim_{\epsilon \to 0} \delta_\epsilon(x - x_0) \tag{7.3}$$

and is not a function in the usual sense. This function is zero everywhere except at the point x_0 where it becomes infinite yet it still subtends an area of unity such that $\int_{-\infty}^{\infty} \delta(x - x_0)\,dx = 1$. It is formally known as belonging to the class of generalized functions from the theory of distributions developed by L. Schwartz[†] around 1950. For our purposes we wish to use some of the operational properties possessed by the Dirac delta function. A very useful property is the sifting property

$$\int_{-\infty}^{\infty} \delta(x - x_0)f(x)\,dx = f(x_0). \tag{7.4}$$

By using the above limit definition we can write the integral (7.4) as a limiting process

$$\int_{-\infty}^{\infty} \delta(x - x_0)f(x)\,dx = \lim_{\epsilon \to 0} \int_{-\infty}^{\infty} \delta_\epsilon(x - x_0)f(x)\,dx$$

$$= \lim_{\epsilon \to 0} \left[\int_{-\infty}^{x_0} 0 \cdot f(x)\,dx + \int_{x_0}^{x_0+\epsilon} \frac{1}{\epsilon}f(x)\,dx + \int_{x_0+\epsilon}^{\infty} 0 \cdot f(x)\,dx \right]$$

[†]   See Appendix C

Using the mean value theorem for integrals we can write for $0 < \theta < 1$

$$\int_{-\infty}^{\infty} \delta(x - x_0) f(x)\, dx = \lim_{\epsilon \to 0} \frac{1}{\epsilon} f(x_0 + \theta\epsilon)\epsilon = f(x_0).$$

Another useful property for the Dirac delta function is the interpretation that the derivative of the Heaviside unit step function produces the Dirac delta function or

$$\frac{dH(x - x_0)}{dx} = \delta(x - x_0). \tag{7.5}$$

This follows directly from the definition of a derivative where

$$H'(x - x_0) = \frac{dH(x - x_0)}{dx} = \lim_{\epsilon \to 0} \frac{H(x - x_0) - H(x - (x_0 + \epsilon))}{\epsilon} = \delta(x - x_0).$$

If the Dirac delta function is integrated there results

$$\int_{-\infty}^{x} \delta(x - x_0)\, dx = \begin{cases} 0 & \text{if} & x < x_0 \\ H(x - x_0) & \text{if} & x > x_0 \end{cases}. \tag{7.6}$$

This result can be achieved by integrating the $\delta_\epsilon(x - x_0)$ impulse function and then taking the limit as $\epsilon \to 0$.

The Dirac delta function also has the derivative properties

$$\int_{-\infty}^{\infty} \delta'(x - x_0) f(x)\, dx = -f'(x_0)$$

$$\int_{-\infty}^{\infty} \delta''(x - x_0) f(x)\, dx = f''(x_0)$$

$$\int_{-\infty}^{\infty} \delta'''(x - x_0) f(x)\, dx = -f'''(x_0) \tag{7.7}$$

$$\cdots$$

$$\int_{-\infty}^{\infty} \delta^{(n)}(x - x_0) f(x)\, dx = (-1)^n f^{(n)}(x_0)$$

which can be achieved using integration by parts.

In two and three dimensions we have the unity properties

$$I_2 = \int_{-\infty}^{\infty} \int_{-\infty}^{\infty} \delta(x - x_0)\delta(y - y_0)\, dx\, dy = 1$$

$$I_3 = \int_{-\infty}^{\infty} \int_{-\infty}^{\infty} \int_{-\infty}^{\infty} \delta(x - x_0)\delta(y - y_0)\delta(z - z_0)\, dx\, dy\, dz = 1$$

which can be used to develop the Dirac delta function representation for other coordinate systems. For example, if we change to polar coordinates using the transformations $x = r\cos\theta$ and $y = r\sin\theta$ we want the product of delta functions to change to polar form. We know $dxdy$ transforms to $r\,dr\,d\theta$ and so we expect there is some value A such that the integral I_2 will transform to the form

$$I_2 = \int_0^{2\pi} \int_0^{\infty} A\delta(r - r_0)\delta(\theta - \theta_0)\,r\,dr\,d\theta = 1,$$

involving the polar point (r_0, θ_0) which corresponds to the (x_0, y_0) point in Cartesian coordinates. The unity property is preserved if we select $A = 1/r$, then we can say that

$$\delta(x - x_0)\delta(y - y_0) \quad \text{transforms to} \quad \frac{\delta(r - r_0)\delta(\theta - \theta_0)}{r}. \tag{7.8}$$

If there is symmetry with respect to θ, then we find that the delta function takes the form

$$\frac{\delta(r - r_0)}{2\pi r}. \tag{7.9}$$

Using similar type of arguments for three dimensions we find that the delta function has the following representations.

$$\frac{\delta(r - r_0)\delta(\theta - \theta_0)\delta(z - z_0)}{r} \quad \{\text{Cylindrical coordinates } (r, \theta, z)$$

$$\frac{\delta(r - r_0)\delta(z - z_0)}{2\pi r} \quad \left\{ \begin{array}{l} \text{Cylindrical coordinates having} \\ \text{symmetry with respect to } \theta.. \end{array} \right.$$

$$\frac{\delta(\rho - \rho_0)\delta(\theta - \theta_0)\delta(\phi - \phi_0)}{\rho^2 \sin\theta} \quad \{\text{Spherical coordinates } (\rho, \theta, \phi). \tag{7.10}$$

$$\frac{\delta(\rho - \rho_0)\delta(\theta - \theta_0)}{2\pi\rho^2 \sin\theta} \quad \left\{ \begin{array}{l} \text{Spherical coordinates } (\rho, \theta) \text{ having} \\ \text{symmetry with respect to } \phi. \end{array} \right.$$

$$\frac{\delta(\rho - \rho_0)}{4\pi\rho^2} \quad \left\{ \begin{array}{l} \text{Spherical coordinates } (\rho) \text{ having symmetry} \\ \text{with respect to both } \theta \text{ and } \phi. \end{array} \right.$$

Note that the transformations

$$x - x_0 = r\cos\theta, \quad y - y_0 = r\sin\theta, \quad z - z_0 = Z$$

to cylindrical coordinates (r, θ, Z) and the transformations

$$x - x_0 = \rho\sin\theta\cos\phi, \quad y - y_0 = \rho\sin\theta\sin\phi, \quad z - z_0 = \rho\cos\theta$$

to spherical coordinates (ρ, θ, ϕ) shifts the impulse function to the origin of the new coordinates.

The Laplace Transform

It is assumed the reader has been exposed to Laplace transforms for the solution of ordinary differential equations where the basic technique for solving linear ordinary differential equations by Laplace transforms is illustrated in the figure 7-3.

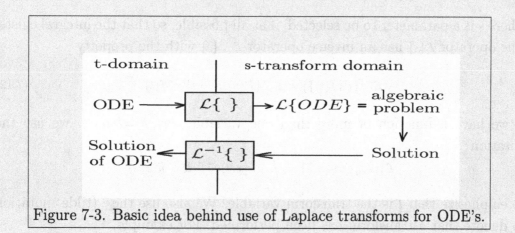

Figure 7-3. Basic idea behind use of Laplace transforms for ODE's.

In figure 7-3 a linear ordinary differential equation (ODE) is transformed from a t-domain to a Laplace transform s-domain where it becomes an algebraic problem. This algebraic problem is solved in the transform domain and inverted back to the t-domain by taking an inverse Laplace transform by either complex variable methods or table look up methods. This same basic idea is applied to solve certain linear partial differential equations (PDE). When linear PDE's are put into the Laplace transform operator box they come out as linear ordinary differential equations or lower order partial differential equations in the transform domain. The transformed equation is solved and then inverted back to the t-domain to obtain the solution to the original PDE. This basic idea works with other transform operators (Laplace, Fourier, etc.). In order for these transform techniques to work, the ordinary differential equation or partial differential equation that the transform is being applied to must be linear. These transform techniques are not applicable to nonlinear equations. Also one must become familiar with the operational properties of the various transform operators that are being used, as this knowledge will enable more extensive use and modifications of transform tables. We begin our study of transform techniques with a review of the Laplace transform and its properties because many of the techniques and methods employed in the use of Laplace transforms will help in understanding the operational properties of other transforms.

Recall the Laplace transform is defined

$$\mathcal{L}\{f(t); t \rightarrow s\} = \int_0^\infty f(t)e^{-st}\, dt = F(s) \tag{7.11}$$

where s is a parameter to be selected, if at all possible, so that the integral exists. The operator $\mathcal{L}\{\ \}$ has an inverse operator $\mathcal{L}^{-1}\{\ \}$ with the property

$$\mathcal{L}^{-1}\{\mathcal{L}\{f(t)\}\} = \mathcal{L}^{-1}\{F(s); s \rightarrow t\} = f(t). \tag{7.12}$$

If we have a function of more than one variable, say $u = u(x,t)$, we use the notation

$$\mathcal{L}\{u(x,t); t \rightarrow s\} = \int_0^\infty u(x,t)e^{-st}\, dt = \widetilde{u}(x,s)$$

to emphasize that t is the transform variable. We also use the ~ (tilde) notation to denote that a transform has been performed. For example,

$$\mathcal{L}\{u(x,t); x \rightarrow \xi\} = \int_0^\infty u(x,t)e^{-\xi x}\, dx = \widetilde{u}(\xi,t)$$

is the notation used to show that the variable x is the variable to be transformed to the variable ξ. In one-dimension we can use capital letters to denote the transform function or one can use the tilde notation. The table 7-1 lists operational properties associated with the Laplace transform. The first property illustrates the Laplace transform notation where one can write $\mathcal{L}\{f(t)\} = F(s)$ if you move across the table from left to right or you can write $f(t) = \mathcal{L}^{-1}\{F(s)\}$ if you move across the table from right to left. The second property is a linearity property which follows from the definition of a Laplace transform and the linearity property of integrals. The integral of a sum is the sum of integrals and also constants can be brought out in front of the integral sign. The property three is a differentiation property and assumes that $f(t)$ is a continuous function. The property follows using integration by parts and the definition of the Laplace transform. We have

$$\mathcal{L}\{f'(t)\} = \int_0^\infty f'(t)e^{-st}\, dt = f(t)e^{-st}\Big]_0^\infty - \int_0^\infty f(t)e^{-st}(-s)\, dt$$

which simplifies to

$$\mathcal{L}\{f'(t)\} = sF(s) - f(0^+)$$

under the assumption $s > 0$ and $\lim_{T \to \infty} f(T)e^{-sT} = 0$. Here we have used the definition $F(s) = \int_0^\infty f(t)e^{-st}\,dt$ and the $f(0^+)$ term is defined as the right-hand limit $f(0^+) = \lim_{\substack{t \to 0 \\ t > 0}} f(t)$. The definition of the Laplace transform implies that for the Laplace transform to exist, the improper integral of $f(t)$ must exist. The function $f(t)$ need only be defined for $t \geq 0$, and the transform parameter s must be selected such that the improper integral exists. In the case where $f(t)$ has a discontinuity the differentiation property in table 7-1, entry number 3, needs to be modified. If we assume $f(t)$ has a single jump discontinuity at the point t_0, then we would write

$$\mathcal{L}\{f'(t)\} = \int_0^\infty f'(t)e^{-st}\,dt = \int_0^{t_0^-} f'(t)e^{-st}\,dt + \int_{t_0^+}^\infty f'(t)e^{-st}\,dt$$

$$\mathcal{L}\{f'(t)\} = f(t)e^{-st}\big]_0^{t_0^-} + s\int_0^{t_0^-} f(t)e^{-st}\,dt + f(t)e^{-st}\big]_{t_0^+}^\infty + s\int_{t_0^+}^\infty f(t)e^{-st}\,dt$$

which simplifies to

$$\mathcal{L}\{f'(t)\} = sF(s) - f(0^+) + \left[f(t_0^-) - f(t_0^+)\right]e^{-st_0}.$$

This illustrates that for each jump discontinuity in the function an additional term, involving the jump, must be added to the derivative property listed in the Laplace transform table 7-1. For the large majority of physics, chemistry and engineering problems we assume continuity for $f(t)$. The hieroglyphics denoting property 3 can be put into words. We would read the symbols denoting property 3 as follows.

"If the Laplace transform of $f(t)$ is $F(s)$, then the Laplace transform of the derivative $f'(t)$ is s-times the transform of the function differentiated minus the initial value of the function differentiated."

Now the properties 4 and 5 follow by repeated applications of this differentiation property. For example, "If the Laplace transform of $f'(t)$ is $sF(s) - f(0^+)$, then the Laplace transform of the derivative $f''(t)$ is s-times the transform of the function differentiated minus the initial value of the function differentiated." In terms of symbols

$$\mathcal{L}\{f''(t)\} = s[sF(s) - f(0^+)] - f'(0^+).$$

Table 7-1		
Laplace Transform Properties		
1. $f(t) = \mathcal{L}^{-1}\{F(s)\}$	$F(s) = \mathcal{L}\{f(t)\}$	Comments
2. $c_1 f(t) + c_2 g(t)$	$c_1 F(s) + c_2 G(s)$	Linearity property
3. $f'(t)$	$sF(s) - f(0^+)$	Derivative property
4. $f''(t)$	$s^2 F(s) - s f(0^+) - f'(0^+)$	Second derivative
5. $f^{(n)}(t)$	$s^n F(s) - s^{n-1} f(0^+) - \cdots$ $- s f^{(n-2)}(0^+) - f^{(n-1)}(0^+)$	nth derivative
6. $\int_0^t f(\tau)\, d\tau$	$\dfrac{F(s)}{s}$	Integration property
7. $t f(t)$	$-F'(s)$	Multiplication by t
8. $t^n f(t)$	$(-1)^n F^{(n)}(s)$	Repeated multiplication by t
9. $\dfrac{f(t)}{t}$	$\int_s^\infty F(s)\, ds$	Division by t
10. $e^{at} f(t)$	$F(s-a)$	First shift property
11. $f(t-a) H(t-a)$	$e^{-as} F(s), \quad a > 0$	Second shift property
12. $H(t-a)$	$\dfrac{e^{-as}}{s}, \quad a > 0$	Heaviside function
13. $\delta(t-a)$	$e^{-as}, \quad a > 0$	Dirac delta function
14. $\dfrac{1}{a} f\!\left(\dfrac{t}{a}\right)$	$F(as), \quad a > 0$	Scaling property
15. $f(t) * g(t)$	$F(s) G(s)$	Convolution property $f(t) * g(t) = \int_0^t f(\tau) g(t-\tau)\, d\tau$ $g(t) * f(t) = \int_0^t g(\tau) f(t-\tau)\, d\tau$
16. $\lvert f(t) \rvert \le M e^{\alpha t}$	$\lim_{s \to \infty} F(s) = 0$	$f(t)$ of exponential order.

	Table 7-2	
	Short Table of Laplace Transforms	
	$f(t) = \mathcal{L}^{-1}\{F(s)\}$	$F(s) = \mathcal{L}\{f(t)\}$
1.	1	$\dfrac{1}{s}$
2.	t	$\dfrac{1}{s^2}$
3.	t^n	$\dfrac{n!}{s^{n+1}}$ n integer, $s > 0$
4.	t^α	$\dfrac{\Gamma(\alpha+1)}{s^{\alpha+1}}$ $\alpha > -1,\, s > 0$
5.	$e^{\omega t}$	$\dfrac{1}{s - \omega}$
6.	$\sin \omega t$	$\dfrac{\omega}{s^2 + \omega^2}$
7.	$\cos \omega t$	$\dfrac{s}{s^2 + \omega^2}$
8.	$\sinh \omega t$	$\dfrac{\omega}{s^2 - \omega^2}$
9.	$\cosh \omega t$	$\dfrac{s}{s^2 - \omega^2}$
10.	$\left(\dfrac{t}{\beta}\right)^{n/2} J_n(2\sqrt{\beta t})$	$\dfrac{1}{s^{n+1}} e^{-\beta/s}$ $n > -1$
11.	$J_0(\omega t)$	$\dfrac{1}{\sqrt{s^2 + \omega^2}}$
12.	$\dfrac{1}{\sqrt{\omega}} e^{\omega t} \operatorname{erf}(\sqrt{\omega t})$	$\dfrac{1}{\sqrt{s}(s - \omega)}$
13.	$\dfrac{1}{\sqrt{\pi t}} e^{-\omega^2/4t}$	$\dfrac{1}{\sqrt{s}} e^{-\omega\sqrt{s}}$
14.	$\dfrac{\omega}{2\sqrt{\pi t^3}} e^{-\omega^2/4t}$	$e^{-\omega\sqrt{s}}$
15.	$\operatorname{erf}\left(\dfrac{\omega}{2\sqrt{t}}\right)$	$\dfrac{1}{s}\left(1 - e^{-\omega\sqrt{s}}\right)$
16.	$\operatorname{erfc}\left(\dfrac{\omega}{2\sqrt{t}}\right)$	$\dfrac{1}{s} e^{-\omega\sqrt{s}}$
17.	$e^{\beta^2 t + \alpha\beta} \operatorname{erfc}\left(\beta\sqrt{t} + \dfrac{\alpha}{2\sqrt{t}}\right)$	$\dfrac{1}{\sqrt{s}(\beta + \sqrt{s})} e^{-\alpha\sqrt{s}}$
18.	$H(t - a)$	$\dfrac{1}{s} e^{-as}$, $a > 0$
19.	$\delta(t - a)$	e^{-as}, $a > 0$
20.	$\dfrac{x}{b} + \dfrac{2}{\pi} \displaystyle\sum_{n=1}^{\infty} \dfrac{(-1)^n}{n} e^{-n^2\pi^2 t/b^2} \sin\dfrac{n\pi x}{b}$	$\dfrac{\sinh x\sqrt{s}}{s \sinh b\sqrt{s}}$

The integration property 6 is derived using the differentiation property 3. Let $g(t) = \int_0^t f(t)\,dt$ with $g'(t) = f(t)$ and $g(0) = 0$, then $\mathcal{L}\{g'(t)\} = \mathcal{L}\{f(t)\}$ gives

$$sG(s) - 0 = \mathcal{L}\{f(t)\} = F(s) \quad \text{or} \quad G(s) = \mathcal{L}\{g(t)\} = \frac{F(s)}{s}.$$

The multiplication by t property, entry 7 of table 7-1, is obtained by differentiating with respect to s the definition

$$F(s) = \mathcal{L}\{f(t)\} = \int_0^\infty f(t)e^{-st}\,dt$$

where the right-hand side is differentiated under the integral sign to obtain

$$F'(s) = \frac{dF}{ds} = \int_0^\infty f(t)e^{-st}(-t)\,dt$$

or

$$-F'(s) = \int_0^\infty tf(t)e^{-st}\,dt = \mathcal{L}\{tf(t)\}$$

or

$$\mathcal{L}^{-1}\{-F'(s)\} = tf(t).$$

By continually applying the multiplication property there results the entry 8 of table 7-1. That is, each time you multiply by t in the t-domain, there results a negative derivative of the transform function in the transform domain.

The division by t property, entry 9 of table 7-1, results by integrating the definition of the Laplace transform from s to ∞. If

$$F(s) = \mathcal{L}\{f(t)\} = \int_0^\infty f(t)e^{-st}\,dt,$$

then $$\int_s^\infty F(s)\,ds = \int_s^\infty \int_0^\infty f(t)e^{-st}\,dt\,ds = \int_0^\infty f(t)\left[\int_s^\infty e^{-st}\,ds\right]dt$$

where we interchange the order of integration to achieve this result. Performing the integration gives

$$\int_s^\infty F(s)\,ds = \int_0^\infty \frac{f(t)}{t}e^{-st}\,dt = \mathcal{L}\{\frac{f(t)}{t}\}$$

or $$\mathcal{L}^{-1}\{\int_s^\infty F(s)\,ds\} = \frac{f(t)}{t}.$$

This property illustrates that division by t in the t-domain corresponds to an integration s to ∞ in the transform domain.

The first shift property indicates that when $f(t)$ is multiplied by e^{at} in the t-domain the corresponding action in the transform s-domain is a shifting of the transform function $F(s)$ to $F(s-a)$. Again, this follows from the definition. If

$$F(s) = \mathcal{L}\{f(t)\} = \int_0^\infty f(t)e^{-st}\,dt,$$

then

$$F(s-a) = \int_0^\infty f(t)e^{-(s-a)t}\,dt = \int_0^\infty e^{at}f(t)e^{-st}\,dt = \mathcal{L}\{e^{at}f(t)\},$$

or

$$\mathcal{L}^{-1}\{F(s-a)\} = e^{at}f(t).$$

The second shift property involves the Heaviside unit step function $H(t-a)$ and involves a shifting and cutting off or chopping of the original function $f(t)$ whenever $F(s)$ is multiplied by e^{-as}. This second shift property is derived as follows. We write the definition of the Laplace transform using a dummy variable of integration

$$F(s) = \mathcal{L}\{f(t)\} = \int_0^\infty f(u)e^{-su}\,du,$$

then

$$e^{-as}F(s) = \int_0^\infty f(u)e^{-su}e^{-as}\,du = \int_0^\infty f(u)e^{-s(u+a)}\,du.$$

Now make the change of variable $t = u+a$ with $dt = du$ and new limits of integration from $t = a$ to ∞ and write the new integral in the form

$$e^{-as}F(s) = \int_{t=a}^\infty f(t-a)e^{-st}\,dt = \int_0^a (0)\cdot f(t-a)e^{-st}\,dt + \int_a^\infty (1)\cdot f(t-a)e^{-st}\,dt.$$

This last integral has the equivalent form

$$e^{-as}F(s) = \int_0^\infty f(t-a)H(t-a)e^{-st}\,dt,$$

since $H(t-a) = 0$ for $t < a$ and $H(t-a) = 1$ for $t > a$. Hence, we find

$$e^{-as}F(s) = \mathcal{L}\{f(t-a)H(t-a)\}$$

$$\text{or} \qquad \mathcal{L}^{-1}\{e^{-as}F(s)\} = f(t-a)H(t-a).$$

The figure 7-4 illustrates the shifting and chopping action associated with the second shift property.

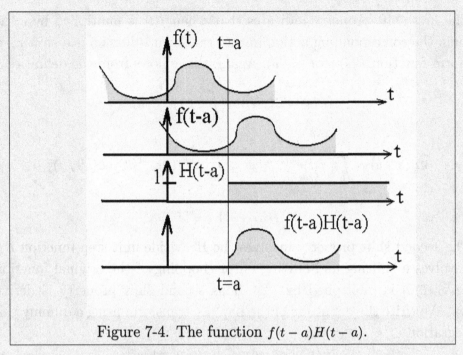

Figure 7-4. The function $f(t-a)H(t-a)$.

A special case results when $f(t) = 1$ and $F(s) = 1/s$. The second shift property gives

$$\mathcal{L}\{H(t-a)\} = \frac{e^{-as}}{s}$$

which is the property 12 listed in table 7-1. The Dirac delta function sifting property gives

$$\mathcal{L}\{\delta(t-a)\} = \int_0^\infty \delta(t-a)e^{-st}\,dt = e^{-sa}$$

or $\qquad \delta(t-a) = \mathcal{L}^{-1}\{e^{-sa}\}.$

which is entry 13 from table 7-1.

The scaling property, entry 14 of table 7-1, results by letting u denote a dummy variable of integration and writing

$$F(s) = \mathcal{L}\{f(t)\} = \int_0^\infty f(u)e^{-su}\,du$$

then $\qquad \dfrac{1}{a}F(\dfrac{s}{a}) = \displaystyle\int_0^\infty \dfrac{1}{a}f(u)e^{-(s/a)u}\,du.$

Make the substitution $t = u/a$ with $dt = du/a$ to obtain

$$\frac{1}{a}F(\frac{s}{a}) = \int_0^\infty f(at)e^{-st}\,dt = \mathcal{L}\{f(at)\}$$

or $\qquad \mathcal{L}^{-1}\{\dfrac{1}{a}F(\dfrac{s}{a})\} = f(at).$

For example, if $f(t) = \sin t$ has the Laplace transform $F(s) = \frac{1}{s^2+1}$, then $g(t) = f(\omega t) = \sin \omega t$ has the Laplace transform

$$G(s) = \frac{1}{\omega} F(\frac{s}{\omega}) = \frac{1}{\omega} \frac{1}{\left(\frac{s}{\omega}\right)^2 + 1} = \frac{\omega}{s^2 + \omega^2}.$$

The convolution of two functions $f(t)$ and $g(t)$ is defined

$$f(t) * g(t) = \int_0^t f(\tau) g(t - \tau)\, d\tau$$

where the right-hand side is called a convolution integral. The convolution property is commutative since

$$g(t) * f(t) = \int_0^t g(\tau) f(t - \tau)\, d\tau$$

is obtained by interchanging the functions f and g in the above definition. By making a change of variable it is an easy exercise to show that

$$f(t) * g(t) = g(t) * f(t)$$

and so one should examine both integrals to see which one is easier to integrate. To prove property 15 of table 7-1 we note that if

$$F(s) = \mathcal{L}\{f(t)\} = \int_0^\infty f(t) e^{-st}\, dt,$$

then by introducing a dummy variable of integration τ we can write

$$G(s)F(s) = \int_0^\infty f(\tau) e^{-s\tau} G(s)\, d\tau. \tag{7.13}$$

By the second shift property, entry 11 of table 7-1, we have

$$e^{-s\tau} G(s) = \mathcal{L}\{g(t - \tau) H(t - \tau)\} = \int_0^\infty g(t - \tau) H(t - \tau) e^{-st}\, dt. \tag{7.14}$$

We substitute equation (7.14) into equation (7.13) and write

$$F(s)G(s) = \int_{\tau=0}^\infty f(\tau) \int_{t=0}^\infty g(t - \tau) H(t - \tau) e^{-st}\, dt d\tau. \tag{7.15}$$

The integrals in equation (7.15) are improper integrals and result from the limiting process

$$F(s)G(s) = \lim_{T \to \infty} \int_{\tau=0}^T f(\tau) \int_{t=0}^T g(t - \tau) H(t - \tau) e^{-st}\, dt d\tau \tag{7.16}$$

This limiting form produces the figure 7-5 illustrating the region of integration associated with the element of area $dtd\tau$.

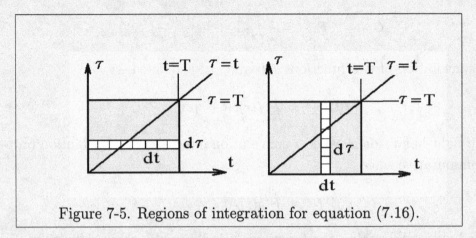

Figure 7-5. Regions of integration for equation (7.16).

The double integral region of integration for equation (7.16), which is illustrated in figure 7-5, shows that an integration is performed first in the t-direction from 0 to T. This forms the horizontal strip illustrated in the figure 7-5. The second integration, in the τ direction, moves the horizontal strip from 0 to T and thereby covers the square with sides T. We assume the integrand in equation (7.16) are continuous functions and so we can interchange the order of the integrations in that equation and write

$$F(s)G(s) = \lim_{T\to\infty} \int_{t=0}^{T} \left[\int_{\tau=0}^{T} f(\tau)g(t-\tau)H(t-\tau)\,d\tau \right] e^{-st}\,dt. \tag{7.17}$$

Here the element of area $d\tau dt$ is integrated first in the τ direction, from 0 to T, to form the vertical strip illustrated in the figure 7-5. The integral inside the brackets can be written in the form

$$\int_{\tau=0}^{T} f(\tau)g(t-\tau)H(t-\tau)\,d\tau = \int_{\tau=0}^{t} f(\tau)g(t-\tau)H(t-\tau)\,d\tau$$
$$+ \int_{\tau=t}^{T} f(\tau)g(t-\tau)H(t-\tau)\,d\tau. \tag{7.18}$$

By definition, the Heaviside step function satisfies $H(\xi) = 0$ if $\xi < 0$ and $H(\xi) = 1$ if $\xi > 0$ and so its value depends upon whether the argument of the function is positive or negative. Examine the two integrals on the right-hand side of equation (7.18). In the first integral we have $0 < \tau < t$ and so $t - \tau > 0$ which makes $H(t-\tau) = 1$. In the second integral we have $t < \tau < T$ and so $t - \tau < 0$ which make $H(t-\tau)=0$. Therefore, the integral given by equation (7.18) reduces to the convolution integral $f(t)*g(t)$ and therefore the integral given by equation (7.17) becomes

$$F(s)G(s)=\int_0^\infty f(t)^*g(t)e^{-st}\,dt=\mathcal{L}\{f(t)^*g(t)\}$$

or $\qquad \mathcal{L}^{-1}\{F(s)G(s)\}=f(t)^*g(t).$

The table 7-2 gives a listing of some Laplace transforms that are useful. For a more complete table see the Bateman tables given in the list of references.

Example 7-1. (Heat equation.)

Use Laplace transforms to solve the boundary value problem for the heat equation with a constant source term q for a semi-infinite rod.

$$\text{PDE:} \qquad \frac{\partial u}{\partial t}=K\frac{\partial^2 u}{\partial x^2}+q, \quad 0<x<\infty, \quad t>0$$
$$\text{BC:} \qquad u(0,t)=0$$
$$\text{IC:} \qquad u(x,0)=0$$

Solution: For problems with infinite domain one usually assumes some type of boundedness condition at infinity. For this problem we assume the flux is zero so that $\lim_{x\to\infty}\frac{\partial u(x,t)}{\partial x}=0$. Define

$$\widetilde{u}(x,s)=\mathcal{L}\{u(x,t); t\to s\}=\int_0^\infty u(x,t)e^{-st}\,dt$$

as the Laplace transform of the temperature $u(x,t)$ with respect to the t variable. We then take the Laplace transform of the PDE and BC to obtain

$$\mathcal{L}\{\frac{\partial u}{\partial t}\}=\mathcal{L}\{K\frac{\partial^2 u}{\partial x^2}\}+\mathcal{L}\{q\}$$
$$\mathcal{L}\{u(0,t)\}=\mathcal{L}\{0\}. \tag{7.19}$$

Using the properties of the Laplace transform the equation (7.19) reduces to an ordinary differential equation

$$s\widetilde{u}-u(x,0)=K\frac{d^2\widetilde{u}}{dx^2}+\frac{q}{s}, \quad 0<x<\infty$$
$$\widetilde{u}(0,s)=0, \qquad \lim_{x\to\infty}\frac{\partial\widetilde{u}(x,s)}{\partial x}=0. \tag{7.20}$$

Observe we have interchanged the operations of integration and differentiation

$$\mathcal{L}\{K\frac{\partial^2 u}{\partial x^2}\}=K\frac{\partial^2\mathcal{L}\{u\}}{\partial x^2}$$

to obtain the above result. The equation (7.20) can be written in the form

$$\frac{d^2\tilde{u}}{dx^2} - \frac{s}{K}\tilde{u} = \frac{-q}{Ks}$$

and treating s as a constant we solve this differential equation to obtain the general solution

$$\tilde{u} = \tilde{u}(x,s) = c_1 e^{x\sqrt{s/K}} + c_2 e^{-x\sqrt{s/K}} + \frac{q}{s^2}.$$

We require $\lim_{x\to\infty} \frac{\partial \tilde{u}(x,s)}{\partial x} = 0$ and so we set $c_1 = 0$. The condition $\tilde{u}(0,s) = 0$ is obtained if we set $c_2 = -q/s^2$. This gives the solution

$$\tilde{u} = \tilde{u}(x,s) = \frac{q}{s^2} - q\left(\frac{e^{-x\sqrt{s/K}}}{s^2}\right)$$

in the transform domain. An examination of the table 7-2 shows that

$$\mathcal{L}^{-1}\{\frac{1}{s^2}\} = t$$

and $\qquad \mathcal{L}^{-1}\{\frac{1}{s}e^{-x\sqrt{s/K}}\} = \text{erfc}\left(\frac{x}{2\sqrt{Kt}}\right).$

Using the property 6 from table 7-1 we can write

$$\mathcal{L}^{-1}\{\frac{1}{s^2}e^{-x\sqrt{s/K}}\} = \int_0^t \text{erfc}\left(\frac{x}{2\sqrt{K\tau}}\right) d\tau.$$

Therefore, the solution of the original boundary value problem can be written as

$$u = u(x,t) = \mathcal{L}^{-1}\{\tilde{u}(x,s); s \to t\} = q\left[t - \int_0^t \text{erfc}\left(\frac{x}{2\sqrt{K\tau}}\right) d\tau\right]. \qquad (7.21)$$

A graph of this solution is illustrated in the figure 7-6.

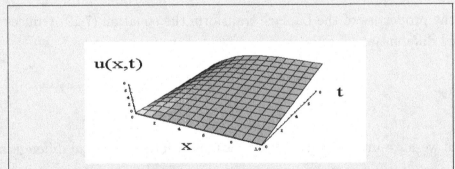

Figure 7-6. Temperature as function of distance and time.

We note from the entries in table 7-2

$$\mathcal{L}^{-1}\{\frac{1}{s^2}\} = t = f(t)$$

and $\quad \mathcal{L}^{-1}\{e^{-x\sqrt{s/K}}\} = \dfrac{x}{2\sqrt{\pi K t^3}} e^{-x^2/4Kt} = g(t).$

Using the convolution property we can write the solution in the forms

$$u(x,t) = qt - q \int_0^t \frac{x}{2\sqrt{\pi K \tau^3}} e^{-x^2/4K\tau} (t - \tau)\, d\tau = q[t - g(t) * f(t)]$$

or $\quad u(x,t) = qt - q \displaystyle\int_0^t \tau \frac{x}{2\sqrt{\pi K(t-\tau)^3}} e^{-x^2/4K(t-\tau)}\, d\tau = q[t - f(t)^* g(t)].$

These are equivalent forms for the solution. It is left as a exercise to integrate the equation (7.21) by parts to obtain the first of the convolution forms above. Note that if one knows the solution of a problem is unique, then if one obtains what appears to be two different solutions to the same problem our only conclusion is that these two solution forms must equal one another. Thus, uniqueness proofs are an important consideration in studying applied problems.

∎

Example 7-2. (Heat equation and shock tubes[‡].)

Shock tubes are used to study hypersonic flow. In a typical shock tube experiment shock waves are created by bursting a diaphragm between a high-pressure chamber and a low pressure chamber. The resulting shock wave travels the length of a test section at supersonic speed and then discharges into a large chamber called a dump tank. This causes the pressure of the entire system to stabilize. When a secondary diaphragm is place between the test section and the dump tank, the shock tube is said to be operating in a "shock expansion" mode. In this mode a high temperature shock wave is created which travels the length of the test tube and bursts the secondary diaphragm and then expands into the low pressure dump tank. This bursting of the secondary diaphragm changes the thermodynamics, and chemical properties of the gas flow as well as increasing the speed of the gas flow. In the study of heat effects on the secondary diaphragm, which occurs over a period of micro seconds, the following mixed boundary value

---

[‡] The examples 7-2 and 7-15 evolved from the research of M.D. Williams, Senior Research Scientist, Hypersonic Airbreathing Propulsion Branch, Aerodynamics, Aerothermodynamics, and Acoustics Competency, NASA Langley Research Center, Hampton, Virginia.

problem arises

$$\text{PDE:} \quad \frac{\partial u}{\partial t} = \beta \frac{\partial^2 u}{\partial x^2}, \quad 0 < x < L, \quad t > 0$$

$$\text{BC:} \quad u(0,t) = T_s$$

$$\frac{\partial u(L,t)}{\partial x} = 0$$

$$\text{IC:} \quad u(x,0) = T_r$$

where $u = u(x,t)$ is the temperature (K) in the diaphragm, x is distance (cm) into the diaphragm, T_s is the shock temperature (4681 K), T_r is the room temperature (300 K), L is the diaphragm thickness (0.005 cm). We also assume T_s, T_r and $\beta > 0$ remain constant.

Solution: Using Laplace transforms we define

$$\widetilde{u}(x,s) = \mathcal{L}\{u(x,t); t \to s\}$$

and take the Laplace transform of the PDE and BC's to obtain the ordinary differential equation

$$s\widetilde{u} - T_r = \beta \frac{d^2\widetilde{u}}{dx^2}$$

$$\widetilde{u}(0,s) = \frac{T_s}{s}$$

$$\frac{d\widetilde{u}(L,s)}{dx} = 0$$

We write the differential equation in the form

$$\frac{d^2\widetilde{u}}{dx^2} - \frac{s}{\beta}\widetilde{u} = -\frac{T_r}{\beta}$$

and find the solution

$$\widetilde{u} = \widetilde{u}(x,s) = c_1 e^{-x\sqrt{s/\beta}} + c_2 e^{x\sqrt{s/\beta}} + \frac{T_r}{s}$$

with derivative $\quad \dfrac{d\widetilde{u}}{dx} = -c_1\sqrt{s/\beta}\, e^{-x\sqrt{s/\beta}} + c_2\sqrt{s/\beta}\, e^{x\sqrt{s/\beta}}$

The end boundary condition requires

$$\widetilde{u}(0,s) = c_1 + c_2 + \frac{T_r}{s} = \frac{T_s}{s} \tag{7.22}$$

and the flux boundary condition requires

$$\frac{d\widetilde{u}(L,s)}{dx} = -c_1\sqrt{s/\beta}\, e^{-\sqrt{s/\beta}L} + c_2\sqrt{s/\beta}\, e^{\sqrt{s/\beta}L} = 0. \tag{7.23}$$

We solve equations (7.22) and (7.23) for the constants c_1 and c_2 and obtain the transform domain solution

$$\tilde{u} = \tilde{u}(x,s) = \frac{1}{s}\frac{T_s - T_r}{1 + e^{-2\sqrt{s/\beta}L}} + \frac{1}{s}\frac{T_s - T_r}{1 + e^{-2\sqrt{s/\beta}L}}e^{-\sqrt{s/\beta}(2L-x)} + \frac{T_r}{s}.$$

The table 7-1 does not contain transforms which we can use to invert this transform directly and so we rearrange the terms of the solution into a form where we can make use of the operational properties in tables 7-1 and 7-2. One method is to expand the function $\dfrac{1}{1 + e^{-2\sqrt{s/\beta}L}}$ into a series using

$$\frac{1}{1+x} = 1 - x + x^2 - x^3 + \cdots$$

with $x = e^{-2\sqrt{s/\beta}L}$ to obtain

$$\frac{1}{1 + e^{-2\sqrt{s/\beta}L}} = \sum_{n=0}^{\infty}(-1)^n e^{-2n\sqrt{s/\beta}L}.$$

The solution in the transform domain can now be expressed in the form

$$\tilde{u} = \frac{T_r}{s} + (T_s - T_r)\sum_{n=0}^{\infty}(-1)^n\left[\frac{e^{-[(2n+1)L-x]\sqrt{s/\beta}}}{s} + \frac{e^{-(2nL+x)\sqrt{s/\beta}}}{s}\right]. \tag{7.24}$$

The table 7-2 gives the inverse transform $\mathcal{L}^{-1}\{\frac{e^{-a\sqrt{s}}}{s}\} = \operatorname{erfc}\left(\frac{a}{2\sqrt{t}}\right)$ which is applicable for finding the inverse Laplace transform associated with each of the series terms in the equation (7.24). Using this result we invert the transform given by equation (7.24) to the form

$$u = \mathcal{L}^{-1}\{\tilde{u}(x,s); s \to t\}$$
$$u = T_r + (T_s - T_r)\sum_{n=0}^{\infty}(-1)^n\left\{\operatorname{erfc}\left[\frac{2(n+1)L - x}{2\sqrt{\beta t}}\right] + \operatorname{erfc}\left[\frac{2nL + x}{2\sqrt{\beta t}}\right]\right\}. \tag{7.25}$$

The figure 7-7 illustrates a graph of temperature vs distance after $t = 100\,\mu$ sec for various values of the ratio $a = L/\sqrt{\beta}$.

374

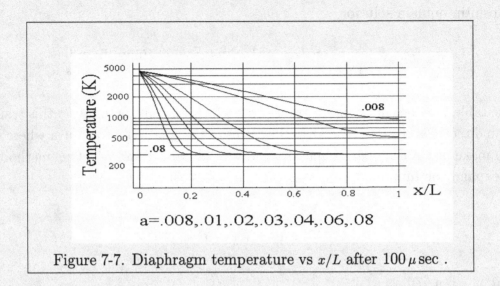

Figure 7-7. Diaphragm temperature vs x/L after $100\,\mu\sec$.

Bending of Beams

Let $w = w(x)$ denote the transverse load distribution per unit length along a beam as illustrated in the adjoining figure. Denote by
$y = y(x)$ the displacement of the center line of
the beam. In simplified beam theory this displacement is assumed to be small so that $\frac{dy}{dx}$ is
also small. This assumption produces the equations

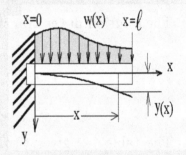

$$EI\frac{d^2y}{dx^2} = M(x) = \text{Internal bending moment at position } x.$$

$$EI\frac{d^3y}{dx^3} = -V(x) = \text{Internal shearing force at position } x.$$

$$EI\frac{d^4y}{dx^4} = w(x) = \text{Transverse load per unit length at position } x.$$

The quantity EI is called the flexural rigidity of the beam and is assumed to
be constant. Here E denotes Young's modulus of elasticity for the beam and I
denotes the moment of inertia of the cross section about the neutral axis. Boundary conditions associated with various types of beams have the following forms.

(1.) Clamped or fixed ends
$$y(0) = y(\ell) = 0$$
$$y'(0) = y'(\ell) = 0$$

(2.) Simply supported ends
$$y(0) = 0, \quad y''(0) = 0$$
$$y(\ell) = 0, \quad y''(\ell) = 0$$

(3.) Free end, cantilever beam
$$y(0) = 0, \quad y'(0) = 0$$
$$y''(\ell) = 0, \quad y'''(\ell) = 0$$

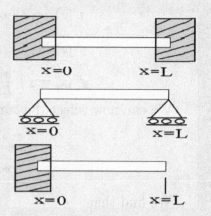

Example 7-3. (Bending of beams.)
Consider a cantilever beam with a uniform load w_0 and a concentrated load P_0 acting at $x = \ell/2$ as illustrated. We represent the concentrated load by a Dirac delta function and solve

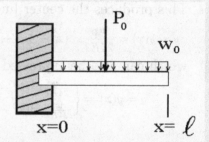

$$\frac{d^4y}{dx^4} = \frac{w_0}{EI} + \frac{P_0}{EI}\delta(x - \ell/2), \quad 0 < x < \ell$$
$$y(0) = 0, \quad y'(0) = 0$$
$$y''(\ell) = 0, \quad y'''(\ell) = 0.$$

We use Laplace transforms and solve the above differential equation over the interval $0 < x < \infty$. Let $\widetilde{y} = \mathcal{L}\{y(x); x \to s\}$ denote the Laplace transform of the beam deflection. We take the Laplace transform of the above differential equation to obtain

$$s^4\widetilde{y} - s^3 y(0) - s^2 y'(0) - sy''(0) - y'''(0) = \frac{w_0}{EIs} + \frac{P_0}{EI}e^{-\ell s/2}. \quad (7.26)$$

We do not know the initial conditions $y''(0)$ and $y'''(0)$ and so we denote these values by unknown constants c_1 and c_2. The equation (7.26) is an algebraic equation and so one can solve for the transform variable $\widetilde{y}$ and find

$$\widetilde{y} = \frac{c_1}{s^3} + \frac{c_2}{s^4} + \frac{w_0}{EIs^5} + \frac{P_0}{EIs^4}e^{-\ell s/2}. \quad (7.27)$$

We take the inverse Laplace transform of the equation (7.27) and obtain the solution in terms of the unknown constants c_1 and c_2. This solution is represented

$$y(x) = \mathcal{L}^{-1}\{\widetilde{y}; s \to x\} = c_1\frac{x^2}{2!} + c_2\frac{x^3}{3!} + \frac{w_0}{EI}\frac{x^4}{4!} + \frac{P_0}{EI}\frac{(x - \ell/2)^3}{3!}H(x - \ell/2). \quad (7.28)$$

This solution has the derivatives

$$y'(x) = c_1 x + c_2 \frac{x^2}{2!} + \frac{w_0}{EI} \frac{x^3}{3!} + \frac{P_0}{EI} \frac{(x - \ell/2)^2}{2!}, \qquad \ell/2 < x < \ell$$

$$y''(x) = c_1 + c_2 x + \frac{w_0}{EI} \frac{x^2}{2!} + \frac{P_0}{EI}(x - \ell/2), \qquad \ell/2 < x < \ell$$

$$y'''(x) = c_2 + \frac{w_0}{EI} x + \frac{P_0}{EI}, \qquad \ell/2 < x < \ell$$

We can now select the constants c_1 and c_2 to satisfy the boundary conditions

$$y''(\ell) = c_1 + c_2 \ell + \frac{w_0}{EI} \frac{\ell^2}{2!} + \frac{P_0}{EI} \ell/2 = 0$$

$$y'''(\ell) = c_2 + \frac{w_0}{EI} \ell + \frac{P_0}{EI} = 0$$

We find that

$$c_1 = \frac{w_0}{EI} \frac{\ell^2}{2} + \frac{P_0}{EI} \frac{\ell}{2} \qquad \text{and} \qquad c_2 = -\frac{w_0}{EI} \ell - \frac{P_0}{EI} \tag{7.29}$$

This produces the center line deflection

$$y = y(x) = \frac{w_0}{24EI} x^2 (x^2 - 4\ell x + 6\ell^2) + \frac{P_0}{12EI} x^2 (3\ell - 2x) + \frac{P_0}{6EI}(x - \ell/2)^3 H(x - \ell/2) \tag{7.30}$$

which can also be expressed in the form

$$y = y(x) = \begin{cases} \frac{w_0}{24EI} x^2 (x^2 - 4\ell x + 6\ell^2) + \frac{P_0}{12EI}(3\ell - 2x), & 0 < x < \ell/2 \\ \frac{w_0}{24EI} x^2 (x^2 - 4\ell x + 6\ell^2) + \frac{P_0}{48EI} \ell^2 (\ell - 6x), & \ell/2 < x < \ell. \end{cases} \tag{7.31}$$

∎

The Fourier Exponential Transform

Consider the Fourier integral formula from equation (3.77) which can be written in the form

$$f(x) = \int_{-\infty}^{\infty} e^{-i\omega x} \left[\frac{1}{2\pi} \int_{-\infty}^{\infty} f(\xi) e^{i\omega \xi} \, d\xi \right] d\omega. \tag{7.32}$$

The Fourier integral formula given by equation (7.32) can be written as the Fourier exponential transform pair

$$\mathcal{F}_e\{f(x); x \to \omega\} = F(\omega) = \frac{1}{2\pi} \int_{-\infty}^{\infty} f(\xi) e^{i\omega \xi} \, d\xi$$

$$\mathcal{F}_e^{-1}\{F(\omega); \omega \to x\} = f(x) = \int_{-\infty}^{\infty} F(\omega) e^{-i\omega x} \, d\omega \tag{7.33}$$

This is the transform pair we will use in this text. All of the tables and properties of Fourier exponential transforms that we develop will come from the above definitions.

Other Definitions

We divert from the above definition and point out that there are many different ways to define a Fourier transform pair. These other definitions can make the subject of Fourier transforms very confusing to beginning students when moving back and forth between different textbooks. We present four ways of defining a Fourier transform pair.

1. The symmetric form[1] for the Fourier transform pair is defined

$$\mathcal{F}_1\{f_1(x); x \to \omega\} = F_1(\omega) = \frac{1}{\sqrt{2\pi}} \int_{-\infty}^{\infty} f_1(x)e^{i\omega x}\, dx$$

$$\mathcal{F}_1^{-1}\{F_1(\omega); \omega \to x\} = f_1(x) = \frac{1}{\sqrt{2\pi}} \int_{-\infty}^{\infty} F_1(\omega)e^{-i\omega x}\, d\omega$$

(7.34)

2. Another form[2] for the definition of the Fourier transform is

$$\mathcal{F}_2\{f_2(x); x \to y\} = F_2(y) = \int_{-\infty}^{\infty} f_2(x)e^{ixy}\, dx$$

$$\mathcal{F}_2^{-1}\{F_2(y); y \to x\} = f_2(x) = \frac{1}{2\pi} \int_{-\infty}^{\infty} F_2(y)e^{-ixy}\, dy$$

(7.35)

3. From the Bateman[†] project we have the Fourier transform definition[3]

$$\mathcal{F}_3\{f_3(x); x \to p\} = F_3(p) = \int_{-\infty}^{\infty} f_3(x)e^{-ipx}\, dx$$

$$\mathcal{F}_3^{-1}\{F_3(p); p \to x\} = f_3(x) = \frac{1}{2\pi} \int_{-\infty}^{\infty} F_3(p)e^{ipx}\, dp$$

(7.36)

At first glance, when comparing equations (7.36) with those of equations (7.35) and (7.34), there seems to be a sign error. However, the definitions given by equations (7.36) are correct.

4. The Campbell and Foster form[4] for the definition of a Fourier transform is

$$\mathcal{F}_4\{f_4(g); g \to f\} = F_4(f) = \int_{-\infty}^{\infty} f_4(g)e^{-i2\pi fg}\, dg$$

$$\mathcal{F}_4^{-1}\{F_4(f); f \to g\} = f_4(g) = \int_{-\infty}^{\infty} F_4(f)e^{i2\pi fg}\, df$$

(7.37)

[1] I.N. Sneddon, The Use of Integral Transforms, McGraw Hill Book Co., N.Y., 1951.

[2] W.Magnus, F. Oberhettinger, Formulas and Theorems for the Functions of Mathematical Physics, Chelsea Publishing Co., N.Y., 1954.

[†] See Appendix C

[3] Erdelyi,A.(Ed.) Tables of Integral Transforms, 2 volumes, Bateman Manuscript Project, McGraw Hill Book Co.,N.Y.,1954.

[4] G.A. Campbell, R.M. Foster, Fourier Integrals for Practical Application, John Wiley & Sons, N.Y., 1948

These other definitions are found scattered among various physics, mathematics, engineering texts and reference books. Most of the definitions result from the Fourier integral representation given by the equation (7.32) when it is written in the form

$$\mathcal{F}_5\{f_5(x); x \to \omega\} = F_5(\omega) = \frac{\beta}{2\pi}\int_{-\infty}^{\infty} f_5(x)e^{i\omega x}\,dx$$

$$\mathcal{F}_5^{-1}\{F_5(\omega); \omega \to x\} = f_5(x) = \frac{1}{\beta}\int_{-\infty}^{\infty} F_5(\omega)e^{-i\omega x}\,d\omega$$

(7.38)

where β can be any constant factor which satisfies $\frac{\beta}{2\pi}\frac{1}{\beta} = \frac{1}{2\pi}$. For example, the symmetric definition above uses $\beta = \sqrt{2\pi}$. The definition 2 above uses $\beta = 2\pi$. By making appropriate changes of the dummy variables of integration one can relate all of the above definitions to the definition we will be using in this text.

Example 7-4. (Relation between Fourier transform definitions.)

How are the Fourier transform definitions given by equations (7.34) and (7.36) related? Explain the difference using the following tabular values.

Table A		Table B	
Fourier Transform		Fourier Transform	
Symmetric Form		Bateman Form	
$f_1(x)$	$F_1(p)$	$f_3(x)$	$F_3(p)$
$\dfrac{\sin(ax)}{x}$	$\sqrt{\frac{\pi}{2}}, \quad \lvert p\rvert < a$ $0, \quad \lvert p\rvert > a$	$\dfrac{1}{2}, \quad \lvert x\rvert < a$ $0, \quad \lvert x\rvert > a$	$\dfrac{\sin(ap)}{p}$
$\sqrt{\frac{\pi}{2}}, \quad \lvert x\rvert < a$ $0, \quad \lvert x\rvert > a$	$\dfrac{\sin(ap)}{p}$	$\dfrac{\sin(ax)}{x}$	$\pi, \quad \lvert p\rvert < a$ $0, \quad \lvert p\rvert > a$

Solution: Assume that we use different transforms on the same function $f(x)$. The symmetric definition would produce the equations

$$\mathcal{F}_1\{f(x)\} = F_1(p) = \frac{1}{\sqrt{2\pi}}\int_{-\infty}^{\infty} f(x)e^{ipx}\,dx$$

(7.39)

$$\mathcal{F}_1^{-1}\{F_1(p)\} = f(x) = \frac{1}{\sqrt{2\pi}}\int_{-\infty}^{\infty} F_1(p)e^{-ipx}\,dp$$

(7.40)

and the Bateman definition would produce

$$\mathcal{F}_3\{f(x)\} = F_3(p) = \int_{-\infty}^{\infty} f(x)e^{-ipx}\,dx \qquad (7.41)$$

$$\mathcal{F}_3^{-1}\{F_3(p)\} = f(x) = \frac{1}{2\pi}\int_{-\infty}^{\infty} F_3(p)e^{ipx}\,dp. \qquad (7.42)$$

Note 1: In equation (7.40) replace x by $-x$ to obtain

$$f(-x) = \frac{1}{\sqrt{2\pi}}\int_{-\infty}^{\infty} F_1(p)e^{ipx}\,dp. \qquad (7.43)$$

Now interchange the variables x and p in equation (7.43) to obtain

$$f(-p) = \frac{1}{\sqrt{2\pi}}\int_{-\infty}^{\infty} F_1(x)e^{ixp}\,dx = \mathcal{F}_1\{F_1(x)\}. \qquad (7.44)$$

The equation (7.44) tells us the table A can be used in two ways. That is,

$$\text{if} \quad \mathcal{F}_1\{f(x)\} = F_1(p), \quad \text{then} \quad \mathcal{F}_1\{F_1(x)\} = f(-p). \qquad (7.45)$$

Note 2: A similar result holds for the Bateman definition of the Fourier transform. Observe that if we replace x by $-x$ in equation (7.42) we get

$$f(-x) = \frac{1}{2\pi}\int_{-\infty}^{\infty} F_3(p)e^{-ipx}\,dx. \qquad (7.46)$$

Now interchange the variables x and p in equation (7.46) to get

$$f(-p) = \frac{1}{2\pi}\int_{-\infty}^{\infty} F_3(x)e^{-ixp}\,dp$$

$$\text{or} \quad 2\pi f(-p) = \int_{-\infty}^{\infty} F_3(x)e^{-ixp}\,dp = \mathcal{F}_3\{F_3(x)\}. \qquad (7.47)$$

The equation (7.47) tells us the table B can be used in two ways. That is,

$$\text{if} \quad \mathcal{F}_3\{f(x)\} = F_3(p), \quad \text{then} \quad \mathcal{F}_3\{F_3(x)\} = 2\pi f(-p). \qquad (7.48)$$

Note 3: The relation between the transform definitions is achieved by making appropriate notation changes. For example, in equations (7.41) and (7.39) we observe that for transforms associated with the same function $f(x)$

$$F_3(-p) = \int_{-\infty}^{\infty} f(x)e^{ipx}\,ds = \sqrt{2\pi}F_1(p). \qquad (7.49)$$

Thus, from table A we find

$$\mathcal{F}_1\{\frac{\sin(ax)}{x}\} = F_1(p) = \begin{cases} \sqrt{\frac{\pi}{2}}, & |p| < a \\ 0, & |p| > a \end{cases}$$

so that

$$\sqrt{2\pi}F_1(-p) = F_3(p) = \begin{cases} \pi, & |p| < a \\ 0, & |p| > a \end{cases}.$$

These properties can be discerned by examining the above tables.

■

Fourier Transforms Continued

We will now concentrate upon developing the Fourier exponential transform properties associated with the defining equations

$$\mathcal{F}_e\{f(x); x \to \omega\} = F(\omega) = \frac{1}{2\pi} \int_{-\infty}^{\infty} f(\xi) e^{i\omega\xi} \, d\xi$$

$$\mathcal{F}_e^{-1}\{F(\omega); \omega \to x\} = f(x) = \int_{-\infty}^{\infty} F(\omega) e^{-i\omega x} \, d\omega \tag{7.50}$$

previously cited in equations (7.33). This Fourier transform pair comes from the Fourier integral formula considered earlier. Recall that for the Fourier integral formula to exist the function $f(x)$ must be sectionally continuous and the absolute integral $\int_{-\infty}^{\infty} |f(x)| \, dx$ must exist.

The table 7-3 lists some important operational properties associated with the Fourier exponential transform while the table 7-4 is a short table of Fourier exponential transforms.

The Fourier transform and the Laplace transform are related. In the special case $f(t) = 0$ for $t < 0$, the Fourier transform becomes $\frac{1}{2\pi}$ times the Laplace transform with transform parameter $-i\omega$ instead of s. This is equivalent to restricting the Laplace transform to have pure imaginary parameters. This restricts the class of functions for which the Laplace transform exist. It is more desirable to expand the class of functions for which a transform exits. For many functions the condition $\int_{-\infty}^{\infty} |f(x)| \, dx < M$ of absolute integrability is not satisfied and so a Fourier transform does not exist. One possible way around such a restriction is to define the Fourier transform in the following limiting sense. If $\mathcal{F}_e\{f(x)\}$ does not exist, then consider the limiting processes

$$\mathcal{F}_e\{f(x)\} = \lim_{\substack{\beta \to 0 \\ \beta > 0}} \mathcal{F}_e\{f(x) e^{-\beta|x|}\}. \tag{7.51}$$

For example, using this type of limiting process we obtain from the Laplace transform table

$$\mathcal{L}\{e^{-\beta t} \sin \alpha t\} = \frac{\alpha}{(s+\beta)^2 + \alpha^2}.$$

Therefore, the Fourier exponential transform can be written

$$\mathcal{F}_e\{e^{-\beta x} \sin \alpha x \, H(x)\} = \frac{1}{2\pi} \frac{\alpha}{(-i\omega + \beta)^2 + \alpha^2}$$

so that

$$\lim_{\beta \to 0} \mathcal{F}_e\{e^{-\beta x} \sin \alpha x \, H(x)\} = \frac{1}{2\pi} \frac{\alpha}{(-i\omega)^2 + \alpha^2} = \mathcal{F}_e\{\sin \alpha x \, H(x)\}.$$

Similarly one can show

$$\mathcal{F}_e\{\cos \alpha x \, H(x)\} = \frac{1}{2\pi} \frac{-i\omega}{(-i\omega)^2 + \alpha^2}.$$

The proof is left as an exercise.

The property 1 from table 7-3 states that the columns in a Fourier exponential transform table can be interchanged if proper changes of variables are made. That is, if

$$F(\omega) = \mathcal{F}_e\{f(x)\} = \frac{1}{2\pi} \int_{-\infty}^{\infty} f(x) e^{i\omega x} \, dx, \qquad (7.52)$$

then upon replacing ω by $-x$ and x by ω simultaneously in the equation (7.52) we obtain

$$2\pi F(-x) = \int_{-\infty}^{\infty} f(\omega) e^{-ix\omega} \, d\omega = \mathcal{F}_e^{-1}\{f(\omega)\}. \qquad (7.53)$$

The equation (7.53) is the column interchange property given as the first entry of table 7-3. See also the entries labeled 1 and 2 in the table 7-4 for an example of this interchange property.

The property 2 in table 7-3 follows from the definition of a Fourier exponential transform since the integral operator is a linear operator. The property 3 in table 7-3 is a differentiation property of the Fourier exponential transform. If $\mathcal{F}_e\{f(x)\} = F(\omega)$, then $\mathcal{F}_e\{f'(x)\} = -i\omega F(\omega)$. This property follows from the definition

$$\mathcal{F}_e\{f'(x)\} = \frac{1}{2\pi} \int_{-\infty}^{\infty} f'(x) e^{i\omega x} \, dx$$

which we integrate by parts to obtain

$$\mathcal{F}_e\{f'(x)\} = \frac{1}{2\pi} \left[f(x) e^{i\omega x} \right]_{-\infty}^{\infty} - (i\omega) \int_{-\infty}^{\infty} f(x) e^{i\omega x} \, dx = -i\omega \mathcal{F}_e\{f(x)\} = -i\omega F(\omega).$$

Here we assume $\lim_{x \to \pm\infty} f(x) = 0$ to obtain the above result.

Table 7-3. Fourier Exponential Transform Properties

$f(x) = \int_{-\infty}^{\infty} F(\omega)e^{-i\omega x}\,d\omega = \mathcal{F}_e^{-1}\{F(\omega)\}$	$F(\omega) = \frac{1}{2\pi}\int_{-\infty}^{\infty} f(x)e^{i\omega x}\,dx = \mathcal{F}_e\{f(x)\}$	
	$f(x) = \mathcal{F}_e^{-1}\{F(\omega)\}$	$F(\omega) = \mathcal{F}_e\{f(x)\}$	Comments
1.	$2\pi F(-x)$	$f(\omega)$	Column interchange
2.	$c_1 f(x) + c_2 g(x)$	$c_1 F(\omega) + c_2 G(\omega)$	Linearity property
3.	$f'(x)$	$-i\omega F(\omega)$	Derivative property
4.	$f''(x)$	$(-i\omega)^2 F(\omega)$	
5.	$f^{(n)}(x)$	$(-i\omega)^n F(\omega)$	
6.	$f(x-\alpha)$	$e^{i\omega\alpha} F(\omega)$	Shift property
7.	$xf(x)$	$-i\dfrac{dF}{d\omega} = -iF'(\omega)$	Multiplication by x property
8.	$x^n f(x)$	$(-i)^n \dfrac{d^n F(\omega)}{d\omega^n}$	
9.	$f{*}g = \dfrac{1}{2\pi}\displaystyle\int_{-\infty}^{\infty} f(\tau)g(x-\tau)\,d\tau$	$F(\omega)G(\omega)$	Convolution property
10.	$\delta(x-x_0)$	$\dfrac{1}{2\pi}e^{i\omega x_0}$	Dirac delta function
11.	$f(ax),\quad a>0$	$\dfrac{1}{a}F\left(\dfrac{\omega}{a}\right)$	Scaling property
12.	$f(ax)e^{ibx},\ a>0$	$\dfrac{1}{a}F\left(\dfrac{\omega+b}{a}\right)$	Shift and scaling
13.	$f(ax)\cos bx$	$\dfrac{1}{2a}\left[F\left(\dfrac{\omega+b}{a}\right) + F\left(\dfrac{\omega-b}{a}\right)\right]$	
14.	$f(ax)\sin bx$	$\dfrac{1}{2ia}\left[F\left(\dfrac{\omega+b}{a}\right) - F\left(\dfrac{\omega-b}{a}\right)\right]$	
15.	$f(x)e^{iax}$	$F(\omega+a)$	Shift property

Table 7-4. Fourier Exponential Transforms

	$f(x) = \mathcal{F}_e^{-1}\{F(\omega)\} = \int_{-\infty}^{\infty} F(\omega)e^{-i\omega x}\,d\omega$	$F(\omega) = \mathcal{F}_e\{f(x)\} = \frac{1}{2\pi}\int_{-\infty}^{\infty} f(x)e^{i\omega x}\,dx$
1.	$e^{-\alpha|x|}$	$\dfrac{\alpha}{\pi(\alpha^2 + \omega^2)}$
2.	$\dfrac{2\alpha}{\alpha^2 + x^2}$	$e^{-\alpha|\omega|}$
3.	$\begin{cases} 1, & |x| < L \\ 0, & |x| > L \end{cases}$	$\dfrac{1}{\pi}\dfrac{\sin \omega L}{\omega}$
4.	$e^{-\alpha x}H(x)$	$\dfrac{1}{2\pi(a - i\omega)}$
5.	$e^{-\alpha x^2}$	$\dfrac{1}{2\sqrt{\pi\alpha}}e^{-\omega^2/4\alpha}$
6.	$\operatorname{sgn}(x) = \begin{cases} 1, & x > 0 \\ 0, & x = 0 \\ -1, & x < 0 \end{cases}$	$\dfrac{i}{\pi\omega}$
7.	$\dfrac{\sinh ax}{\sinh \pi x}, \quad -\pi < a < \pi$	$\dfrac{1}{2\pi}\dfrac{\sin a}{\cosh \omega + \cos a}$
8.	$\dfrac{x}{x^2 + \alpha^2}$	$\dfrac{i}{2}e^{-\alpha|\omega|}\operatorname{sgn}(\omega)$
9.	$\sin \alpha x\, H(x)$	$\dfrac{1}{2\pi}\dfrac{\alpha}{(-i\omega)^2 + \alpha^2}$
10.	$\cos \alpha x\, H(x)$	$\dfrac{1}{2\pi}\dfrac{-i\omega}{(-i\omega)^2 + \alpha^2}$
11.	$\dfrac{2\alpha x}{(x^2 + \alpha^2)^2}$	$\dfrac{i\omega}{2}e^{-\alpha|\omega|}$
12.	$\dfrac{1}{|x|}$	$\dfrac{1}{\sqrt{2\pi}}\dfrac{1}{|\omega|}$
13.	$xe^{-\alpha|x|} \quad \alpha > 0$	$\dfrac{2}{\pi}\dfrac{2\alpha i\omega}{(\alpha^2 + \omega^2)^2} \quad \omega > 0$
14.	$\dfrac{2\sin \alpha x}{x}$	$\begin{cases} 1 & |\omega| < \alpha \\ 0 & |\omega| > \alpha \end{cases}$
15.	$\operatorname{erf}\left(\dfrac{x}{2\sqrt{Kt}}\right)$	$\dfrac{i}{\pi\omega}e^{-Kt\omega^2}$
16.	$\begin{cases} (a^2 - x^2)^{-1/2}, & |x| < a \\ 0, & |x| > a \end{cases}$	$\dfrac{1}{2}J_0(a\omega)$

The properties 4 and 5 in table 7-3 follow from the property 3. That is, each differentiation in the x-domain corresponds to a multiplication by $-i\omega$ in the transform domain. The property 6 is a shifting property. If $\mathcal{F}_e\{f(x)\} = F(\omega)$, then $\mathcal{F}_e\{f(x-\alpha)\} = e^{i\omega\alpha}F(\omega)$. This also follows from the definition since

$$\mathcal{F}_e\{f(x)\} = F(\omega) = \frac{1}{2\pi}\int_{-\infty}^{\infty} f(x)e^{i\omega x}\,dx$$

so that

$$e^{i\omega\alpha}F(\omega) = \frac{1}{2\pi}\int_{-\infty}^{\infty} f(x)e^{i\omega x}e^{i\omega\alpha}\,dx = \frac{1}{2\pi}\int_{-\infty}^{\infty} f(x)e^{i\omega(x+\alpha)}\,dx.$$

Now make the change of variable $\xi = x + \alpha$, or $x = \xi - \alpha$ with $dx = d\xi$ to obtain

$$e^{i\omega\alpha}F(\omega) = \frac{1}{2\pi}\int_{-\infty}^{\infty} f(\xi-\alpha)e^{i\omega\xi}\,d\xi = \mathcal{F}_e\{f(x-\alpha)\}.$$

In a similar fashion the property 12 of table 7-3 is established. If $\mathcal{F}_e\{f(x)\} = F(\omega)$, then $\mathcal{F}_e\{f(ax)e^{ibx}\} = \frac{1}{a}F\left(\frac{\omega+b}{a}\right)$. The proof follows from the definition. If

$$F(\omega) = \frac{1}{2\pi}\int_{-\infty}^{\infty} f(x)e^{i\omega x}\,dx, \qquad \text{then}$$

$$\frac{1}{a}F\left(\frac{\omega+b}{a}\right) = \frac{1}{2\pi a}\int_{-\infty}^{\infty} f(x)e^{i\left(\frac{\omega+b}{a}\right)x}\,dx$$

where the change of variable $x = a\xi$, with $dx = a\,d\xi$ produces the result

$$\frac{1}{a}F\left(\frac{\omega+b}{a}\right) = \frac{1}{2\pi}\int_{-\infty}^{\infty} f(a\xi)e^{ib\xi}e^{i\omega\xi}\,d\xi = \mathcal{F}_e\{f(ax)e^{ibx}\}.$$

Note entry number 11 of table 7-3 is a special case of the above scaling and shifting property.

The entry number 7 in table 7-3 states that a multiplication by x in the x-domain corresponds to a differentiation with respect to ω in the transform domain followed by a (-i) multiplication. Again this result follows from the definition. If

$$F(\omega) = \mathcal{F}_e\{f(x)\} = \frac{1}{2\pi}\int_{-\infty}^{\infty} f(x)e^{i\omega x}\,dx, \qquad \text{then}$$

$$\frac{dF(\omega)}{d\omega} = \frac{1}{2\pi}\int_{-\infty}^{\infty} f(x)\frac{d}{d\omega}\left(e^{i\omega x}\right)\,dx = \frac{1}{2\pi}\int_{-\infty}^{\infty} f(x)(ix)e^{i\omega x}\,dx$$

or $\qquad -i\dfrac{dF(\omega)}{d\omega} = \dfrac{1}{2\pi}\displaystyle\int_{-\infty}^{\infty} xf(x)e^{i\omega x}\,dx = \mathcal{F}_e\{xf(x)\}.$

The property number 8 results from a repeated application of the property 7. The property 10 of table 7-3 follows from the sifting property of the Dirac delta function. The property 9 is a convolution property of the Fourier exponential transform where the convolution of two functions $f(x)$ and $g(x)$ over the domain $-\infty < x < \infty$ is defined

$$f(x) * g(x) = \frac{1}{2\pi} \int_{-\infty}^{\infty} f(\tau) g(x - \tau) \, d\tau.$$

The property 9 states that if $\mathcal{F}_e\{f(x)\} = F(\omega)$ and $\mathcal{F}_e\{g(x)\} = G(\omega)$, then

$$\mathcal{F}_e^{-1}\{F(\omega)G(\omega)\} = f(x)^* g(x) = \frac{1}{2\pi} \int_{-\infty}^{\infty} f(\tau) g(x - \tau) \, d\tau.$$

To prove this property, let $H(\omega) = F(\omega)G(\omega)$ so that

$$h(x) = \mathcal{F}_e^{-1}\{H(\omega)\} = \int_{-\infty}^{\infty} H(\omega) e^{-i\omega x} \, d\omega = \int_{-\infty}^{\infty} F(\omega)G(\omega) e^{-i\omega x} \, d\omega. \tag{7.54}$$

We substitute into the equation (7.54) the definition of $G(\omega)$ given by

$$G(\omega) = \mathcal{F}_e\{g(x)\} = \frac{1}{2\pi} \int_{-\infty}^{\infty} g(\xi) e^{i\omega\xi} \, d\xi.$$

This produces

$$h(x) = \mathcal{F}_e^{-1}\{H(\omega)\} = \int_{-\infty}^{\infty} F(\omega) \frac{1}{2\pi} \int_{-\infty}^{\infty} g(\xi) e^{i\omega\xi} \, d\xi \, e^{-i\omega x} \, d\omega$$

where we can change the order of integration and write

$$h(x) = \mathcal{F}_e^{-1}\{H(\omega)\} = \frac{1}{2\pi} \int_{-\infty}^{\infty} g(\xi) \left[\int_{-\infty}^{\infty} F(\omega) e^{-i\omega(x - \xi)} \, d\omega \right] d\xi$$
$$= \frac{1}{2\pi} \int_{-\infty}^{\infty} g(\xi) f(x - \xi) \, d\xi$$
$$= g(x) * f(x)$$

The change of variable $x - \xi = r$ with $\xi = x - r$ and $d\xi = -dr$ further reduces the above result to the form

$$h(x) = \mathcal{F}_e^{-1}\{H(\omega)\} = \frac{1}{2\pi} \int_{\infty}^{-\infty} g(x - r) f(r)(-dr) = \frac{1}{2\pi} \int_{-\infty}^{\infty} g(x - r) f(r) \, dr = f(x) * g(x).$$

Note that the convolution is commutative since $f^* g = g^* f$.

Example 7-5. (Heat equation.)

Solve the heat equation $\frac{\partial u}{\partial t} = K\frac{\partial^2 u}{\partial x^2}$ for the temperature $u = u(x,t)$ over an infinite rod $-\infty < x < \infty$ subject to the initial condition $u(x,0) = \delta(x - x_0)$. This represents diffusion of an initial "squirt" of heat injected into the rod at a point x_0 at time $t = 0$.

Solution: The infinite domain over which the solution is desired suggests a Fourier exponential transform. Define the Fourier exponential transform of the temperature

$$\widetilde{u}(\omega,t) = \mathcal{F}_e\{u(x,t); x \to \omega\} = \frac{1}{2\pi}\int_{-\infty}^{\infty} u(x,t)e^{i\omega x}\,dx$$

and take the Fourier exponential transform of the partial differential equation and initial condition to obtain

$$\mathcal{F}_e\{\frac{\partial u}{\partial t}\} = \mathcal{F}_e\{K\frac{\partial^2 u}{\partial x^2}\}$$
$$\mathcal{F}_e\{u(x,0)\} = \mathcal{F}_e\{\delta(x - x_0)\}.$$

Interchange the order of integration and differentiation and write

$$\frac{\partial \mathcal{F}_e\{u\}}{\partial t} = K\mathcal{F}_e\{\frac{\partial^2 u}{\partial x^2}\}$$
$$\frac{\partial \widetilde{u}}{\partial t} = K(-i\omega)^2\widetilde{u} = -K\omega^2\widetilde{u}, \qquad \widetilde{u}(\omega,0) = \frac{1}{2\pi}e^{i\omega x_0} \tag{7.55}$$

Here the Fourier exponential transform of the original partial differential equation plus initial condition produced an ordinary differential equation with initial condition in the transform space. The solution of the ordinary differential equation plus initial condition is given by

$$\widetilde{u} = \widetilde{u}(\omega,t) = \frac{1}{2\pi}e^{i\omega x_0}e^{-K\omega^2 t}.$$

The inverse Fourier exponential transform gives

$$u(x,t) = \mathcal{F}_e^{-1}\{\widetilde{u}(\omega,t)\} = \int_{-\infty}^{\infty} \frac{1}{2\pi}e^{i\omega x_0}e^{-K\omega^2 t}e^{-i\omega x}\,d\omega$$

or

$$u(x,t) = \frac{1}{2\pi}\int_{-\infty}^{\infty} e^{-K\omega^2 t}e^{i\omega(x-x_0)}\,d\omega.$$

From the table 7-4 we find

$$e^{-\alpha x^2} = \int_{-\infty}^{\infty} \frac{1}{2\sqrt{\pi\alpha}}e^{-\omega^2/4\alpha}e^{-i\omega x}\,d\omega. \tag{7.56}$$

Let $\dfrac{1}{4\alpha} = Kt$ or $\alpha = 1/4Kt$ then the integral given by equation (7.56) becomes

$$\frac{1}{\sqrt{4\pi Kt}}e^{-x^2/4Kt} = \frac{1}{2\pi}\int_{-\infty}^{\infty} e^{-K\omega^2 t}e^{-i\omega x}\,d\omega = g(x) \tag{7.57}$$

Replacing x by $x - x_0$ gives

$$g(x - x_0) = \frac{1}{\sqrt{4\pi Kt}}e^{-(x-x_0)^2/4Kt} = \frac{1}{2\pi}\int_{-\infty}^{\infty} e^{-K\omega^2 t}e^{-i\omega(x-x_0)}\,d\omega.$$

The solution of the original problem can now be expressed in the form

$$u = u(x,t) = \frac{1}{\sqrt{4\pi Kt}}e^{-(x-x_0)^2/4Kt}.$$

A graph of the solution response function due to a Dirac delta function initial condition is plotted for various values of t in the figure 7-8. Note also this result is a special case of equation (4.134) considered earlier.

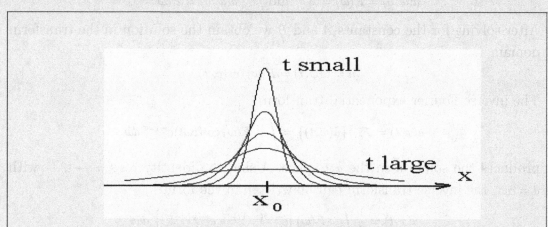

Figure 7-8. Response function due to Dirac delta function initial condition.

■

Example 7-6.   (Wave equation.)

   Solve the one-dimensional wave equation for the displacement $u = u(x,t)$ in an infinite string modeled by the wave equation $\dfrac{\partial^2 u}{\partial t^2} = c^2\dfrac{\partial^2 u}{\partial x^2}$, $-\infty < x < \infty$ and subject to the initial conditions $u(x,0) = f(x)$ and $\dfrac{\partial u(x,0)}{\partial t} = 0$.

Solution: Let $\widetilde{u} = \widetilde{u}(\omega,t) = \mathcal{F}_e\{u(x,t); x \to \omega\} = \dfrac{1}{2\pi}\displaystyle\int_{-\infty}^{\infty} u(x,t)e^{i\omega x}\,dx$, denote the

Fourier exponential transform of $u(x,t)$ and take the Fourier exponential transform of the partial differential equation and initial conditions to obtain

$$\mathcal{F}_e\{\frac{\partial^2 u}{\partial t^2}\} = \mathcal{F}_e\{c^2\frac{\partial^2 u}{\partial x^2}\}$$

$$\frac{\partial^2 \mathcal{F}_e\{u\}}{\partial t^2} = c^2\mathcal{F}_e\{\frac{\partial^2 u}{\partial x^2}\}$$

$$\frac{d^2\widetilde{u}}{dt^2} = c^2(-i\omega)^2\widetilde{u} = -c^2\omega^2\widetilde{u}$$

$$\mathcal{F}_e\{u(x,0)\} = \mathcal{F}_e\{f(x)\} = F(\omega) = \widetilde{u}(\omega,0)$$

$$\mathcal{F}_e\{\frac{\partial u(x,0)}{\partial t}\} = \frac{d}{dt}\mathcal{F}_e\{u(x,0)\} = \frac{d}{dt}\widetilde{u}(\omega,0) = \mathcal{F}_e\{0\}$$

The Fourier exponential transform converts the partial differential equation into an ordinary differential equation with initial conditions. The differential equation in the transform space is easily solved to obtain

$$\widetilde{u} = \widetilde{u}(\omega,t) = A\cos(c\omega t) + B\sin(c\omega t)$$

with derivative $\dfrac{d\widetilde{u}}{dt} = -Ac\omega\sin(c\omega t) + Bc\omega\cos(c\omega t).$

The initial conditions require

$$\widetilde{u}(\omega,0) = F(\omega) = A \quad \text{and} \quad \frac{d}{dt}\widetilde{u}(\omega,0) = Bc\omega = 0.$$

After solving for the constants A and B we obtain the solution in the transform domain

$$\widetilde{u} = \widetilde{u}(\omega,t) = F(\omega)\cos(c\omega t).$$

The inverse Fourier exponential transform

$$u(x,t) = \mathcal{F}_e^{-1}\{\widetilde{u}(\omega,t)\} = \int_{-\infty}^{\infty} F(\omega)\cos(c\omega t)e^{-i\omega x}\,d\omega$$

produces the solution in the x,t space. Using the identity $\cos\theta = \frac{e^{i\theta}+e^{-i\theta}}{2}$ with $\theta = c\omega t$, the inverse transform can be written in the form

$$u(x,t) = \frac{1}{2}\int_{-\infty}^{\infty} F(\omega)\left[e^{-i\omega(x-ct)} + e^{-i\omega(x+ct)}\right]d\omega.$$

Now apply the shifting property of the Fourier exponential transform. If

$$f(x) = \int_{-\infty}^{\infty} F(\omega)e^{-i\omega x}\,d\omega,$$

then

$$f(x\pm ct) = \int_{-\infty}^{\infty} F(\omega)e^{-i\omega(x\pm ct)}\,d\omega.$$

The solution can therefore be written in the form

$$u(x,t) = \frac{1}{2}\left[f(x-ct) + f(x+ct)\right].$$

Example 7-7. (Steady state heat equation.)

Solve for the steady state temperature in the infinite strip $-\infty < x < \infty$ with $0 < y < y_0$ and subject to the boundary conditions $u(x, 0) = f(x)$ and $u(x, y_0) = g(x)$.
Solution: The given problem is modeled by the partial differential equation and boundary conditions

$$PDE: \qquad \frac{\partial^2 u}{\partial x^2} + \frac{\partial^2 u}{\partial y^2} = 0, \quad -\infty < x < \infty, \quad 0 < y < y_0$$

$$BC: \qquad u(x,0) = f(x), \quad u(x, y_0) = g(x) \quad -\infty < x < \infty$$

Note that for problems over an infinite domain there is always implied some kind of limit condition at infinity. Here we assume boundedness for these conditions. The geometry of the problem suggests a Fourier exponential transform and so we define

$$\widetilde{u} = \widetilde{u}(\omega, y) = \mathcal{F}_e\{u(x, y); x \to \omega\} = \frac{1}{2\pi} \int_{-\infty}^{\infty} u(x, y) e^{i\omega x}\, dx.$$

We take the Fourier exponential transform of the given partial differential equations and boundary conditions to obtain

$$\mathcal{F}_e\{\frac{\partial^2 u}{\partial x^2}\} + \mathcal{F}_e\{\frac{\partial^2 u}{\partial y^2}\} = \mathcal{F}_e\{0\}$$

$$\mathcal{F}_e\{u(x,0)\} = \mathcal{F}_e\{f(x)\}, \qquad \mathcal{F}_e\{u(x, y_0)\} = \mathcal{F}_e\{g(x)\}$$

Then interchange the order of differentiation and integration to obtain the transformation domain problem

$$\frac{d^2\widetilde{u}}{dy^2} - \omega^2 \widetilde{u} = 0$$

$$\widetilde{u}(\omega, 0) = F(\omega), \qquad \widetilde{u}(\omega, y_0) = G(\omega) \tag{7.58}$$

where $F(\omega) = \mathcal{F}_e\{f(x)\}$ and $G(\omega) = \mathcal{F}_e\{g(x)\}$. We write the solution in the transform domain as

$$\widetilde{u} = \widetilde{u}(\omega, y) = C_1 \sinh \omega y + C_2 \sinh \omega(y_0 - y) \tag{7.59}$$

where C_1, C_2 are arbitrary constants and must be chosen such that the solution satisfies the transformed boundary conditions. This form of the solution is suggested by the boundary conditions and will make algebra easier. We find C_1, C_2 must be selected to satisfy the conditions

$$\widetilde{u}(\omega, 0) = C_2 \sinh \omega y_0 = F(\omega)$$

$$\widetilde{u}(\omega, y_0) = C_1 \sinh \omega y_0 = G(\omega).$$

We solve for C_1, C_2 and obtain the solution

$$\widetilde{u} = \widetilde{u}(\omega, y) = G(\omega)\frac{\sinh \omega y}{\sinh \omega y_0} + F(\omega)\frac{\sinh \omega (y_0 - y)}{\sinh \omega y_0}. \tag{7.60}$$

Note that one could write the solution to the differential equation (7.58) in the alternate form

$$\widetilde{u} = \widetilde{u}(\omega, y) = A \sinh \omega y + B \cosh \omega y$$

where A, B are arbitrary constants. The boundary conditions then require A and B be selected to satisfy the conditions

$$\widetilde{u}(\omega, 0) = B = F(\omega)$$

$$\widetilde{u}(\omega, y_0) = A \sinh \omega y_0 + B \cosh \omega y_0 = G(\omega).$$

Solving for A, B gives the solution

$$\widetilde{u} = \widetilde{u}(\omega, y) = \left[\frac{G(\omega) - F(\omega) \cosh \omega y_0}{\sinh \omega y_0}\right] \sinh \omega y + F(\omega) \cosh \omega y.$$

Some algebra reduces this solution to the form given by equation (7.59). To avoid all the algebra, a little thought concerning the boundary conditions and an appropriate form for the solution to the differential equation can save you a lot of work.

The solution to the original partial differential equation and boundary conditions is obtained by taking the inverse transform of equation (7.60) and can be represented

$$u = u(x, y) = \mathcal{F}_e^{-1}\{\widetilde{u}(\omega, y)\} = \int_{-\infty}^{\infty} \widetilde{u}(\omega, y) e^{-i\omega x}\, d\omega$$

$$u = u(x, y) = \int_{-\infty}^{\infty} \left[G(\omega)\frac{\sinh \omega y}{\sinh \omega y_0} + F(\omega)\frac{\sinh \omega (y_0 - y)}{\sinh \omega y_0}\right] e^{-i\omega x}\, d\omega$$

A good exercise to test your knowledge of the Fourier transform properties is to use the convolution integral and express the above solution in the alternate form

$$u = u(x, y) = \frac{1}{2y_0}\int_{-\infty}^{\infty}\left[g(\tau)\frac{\sin \frac{\pi}{y_0} y}{\cosh \frac{\pi}{y_0}(x - \tau) + \cos \pi \frac{y}{y_0}} + f(\tau)\frac{\sin \frac{\pi}{y_0}(y_0 - y)}{\cosh \frac{\pi}{y_0}(x - \tau) + \cos \frac{\pi}{y_0}(y_0 - y)}\right] d\tau.$$

Example 7-8. (Steady state heat equation.)

Solve for the steady state temperature $u = u(x, y)$ for the half-plane $y > 0$ subject to the boundary condition $u(x, 0) = f(x)$. Evaluate the solution for the special case $f(x) = \begin{cases} 0, & x < 0 \\ 1, & x > 0 \end{cases}$

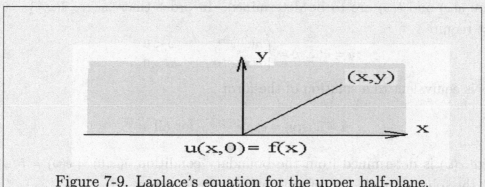

Figure 7-9. Laplace's equation for the upper half-plane.

Solution: The given problem is to solve

$$\nabla^2 u = \frac{\partial^2 u}{\partial x^2} + \frac{\partial^2 u}{\partial y^2} = 0, \quad y > 0, \qquad u(x, 0) = f(x), \quad -\infty < x < \infty$$

with the implied boundary conditions

$$\lim_{|x| \to \infty} u(x, y) = 0, \qquad \lim_{y \to \infty} u(x, y) = 0.$$

Define

$$\widetilde{u}(\omega, y) = \mathcal{F}_e\{u(x, y); x \to \omega\} = \frac{1}{2\pi} \int_{-\infty}^{\infty} u(x, y) e^{i\omega x} \, dx$$

as the Fourier exponential transform of $u(x, y)$ and then take the Fourier exponential transform of Laplace's equation and the boundary condition to obtain

$$\mathcal{F}_e\{\frac{\partial^2 u}{\partial x^2}\} + \mathcal{F}_e\{\frac{\partial^2 u}{\partial y^2}\} = \mathcal{F}_e\{0\}, \qquad \mathcal{F}_e\{u(x, 0)\} = \mathcal{F}_e\{f(x)\}.$$

From the table of properties for the Fourier exponential transform we find

$$\mathcal{F}_e\{\frac{\partial^2 u}{\partial x^2}\} = (-i\omega)^2 \widetilde{u}(\omega, y), \qquad \mathcal{F}_e\{\frac{\partial^2 u}{\partial y^2}\} = \frac{d^2}{dy^2} \mathcal{F}_e\{u\} = \frac{d^2 \widetilde{u}}{dy^2}$$

so that in the transform domain we are required to solve the problem

$$\frac{d^2 \widetilde{u}}{dy^2} - \omega^2 \widetilde{u} = 0$$

subject to the boundary conditions $\lim_{y \to \infty} \tilde{u}(\omega, y) = 0$ and $\tilde{u}(\omega, y) = F(\omega)$, where $F(\omega) = \mathcal{F}_e\{f(x)\}$. The solution in the transform domain is given by

$$\tilde{u} = \tilde{u}(\omega, y) = a(\omega)e^{\omega y} + b(\omega)e^{-\omega y}$$

where $a(\omega)$ and $b(\omega)$ are to be determined. In order that $\lim_{y \to \infty} \tilde{u}(\omega, y) = 0$ we must require

$$\tilde{u} = \tilde{u}(\omega, y) = \begin{cases} a(\omega)e^{\omega y}, & \omega < 0 \\ b(\omega)e^{-\omega y}, & \omega > 0 \end{cases}.$$

This is equivalent to a solution of the form

$$\tilde{u} = \tilde{u}(\omega, y) = c(\omega)e^{-|\omega|y} \quad \text{for all } \omega$$

where $c(\omega)$ is determined from the boundary condition $\tilde{u}(\omega, 0) = c(\omega) = F(\omega)$ so that the solution in the transform domain can be expressed

$$\tilde{u} = \tilde{u}(\omega, y) = F(\omega)e^{-|\omega|y}.$$

This solution is a product of two Fourier transforms and suggests that the convolution property for the inverse Fourier transform can be used to express the solution to the original problem. That is, if $\mathcal{F}_e^{-1}\{F(\omega)\} = f(x)$ and $\mathcal{F}_e^{-1}\{e^{-|\omega|y}\} = g(x, y)$, then

$$\mathcal{F}_e^{-1}\{\tilde{u}(\omega, y)\} = \mathcal{F}_e^{-1}\{F(\omega)e^{-|\omega|y}\} = \frac{1}{2\pi} \int_{-\infty}^{\infty} f(\xi)g(x - \xi, y)\, d\xi.$$

Note

$$\mathcal{F}_e^{-1}\{e^{-|\omega|y}\} = g(x, y) = \int_{-\infty}^{\infty} e^{-|\omega|y} e^{-i\omega x}\, d\omega$$

can be evaluated directly. We find

$$g(x, y) = \int_{-\infty}^{0} e^{\omega y} e^{-i\omega x}\, d\omega + \int_{0}^{\infty} e^{-\omega y} e^{-i\omega x}\, d\omega$$

$$g(x, y) = \frac{e^{\omega(y-ix)}}{y - ix}\Big|_{-\infty}^{0} + \frac{e^{-\omega(y+ix)}}{-(y + ix)}\Big|_{0}^{\infty}$$

$$g(x, y) = \frac{1}{y - ix} + \frac{1}{y + ix} = \frac{2y}{x^2 + y^2}$$

Therefore, the solution to the original problem can be represented

$$u(x, y) = \frac{1}{2\pi} \int_{-\infty}^{\infty} f(\xi) \frac{2y}{(x - \xi)^2 + y^2}\, d\xi. \tag{7.61}$$

In the special case $f(x) = \begin{cases} 0, & x < 0 \\ 1, & x > 0 \end{cases}$ the integral given by equation (7.61) reduces to

$$u(x,y) = \frac{1}{\pi} \int_0^\infty \frac{y}{(x-\xi)^2 + y^2} \, d\xi = \frac{1}{\pi} \arctan\left(\frac{\xi - x}{y}\right)\Big|_0^\infty = \frac{1}{\pi}\left[\frac{\pi}{2} + \arctan\left(\frac{x}{y}\right)\right].$$

It is left as an exercise to show that an alternate form of this solution is given by

$$u(x,y) = 1 - \frac{1}{\pi} \arctan\left(\frac{y}{x}\right).$$

Example 7-9. (Fluid flow.)

The two-dimensional irrotational flow of a perfect fluid with velocity components $\{u(x,y), v(x,y)\}$ in the x- and y-directions are represented by the Cauchy-Riemann equations

$$\frac{\partial u}{\partial y} - \frac{\partial v}{\partial x} = 0, \qquad \frac{\partial u}{\partial x} + \frac{\partial v}{\partial y} = 0.$$

Consider the problem of fluid flowing through a slot $|x| \le a$ into the upper half plane $y \ge 0$ such that on the boundary $y = 0$ we have

$$v(x,0) = f(x)H(a - |x|)$$

where $H(\xi)$ is the Heaviside step function. The situation is illustrated in the figure 7-10.

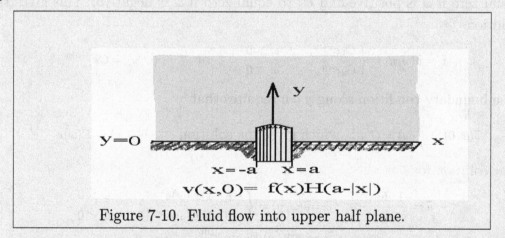

Figure 7-10. Fluid flow into upper half plane.

Solution: We assume that the velocity components u and v approach zero as y increases without bound. Define

$$\tilde{u} = \tilde{u}(\omega,y) = \mathcal{F}_e\{u(x,y); x \to \omega\}, \qquad \tilde{v} = \tilde{v}(\omega,y) = \mathcal{F}_e\{v(x,y); x \to \omega\}$$

and take the Fourier exponential transform of the above Cauchy-Riemann equations to obtain

$$\mathcal{F}_e\{\frac{\partial u}{\partial y}\} - \mathcal{F}_e\{\frac{\partial v}{\partial x}\} = \mathcal{F}_e\{0\} \qquad \text{and} \qquad \mathcal{F}_e\{\frac{\partial u}{\partial x}\} + \mathcal{F}_e\{\frac{\partial v}{\partial y}\} = \mathcal{F}_e\{0\}$$

$$\frac{d\widetilde{u}}{dy} + i\omega\widetilde{v} = 0 \qquad\qquad -i\omega\widetilde{u} + \frac{d\widetilde{v}}{dy} = 0.$$

Define $F(\omega) = \mathcal{F}_e\{f(x)H(a - |x|)\} = \mathcal{F}_e\{v(x,0); x \to \omega\} = \widetilde{v}(\omega,0)$ as the boundary condition associated with the above transformed equations. Differentiating the right equation gives

$$-i\omega\frac{d\widetilde{u}}{dy} + \frac{d^2\widetilde{v}}{dy^2} = 0$$

and substituting for the derivative $\frac{d\widetilde{u}}{dy}$ from the left equation gives

$$\frac{d^2\widetilde{v}}{dy^2} - \omega^2\widetilde{v} = 0, \qquad \widetilde{v}(\omega,0) = F(\omega) \quad \text{and} \quad \lim_{y\to\infty}\widetilde{v} = 0$$

The general solution of this equation is given by

$$\widetilde{v} = C_1 e^{-\omega y} + C_2 e^{\omega y}.$$

The condition the $\widetilde{v}$ tend to zero as y increases without bound requires C_2 to equal zero if ω is positive and C_1 to equal zero if ω is negative. This gives the solution

$$\widetilde{v} = \widetilde{v}(\omega, y) = \begin{cases} C_1 e^{-\omega y}, & \omega > 0 \\ C_2 e^{\omega y}, & \omega < 0 \end{cases} \qquad \text{or} \qquad \widetilde{v}(\omega, y) = Ce^{-|\omega|y}.$$

The boundary condition along $y = 0$ requires that

$$\widetilde{v}(\omega,0) = F(\omega) = C \qquad \text{which gives the solution} \qquad \widetilde{v}(\omega, y) = F(\omega)e^{-|\omega|y}.$$

The solution for $\widetilde{u}$ is

$$\widetilde{u} = \frac{1}{i\omega}\frac{d\widetilde{v}}{dy} \quad \text{or} \quad \widetilde{u} = \widetilde{u}(\omega, y) = \begin{cases} -i(-1)F(\omega)e^{-\omega y}, & \omega > 0 \\ -i(+1)F(\omega)e^{\omega y}, & \omega < 0 \end{cases}$$

$$\widetilde{u} = i\,\text{sgn}(\omega)\,F(\omega)e^{-|\omega|y}.$$

We can now use the convolution property of the Fourier exponential transform to obtain the solutions

$$u(x,y) = \mathcal{F}_e^{-1}\{F(\omega)G_1(\omega)\}, \qquad v(x,y) = \mathcal{F}_e^{-1}\{F(\omega)G_2(\omega)\}$$

where

$$G_1(\omega) = i\,\mathrm{sgn}(\omega)\,e^{-|\omega|y} \qquad \text{and} \qquad G_2(\omega) = e^{-|\omega|y}.$$

One can write

$$\mathcal{F}_e^{-1}\{G_1(\omega)\} = \int_{-\infty}^{\infty} i\,\mathrm{sgn}(\omega)\,e^{-|\omega|y}e^{-i\omega x}\,d\omega$$

$$= \int_{-\infty}^{0} i(-1)e^{\omega y}e^{-i\omega x}\,d\omega + \int_0^{\infty} i(+1)e^{-\omega y}e^{-i\omega x}\,d\omega$$

Make a change of variable in the first integral and replace ω by $-\omega$ to obtain

$$\mathcal{F}_e^{-1}\{G_1(\omega)\} = \int_0^{\infty} -ie^{-\omega y}e^{i\omega x}\,d\omega + \int_0^{\infty} ie^{-\omega y}e^{-i\omega x}\,d\omega$$

$$= \int_0^{\infty} 2e^{-\omega y}\left(\frac{e^{i\omega x}-e^{-i\omega x}}{2i}\right)d\omega = 2\int_0^{\infty} e^{-\omega y}\sin\omega x\,d\omega = \frac{2x}{x^2+y^2}$$

This last result is obtained from the Laplace transform table. From the Fourier exponential transform table we find by direct table look up that

$$\mathcal{F}_e^{-1}\{G_2(\omega)\} = \mathcal{F}_e^{-1}\{e^{-|\omega|y}\} = \frac{2y}{x^2+y^2}.$$

Therefore, the convolution integrals produce the solutions

$$u(x,y) = \mathcal{F}_e^{-1}\{\widetilde{u}\} = \frac{1}{2\pi}\int_{-\infty}^{\infty} f(t)H(a-|t|)\frac{2(x-t)}{(x-t)^2+y^2}\,dt = \frac{1}{\pi}\int_{-a}^{a}\frac{f(t)(x-t)}{(x-t)^2+y^2}\,dt$$

$$v(x,y) = \mathcal{F}_e^{-1}\{\widetilde{v}\} = \frac{1}{2\pi}\int_{-\infty}^{\infty} f(t)H(a-|t|)\frac{2y}{(x-t)^2+y^2}\,dt = \frac{y}{\pi}\int_{-a}^{a}\frac{f(t)}{(x-t)^2+y^2}\,dt$$

∎

Fourier sine and cosine transform

From the Fourier integral representation (3.72) one can write

$$f(x) = \frac{1}{\pi}\int_0^{\infty}\cos\omega x\int_{-\infty}^{\infty} f(\xi)\cos\omega\xi\,d\xi d\omega$$
$$+\frac{1}{\pi}\int_0^{\infty}\sin\omega x\int_{-\infty}^{\infty} f(\xi)\sin\omega\xi\,d\xi d\omega. \tag{7.62}$$

If $f(x)$ is an odd function $f(-x) = -f(x)$, then the first integral in equation (7.62) vanishes and one obtains

$$f(x) = \frac{2}{\pi}\int_0^{\infty}\sin\omega x\int_0^{\infty} f(\xi)\sin\omega x\,d\xi d\omega \tag{7.63}$$

which can be used to define the Fourier sine transform pair.

Define the Fourier sine transform

$$\mathcal{F}_s\{f(x)\} = F_s(\omega) = \frac{2}{\pi} \int_0^\infty f(x) \sin \omega x \, dx \qquad (7.64)$$

with inverse transform

$$\mathcal{F}_s^{-1}\{F_s(\omega)\} = f(x) = \int_0^\infty F_s(\omega) \sin \omega x \, d\omega. \qquad (7.65)$$

Table 7-5. Fourier Sine Transform

$f(x) = \mathcal{F}_s^{-1}\{F_s(\omega)\} = \int_0^\infty F_s(\omega) \sin \omega x \, d\omega$		$F_s(\omega) = \mathcal{F}_s\{f(x)\} = \frac{2}{\pi} \int_0^\infty f(x) \sin \omega x \, dx$
1.	$f'(x)$	$-\omega \mathcal{F}_c\{f(x)\}$
2.	$f''(x)$	$-\omega^2 F_s(\omega) + \dfrac{2\omega}{\pi} f(0)$
3.	$f(ax), \quad a > 0$	$\dfrac{1}{a} F_s\left(\dfrac{\omega}{a}\right)$
4.	$f(ax) \cos bx \quad a > 0, b > 0$	$\dfrac{1}{2a}\left[F_s\left(\dfrac{\omega+b}{a}\right) + F_s\left(\dfrac{\omega-b}{a}\right)\right]$
5.	1	$\dfrac{2}{\pi} \cdot \dfrac{1}{\omega}$
6.	$\dfrac{1}{x}$	$\operatorname{sgn}(\omega)$
7.	$e^{-\beta x} \quad \beta > 0$	$\dfrac{2}{\pi} \dfrac{\omega}{\beta^2 + \omega^2}$
8.	$\dfrac{e^{-\beta x}}{x} \quad \beta > 0$	$\dfrac{2}{\pi} \arctan\left(\dfrac{\omega}{\beta}\right)$
9.	$\dfrac{1}{2}\operatorname{erfc}\left(\dfrac{x}{2\sqrt{Kt}}\right), \quad x > 0$	$\dfrac{1 - e^{-Kt\omega^2}}{\pi \omega}$
10.	$\dfrac{1}{\pi}\displaystyle\int_0^\infty f(\xi)\left[g(x-\xi) - g(x+\xi)\right] d\xi$	$F_s(\omega) G_c(\omega)$
11.	$\dfrac{1}{4\beta}\sqrt{\dfrac{\pi}{\beta}} x e^{-x^2/4\beta}$	$\omega e^{-\beta \omega^2}$

If $f(x)$ is an even function $f(-x) = f(x)$, then the second integral in equation (7.62) vanishes and one obtains

$$f(x) = \frac{2}{\pi} \int_0^\infty \cos \omega x \int_0^\infty f(\xi) \cos \omega \xi \, d\xi d\omega. \qquad (7.66)$$

The integral (7.66) is used to define the Fourier cosine transform pair.

Define the Fourier cosine transform

$$\mathcal{F}_c\{f(x)\} = F_c(\omega) = \frac{2}{\pi} \int_0^\infty f(x) \cos \omega x \, dx \qquad (7.67)$$

with inverse transform

$$\mathcal{F}_c^{-1}\{F_c(\omega)\} = f(x) = \int_0^\infty F_c(\omega) \cos \omega x \, d\omega. \qquad (7.68)$$

Table 7-6. Fourier Cosine Transform

	$f(x) = \mathcal{F}_c^{-1}\{F_c(\omega)\} = \int_0^\infty F_c(\omega)\cos\omega x\,d\omega$	$F_c(\omega) = \mathcal{F}_c\{f(x)\} = \frac{2}{\pi}\int_0^\infty f(x)\cos\omega x\,dx$
1.	$f'(x)$	$\omega \mathcal{F}_s\{f(x)\} - \dfrac{2}{\pi} f(0)$
2.	$f''(x)$	$-\omega^2 F_c(\omega) - \dfrac{2}{\pi} f'(0)$
3.	$f(ax)$	$\dfrac{1}{a} F_c\left(\dfrac{\omega}{a}\right)$
4.	$f(ax)\cos bx \quad a>0,\, b>0$	$\dfrac{1}{2a}\left[F_c\left(\dfrac{\omega+b}{a}\right) + F_c\left(\dfrac{\omega-b}{a}\right)\right]$
5.	$e^{-\beta x} \quad \beta>0$	$\dfrac{2}{\pi}\dfrac{\beta}{\beta^2+\omega^2}$
6.	$e^{-\beta^2 x^2}$	$\dfrac{1}{\sqrt{\pi}}\dfrac{e^{-\omega^2/4\beta^2}}{|\beta|}$
7.	$\dfrac{\beta}{x^2+\beta^2}$	$e^{-\omega\beta}$
8.	$\begin{cases} 1 & 0<x<L \\ 0 & L<x \end{cases}$	$\dfrac{2}{\pi}\dfrac{\sin\omega L}{\omega}$
9.	$\sqrt{\dfrac{Kt}{\pi}}\,e^{-x^2/4Kt} - \dfrac{x}{2}\mathrm{erfc}\left(\dfrac{x}{2\sqrt{Kt}}\right)$	$\dfrac{1-e^{-\omega^2 Kt}}{\pi\omega^2}$
10.	$\dfrac{1}{\pi}\displaystyle\int_0^\infty g(\xi)\left[f(x-\xi)+f(x+\xi)\right]d\xi$	$G_c(\omega)F_c(\omega)$
11.	$\dfrac{1}{2}\sqrt{\dfrac{\pi}{\beta}}\,e^{-x^2/4\beta}$	$e^{-\beta\omega^2}$

The tables 7-5 and 7-6 are short tables for values and properties of the Fourier sine transform and cosine transform. The entry numbers 1 and 2 of table 7-5 are operational properties of the Fourier sine transform. These results can be written $\mathcal{F}_s\{f'(x)\} = -\omega\mathcal{F}_c\{f(x)\}$ and $\mathcal{F}_s\{f''(x)\} = -\omega^2\mathcal{F}_s\{f(x)\} + \dfrac{2\omega}{\pi}f(0)$ and are obtained from the definition and using integration by parts.

The properties 3 and 4 of table 7-5 are scaling properties and follow from the definition. Note entry 5 in table 7-5 is a limiting value of entry number 7, in the limit as $\beta \to 0$. Similarly, the entry number 6 is a limiting value of entry number 8, in the limit as $\beta \to 0$. The entry number 10 of table 7-5 is a convolution property of the Fourier sine transform. By using $G_c(\omega)$ as the Fourier cosine transform of $g(x)$ one can write entry number 10 as

$$\mathcal{F}_s^{-1}\{F_s(\omega)G_c(\omega)\} = \int_0^\infty F_s(\omega)G_c(\omega)\sin\omega x\,d\omega$$

$$= \int_0^\infty \frac{2}{\pi}\int_0^\infty f(\xi)\sin\omega\xi\,d\xi G_c(\omega)\sin\omega x\,d\omega.$$

Using the identity $\sin\omega\xi\sin\omega x = \dfrac{1}{2}[\cos\omega(x-\xi) - \cos\omega(x+\xi)]$ together with an interchange in the order of integration the above result can be written in the form

$$\mathcal{F}_s^{-1}\{F_s(\omega)G_c(\omega)\} = \frac{1}{\pi}\int_0^\infty f(\xi)\int_0^\infty G_c(\omega)\left[\cos\omega(x-\xi) - \cos\omega(x+\xi)\right]d\omega d\xi$$

$$= \frac{1}{\pi}\int_0^\infty f(\xi)\left[g(x-\xi) - g(x+\xi)\right]d\xi \tag{7.69}$$

Here $f(x)$ and $g(x)$ are understood to represent the odd and even extension over the interval $-\infty < x < \infty$. That is, it is assumed that the Fourier sine transform is associated with odd functions and the Fourier cosine transform is associated with even functions. Hence, when necessary these functions can be extended to the interval $-\infty < x < \infty$.

The entries 1 and 2 of the table 7-6 are operational properties of the Fourier cosine transform and follow from the definition and integration by parts. The entries 3,4 of table 7-6 are scaling properties of the Fourier cosine transform. The property 10 is a convolution property of the Fourier cosine transform. Note that

$$\mathcal{F}_c^{-1}\{G_c(\omega)F_c(\omega)\} = \int_0^\infty G_c(\omega)F_c(\omega)\cos\omega x\,d\omega$$

$$= \int_0^\infty F_c(\omega)\frac{2}{\pi}\int_0^\infty g(\xi)\cos\omega\xi\,d\xi\,\cos\omega x d\omega$$

$$= \frac{1}{\pi}\int_0^\infty g(\xi)\int_0^\infty F_c(\omega)\left[\cos\omega(x-\xi) + \cos\omega(x+\xi)\right]d\omega d\xi$$

$$= \frac{1}{\pi}\int_0^\infty g(\xi)\left[f(x-\xi) + f(x+\xi)\right]d\xi$$

where f, g are understood to be the even extensions over the interval $-\infty < x < \infty$.

Example 7-10. (Heat equation.)

Solve for the temperature $u = u(x,t)$, in a long thin rod subject to the initial condition $u(x,0) = 0$, $x > 0$, and boundary condition $u(0,t) = T_0$, $t > 0$ where T_0 is constant. This problem is modeled

$$\text{PDE:} \qquad \frac{\partial^2 u}{\partial x^2} = \frac{1}{K}\frac{\partial u}{\partial t}$$

$$\text{BC:} \qquad u(0,t) = T_0, \quad t > 0$$

$$\text{IC:} \qquad u(x,0) = 0, \quad x > 0$$

Solution: The geometry of the problem, together with the fact $u(0,t)$ is known, suggests a Fourier sine transform. We define

$$\tilde{u}(\omega, t) = \mathcal{F}_s\{u(x,t); x \to \omega\} = \frac{2}{\pi}\int_0^\infty u(x,t)\sin\omega x\, dx$$

and take the Fourier sine transform of the PDE and IC to obtain

$$\mathcal{F}_s\{\frac{\partial^2 u}{\partial x^2}\} = \frac{1}{K}\mathcal{F}_s\{\frac{\partial u}{\partial t}\} \qquad\qquad \mathcal{F}_s\{u(x,0)\} = \mathcal{F}_s\{0\}$$

$$-\omega^2\tilde{u}(\omega,t) + \frac{2}{\pi}\omega u(0,t) = \frac{1}{K}\frac{\partial}{\partial t}(\mathcal{F}_s\{u\}) \qquad \tilde{u}(\omega, 0) = 0$$

$$-\omega^2\tilde{u} + \frac{2}{\pi}\omega T_0 = \frac{1}{K}\frac{d\tilde{u}}{dt}$$

Here we have used the result number 2 from table 7-5 involving the Fourier sine transform of a second derivative. Also note we have interchanged the order of differentiation and integration to obtain the above result. The heat equation is transformed to the ordinary differential equation

$$\frac{1}{K}\frac{d\tilde{u}}{dt} = -\omega^2\tilde{u} + \frac{2}{\pi}\omega T_0, \qquad \tilde{u}(\omega,0) = 0$$

with solution in the transform space given by $\tilde{u} = \tilde{u}(\omega, t) = \dfrac{2T_0}{\pi\omega}\left(1 - e^{-\omega^2 Kt}\right)$. Taking the inverse sine transform of this result produces the solution to the original problem. We find

$$u = u(x,t) = \mathcal{F}_s^{-1}\{\tilde{u}(\omega,t); \omega \to x\} = \int_0^\infty \frac{2T_0}{\pi\omega}\left(1 - e^{-\omega^2 Kt}\right)\sin\omega x\, d\omega \qquad (7.70)$$

which represents one form for the solution. Whenever possible we try to evaluate the integrals resulting from the inverse sine transform. If the integrals cannot be evaluated one can always use numerical methods to evaluate the integral. The equation (7.70) can also be written as

$$u = u(x, t) = \frac{2T_0}{\pi} \int_0^\infty \frac{\sin \omega x}{\omega} \, d\omega - \frac{2T_0}{\pi} \int_0^\infty e^{-\omega^2 K t} \frac{\sin \omega x}{\omega} \, d\omega. \tag{7.71}$$

From the Fourier sine transform table 7-5, entry number 5, we find

$$\mathcal{F}_s\{1\} = F(\omega) = \frac{2}{\pi}\frac{1}{\omega} \quad \text{or} \quad \mathcal{F}_s^{-1}\{\frac{2}{\pi}\frac{1}{\omega}\} = \frac{2}{\pi} \int_0^\infty \frac{\sin \omega x}{\omega} \, d\omega = 1$$

and so the first integral in equation (7.71) can be evaluated. The second integral in equation (7.71) can be written

$$\frac{2T_0}{\pi} \int_0^\infty e^{-\omega^2 K t} \frac{\sin \omega x}{\omega} \, d\omega = \frac{2T_0}{\pi} \int_0^\infty e^{-\omega^2 K t} \left[\int_0^x \cos \omega u \, du \right] d\omega$$

$$= \frac{2T_0}{\pi} \int_0^x \left[\int_0^\infty e^{-\omega^2 K t} \cos \omega u \, d\omega \right] du$$

where we have interchanged the order of integration to achieve this result. The integral of the term in brackets is a special case of the integral given by equation (4.132) and so the solution given by equation (7.70) can be expressed in the form

$$u = u(x, t) = T_0 - \frac{2T_0}{\sqrt{\pi}} \int_0^x \frac{e^{-u^2/4Kt}}{2\sqrt{Kt}} \, du.$$

If we make the change of variable $\nu^2 = u^2/4Kt$ with $d\nu = du/2\sqrt{Kt}$, the solution can be written in terms of the complementary error function previously defined by the equation (4.142). This alternate form for the solution is

$$u = u(x, t) = T_0 \operatorname{erfc}\left(\frac{x}{2\sqrt{Kt}}\right). \tag{7.72}$$

∎

Hankel transforms

A transform which is related to the Fourier transform is the Hankel[†] transform. The Hankel transform of order ν is defined

$$H_\nu\{f(r); r \to p\} = F(p) = \int_0^\infty f(r) r J_\nu(pr) \, dr \tag{7.73}$$

[†] See Appendix C

with inverse transform

$$H_\nu^{-1}\{F(p); p \to r\} = f(r) = \int_0^\infty F(p)pJ_\nu(rp)\, dp. \tag{7.74}$$

This transform is applicable for problems in polar or cylindrical coordinates where $0 < r < \infty$. The Hankel transform has the operational property

$$H_n\{\frac{d^2y}{dr^2} + \frac{1}{r}\frac{dy}{dr} - \frac{n^2}{r^2}y\} = -p^2 H_n\{y(r)\} = -p^2 Y(p) \tag{7.75}$$

where $Y(p)$ is the Hankel transform of $y(r)$ and it is assumed that $\lim_{r \to \infty} ry(r) = 0$. The table 7-7 is a short table of Hankel transforms. Note that the column information can be interchanged in the Hankel transform table. That is, if $\mathcal{H}_\nu\{f(r)\} = F(p)$, then $\mathcal{H}_\nu\{F(r)\} = f(p)$.

	ν	$f(r) = H_\nu^{-1}\{F(p)\} = \int_0^\infty F(p)pJ_\nu(rp)\,dp$	$F(p) = H_\nu\{f(r)\} = \int_0^\infty f(r)rJ_\nu(pr)\,dr$
		Table 7-7. Short Table of Hankel Transforms	
1.	0	$\begin{cases} 0 & b < r \\ \frac{1}{\sqrt{b^2-r^2}} & b > r \end{cases}$	$\frac{1}{p}\sin bp$
2.	0	$H(b-r)$	$\frac{b}{p}J_1(bp)$
3.	0	e^{-br}	$b(b^2+p^2)^{-3/2}$
4.	0	$\frac{1}{r}e^{-br}$	$(b^2+p^2)^{-1/2}$
5.	0	e^{-r^2/b^2}	$\frac{b^2}{2}e^{-b^2p^2/4}$
6.	1	$rH(b-r)$	$\frac{b^2}{p}J_2(bp)$
7.	1	e^{-br}	$p(b^2+p^2)^{-3/2}$
8.	1	$\frac{1}{r}e^{-br}$	$p(b^2+p^2)^{-1/2}(b+\sqrt{b^2+p^2})^{-1}$
9.	1	re^{-r^2/b^2}	$\frac{bp^4}{4}e^{-p^2b^2/4}$

Sturm-Liouville Transforms

In this section we introduce transforms that are applicable for use with certain partial differential equations over finite domains as well as infinite domains. In particular, we will study the finite sine transform, the finite cosine transform, the finite Hankel transform and general Sturm-Liouville transforms and point out how one can construct special types of transforms for special equations.

We have shown that Sturm-Liouville problems of the form

$$L(y) = \frac{d}{dx}\left(p(x)\frac{dy}{dx}\right) + q(x)y = -\lambda r(x)y \tag{7.76}$$

defined over an interval $a \le x \le b$ with parameter λ, and subject to the boundary conditions

$$\beta_1 y(a) + \beta_2 \frac{dy(a)}{dx} = 0, \qquad \beta_3 y(b) + \beta_4 \frac{dy(b)}{dx} = 0 \tag{7.77}$$

where $\beta_1, \beta_2, \beta_3, \beta_4$ are real constants independent of λ, give rise to sets of functions $\{\phi_n(x)\}$, for $n = 1, 2, 3, \ldots$ which are orthogonal over an interval (a, b) with respect to a weight function $r(x) > 0$. The inner product of any two of these functions satisfies

$$(\phi_n, \phi_m) = \int_a^b r(x)\phi_n(x)\phi_m(x)\,dx = \|\phi_n\|^2 \delta_{nm} \tag{7.78}$$

where δ_{nm} is the Kronecker delta and $\|\phi_n\|^2$ is the norm squared which is defined

$$\|\phi_n\|^2 = (\phi_n, \phi_n) = \int_a^b r(x)\phi_n^2(x)\,dx \tag{7.79}$$

and is assumed to be nonzero. Functions continuous on the interval (a, b) can be written in the form of a series of these orthogonal functions in the form

$$f(x) = \sum_{n=1}^{\infty} C_n \phi_n(x) \tag{7.80}$$

called a generalized Fourier series with Fourier coefficients C_n given by an inner product divided by a norm squared

$$C_n = \frac{(f, \phi_n)}{\|\phi_n\|^2} = \frac{\int_a^b r(x)f(x)\phi_n(x)\,dx}{\int_a^b r(x)\phi_n^2(x)\,dx}, \quad n = 1, 2, 3\ldots. \tag{7.81}$$

The above results can be written in a language and notation very similar to the Laplace transform notation. The equations (7.80) and (7.81) can be written as a general transform pair. Define the Sturm-Liouville transform of $f(x)$

$$T_{SL}\{f(x); x \to n\} = \widetilde{F}_{SL}(n) = \int_a^b r(x)f(x)\phi_n(x)\,dx \quad n = 1, 2, 3, \ldots \tag{7.82}$$

with inverse transform

$$T_{SL}^{-1}\{\widetilde{F}_{SL}(n)\} = f(x) = \sum_{n=1}^{\infty} \widetilde{F}_{SL}(n) \frac{\phi_n(x)}{\| \phi_n \|^2} \qquad (7.83)$$

which represents the generalized Fourier series expansion of $f(x)$ in terms of the orthogonal set of functions $\{\phi_n(x)\}$. An important operational property associated with this type of transform comes from the differential equation which defines the orthogonal set. Consider the differential operator

$$L_T(y) = \frac{1}{r(x)} \left[\frac{d}{dx} \left(p(x) \frac{dy}{dx} \right) + q(x)y \right] \qquad (7.84)$$

obtained from the differential equation (7.76) by dividing each term by $r(x)$. We use the Sturm-Liouville transform as defined by equation (7.82) and multiply equation (7.84) by $r(x)\phi_n(x)\,dx$ and then integrate from a to b. This gives the transform of the operator L_T

$$T_{SL}\{L_T(y)\} = \int_a^b \frac{d}{dx} \left(p(x) \frac{dy}{dx} \right) \phi_n(x)\,dx + \int_a^b q(x)y(x)\phi_n(x)\,dx \qquad (7.85)$$

Integrate the first term on the right-hand side of equation (7.85) using integration by parts to obtain

$$T_{SL}\{L_T(y)\} = \phi_n(x)p(x)\frac{dy}{dx}\Big]_a^b - \int_a^b p(x)\phi_n'(x)\frac{dy}{dx}\,dx + \int_a^b q(x)y(x)\phi_n(x)\,dx. \qquad (7.86)$$

Now integrate by parts a second time to obtain

$$T_{SL}\{L_T(y)\} = [p(x)(\phi_n(x)y'(x) - \phi_n'(x)y(x))]_a^b$$
$$+ \int_a^b \left[\frac{d}{dx}(p(x)\phi_n'(x)) + q(x)\phi_n(x) \right] y(x)\,dx \qquad (7.87)$$

Recall that each function $\phi_n(x)$ satisfies the differential equation

$$\frac{d}{dx}(p(x)\phi_n'(x)) + q(x)\phi_n(x) = -\lambda_n r(x)\phi_n(x)$$

and so the transform given by equation (7.87) reduces to

$$T_{SL}\{L_T(y)\} = [p(x)(\phi_n(x)y'(x) - \phi_n'(x)y(x))]_a^b - \lambda_n \int_a^b r(x)\phi_n(x)y(x)\,dx \qquad (7.88)$$

or

$$T_{SL}\{L_T(y)\} = -\lambda_n T_{SL}\{y(x)\} + [p(x)(\phi_n(x)y'(x) - \phi_n'(x)y(x))]_a^b. \qquad (7.89)$$

This shows the Sturm-Liouville transform, as defined by equation (7.82), can be applied to differential operators having the form $L_T(y)$ to obtain a transform of the dependent variable times the eigenvalue associated with $\phi_n(x)$ together with terms involving the boundary conditions at the end points a and b. That is, in order for the Sturm-Liouville transformation of the operator $L_T(y)$ to be applicable, one must know how to evaluate the boundary terms

$$[p(x)\,(\phi_n(x)y'(x) - \phi_n'(x)y(x))]_a^b = p(b)(\phi_n(b)y'(b) - \phi_n'(b)y(b))$$
$$-p(a)(\phi_n(a)y'(a) - \phi_n'(a)y(a)). \tag{7.90}$$

In some situations, where two or more of the boundary conditions $y(a), y'(a), y(b), y'(b)$ are unknown, then one can replace the unknown values with constants terms, then after the solution is obtained in terms of these constants, one can use some other conditions to solve for the unknown constant. The boundary conditions are an important part of any finite Sturm-Liouville transform.

There are many finite transform pairs. We will use the above ideas to concentrate upon developing finite sine and cosine transforms and their operational properties. Examples will be given of how Bessel functions can be used to construct finite Hankel transforms. These examples should then suggest how the many other finite transforms can be developed. The development and use of some of these other types will be given as exercises at the end of this chapter.

Finite Sine Transforms

We have shown the set of functions $\{\sin \frac{n\pi x}{L}\}$, $n = 1, 2, 3, \ldots$ are orthogonal over the interval $(0, L)$ with respect to the weight function $r(x) = 1$. A sectionally continuous function $f(x)$ defined on the interval $(0, L)$ can be represented as a Fourier sine series

$$f(x) = \sum_{n=1}^{\infty} b_n \sin \frac{n\pi x}{L} \tag{7.91}$$

with Fourier coefficients

$$b_n = \frac{(f, \sin \frac{n\pi x}{L})}{\| \sin \frac{n\pi x}{L} \|^2} = \frac{2}{L} \int_0^L f(x) \sin \frac{n\pi x}{L}\, dx. \tag{7.92}$$

The resulting series represents $f(x)$ on the interval $(-L, L)$ as an odd function and outside the full Fourier interval the series gives the periodic extension of $f(x)$. We now rewrite the equations (7.91) and (7.92) in the form of a transform pair. Define the finite sine transform of $f(x)$ as

$$S_n\{f(x); x \to n\} = \widetilde{F}_s(n) = \int_0^L f(x) \sin \frac{n\pi x}{L}\, dx \qquad n = 1, 2, 3, \ldots \tag{7.93}$$

with inverse transform

$$S_n^{-1}\{\widetilde{F}_s(n)\} = f(x) = \frac{2}{L} \sum_{n=1}^{\infty} \widetilde{F}_s(n) \sin \frac{n\pi x}{L}. \tag{7.94}$$

Note that the inverse transform is a Fourier sine series and so only represents $f(x)$ on the interval $(0, L)$. The equations (7.93) and (7.94) are just another way of representing the equations (7.91) and (7.92) using a different notation. The table 7-8 is a short table for the finite sine transform.

An important operational property of the finite sine transform is given in entry number 2 of table 7-8. If $S_n\{f(x)\} = \widetilde{F}_s(n)$, then

$$S_n\{f''(x)\} = -\frac{n^2\pi^2}{L^2}\widetilde{F}_s(n) + \frac{n\pi}{L}[f(0^+) - (-1)^n f(L^-)]. \tag{7.95}$$

This property requires that Dirichlet conditions be known at the end points $x = 0$ and $x = L$. The proof of the above property follows using integration by parts. By definition

$$S_n\{f''(x)\} = \int_0^L f''(x) \sin \frac{n\pi x}{L}\, dx$$

$$= f'(x) \sin \frac{n\pi x}{L}\Big]_0^L - \frac{n\pi}{L} \int_0^L f'(x) \cos \frac{n\pi x}{L}\, dx.$$

Another integration by parts gives

$$S_n\{f''(x)\} = -\frac{n\pi}{L}\left[f(x) \cos \frac{n\pi}{L}\right]_0^L - \frac{n^2\pi^2}{L^2} \int_0^L f(x) \sin \frac{n\pi x}{L}\, dx$$

which simplifies to the result given by equation (7.95). Note that this is a special case of equation (7.89) developed earlier.

Example 7-11. (Fourier sine series.)
Find the Fourier sine series representation of the function
$$f(x) = x, \qquad 0 < x < L.$$
Solution: From the entry number 4, table 7-8, we find

$$S_n\{x\} = \widetilde{F}_s(n) = (-1)^{n+1}\frac{L^2}{n\pi}.$$

	Table 7-8	
	Short Table of Finite Sine Transforms	
	$f(x) = S_n^{-1}\{\widetilde{F}_s(n)\}$	$\widetilde{F}_s(n) = S_n\{f(x)\}$
1.	$f(x) = \frac{2}{L}\sum_{n=1}^{\infty}\widetilde{F}_s(n)\sin\frac{n\pi x}{L}$	$\widetilde{F}_s(n) = \int_0^L f(x)\sin\frac{n\pi x}{L}\,dx$
2.	$f''(x)$	$-\frac{n^2\pi^2}{L^2}\widetilde{F}_s(n) + \frac{n\pi}{L}\left[f(0^+) - (-1)^n f(L^-)\right]$
3.	1	$\frac{L}{n\pi}\left[1 - (-1)^n\right]$
4.	x	$(-1)^{n+1}\frac{L^2}{n\pi}$
5.	x^2	$\frac{L^3}{n\pi}(-1)^{n-1} - \frac{2L^3}{n^3\pi^3}\left[1 - (-1)^n\right]$
6.	x^3	$(-1)^n\frac{L^4}{\pi^3}\left(\frac{6}{n^3} - \frac{\pi^2}{n}\right)$
7.	$e^{\alpha x}$	$\frac{n\pi L}{n^2\pi^2 + \alpha^2 L^2}\left[1 - (-1)^n\right]$
8.	$\begin{cases} x, & 0 \leq x \leq L/2 \\ L-x, & L/2 \leq x \leq L \end{cases}$	$\frac{2L^2}{n^2\pi^2}\sin\frac{n\pi}{2}$
9.	$\cos\alpha x$	$\frac{n\pi L}{n^2\pi^2 - \alpha^2 L^2}\left[1 - (-1)^n\cos\alpha L\right], \quad n \neq \frac{\alpha L}{\pi}$
10.	$\begin{cases} x, & 0 < x < L/2 \\ 0, & L/2 < x < L \end{cases}$	$\frac{L^2}{n^2\pi^2}\left[\sin\frac{n\pi}{2} - \frac{n\pi}{2}\cos\frac{n\pi}{2}\right]$
11.	$x(L-x)$	$\frac{2L^3}{n^3\pi^3}\left[1 - (-1)^n\right]$
12.	$\begin{cases} 1, & 0 < x < L/2 \\ -1 & L/2 < x < L \end{cases}$	$\frac{L}{n\pi}\left[1 + (-1)^n\right] - \frac{2L}{n\pi}\cos\frac{n\pi}{2}$
13.	$\begin{cases} 2x/L, & 0 < x < L/2 \\ 2(1-x/L) & L/2 < x < L \end{cases}$	$\frac{4L}{n^2\pi^2}\sin\frac{n\pi}{2}$
14.	$\begin{cases} 2x/L, & 0 < x < L/2 \\ 2(x/L - 1) & L/2 < x < L \end{cases}$	$-\frac{2L}{n\pi}\cos\frac{n\pi}{2}$
15.	$L - x$	$\frac{L^2}{n\pi}$
16.	$f_o(x)^* g_e(x)$	$2\widetilde{F}_s(n)\widetilde{G}_c(n)$

The inverse finite sine transform gives the Fourier sine series

$$S_n^{-1}\{\widetilde{F}_s(n)\} = f(x) = x = \frac{2}{L}\sum_{n=1}^{\infty}\widetilde{F}_s(n)\sin\frac{n\pi x}{L} = \frac{2L}{\pi}\sum_{n=1}^{\infty}\frac{(-1)^{n+1}}{n}\sin\frac{n\pi x}{L}$$

which is the same result obtained in example 3-4.

Example 7-12. (Heat equation.)

Use finite sine transforms to solve the boundary value problem for the heat equation.

$$\text{PDE:} \qquad \frac{\partial u}{\partial t} = K \frac{\partial^2 u}{\partial x^2}, \qquad 0 < x < L, \quad t > 0$$

$$\text{BC:} \qquad u(0,t) = g_1(t), \qquad u(L,t) = g_2(t)$$

$$\text{IC:} \qquad u(x,0) = f(x)$$

Solution: Let $\tilde{u}(n,t) = S_n\{u(x,t); x \to n\} = \displaystyle\int_0^L u(x,t)\sin\frac{n\pi x}{L}\, dx$ denote the finite sine transform of the temperature $u(x,t)$. We then take the finite sine transform of the PDE and IC to obtain

$$S_n\{\frac{\partial u}{\partial t}\} = S_n\{K\frac{\partial^2 u}{\partial x^2}\}$$

$$S_n\{u(x,0)\} = \tilde{u}(n,0) = S_n\{f(x)\} = F_s(n).$$

Interchange the order of integration and differentiation and make use of the operational property from entry 2, table 7-8, to find

$$\frac{\partial}{\partial t}S_n\{u\} = K\, S_n\left\{\frac{\partial^2 u}{\partial x^2}\right\}$$

$$\frac{d\tilde{u}}{dt} = K\left(-\frac{n^2\pi^2}{L^2}\tilde{u} + \frac{n\pi}{L}\left[u(0,t) - (-1)^n u(L,t)\right]\right)$$

$$\tilde{u}(n,0) = F_s(n).$$

Substituting the given boundary conditions and simplifying we find

$$\frac{d\tilde{u}}{dt} + \frac{Kn^2\pi^2}{L^2}\tilde{u} = \frac{Kn\pi}{L}[g_1(t) - (-1)^n g_2(t)]$$

$$\tilde{u}(n,0) = F_s(n)$$

This differential equation has the solution

$$\tilde{u}(n,t) = F_s(n)e^{-\frac{Kn^2\pi^2}{L^2}t} + e^{-\frac{Kn^2\pi^2}{L^2}t}\int_0^t \frac{Kn\pi}{L}[g_1(\tau) - (-1)^n g_2(\tau)]e^{\frac{Kn^2\pi^2}{L^2}\tau}\, d\tau. \qquad (7.96)$$

The solution of the original partial differential equation is then

$$u = u(x,t) = S_n^{-1}\{\tilde{u}(n,t); n \to x\} = \frac{2}{L}\sum_{n=1}^{\infty}\tilde{u}(n,t)\sin\frac{n\pi x}{L}$$

where $\tilde{u}(n,t)$ is given by equation (7.96). ∎

408

Finite Cosine transform

We have shown the set of functions $\{\cos \frac{n\pi x}{L}\}$ for $n = 0, 1, 2, 3, \ldots$ are orthogonal over the interval $(0, L)$ with respect to the weight function $r(x) = 1$. For even extensions of $f(x)$ to the interval $(-L, 0)$ there results the Fourier cosine series

$$f(x) = a_0 + \sum_{n=1}^{\infty} a_n \cos \frac{n\pi x}{L} \tag{7.97}$$

with Fourier coefficients given by an inner product divided by a norm squared

$$a_0 = \frac{(f, 1)}{\| 1 \|^2} = \frac{1}{2L} \int_{-L}^{L} f(x)\, dx = \frac{1}{L} \int_0^L f(x)\, dx$$

$$a_n = \frac{(f, \cos \frac{n\pi x}{L})}{\| \cos \frac{n\pi x}{L} \|^2} = \frac{1}{L} \int_{-L}^{L} f(x) \cos \frac{n\pi x}{L}\, dx = \frac{2}{L} \int_0^L f(x) \cos \frac{n\pi x}{L}\, dx. \tag{7.98}$$

The equations (7.97) and (7.98) can also be written as a transform pair. Define the finite cosine transform of $f(x)$ as

$$C_n\{f(x); x \to n\} = \widetilde{F}_c(n) = \int_0^L f(x) \cos \frac{n\pi x}{L}\, dx \tag{7.99}$$

for $n = 0, 1, 2, \ldots$ and write the inverse transform

$$C_n^{-1}\{\widetilde{F}_c(n); n \to x\} = f(x) = \frac{F_c(0)}{L} + \frac{2}{L} \sum_{n=1}^{\infty} \widetilde{F}_c(n) \cos \frac{n\pi x}{L}\, dx \tag{7.100}$$

The table 7-9 is a short table of finite cosine transforms. The entry number 1 in table 7-9 is an important operational property of the finite cosine transform. This property states that if $C_n\{f(x)\} = \widetilde{F}_c(n)$, then

$$C_n\{f''(x)\} = -\frac{n^2\pi^2}{L^2} \widetilde{F}_c(n) - f'(0) + (-1)^n f'(L). \tag{7.101}$$

Observe that in order to use this property the flux boundary conditions at $f'(0)$ and $f'(L)$ must be known. To prove this property we use the definition of a finite cosine transform and write

$$C_n\{f''(x)\} = \int_0^L f''(x) \cos \frac{n\pi x}{L}\, dx$$

and integrate by parts to obtain

$$C_n\{f''(x)\} = \left[f'(x) \cos \frac{n\pi x}{L} \right]_0^L + \int_0^L \frac{n\pi}{L} f'(x) \sin \frac{n\pi x}{L}\, dx.$$

Another integration by parts produces the result given by equation (7.101). These results are valid provided $f(x), f'(x), f''(x)$ are continuous functions. Again one should note that this is a special case of the more general result given by equation (7.89) which we developed earlier.

Example 7-13. (Cosine Series)

Find the Fourier cosine series expansion for the function $f(x) = x^2$, $0 < x < L$.

Solution: From the table 7-9, entry number 5, we find

$$C_n\{x^2\} = F_c(n) = \frac{2L^3}{n^2\pi^2}(-1)^n \quad \text{with} \quad F_c(0) = \frac{L^3}{3}.$$

This gives the inverse transform

$$C_n^{-1}\{F_c(n)\} = f(x) = x^2 = \frac{L^2}{3} + \frac{4L^2}{\pi^2}\sum_{n=1}^{\infty}\frac{(-1)^n}{n^2}\cos\frac{n\pi x}{L}$$

which is the Fourier cosine series representation for the function $f(x) = x^2$. This is the same result obtain in the previous example 3-7.

■

Example 7-14. (Heat equation finite rod.)

Use finite cosine transforms to solve the following boundary value problem for the temperature in a finite length rod with zero flux boundary conditions and given initial temperature distribution.

$$\text{PDE:} \qquad \frac{\partial u}{\partial t} = K\frac{\partial^2 u}{\partial x^2}, \quad 0 < x < L, \quad t > 0$$

$$\text{BC:} \qquad \frac{\partial u(0,t)}{\partial x} = 0$$

$$\frac{\partial u(L,t)}{\partial x} = 0$$

$$\text{IC:} \qquad u(x,0) = f(x)$$

Solution: The boundary conditions suggest a finite cosine transform and so we define

$$\widetilde{u}(n,t) = C_n\{u(x,t); x \to n\} = \int_0^L u(x,t)\cos\frac{n\pi x}{L}\,dx$$

and take the finite transform of both the PDE and IC to obtain

$$C_n\{\frac{\partial u}{\partial t}\} = \frac{\partial C_n\{u\}}{\partial t} = KC_n\{\frac{\partial^2 u}{\partial x^2}\}$$

$$\frac{d\widetilde{u}}{dt} = K\left(-\frac{n^2\pi^2}{L^2}\right)\widetilde{u} - \frac{\partial u(0,t)}{\partial x} + (-1)^n\frac{\partial u(L,t)}{\partial x}$$

$$\frac{d\widetilde{u}}{dt} = K\left(-\frac{n^2\pi^2}{L^2}\right)\widetilde{u}$$

$$\widetilde{u}(n,0) = C_n\{f(x)\} = \widetilde{F}_c(n) = \int_0^L f(x)\cos\frac{n\pi x}{L}\,dx.$$

410

Table 7-9		
Short Table of Finite Cosine Transforms		
$f(x) = \frac{\widetilde{F}_c(0)}{L} + \frac{2}{L}\sum_{n=1}^{\infty}\widetilde{F}_c(n)\cos\frac{n\pi x}{L}$	$\widetilde{F}_c(n) = \int_0^L f(x)\cos\frac{n\pi x}{L}\,dx$	
1.	$f''(x)$	$-\frac{n^2\pi^2}{L^2}\widetilde{F}_c(n) - f'(0) + (-1)^n f'(L)$	
2.	$f(x) = C_n^{-1}\{\widetilde{F}_c(n)\}$	$\widetilde{F}_c(0)$	$\widetilde{F}_c(n) = C_n\{f(x)\}\quad n\neq 0$
3.	1	L	0
4.	x	$\frac{1}{2}L^2$	$\left(\frac{L}{n\pi}\right)^2[(-1)^n - 1]$
5.	x^2	$\frac{1}{3}L^3$	$\frac{2L^3}{n^2\pi^2}(-1)^n$
6.	x^3	$\frac{1}{4}L^4$	$\frac{3L^4}{\pi^2 n^2}(-1)^n + \frac{6L^4}{\pi^4 n^4}[(-1)^n - 1]$
7.	$e^{\alpha x}$	$\frac{1}{\alpha}(e^{\alpha L} - 1)$	$\frac{\alpha L^2}{\alpha^2 L^2 + n^2\pi^2}[(-1)^n e^{\alpha L} - 1]$
8.	$\begin{cases} x, & 0\leq x\leq L/2 \\ L-x, & L/2\leq x\leq L\end{cases}$	$\frac{L^2}{4}$	$\frac{4L^2}{n^2\pi^2}\cos\frac{n\pi}{2}\sin^2\frac{n\pi}{4}$
9.	$\sin\alpha x \quad n\neq\frac{\alpha L}{\pi}$	$\frac{1}{\alpha}(1-\cos\alpha L)$	$\frac{\alpha L^2}{n^2\pi^2 - \alpha^2 L^2}[(-1)^n\cos\alpha L - 1]$
10.	$\begin{cases} x, & 0<x<L/2 \\ 0, & L/2<x<L\end{cases}$	$\frac{L^2}{8}$	$\frac{L^2}{n^2\pi^2}[-1 + \cos\frac{n\pi}{2} + \frac{n\pi}{2}\sin\frac{n\pi}{2}]$
11.	$x(L-x)$	$\frac{L^3}{6}$	$-\frac{L^3}{n^2\pi^2}[1+(-1)^n]$
12.	$\begin{cases} 1, & 0<x<L/2 \\ -1 & L/2<x<L\end{cases}$	0	$\frac{2L}{n\pi}\sin\frac{n\pi}{2}$
13.	$\left(1-\frac{x}{L}\right)^2$	$\frac{L}{3}$	$\frac{2L^3}{n^2\pi^2}$
14.	$\begin{cases} 2x/L, & 0<x<L/2 \\ 2(x/L-1) & L/2<x<L\end{cases}$	0	$\frac{2L}{n^2\pi^2}[-1+(-1)^n + n\pi\sin\frac{n\pi}{2}]$
15.	$L-x$	$\frac{L^2}{2}$	$\frac{L^2}{n^2\pi^2}[1-(-1)^n]$
16.	$f_o(x)^* g_o(x)$		$-2\widetilde{F}_s(n)\widetilde{G}_s(n)$
17.	$f_e(x)^* g_e(x)$		$2\widetilde{F}_c(n)\widetilde{G}_c(n)$

This gives the first order differential equation

$$\frac{d\widetilde{u}}{dt} + \frac{Kn^2\pi^2}{L^2}\widetilde{u} = 0, \qquad \widetilde{u}(n,0) = \widetilde{F}_c(n)$$

with bounded solution in the transform domain

$$\widetilde{u}(n,t) = \widetilde{F}_c(n)e^{-\frac{Kn^2\pi^2 t}{L^2}}.$$

The inverse transform gives the solution to the original problem as

$$u = u(x,t) = \frac{\widetilde{F}_c(0)}{L} + \frac{2}{L}\sum_{n=1}^{\infty}\widetilde{F}_c(n)e^{-\frac{Kn^2\pi^2 t}{L^2}}\cos\frac{n\pi x}{L}.$$

∎

Convolution Integrals

The convolution of two functions $f(x)$ and $g(x)$ on an interval $(-L, L)$ is defined as

$$f(x)^*g(x) = \int_{-L}^{L} f(x-\tau)g(\tau)\,d\tau. \tag{7.102}$$

In the case $f(x)$ or $g(x)$ is only defined on the interval $(0, L)$, then one must use either the odd or even extensions of these functions to calculate the convolution integral. For example, let $f_o(x), g_o(x)$ and $f_e(x), g_e(x)$ denote respectively the odd and even extensions of $f(x)$ and $g(x)$, then the following transform properties can be verified[1]

$$\begin{aligned}
S_n\{f_o(x)^*g_e(x)\} &= 2\widetilde{F}_s(n)\widetilde{G}_c(n) \\
C_n\{f_e(x)^*g_e(x)\} &= 2\widetilde{F}_c(n)\widetilde{G}_c(n) \\
C_n\{f_o(x)^*g_o(x)\} &= -2\widetilde{F}_s(n)\widetilde{G}_s(n)
\end{aligned} \tag{7.103}$$

These properties are listed in the tables 7-8 and 7-9.

Finite Hankel Transforms

Consider the singular Sturm-Liouville problem

$$\frac{d}{dx}\left(x\frac{dy}{dx}\right) + (\lambda^2 x - \frac{\nu^2}{x})y = 0, \qquad 0 < x < b$$

subject to the boundary condition $y(b) = 0$. This is a Sturm-Liouville problem involving Bessel's differential equation where the eigenvalues λ_n are determined from the roots of the equation $J_\nu(\lambda_n b) = 0$ for $n = 1, 2, 3, \ldots$ and the eigenfunctions are $\phi_n(x) = J_\nu(\lambda_n x)$. These eigenfunctions are orthogonal over the interval $(0, b)$ with respect to the weight function $r(x) = x$. For this Sturm-Liouville problem we can define the finite Hankel transform of order ν

$$\mathcal{H}_\nu\{f(x); x \to n\} = \widetilde{F}_H(n) = \int_0^b xf(x)J_\nu(\lambda_n x)\,dx \qquad n = 1, 2, 3, \ldots \tag{7.104}$$

---

[1] R.V. Churchill, Operational Mathematics, McGraw Hill Book Co., N.Y, 1958.

412

with inverse transform

$$\mathcal{H}_\nu^{-1}\{\widetilde{F}_H(n)\} = f(x) = \frac{2}{b^2}\sum_{n=1}^{\infty}\widetilde{F}_H(n)\frac{J_\nu(\lambda_n x)}{[J_\nu'(\lambda_n b)]^2} \qquad (7.105)$$

which represents the Fourier-Bessel series expansion for the function $f(x)$. This transform pair has the operational property

$$\mathcal{H}_\nu\{\frac{d^2 y}{dx^2} + \frac{1}{x}\frac{dy}{dx} - \frac{\nu^2}{x^2}y\} = -b\lambda_n y(b)J_\nu'(\lambda_n b) - \lambda_n^2\widetilde{F}_H(n) \qquad (7.106)$$

which is a special case of the equation (7.89). The table 7-10 is a short table of finite Hankel transforms.

Table 7-10		
Short Table of Finite Hankel Transforms $\mathcal{H}_\nu\{\quad\}$ $\quad\nu = 0, 1$ $\quad J_\nu(\lambda_n b) = 0$		
$f(x) = \mathcal{H}_\nu^{-1}\{\widetilde{F}_{H_\nu}(n); n \to x\}$ $f(x) = \frac{2}{b^2}\sum_{n=1}^{\infty}\widetilde{F}_{H_\nu}(n)\frac{J_\nu(\lambda_n x)}{[J_\nu'(\lambda_n b)]^2}$		$\widetilde{F}_{H_\nu}(n) = \mathcal{H}_\nu\{f(x); x \to n\}$ $\widetilde{F}_{H_\nu}(n) = \int_0^b xf(x)J_\nu(\lambda_n x)\,dx$
1.	$f''(x) + \frac{1}{x}f'(x) - \frac{\nu^2}{x^2}f(x)$	$-b\lambda_n f(b)J_\nu'(\lambda_n b) - \lambda_n^2\widetilde{F}_{H_\nu}(n)$
2. $\nu = 0$	1	$\widetilde{F}_{H_0}(n) = \frac{b}{\lambda_n}J_1(\lambda_n b)$
3. $\nu = 0$	x^2	$\widetilde{F}_{H_0}(n) = \left(\frac{-4b}{\lambda_n^3} + \frac{b^3}{\lambda_n}\right)J_1(\lambda_n b)$
4. $\nu = 0$	$\left(1 - \frac{x^2}{b^2}\right)$	$\widetilde{F}_{H_0}(n) = \frac{4}{b\lambda_n^3}J_1(\lambda_n b)$
5. $\nu = 0$	$J_0(\alpha x)\quad \alpha \neq \lambda_n$	$\widetilde{F}_{H_0}(n) = \frac{b}{\lambda_n^2 - \alpha^2}[\lambda_n J_1(\lambda_n b)J_0(\alpha b)]$
6. $\nu = 1$	x	$\widetilde{F}_{H_1}(n) = \frac{b^2}{\lambda_n}J_2(\lambda_n b)$
7. $\nu = 1$	x^3	$\widetilde{F}_{H_1}(n) = \left(\frac{b^4}{\lambda_n} - \frac{8b^2}{\lambda_n^3}\right)J_2(\lambda_n b)$
8. $\nu = 1$	$x\left(1 - \frac{x^2}{b^2}\right)$	$\widetilde{F}_{H_1}(n) = \frac{8}{\lambda_n^3}J_2(\lambda_n b)$
9. $\nu = 1$	$J_1(\alpha x)\quad \alpha \neq \lambda_n$	$\widetilde{F}_{H_1}(n) = \frac{\lambda_n b}{\alpha^2 - \lambda_n^2}J_0(\lambda_n b)J_1(\alpha b)$

Example 7-15. (Hankel Transform.)

In the shock tube experiments discussed in example 7-2, the circular membrane placed in the test section of the tube is subjected to a uniform load per unit area by the dynamic pressure. The axial displacement of the circular membrane $u(r,t)$ is found to satisfy the nonhomogeneous wave equation with boundary and initial conditions given by the boundary value problem

$$
\begin{aligned}
\text{PDE:} \quad & \frac{\partial^2 u}{\partial r^2} + \frac{1}{r}\frac{\partial u}{\partial r} - \frac{1}{\alpha^2}\frac{\partial^2 u}{\partial t^2} = -P_0, \qquad 0 < r < b \\
\text{BC:} \quad & u(b,t) = 0 \\
& u(r,0) = \ell\left[1 - (r/b)^2\right] \\
\text{IC:} \quad & \frac{\partial u(r,0)}{\partial t} = 0
\end{aligned}
\tag{7.107}
$$

where α, ℓ, b, P_0 are constants.

Solution: We use the finite Hankel transform of order zero and define

$$
\mathcal{H}_0\{u(r,t); r \to n\} = \widetilde{u}(n,t) = \int_0^b u(r,t)\, r J_0(\lambda_n r)\, dr
\tag{7.108}
$$

where λ_n are the zero's of the equation $J_0(\lambda_n b) = 0$. The zeroth order Hankel transform of the given partial differential equation gives

$$
\mathcal{H}_0\left\{\frac{\partial^2 u}{\partial r^2} + \frac{1}{r}\frac{\partial u}{\partial r}\right\} - \frac{1}{\alpha^2}\mathcal{H}_0\left\{\frac{\partial^2 u}{\partial t^2}\right\} = \mathcal{H}_0\{-P_0\}.
\tag{7.109}
$$

We use the operational property

$$
\mathcal{H}_0\left\{\frac{\partial^2 u}{\partial r^2} + \frac{1}{r}\frac{\partial u}{\partial r}\right\} = -b\lambda_n u(b,t) J_0'(\lambda_n b) - \lambda_n^2 \widetilde{u}(n,t),
\tag{7.110}
$$

with the boundary condition $u(b,t) = 0$, and then we interchange the order of differentiation and integration to obtain from equation (7.109) the transformed equation

$$
-\lambda_n^2 \widetilde{u}(n,t) - \frac{1}{\alpha^2}\frac{d^2\widetilde{u}(n,t)}{dt^2} = -P_0 \widetilde{f}(n)
\tag{7.111}
$$

where

$$
\widetilde{f}(n) = \mathcal{H}_0\{1\} = \int_0^b r J_0(\lambda_n r)\, dr = \frac{b}{\lambda_n} J_1(\lambda_n b).
\tag{7.112}
$$

The zeroth order Hankel transform of the initial conditions produces

$$
\mathcal{H}_0\{u(r,0)\} = \widetilde{u}(n,0) = \widetilde{F}(n) = \mathcal{H}_0\{\ell\left[1 - (r/b)^2\right]\} = \frac{4\ell}{b\lambda_n^3} J_1(\lambda_n b)
$$

$$
\mathcal{H}_0\left\{\frac{\partial u(r,0)}{\partial t}\right\} = \frac{\partial \widetilde{u}(n,0)}{\partial t} = \mathcal{H}_0\{0\} = 0
$$

The zeroth order finite Hankel transform of the given partial differential equation and initial conditions produces the ordinary differential equation

$$\frac{d^2\widetilde{u}}{dt^2} + \alpha^2\lambda_n^2\widetilde{u} = P_0\alpha^2\widetilde{f}(n) \tag{7.113}$$

subject to the initial conditions

$$\widetilde{u}(n,0) = \frac{4\ell}{b\lambda_n^3}J_1(\lambda_n b)$$
$$\frac{\partial\widetilde{u}(n,0)}{\partial t} = 0 \tag{7.114}$$

The differential equation (7.113) has the general solution

$$\widetilde{u} = \widetilde{u}(n,t) = A\cos\alpha\lambda_n t + B\sin\alpha\lambda_n t + \frac{P_0}{\lambda_n^2}\widetilde{f}(n). \tag{7.115}$$

We apply the initial conditions from equation (7.114) and find the solution

$$\widetilde{u}(n,t) = \left[\widetilde{F}(n) - \frac{P_0}{\lambda_n^2}\right]\cos\alpha\lambda_n t + \frac{P_0}{\lambda_n^2}\widetilde{f}(n) \tag{7.116}$$

which satisfies the differential equation (7.113) and the initial conditions (7.114). We take the inverse Hankel transform to obtain the solution to the original problem. We find

$$u(r,t) = \mathcal{H}_0^{-1}\{\widetilde{u}(n,t)\}$$
$$u(r,t) = \frac{2}{b^2}\sum_{n=1}^{\infty}\left\{\left[\frac{4\ell}{b\lambda_n^3}J_1(\lambda_n b) - \frac{P_0 b}{\lambda_n^3}J_1(\lambda_n b)\right]\cos\alpha\lambda_n t + \frac{P_0 b}{\lambda_n^3}J_1(\lambda_n b)\right\}\frac{J_0(\lambda_n r)}{[J_1(\lambda_n b)]^2}$$

which simplifies to

$$u(r,t) = \frac{2}{b^2}\sum_{n=1}^{\infty}\left[\frac{4\ell}{b}\cos\alpha\lambda_n t + P_0 b(1 - \cos\alpha\lambda_n t)\right]\frac{J_0(\lambda_n r)}{\lambda_n^3 J_1(\lambda_n b)}. \tag{7.117}$$

∎

We conclude this chapter with tables 7-11,7-12,7-13 and 7-14 which illustrate solutions to selected Sturm-Liouville problems over discrete and infinite domains for different types of boundary conditions. These tables illustrate the analogies that exist between discrete and continuous spectrum systems which are associated with periodic, Dirichlet, Neumann and Robin type boundary conditions.

Table 7-11		
Periodic Solutions		
Discrete System		Continuous System
$\frac{d^2y}{dx^2} = \lambda y$ $-L < x < L$ $y(-L) = y(L)$ $y'(-L) = y'(L)$	Differential Equation BC	$\frac{d^2y}{dx^2} = \lambda y$ $-\infty < x < \infty$ $\lim_{x \to \pm\infty} y(x)$ bounded
$\lambda_n = -\left(\frac{n\pi}{L}\right)^2, \quad n = 0, \pm 1, \pm 2, \ldots$	Eigenvalues	$\lambda_\omega = -\omega^2, \quad -\infty < \omega < \infty$
$y_n(x) = e^{i\frac{n\pi x}{L}}, \quad n = 0, \pm 1, \pm 2, \ldots$	Eigenfunctions	$y_\omega(x) = e^{i\omega x}, \quad -\infty < \omega < \infty$
$\int_{-L}^{L} e^{i\frac{n\pi x}{L}} e^{-i\frac{m\pi x}{L}} \, dx = 2L\delta_{nm}$	Orthogonality	$\int_{-\infty}^{\infty} e^{i(\xi-\omega)x} \, dx = 2\pi\delta(\omega - \xi)$
$c_n = \frac{1}{2L} \int_{-L}^{L} f(x) e^{i\frac{n\pi x}{L}} \, dx$ $f(x) = \sum_{n=-\infty}^{\infty} c_n e^{-i\frac{n\pi x}{L}}$	Fourier Expansion	$F(\omega) = \frac{1}{2\pi} \int_{-\infty}^{\infty} f(x) e^{i\omega x} \, dx$ $f(x) = \int_{-\infty}^{\infty} F(\omega) e^{-i\omega x} \, d\omega$

Table 7-12		
Dirichlet Conditions		
Discrete System		Continuous System
$\frac{d^2y}{dx^2} = \lambda y$ $0 < x < L$ $y(0) = 0$ $y(L) = 0$	Differential Equation BC	$\frac{d^2y}{dx^2} = \lambda y$ $0 < x < \infty$ $y(0) = 0$ $\lim_{x \to \infty} y(x)$ bounded
$\lambda_n = -\left(\frac{n\pi}{L}\right)^2, \quad n = 1, 2, 3, \ldots$	Eigenvalues	$\lambda_\omega = -\omega^2, \quad 0 < \omega < \infty$
$y_n(x) = \sin \frac{n\pi x}{L}$	Eigenfunctions	$y_\omega(x) = \sin \omega x$
$\int_0^L \sin \frac{n\pi x}{L} \sin \frac{m\pi x}{L} \, dx = \frac{L}{2}\delta_{nm}$	Orthogonality	$\int_0^\infty \sin \omega x \sin \xi x \, dx = \frac{\pi}{2}\delta(\omega - \xi)$
$b_n = \frac{2}{L} \int_0^L f(x) \sin \frac{n\pi x}{L} \, dx$ $f(x) = \sum_{n=1}^{\infty} b_n \sin \frac{n\pi x}{L}$	Fourier Expansion	$F_s(\omega) = \frac{2}{\pi} \int_0^\infty f(x) \sin \omega x \, dx$ $f(x) = \int_0^\infty F_s(\omega) \sin \omega x \, d\omega$

	Table 7-13	
	Neumann Conditions	
Discrete System		Continuous System
$\frac{d^2y}{dx^2} = \lambda y, \quad 0 < x < L$ $y'(0) = 0, \quad y'(L) = 0$	Differential Eq. BC	$\frac{d^2y}{dx^2} = \lambda y, \quad 0 < x < \infty$ $y'(0) = 0, \quad \lim_{x\to\infty} y'(x)$
$\lambda_n = -\left(\frac{n\pi}{L}\right)^2, \quad n = 0, 1, 2, \ldots$	Eigenvalues	$\lambda_\omega = -\omega^2, \quad 0 < \omega < \infty$
$y_0(x) = 1, \quad y_n(x) = \cos\frac{n\pi x}{L}$	Eigenfunctions	$y_\omega(x) = \cos\omega x$
$\int_0^L \cos\frac{n\pi x}{L}\cos\frac{m\pi x}{L}\,dx = \|y_n\|^2\,\delta_{nm}$ $\|y_n\|^2 = \begin{cases} L, & m = n = 0 \\ L/2, & m = n \neq 0 \end{cases}$	Orthogonality	$\int_0^\infty \cos\omega x \cos\xi x\,dx = \frac{\pi}{2}\delta(\omega - \xi)$
$a_0 = \frac{1}{L}\int_0^L f(x)\,dx$ $a_n = \frac{2}{L}\int_0^L f(x)\cos\frac{n\pi x}{L}\,dx$ $f(x) = a_0 + \sum_{n=1}^\infty a_n \cos\frac{n\pi x}{L}$	Fourier Expansion	$F_c(\omega) = \frac{2}{\pi}\int_0^\infty f(x)\cos\omega x\,dx$ $f(x) = \int_0^\infty F_c(\omega)\cos\omega x\,d\omega$

	Table 7-14	
	Robin Conditions	
Discrete System		Continuous System
$\frac{d^2y}{dx^2} = \lambda y, \quad 0 < x < L$ $y'(0) + hy(0) = 0, \quad h > 0$ $y(L) = 0$	Differential Eq. BC	$\frac{d^2y}{dx^2} = \lambda y, \quad 0 < x < \infty$ $y'(0) + hy(0) = 0, \quad h > 0$ $\lim_{x\to\infty} y(x) \quad$ bounded
$\lambda_n = -\omega_n^2 \quad \tan\omega_n L = \omega_n/h$ $n = 1, 2, \ldots$ and $h > 0$ $\lambda_0 = \omega_0^2$ where $\tanh\omega_0 L = \omega_0/h$	Eigenvalues	$\lambda_\omega = -\omega^2, \quad 0 \le \omega < \infty$ $\lambda_0 = h^2, \quad h > 0$
$y_0(x) = e^{-\omega_0 x} - e^{-\omega_0 L}\cosh[\omega_0(x - L)]$ $y_n(x) = \sin[\omega_n x - \arctan(\omega_n/h)]$	Eigenfunctions	$y_0(x) = e^{-hx}$ $y_\omega(x) = \sin[\omega x - \arctan(\omega/h)]$
$(y_n, y_0) = \int_0^L y_n(x)y_0(x)\,dx = 0$ $(y_n, y_m) = \int_0^L y_n(x)y_m(x)\,dx = \|y_n\|^2\,\delta_{nm}$ $\|y_0\|^2 = -\frac{L}{2}e^{-2\omega_0 L} + \frac{1 - e^{-4\omega_0 L}}{8\omega_0}$ $\|y_n\|^2 = \frac{1}{2}\left[L - \frac{h}{\omega_n^2 + h^2}\right]$	Orthogonality	$\int_0^\infty y_\omega(x)y_\xi(x)\,dx = \frac{\pi}{2}\delta(\omega - \xi)$ $\int_0^\infty y_\omega(x)e^{-hx}\,dx = 0, \quad h > 0$ $\int_0^\infty y_0(x)y_0(x)\,dx = \frac{1}{2h}, \quad h > 0$
$f(x) = a_0 y_0(x) + \sum_{n=1}^\infty a_n y_n(x)$ $a_0 = \frac{(f, y_0)}{\|y_0\|^2} = \frac{1}{\|y_0\|^2}\int_0^L f(x)y_0(x)\,dx$ $a_n = \frac{(f, y_n)}{\|y_n\|^2} = \frac{1}{\|y_n\|^2}\int_0^L f(x)y_n(x)\,dx$	Fourier Expansion	$f(x) = A_0 y_0(x) + \int_0^\infty A(\omega)y_\omega(x)\,d\omega$ $A_0 = 2h\int_0^\infty f(x)e^{-hx}\,dx, \quad h > 0$ $A(\omega) = \frac{2}{\pi}\int_0^\infty f(x)y_\omega(x)\,dx$

Exercises 7

▶ 1.

(a) When rectangular coordinates (x, y, z) are converted by way of transformation equations $x = x(u, v, w)$, $y = y(u, v, w)$, $z = z(u, v, w)$ to a (u, v, w) coordinate system, where the transformation equations are single-valued and continuous with Jacobian different from zero, then we can solve these equations for the inverse transformations $u = u(x, y, z)$, $v = v(x, y, z)$, $w = w(x, y, z)$. The vector $\vec{r} = \vec{r}(u, v, w) = x(u, v, w)\,\widehat{e}_1 + y(u, v, w)\,\widehat{e}_2 + z(u, v, w)\,\widehat{e}_3$ can be used to construct curvilinear coordinate curves through a general point $\vec{r}(c_1, c_2, c_3)$. The surfaces $u(x, y, z) = c_1$, $v(x, y, z) = c_2$ and $w(x, y, z) = c_3$ are called coordinate surfaces. These surfaces intersect in curves called coordinate curves. The vectors $\vec{r}(c_1, c_2, w)$, $\vec{r}(c_1, v, c_3)$ and $\vec{r}(u, c_2, c_3)$ generate these coordinate curves which have the unit tangent vectors

$$\widehat{e}_u = \frac{\frac{\partial \vec{r}}{\partial u}}{\left|\frac{\partial \vec{r}}{\partial u}\right|}, \qquad \widehat{e}_v = \frac{\frac{\partial \vec{r}}{\partial v}}{\left|\frac{\partial \vec{r}}{\partial v}\right|}, \qquad \widehat{e}_w = \frac{\frac{\partial \vec{r}}{\partial w}}{\left|\frac{\partial \vec{r}}{\partial w}\right|}$$

where the quantities $h_u = \left|\dfrac{\partial \vec{r}}{\partial u}\right|$, $h_v = \left|\dfrac{\partial \vec{r}}{\partial v}\right|$, $h_w = \left|\dfrac{\partial \vec{r}}{\partial w}\right|$ are called scale factors. Note $d\vec{r} = \dfrac{\partial \vec{r}}{\partial u}du + \dfrac{\partial \vec{r}}{\partial v}dv + \dfrac{\partial \vec{r}}{\partial w}dw$ is composed of the three vectors $\vec{A} = \dfrac{\partial \vec{r}}{\partial u}du$, $\vec{B} = \dfrac{\partial \vec{r}}{\partial v}dv$, and $\vec{C} = \dfrac{\partial \vec{r}}{\partial w}dw$ which form a parallelepiped volume element in this coordinate system. Show the volume associated with this element is given by $d\tau = h_u h_v h_w\, du\, dv\, dw$.

Hint: Consider the volume element with sides $\vec{A}, \vec{B}, \vec{C}$ having
　　　the volume $d\tau = |\,\vec{A} \cdot (\vec{B} \times \vec{C})\,|$.

(b) If (u_0, v_0, w_0) denotes an arbitrary point in the curvilinear coordinate system, show

$$\delta(x - x_0)\delta(y - y_0)\delta(z - z_0) = \frac{1}{h_u h_v h_w}\delta(u - u_0)\delta(v - v_0)\delta(w - w_0).$$

(c) Verify your results in part (b) for cylindrical coordinates where $(u, v, w) = (r, \theta, z)$ are determined from the transformation equations

$$x = r\cos\theta, \qquad y = r\sin\theta, \qquad z = z$$

(d) Verify your results in part (b) for spherical coordinates where $(u, v, w) = (\rho, \theta, \phi)$ are determined from the transformation equations

$$x = \rho\sin\theta\cos\phi, \qquad y = \rho\sin\theta\sin\phi, \qquad z = \rho\cos\theta.$$

▶ **2.**

For H the Heaviside unit step function and β constant

(a) Graph the function $y = y(x) = H(x^2 - \beta^2)$.

(b) Graph the function $y = y(x) = 3\left[1 - \left(\frac{x}{\beta}\right)^2\right] H(\beta - |x|)$

(c) Graph the function $y = y(x) = \sin(x - 3)H(x - 3)$.

(d) Graph the function $y = y(x) = x^2[H(x - 1) - H(x - 2)]$.

(e) Graph the function $y = y(x) = 1 + H(x-1) + H(x-2) - H(x-3) - H(x-4) - H(x-5)$

▶ **3.**

Show $\delta(ax) = \dfrac{1}{|a|}\delta(x)$ for nonzero constant a.

▶ **4.**

Show $\delta(x^2 - a^2) = \frac{1}{2a}[\delta(x + a) + \delta(x - a)]$ for $a > 0$.

▶ **5.**

Show $\int_{-\infty}^{\infty} \delta(ax - b)f(x)\,dx = \frac{1}{a}f(\frac{b}{a})$.

▶ **6.**

Show $f(x)\delta(x - a) = f(a)\delta(x - a)$ in the sense that

$$\int_{-\infty}^{\infty} f(x)\delta(x - a)\,dx = \int_{-\infty}^{\infty} f(a)\delta(x - a)\,dx = f(a).$$

▶ **7.**

Find the center line deflection associated with the following beam loadings.

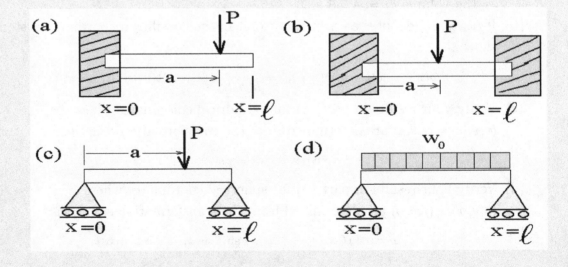

▶ **8.**

(a) If $F(\omega)$ is the Fourier exponential transform of $f(x)$, then show that $\dfrac{1}{2\pi}f(\omega)$ is the Fourier exponential transform of $F(-x)$.

(b) If $F_s(\omega)$ is the Fourier sine transform of $f(x)$, then show the Fourier sine transform of $F_s(x)$ is $\dfrac{2}{\pi}f(\omega)$.

(c) If $F_c(\omega)$ is the Fourier cosine transform of $f(x)$, then show the Fourier cosine transform of $F_c(x)$ is $\dfrac{2}{\pi}f(\omega)$.

▶ **9.**

Show that the result given by equation (7.69) can also be written in the form

$$\mathcal{F}_s^{-1}\{F_s(\omega)G_c(\omega)\} = \frac{1}{\pi}\int_0^\infty g(\xi)\left[f(\xi+x) - f(\xi-x)\right]\,d\xi.$$

▶ **10.**

(a) Evaluate $\mathcal{F}_s\left\{\dfrac{e^{-\beta x}}{x}\right\},\quad \beta > 0$

(b) Evaluate $\mathcal{F}_c\{e^{-\beta^2 x^2}\}$

Hint: Use differentiation with respect to a parameter.

▶ **11.**

Define $\tilde{f}(x) = \mathcal{F}_c^{-1}\{F_s(\omega)\}$. That is, take the sine transform of $f(x)$ to get $F_s(\omega)$, then invert $F_s(\omega)$ using a cosine transform to obtain $\tilde{f}(x)$. Show that

$$\mathcal{F}_s^{-1}\{F_s(\omega)G_s(\omega)\} = \frac{1}{\pi}\int_0^\infty g(\xi)\left[\tilde{f}(x-\xi) - \tilde{f}(x+\xi)\right]\,d\xi.$$

▶ **12.**

(a) If $f(x)$ is an even function of x, show $\mathcal{F}_e\{f(x)\} = \dfrac{1}{2}\mathcal{F}_c\{f(x)\}$.

(b) If $f(x)$ is an odd function of x, show $\mathcal{F}_e\{F(x)\} = \dfrac{i}{2}\mathcal{F}_s\{f(x)\}$

▶ **13.**

Use property 11 from table 7-3 to find

(a) $\mathcal{F}_e\{f(ax)\cos bx\}$

(b) $\mathcal{F}_e\{f(ax)\sin bx\}$ Hint: $\cos bx = \dfrac{e^{ibx} + e^{-ibx}}{2}\qquad \sin bx = \dfrac{e^{ibx} - e^{-ibx}}{2i}$

▶ **14.**

Find the Fourier exponential transform of $f(x) = \begin{cases} 1, & |x| < L \\ 0, & |x| > L. \end{cases}$

▶ **15.**

Use Laplace transforms to solve for the temperature $u = u(x,t)$ in a long thin rod $0 < x < \infty$ modeled by the heat equation with initial and boundary conditions given by

$$\frac{\partial^2 u}{\partial x^2} = \frac{1}{K}\frac{\partial u}{\partial t}, \qquad 0 < x < \infty$$

$$u(x,0) = 0, \qquad \lim_{x \to \infty} u(x,t) = 0$$

$$u(0,t) = f(t)$$

and show the solution is given by $u = u(x,t) = \dfrac{x}{\sqrt{4\pi K}} \displaystyle\int_0^t \dfrac{f(\xi)}{(t-\xi)^{3/2}} e^{-x^2/4K(t-\xi)}\, d\xi$

Show in the special case $f(t) = T_0$, where T_0 is a constant temperature, the solution reduces to $u(x,t) = T_0\, \mathrm{erfc}\left(\dfrac{x}{2\sqrt{Kt}}\right)$.

▶ **16.**

Use the Fourier exponential transform to solve the heat equation for the temperature $u = u(x,t)$ in a long thin rod laterally insulated

$$\frac{\partial u}{\partial t} = K\frac{\partial^2 u}{\partial x^2}, \qquad -\infty < x < \infty$$

subject to the initial condition $u(x,0) = f(x)$.

▶ **17.**

(a) Solve the heat equation for the temperature $u = u(x,t)$ in a long thin rod if

PDE: $\quad \dfrac{\partial^2 u}{\partial x^2} = \dfrac{1}{K}\dfrac{\partial u}{\partial t}, \quad 0 < x < \infty,\ t > 0$

BC: $\quad u(x,0) = T_0, \quad T_0$ constant and $x > 0$

IC: $\quad -\dfrac{\partial u(0,t)}{\partial x} + hu(0,t) = 0, \quad t > 0, \quad h > 0$ is constant.

▶ **18.**

Show that if $\mathcal{L}\{f(t)\} = F(s)$, then $\mathcal{L}\{b\, f(bt)\} = F\left(\dfrac{s}{b}\right)$.

► **19.**

(Duhamel's[†] principle for heat equation)

(a) Use Laplace transforms to show the solution of the boundary-initial value problem for heat flow in an insulated rod

$$\text{PDE:} \qquad \frac{\partial v}{\partial t} = K \frac{\partial^2 v}{\partial x^2}, \qquad 0 < x < L, \ t > 0$$

$$\text{BC:} \qquad v(0,t) = 0, \qquad v(L,t) = 1, \qquad t > 0$$

$$\text{IC:} \qquad v(x,0) = 0, \qquad 0 < x < L$$

is given by

$$v(x,t) = \mathcal{L}^{-1}\{\frac{1}{s}\frac{\sinh\sqrt{\frac{s}{K}}x}{\sinh\sqrt{\frac{s}{K}}L}\} = \frac{x}{L} + \frac{2}{\pi}\sum_{n=1}^{\infty}\frac{(-1)^n}{n}exp\left[-\left(\frac{n\pi}{L}\right)^2 Kt\right]\sin\frac{n\pi x}{L}$$

(b) Use Laplace transforms to show the solution of the boundary-initial value problem for heat flow in an insulated rod

$$\text{PDE:} \qquad \frac{\partial u}{\partial t} = K \frac{\partial^2 u}{\partial x^2}, \qquad 0 < x < L, \ t > 0$$

$$\text{BC:} \qquad u(0,t) = 0, \qquad t > 0$$

$$u(L,t) = f(t)$$

$$\text{IC:} \qquad u(x,0) = 0, \qquad 0 < x < L$$

is given by

$$u(x,t) = \mathcal{L}^{-1}\{F(s)\mathcal{L}\{\frac{\partial v}{\partial t}\}\} = f(t)*\frac{\partial v}{\partial t} = \int_0^t f(\tau)\frac{\partial v(x,t-\tau)}{\partial t}\,d\tau$$

where $v = v(x,t)$ is from part (a). Note that the derivative $\frac{\partial v}{\partial t}$ results in a divergent series and so use integration by parts to represent the solution in the form

$$u(x,t) = \int_0^t f'(\tau)v(x,t-\tau)\,d\tau + f(0)v(x,t).$$

This shows that the solution to the time varying boundary condition problem (part (b)) can be represented in terms of the boundary value problem with constant end conditions (part (a)).

[†] See Appendix C

▶ **20.**

Use Fourier cosine transforms to solve the given boundary value problem.

$$PDE: \qquad \frac{\partial u}{\partial t} = \frac{\partial^2 u}{\partial x^2}, \qquad 0 < x < \infty$$

$$BC: \qquad -\frac{\partial u(0,t)}{\partial x} = q_0, \quad q_0 \text{ constant}, \quad t > 0$$

$$IC: \qquad u(x,0) = 0$$

(a) Give a physical interpretation to the given problem.

(b) Show the solution can be represented in the form

$$u = u(x,t) = -q_0 x \operatorname{erfc}\left(\frac{x}{2\sqrt{t}}\right) + 2q_0 \sqrt{\frac{t}{\pi}} e^{-x^2/4t}.$$

▶ **21.**

Solve for the steady state temperature $u = u(x,y)$ in the semi-infinite strip $0 < x < \pi, y > 0$, where the temperature is bounded and satisfies the boundary conditions $u(0,y) = 0$, $u(\pi,y) = e^{-y}$, $\frac{\partial u(x,0)}{\partial y} = 0$. Here the temperature is specified as the solution of the boundary value problem

$$\frac{\partial^2 u}{\partial x^2} + \frac{\partial^2 u}{\partial y^2} = 0, \qquad 0 < x < \pi, \ y > 0$$

$$u(0,y) = 0, \qquad u(\pi,y) = e^{-y}, \qquad y > 0, \qquad \frac{\partial u(x,0)}{\partial y} = 0, \qquad 0 < x < \pi$$

▶ **22.** Solve for the potential $u = u(x,y)$ satisfying

$$\nabla^2 u = \frac{\partial^2 u}{\partial x^2} + \frac{\partial^2 u}{\partial y^2} = 0, \qquad x, y \in R$$

where $R = \{x,y \mid 0 < x < L, \ -\infty < y < \infty\}$ and u is subject to the boundary conditions of $u(0,y) = f_1(y)$, $u(L,y) = f_2(y)$ with u remaining bounded as $|y| \to \infty$.

▶ **23.**

(a) Solve the heat equation $\frac{\partial u}{\partial t} = K \frac{\partial^2 u}{\partial x^2}$, $-\infty < x < \infty$, and subject to the initial condition $u(x,0) = f(x)$.

(b) Show in the special case $f(x) = \begin{cases} 3, & |x| < 1 \\ 0, & |x| > 1 \end{cases}$ the solution can be represented

$$u(x,t) = \frac{3}{2}\left[\operatorname{erf}\left(\frac{x+1}{2\sqrt{Kt}}\right) - \operatorname{erf}\left(\frac{x-1}{2\sqrt{Kt}}\right)\right]$$

▶ **24.**

(a) What must been done in order to use separation of variables to solve the boundary value problem

$$\text{PDE:} \qquad \frac{\partial u}{\partial t} = K \frac{\partial^2 u}{\partial x^2}, \qquad 0 < x < L, \quad t > 0$$

$$\text{BC:} \qquad u(0,t) = T_0, \qquad u(L,t) = T_1, \quad T_0, T_1 \text{ constants}$$

$$\text{IC:} \qquad u(x,0) = f(x)$$

Make appropriate substitutions and then solve by the method of separation of variables.

(b) Use transform methods to solve the above boundary value problem.

▶ **25.**

(a) Use separation of variables to solve the boundary value problem

$$\text{PDE:} \qquad \frac{\partial u}{\partial t} = K \frac{\partial^2 u}{\partial x^2}, \qquad 0 < x < L, \quad t > 0$$

$$\text{BC:} \qquad \frac{\partial u(0,t)}{\partial x} = 0, \qquad \frac{\partial u(L,t)}{\partial x} = 0$$

$$\text{IC:} \qquad u(x,0) = f(x)$$

(b) Use transform methods to solve the above boundary value problem.

▶ **26.**

Assume $f(x), f'(x), f''(x), f'''(x), f^{iv}(x)$ are continuous functions.

(a) Show

$$C_n\{f^{iv}(x)\} = \frac{n^4\pi^4}{L^4} \widetilde{F}_c(n) + \frac{n^2\pi^2}{L^2}[f'(0) + (-1)^n f'(L)] - f'''(0) + (-1)^n f'''(L)$$

(b) Show

$$S_n\{f^{iv}(x)\} = \frac{n^4\pi^4}{L^4} \widetilde{F}_s(n) - \frac{n^3\pi^3}{L^3}[f(0) - (-1)^n f(L)] + \frac{n\pi}{L}[f''(0) - (-1)^n f''(L)]$$

▶ **27.**

Use transform methods to solve for the potential $u = u(x,y)$ satisfying

$$\nabla^2 u = \frac{\partial^2 u}{\partial x^2} + \frac{\partial^2 u}{\partial y^2} = 0, \qquad x, y \in R$$

where $R = \{x, y \mid 0 < x < \infty, 0 < y < \pi\}$ and subject to the boundary conditions

$$u(0,y) = y, \qquad \frac{\partial u(x,0)}{\partial y} = 0, \qquad u(x,\pi) = g(x).$$

▶ **28.**

Use transform methods to solve

$$\nabla^2 u = \frac{\partial^2 u}{\partial x^2} + \frac{\partial^2 u}{\partial y^2} = 0, \qquad x, y \in R$$

where $R = \{x, y \mid 0 < x < \infty, \; 0 < y < \infty\}$ subject to the boundary conditions $u(0, y) = 0$ and $u(x, 0) = f(x)$.

▶ **29.**

Solve the nonhomogeneous heat equation

$$\frac{\partial u}{\partial t} = K \frac{\partial^2 u}{\partial x^2} + g(t), \quad 0 < x < L, \; t > 0$$

subject to the end conditions $\dfrac{\partial u(0, t)}{\partial x} = 0$ and $\dfrac{\partial u(L, t)}{\partial x} = 0$ and initial condition $u(x, 0) = f(x)$.

▶ **30.**

Use finite Fourier sine transforms to solve

$$\frac{d^2 y}{dx^2} = F(x), \quad 0 < x < L, \quad y(0) = \alpha, \quad y(L) = \beta.$$

▶ **31.**

Use finite Fourier sine transforms to solve the boundary value problem

$$
\begin{aligned}
\text{PDE:} \quad & \frac{\partial u}{\partial t} = K \frac{\partial^2 u}{\partial x^2} + g(t), \quad 0 < x < L, \; t > 0 \\
\text{BC:} \quad & u(0, t) = \alpha \\
& u(L, t) = \beta \\
\text{IC:} \quad & u(x, 0) = f(x)
\end{aligned}
$$

▶ **32.**

(a) Verify the special trigonometric transform $T\{f(x)\} = F(p_n) = \displaystyle\int_0^1 f(x) \cos p_n x \, dx$ where $p_1, p_2, p_3, \ldots, p_n, \ldots$ are solutions of the equation $p \tan p = h$, where $h > 0$ is a constant, has the inverse transform

$$T^{-1}\{F(p_n)\} = f(x) = 2 \sum_{n=1}^{\infty} \frac{p_n^2 + h^2}{p_n^2 + h^2 + h} F(p_n) \cos p_n x.$$

(b) What operational property does this transform possess?

▶ **33.**

(a) Verify the Chebyshev transform $T\{f(x)\} = F(n) = \int_{-1}^{1} \frac{f(x)T_n(x)}{\sqrt{1-x^2}}\, dx$ has the inverse transform

$$T^{-1}\{F(n)\} = f(x) = \frac{2}{\pi}F(0) + \frac{1}{\pi}\sum_{n=1}^{\infty} F(n)T_n(x)$$

(b) Verify the operational property $T\{\left[(1-x^2)\dfrac{d^2 f}{dx^2} - x\dfrac{df}{dx}\right]\} = -n^2 F(n).$

▶ **34.** Find the inverse Laplace transform $f(t) = \mathcal{L}^{-1}\{\dfrac{1}{s^3(1+s^2)}\}$

(a) By use of partial fractions and table look up.

(b) By use of convolution integrals and integration by parts.

▶ **35.** Use Laplace transforms to solve the ordinary differential equation

$$\frac{d^2 y}{dt^2} + 3\frac{dy}{dt} + 2y = te^{-t}$$

subject to the initial conditions $y(0) = 1$ and $\dfrac{dy(0)}{dt} = 0$.

▶ **36.** Use Laplace transforms to solve

$$\frac{\partial^2 u}{\partial\rho^2} + \frac{2}{\rho}\frac{\partial u}{\partial\rho} = \frac{1}{K}\frac{\partial u}{\partial t}, \qquad \rho > \rho_0,\ t > 0$$

subject to the boundary condition $u(\rho_0, t) = T_0$. Assume the initial condition is given by $u(\rho, 0) = 0,\quad \rho > \rho_0$ and $u(\rho, t) \to 0$ as $\rho \to \infty$.
Hint: Let $\bar{u} = \mathcal{L}\{u(\rho, t); t \to s\}$ and make the substitution $w = \rho\bar{u}$

▶ **37.**

(a) Use the Fourier cosine transform to solve the partial differential equation

$$\frac{\partial^2 u}{\partial t^2} = K\frac{\partial^2 u}{\partial x^2} - \beta u, \quad 0 < x < \infty,\ t > 0$$

subject to the boundary condition $\dfrac{\partial u(0, t)}{\partial x} = -\gamma_0$, where γ_0 is a constant.

(b) Use Laplace transforms to solve the above partial differential equation.

(c) Show the solutions from cases (a) and (b) are the same.

Hint: In part (c) use the result $\dfrac{1 - e^{-(K\omega^2 + \beta)t}}{K\omega^2 + \beta} = \int_0^t e^{-(K\omega^2 + \beta)\tau}\, d\tau$

▶ **38.** Use Fourier sine transforms to solve
$$\frac{\partial^2 u}{\partial t^2} = K \frac{\partial^2 u}{\partial x^2}, \qquad 0 < x < \infty, \ t > 0$$
with boundary condition $u(0,t) = 0$ and initial condition $u(x,0) = xe^{-x^2/4\alpha^2}$.

▶ **39.**

(a) Use the Fourier exponential transform to solve
$$\frac{\partial^2 u}{\partial t^2} = c^2 \frac{\partial^2 u}{\partial x^2}, \qquad -\infty < x < \infty, \ t > 0$$
subject to the initial conditions $u(x,0) = f(x)$ and $\dfrac{\partial u(x,0)}{\partial t} = 0$.

(b) Find the solution in the special case $f(x) = e^{-\alpha x^2}$.

▶ **40.** Use the Fourier exponential transform to solve
$$\frac{\partial G}{\partial t} = K \frac{\partial^2 G}{\partial x^2}, \qquad -\infty < x < \infty$$
subject to the initial condition $G(x,0) = \delta(x-\xi)$ where δ is the Dirac delta function.

▶ **41.** Solve the initial value problem
$$\frac{\partial^2 u}{\partial r^2} + \frac{1}{r}\frac{\partial u}{\partial r} - \frac{1}{r^2}u = \frac{\partial u}{\partial t}, \qquad 0 < r < \infty, \ t > 0$$
subject to the initial condition $u(r,0) = f(r)$.

▶ **42.** The solution of the Bessel differential equation
$$\frac{d^2 y}{dx^2} + \frac{1}{x}\frac{dy}{dx} + \left(\alpha^2 - \frac{k^2}{x^2}\right)y = 0, \qquad k = 0,1,2,3,\ldots, \qquad a < x < b$$
which satisfies the end conditions $y(a) = 0$ and $y(b) = 0$ is given by
$$\phi_n(x) = U_k(\alpha_n x) = J_k(\alpha_n x)Y_k(\alpha_n b) - Y_k(\alpha_n x)J_k(\alpha_n b)$$
where α_n, $n = 1,2,3,\ldots$, are the roots satisfying $U_k(\alpha_n a) = 0$. These functions are orthogonal over the interval (a,b) with inner product given by
$$(\phi_n, \phi_m) = \int_a^b x\phi_n(x)\phi_m(x)\,dx = \begin{cases} 0, & m \neq n \\ \frac{2\left(J_k^2(\alpha_n a) - J_k^2(\alpha_n b)\right)}{\pi^2 \alpha_n^2 J_k^2(\alpha_n a)}, & m = n \end{cases}$$

(a) Find the transform $\mathcal{T}$ and inverse transform $\mathcal{T}^{-1}$ associated with this Sturm-Liouville problem

(b) Find the operational property associated with the transform in part (a).

(c) Use your transform to solve the boundary value problem
$$\frac{\partial^2 u}{\partial r^2} + \frac{1}{r}\frac{\partial u}{\partial r} = \frac{1}{K}\frac{\partial u}{\partial t}, \qquad a < r < b, \ t > 0$$
which satisfies $u(a,t) = 0$, $u(b,t) = 0$ and $u(r,0) = f(r)$, $a < r < b$.

Chapter 8
Green's Functions for ODE's

To introduce the concept of Green's functions we begin with Green's identity associated with linear second order differential operators. Every second order linear differential operator

$$L_x(y) = a_0(x)\frac{d^2y}{dx^2} + a_1(x)\frac{dy}{dx} + a_2(x)y \tag{8.1}$$

has associated with it an adjoint operator

$$L_x^*(y) = \frac{d^2}{dx^2}[a_0(x)y] - \frac{d}{dx}[a_1(x)y] + a_2(x)y$$
$$L_x^*(y) = a_0(x)\frac{d^2y}{dx^2} + (2a_0'(x) - a_1(x))\frac{dy}{dx} + (a_0''(x) - a_1'(x) + a_2(x))y. \tag{8.2}$$

Let $u = u(x)$ and $v = v(x)$ denote two arbitrary continuous functions with first and second derivatives, then one can use the above operators to verify the Lagrange identity

$$vL_x(u) - uL_x^*(v) = \frac{d}{dx}[P(u,v)], \tag{8.3}$$

where

$$P(u,v) = \left[a_0(x)\frac{du}{dx} + a_1(x)u\right]v - \left[a_0(x)\frac{dv}{dx} + a_0'(x)v\right]u \tag{8.4}$$

is called the bilinear concomitant. The Lagrange identity (8.3) holds for all continuous differentiable functions $u(x)$ and $v(x)$ defined over some solution domain $I = \{x| \; a \le x \le b\}$. The functions u and v must be differentiable in order that $L_x(u)$ and $L_x^*(v)$ exist. This is the only restriction we place upon these functions. The integral of the Lagrange identity (8.3) produces the Green's identity

$$\int_a^b [vL_x(u) - uL_x^*(v)]\,dx = \left[P(u,v)\right]_a^b \tag{8.5}$$

where

$$\left[P(u,v)\right]_a^b = \left[a_0(b)u'(b) + a_1(b)u(b)\right]v(b) - \left[a_0(b)v'(b) + a_0'(b)v(b)\right]u(b)$$
$$- \left[a_0(a)u'(a) + a_1(a)u(a)\right]v(a) + \left[a_0(a)v'(a) + a_0'(a)v(a)\right]u(a) \tag{8.6}$$

is an important right-hand side of boundary terms that must be analyzed for each differential operator $L_x(\;)$. An alternative derivation of the Green's identity is to define the inner product

$$(v, L_x(u)) = \int_a^b vL_x(u)\,dx$$

and then integrate this inner product by parts to obtain

$$(v, L_x(u)) = (u, L_x^*(v)) + \text{Boundary terms.}$$

In the sections that follow we shall find out how to use the Green's identity and appropriately choose the functions u and v in order to construct solutions to the two point boundary value problem

$$L_x(y) = f(x), \quad a \le x \le b$$
$$B_1[y] = g_1 \quad \text{and} \quad B_2[y] = g_2, \tag{8.7}$$

where L_x is the linear operator given by equation (8.1), g_1, g_2 are given constants, and B_1, B_2 are linear boundary operators of the Robin form

$$B_1[y] = \left[\alpha_1 \frac{dy(x)}{dx} + \beta_1 y(x)\right]_{x=a}$$
$$B_2[y] = \left[\alpha_2 \frac{dy(x)}{dx} + \beta_2 y(x)\right]_{x=b} \tag{8.8}$$

with α_1, α_2, β_1, β_2 constants, not all zero, and $f(x)$ is an arbitrary, but known, function. In the special case that $\alpha_1 = \alpha_2 = 0$ the boundary operators dictate that y be specified on the boundary of the interval I. In this case the boundary conditions are said to be of the Dirichlet type. In the special case that $\beta_1 = \beta_2 = 0$ the boundary conditions are of the Neumann type. In the special case that $g_1 = g_2 = 0$ the boundary conditions are homogeneous otherwise they are termed nonhomogeneous.

In the following discussions of Green's functions, note that we shall be changing variables in the middle of a problem. In order to avoid confusion as to what variables are being used in the operators $L_x()$ and $L_x^*()$ just check the dummy variable used in the integration of the Green's identity as well as checking the subscript on the differential operators.

As an example of how one can use the Green's identity to solve boundary value problems, consider the Dirichlet type boundary value problem to solve

$$L_x(y) = f(x), \quad a \le x \le b, \tag{8.9}$$

where the solution is subject to the Dirichlet homogeneous boundary conditions

$$B_1[y] = y(a) = 0 \quad \text{and} \quad B_2[y] = y(b) = 0. \tag{8.10}$$

In order to solve this boundary value problem we change the variable x in equations (8.1) and (8.2) to some new variable, ξ and write the Green's identity as

$$\int_a^b \left[v(\xi) L_\xi(u) - u(\xi) L_\xi^*(v) \right] d\xi = \left[P(u(\xi), v(\xi)) \right]_a^b. \tag{8.11}$$

Notice in the equation (8.11) the variable ξ is used as a dummy variable of integration and so the operators L_ξ and L_ξ^* involve derivatives with respect to this variable. We now choose the functions u and v and use Green's identity to solve the boundary value problem given by equations (8.9), and (8.10). We let $u(\xi) = y(\xi)$ be the solution of the boundary value problem (8.9) (with x replaced by ξ) and replace u by y in the above Green's identity. That is, in the Green's identity, given by equation (8.11), we let $L_\xi(y) = f(\xi)$ to obtain

$$\int_a^b v(\xi) f(\xi) \, d\xi - \int_a^b y(\xi) L_\xi^*(v) \, d\xi = \left[P(y(\xi), v(\xi)) \right]_a^b, \tag{8.12}$$

where

$$\begin{aligned} \left[P(y, v) \right]_a^b = {} & \left[a_0(b) y'(b) + a_1(b) y(b) \right] v(b) - \left[a_0(b) v'(b) + a_0'(b) v(b) \right] y(b) \\ & - \left[a_0(a) y'(a) + a_1(a) y(a) \right] v(a) + \left[a_0(a) v'(a) + a_0'(a) v(a) \right] y(a). \end{aligned} \tag{8.13}$$

We now choose $v(\xi) = G^*(\xi; x)$ as the Green's function which satisfies

$$L_\xi^*(G^*(\xi; x)) = \delta(\xi - x), \quad a \leq \xi \leq b \tag{8.14}$$

involving the adjoint equation with the derivatives in L_ξ^* with respect to ξ and where $\delta(\xi - x)$ is the Dirac delta function introduced earlier which has the sifting property

$$\int_{\xi = x - \epsilon}^{\xi = x + \epsilon} y(\xi) \delta(\xi - x) \, d\xi = y(x). \tag{8.15}$$

This substitution for v reduces Green's identity (8.12) to the form

$$\int_a^b G^*(\xi; x) f(\xi) \, d\xi - y(x) = P(y(b), G^*(b; x)) - P(y(a), G^*(a; x)). \tag{8.16}$$

Thus, the solution $y(x)$ to our boundary value problem can be obtained from the equation (8.16) provided that the remaining terms in equation (8.16) can all be evaluated. Let us examine the terms in the equation (8.16). In the first term $f(\xi)$ is given or known from the boundary value problem. The function G^* is the Green's function which must be obtained by solving another differential equation (8.14) involving the adjoint operator $L_\xi^*(\ )$. In the second term we have $y(x)$ as the solution we desire and this solution can be obtained provided we can solve for the Green's function and also that we can evaluate the right hand side of equation (8.16). The right hand side of equation (8.16) can be obtained from equation (8.13) by replacing $v(\xi)$ by $G^*(\xi; x)$ and by setting $B_1[y] = y(a) = 0$ and $B_2[y] = y(b) = 0$ as these are the boundary conditions required by equation (8.10). This gives for the right hand side of equation (8.16) the expression

$$P(y(\xi), G^*(\xi; x))|_a^b = [a_0(b)y'(b)]G^*(b; x) - [a_0(a)y'(a)]G^*(a; x). \tag{8.17}$$

The terms inside the brackets involve derivatives of the unknown solution y and these terms are not known. However, we can remove these unknown terms by requiring that the Green's function satisfy the boundary conditions

$$B_1^*[G^*] = G^*(\xi; x)\big|_{\xi=b} = 0 \quad \text{and} \quad B_2^*[G^*] = G^*(\xi; x)\big|_{\xi=a} = 0. \tag{8.18}$$

These boundary conditions are called adjoint boundary conditions. In general, the adjoint boundary conditions are determined when we set the bilinear concomitant to zero and require $P(y, G^*)\big|_a^b = 0$. The above boundary conditions reduce the Green's identity given by equation (8.16) to

$$y(x) = \int_a^b G^*(\xi; x) f(\xi)\, d\xi, \tag{8.19}$$

where G^* is the Green's function satisfying

$$L_\xi^*(G^*(\xi; x)) = \delta(\xi - x), \quad a \le \xi \le b$$

with the boundary conditions

$$B_1^*[G^*] = G^*(a; x) = 0, \qquad B_2^*[G^*] = G^*(b; x) = 0. \tag{8.20}$$

The function $y(x)$ given by equation (8.19) is a solution of the boundary value problem (8.9) which satisfies the boundary conditions given by equation (8.10). Before we can actually prove it is a solution we must find out more about Green's functions.

Green's Functions and Adjoint Green's Functions

Define two different Green's functions G and G^* associated with the operators L_x and L_x^* given by the equations (8.1) and (8.2). We require

$$G = G(x;\xi) \quad \text{satisfy} \quad L_x(G(x;\xi)) = \delta(x - \xi), \quad a \le x \le b \qquad (8.21)$$

with boundary conditions $B_1[G] = 0$ and $B_2[G] = 0$. The adjoint Green's function $G^* = G^*(x;\xi)$ is required to satisfy

$$L_x^*(G^*(x;\xi)) = \delta(x - \xi), \quad a \le x \le b \qquad (8.22)$$

with adjoint boundary conditions $B_1^*[G^*] = 0$ and $B_2^*[G^*] = 0$ which are determined by an analysis of the bilinear concomitant in the Green's identity. In the above equations, all differentiation is with respect to x.

The operators (L_x, B_1, B_2) and (L_x^*, B_1^*, B_2^*) are called adjoint operators. The adjoint boundary conditions are chosen such that $P(G, G^*)\big|_a^b = 0$. Under these circumstances the Green's functions given by equations (8.21) and (8.22) will satisfy the relation

$$G^*(x;\xi) = G(\xi;x) \qquad (8.23)$$

of symmetry with respect to the ξ and x variables. To prove this result we first multiply equation (8.21) by $G^*(x;t)$ and then change the variable ξ in equation (8.22) to t. We can then multiply equation (8.22) by $G(x;\xi)$ to obtain

$$\begin{aligned}
G^*(x;t)L_x(G(x;\xi)) &= G^*(x;t)\delta(x - \xi) \qquad \text{and} \\
G(x;\xi)L_x^*(G^*(x;t)) &= G(x;\xi)\delta(x - t).
\end{aligned} \qquad (8.24)$$

We subtract these two equations and then integrate from $x = a$ to $x = b$. We thus produce the Green's identity

$$\begin{aligned}
P(G, G^*)\big|_a^b &= \int_a^b [G^*(x;t)L_x(G(x;\xi)) - G(x;\xi)L_x^*(G^*(x;t))]\, dx \\
&= \int_a^b [G^*(x,t)\delta(x - \xi) - G(x;\xi)\delta(x - t)]\, dx \\
&= G^*(\xi;t) - G(t;\xi).
\end{aligned} \qquad (8.25)$$

Observe that the bilinear concomitant $P(G, G^*)\big|_a^b$ equals zero because of our choice of the adjoint boundary conditions and consequently, the relation given by equation (8.23) follows. The symmetry of the Green's function allows one to express the solution to the previous Dirichlet problem given by equation (8.19) in either of the forms

$$y(x) = \int_a^b G^*(\xi;x)f(\xi)\,d\xi = \int_a^b G(x;\xi)f(\xi)\,d\xi. \tag{8.26}$$

Green's Function by Laplace Transform

Despite the fact that some Green's functions are defined by differential equations over finite intervals, one can still use the Laplace transform to solve for the Green's function. The following example will illustrate how this is done.

Example 8-1. (Laplace Transforms)
Using the Laplace transform find the Green's function for the operator

$$L_x(y) = \frac{d^2y}{dx^2}, \quad a \le x \le b \quad \text{with boundary conditions } y(a) = y(b) = 0.$$

Solution: The Green's function associated with the given differential operator is obtain as a solution of the differential equation

$$\frac{d^2G}{dx^2} = \delta(x - \xi), \quad a \le x \le b, \quad G(a) = G(b) = 0 \tag{8.27}$$

We can still treat this differential equation over the interval $0 < x < \infty$, however, we will only be interested in the solution over the interval $a < x < b$. We let

$$\mathcal{L}\{G\} = \int_0^\infty G\,e^{-sx}\,dx = \overline{G}(s)$$

denote the Laplace transform of G. Upon taking the Laplace transform of the equation (8.27) there is obtained

$$\mathcal{L}\{G''\} = \mathcal{L}\{\delta(x - \xi)\} \qquad \text{or}$$
$$s^2\overline{G} - sG(0) - G'(0) = e^{-s\xi}. \tag{8.28}$$

Now $G(0)$ and $G'(0)$ are unknowns and so we replace these values with constants A and B to be determined at a later time. Solving the equation (8.28) for $\overline{G}$ gives

$$\overline{G} = \frac{A}{s} + \frac{B}{s^2} + \frac{e^{-s\xi}}{s^2}.$$

Taking the inverse Laplace transform of this equation produces

$$G = G(x;\xi) = A + Bx + (x - \xi)H(x - \xi), \tag{8.29}$$

where $H(x - \xi)$ is the Heaviside unit step function and A and B are constants to be determined. This equation implies that for $a \leq x < \xi$ we have $G = A + Bx$ and the condition $G(a) = 0$ requires that

$$A + Ba = 0. \tag{8.30}$$

Similarly, for $\xi < x \leq b$, the equation (8.29) becomes $G = A + Bx + x - \xi$ and the condition $G(b) = 0$ requires that

$$A + Bb + b - \xi = 0. \tag{8.31}$$

Solving the simultaneous equations (8.30) and (8.31) for the constants A and B we find the Green's function can be represented

$$G = G(x; \xi) = \frac{(\xi - b)(x - a)}{b - a} + (x - \xi)H(x - \xi). \tag{8.32}$$

■

Example 8-2. (Greens function.)
Consider the Green's function problem to solve

$$L_x(G) = \frac{d^2 G}{dx^2} + 3\frac{dG}{dx} + 2G = \delta(x - \xi), \qquad 0 < x < \ell \tag{8.33}$$

subject to the homogeneous boundary conditions $G(0) = 0$ and $G(\ell) = 0$. This equation can be solved by the method of Laplace transforms to produce the solution

$$G = G(x; \xi) = \left(\frac{e^{-2(\ell-\xi)} - e^{-(\ell-\xi)}}{e^{-\ell} - e^{-2\ell}} \right) \left(e^{-x} - e^{-2x} \right) + \left(e^{-(x-\xi)} - e^{-2(x-\xi)} \right) H(x - \xi) \tag{8.34}$$

which can also be expressed in the form

$$G = G(x; \xi) = \begin{cases} G_1(x; \xi) = e^{-2x+\xi} \left[\frac{e^x - 1}{e^\ell - 1} \right] (e^\xi - e^\ell), & x < \xi \\ G_2(x; \xi) = e^{-2x+\xi} \left[\frac{e^\xi - 1}{e^\ell - 1} \right] (e^x - e^\ell), & \xi < x \end{cases} \tag{8.35}$$

Consider also the adjoint Green's function problem to solve

$$L_x^*(G^*) = \frac{d^2 G^*}{dx^2} - 3\frac{dG^*}{dx} + 2G^* = \delta(x - \xi), \qquad 0 < x < \ell \tag{8.36}$$

subject to the homogeneous boundary conditions $G^*(0) = 0$ and $G^*(\ell) = 0$. One can use the method of Laplace transforms and readily verify that the above boundary value problem has the solution

$$G^* = G^*(x;\xi) = \left(\frac{e^{\ell-\xi} - e^{2(\ell-\xi)}}{e^{2\ell} - e^\ell}\right)(e^{2x} - e^x) + \left[e^{2(x-\xi)} - e^{(x-\xi)}\right]H(x-\xi). \qquad (8.37)$$

which can also be expressed in the form

$$G^* = G^*(x;\xi) = \begin{cases} G_1^*(x;\xi) = e^{x-2\xi}\left[\frac{e^x-1}{e^\ell-1}\right](e^\xi - e^\ell), & x < \xi \\ G_2^*(x;\xi) = e^{x-2\xi}\left[\frac{e^\xi-1}{e^\ell-1}\right](e^x - e^\ell), & \xi < x \end{cases} \qquad (8.38)$$

Observe that if the variables x and ξ are interchanged in the above equation (8.35), then the inequalities $x < \xi$ and $\xi < x$ must also be interchanged. Note that by interchanging x and ξ in G_1 we obtain G_2^*. Similarly, if we interchange x and ξ in G_2 we obtain G_1^*. In this sense we can write $G(x;\xi) = G^*(\xi;x)$.

We purposely confuse the issue the solving the equations

$$L_\xi\left(G(\xi;x)\right) = \frac{d^2G}{d\xi^2} + 3\frac{dG}{d\xi} + 2G = \delta(\xi - x), \quad 0 < \xi < \ell \qquad (8.39)$$

subject to the boundary conditions $G(0;x) = 0$ and $G(\ell;x) = 0$ to obtain the solution

$$G(\xi;x) = \begin{cases} e^{-2\xi+x}\left[\frac{e^\xi-1}{e^\ell-1}\right](e^x - e^\ell), & \xi < x \\ e^{-2\xi+x}\left[\frac{e^x-1}{e^\ell-1}\right](e^\xi - e^\ell), & x < \xi \end{cases} \qquad (8.40)$$

Now note that $G(\xi;x) = G^*(x;\xi)$. Similarly, the equation

$$L_\xi^*\left(G^*(\xi;x)\right) = \frac{d^2G^*}{d\xi^2} - 3\frac{dG^*}{d\xi} + 2G^* = \delta(\xi - x), \quad 0 < x < \ell \qquad (8.41)$$

subject to the boundary conditions $G^*(0;x) = 0$ and $G^*(\ell;x) = 0$ has the solution

$$G^*(\xi;x) = \begin{cases} e^{\xi-2x}\left[\frac{e^\xi-1}{e^\ell-1}\right](e^x - e^\ell), & \xi < x \\ e^{\xi-2x}\left[\frac{e^x-1}{e^\ell-1}\right](e^\xi - e^\ell), & x < \xi \end{cases} \qquad (8.42)$$

Now note that $G^*(\xi;x) = G(x;\xi)$.

In summary, we observe that

$$G = G(x;\xi) \quad \text{satisfies } L_x(G) = \delta(x-\xi) \text{ in the } x\text{-variable}$$
$$G^* = G^*(\xi;x) = G(x;\xi) \quad \text{satisfies } L_\xi^*(G) = \delta(\xi-x) \text{ in the } \xi\text{-variable}$$
$$G^* = G^*(x;\xi) \quad \text{satisfies } L_x^*(G^*) = \delta(x-\xi) \text{ in the } x\text{-variable}$$
$$G = G(\xi;x) = G^*(x;\xi) \quad \text{satisfies } L_\xi(G^*) = \delta(\xi-x) \text{ in the } \xi\text{-variable}.$$

One can therefore consider either of the equations

$$L_x(G) = \delta(x-\xi) \quad \text{or} \quad L_\xi^*(G) = \delta(\xi-x)$$

as defining the Green's function $G(x;\xi)$.

The solution to equation (8.21) is called a direct Green's function and the solution to equation (8.22) is call the adjoint Green's function. You can interpret these Green's functions as follows: $G(x;\xi)$ satisfies the direct problem given by equation (8.21) in the x variable and it satisfies the adjoint Green's function problem in the ξ variable. We then obtain the equations

$$L_x(G(x;\xi)) = \delta(x-\xi), \qquad a \le x \le b$$
$$L_\xi^*(G(x;\xi)) = \delta(\xi-x), \qquad a \le \xi \le b.$$

One may use either of these equations to calculate the Green's functions.

In the special case L_x is a linear second order self-adjoint differential operator, we have $L_x = L_x^*$ and therefore $G = G^*$. In this case the Green's function is symmetric in ξ and x. That is, since the equations are self-adjoint, the condition $G^*(x;\xi) = G(\xi;x)$ reduces to $G(x;\xi) = G(\xi;x)$. This is particularly useful as all second order linear differential operators can be written in self-adjoint form.

Example 8-3. (Greens function.)
In the case of self-adjoint operators the Green's function representation becomes simpler since $G = G^*$. We have shown that every second order differential equation

$$L_{1x}[y] = a_0(x)\frac{d^2y}{dx^2} + a_1(x)\frac{dy}{dx} + a_2(x)y = 0, \quad a_0(x) \ne 0 \tag{8.43}$$

can be written in the self-adjoint form

$$L_x[y] = \frac{d}{dx}\left(p(x)\frac{dy}{dx}\right) + q(x)y = 0. \tag{8.44}$$

This is accomplished by multiplying the equation (8.43) by the function

$$\mu(x) = \frac{1}{a_0(x)}e^{\int \frac{a_1(x)}{a_0(x)}dx}, \quad a_0(x) \ne 0. \tag{8.45}$$

For example, consider the differential equation

$$L_{1x}[y] = \frac{d^2y}{dx^2} + 3\frac{dy}{dx} + 2y = 0, \quad 0 < x < \ell \tag{8.46}$$

with homogeneous boundary conditions $y(0) = 0$ and $y(\ell) = 0$. We multiply the equation (8.46) by $\mu(x) = e^{3x}$ to obtain the self-adjoint form

$$L_x[y] = \frac{d}{dx}\left(e^{3x}\frac{dy}{dx}\right) + 2e^{3x}y = 0. \tag{8.47}$$

The Green's function associated with this operator satisfies

$$L_x[G] = \frac{d}{dx}\left(e^{3x}\frac{dG}{dx}\right) + 2e^{3x}G = \delta(x - \xi). \tag{8.48}$$

Observe that if we make the change of variable $u = e^{3x}G$, the differential equation (8.48) becomes

$$\frac{d^2u}{dx^2} - 3\frac{du}{dx} + 2u = \delta(x - \xi) \tag{8.49}$$

which is the same differential equation solved in the previous example. We find the solution

$$u = e^{3x}G = \begin{cases} e^{x-2\xi}\left[\frac{e^x-1}{e^\ell-1}\right](e^\xi - e^\ell), & x < \xi \\ e^{x-2\xi}\left[\frac{e^\xi-1}{e^\ell-1}\right](e^x - e^\ell), & \xi < x \end{cases}$$

and consequently we obtain the Green's function

$$G = G(x;\xi) = \begin{cases} G_1(x;\xi) = e^{-2x-2\xi}\left[\frac{e^x-1}{e^\ell-1}\right](e^\xi - e^\ell), & x < \xi \\ G_2(x;\xi) = e^{-2x-2\xi}\left[\frac{e^\xi-1}{e^\ell-1}\right](e^x - e^\ell), & \xi < x \end{cases} \tag{8.50}$$

Observe that when the operator L_x is self-adjoint we have the symmetry conditions $G_2(x;\xi) = G_1(\xi;x)$ and similarly $G_1(x;\xi) = G_2(\xi;x)$. This is an example of a self-adjoint operator where the associated Green's function is symmetric and satisfies $G(x;\xi) = G(\xi;x)$. The symmetry relation $G(x;\xi) = G(\xi;x)$ is known as Maxwell's reciprocity principle which can be interpreted as the response at point x due to an impulse at the point ξ is the same as the response at ξ due to an impulse at the point x. For example, $G(\frac{1}{3}, \frac{2}{3}) = G(\frac{2}{3}, \frac{1}{3})$. To visualize this we plot the two functions $y_1 = G(x; \frac{2}{3})$ and $y_2 = G(x; \frac{1}{3})$ in the figure 8-1 for the special case where $\ell = 1$.

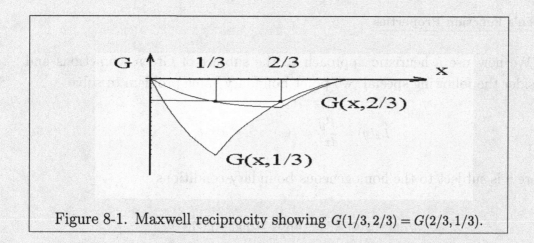

Figure 8-1. Maxwell reciprocity showing $G(1/3, 2/3) = G(2/3, 1/3)$.

Using the result from equation (8.23) we can now show the boundary value problem defined by the equations (8.9) and (8.10) has the solution given by equation (8.19) which can be written as

$$y(x) = \int_a^b G^*(\xi; x) f(\xi)\, d\xi = \int_a^b G(x; \xi) f(\xi)\, d\xi.$$

Let $L_x()$ operate on the above function $y(x)$. We obtain

$$L_x(y) = \int_a^b L_x(G(x; \xi)) f(\xi)\, d\xi = \int_a^b \delta(x - \xi) f(\xi)\, d\xi = f(x)$$

and further the boundary conditions are satisfied since G was chosen to satisfy these boundary conditions.

To summarize, we have shown from an analysis of Green's identity that if we choose G to be a solution of the differential equation

$$L_x(G(x; \xi)) = \delta(x - \xi) \quad a \le x \le b$$

and if we choose boundary conditions for $G(x; \xi)$ to simplify the Green's identity, then a solution of a related boundary value problem can be constructed.

The differential equation (8.21) defining G over the interval (a, b) is a short hand notation for the Dirac delta function dividing the interval (a, b) into two parts. We may write this differential equation as two differential equations

$$L_x(G) = 0, \quad a \le x < \xi \qquad \text{and}$$

$$L_x(G) = 0, \quad \xi < x \le b.$$

Our goal is to solve equations like those given above. We consider this problem in the next section.

Green's Function Properties

We now use a heuristic approach to the subject of Green's functions and consider the following special two point boundary value problem to solve

$$L_x(y) = \frac{d^2y}{dx^2} = f(x), \quad a \le x \le b, \tag{8.51}$$

where y is subject to the homogeneous boundary conditions

$$B_1[y] = y(a) = 0, \quad \text{and} \quad B_2[y] = y(b) = 0,$$

where $f(x)$ can be any arbitrary but known function. In equation (8.51), the right-hand side $f(x)$ can be thought of as an input to a system and the solution $y(x)$ of the differential equation can be thought of as an output or response of the system to the input $f(x)$. The Green's function associated with the above differential operator satisfies the differential equation

$$\frac{d^2G}{dx^2} = \delta(x - \xi), \tag{8.52}$$

where G is interpreted as the response at point x due to a unit impulse at the point ξ. In equation (8.52), the function $\delta(x - \xi)$ is the Dirac delta function satisfying $\delta(x - \xi) = 0$ for $x \ne \xi$ and the sifting property $\int_{-\infty}^{\infty} f(\xi)\delta(x - \xi)\, d\xi = f(x)$. This input is a shorthand notation for the function

$$\delta_\epsilon = \frac{1}{\epsilon}[H(x - \xi) - H(x - (\xi + \epsilon))]$$

in the limit as ϵ tends toward zero. We actually solve the differential equation $\frac{d^2y}{dx^2} = \delta_\epsilon$ and the solution is denoted $y = y(x, \xi; \epsilon)$. If the limit $\lim_{\epsilon \to 0} y(x, \xi; \epsilon)$ exists, this limit is denoted $G = G(x; \xi)$ and called the Green's function associated with the operator $L_x = \frac{d^2}{dx^2}$.

In equation (8.52) the input $\delta(x - \xi)$ is certainly not a continuous function as it is not defined at $x = \xi$. It is treated as a unit impulse occurring at the point $x = \xi$ which is placed somewhere between the end points a and b as illustrated in figure 8-2.

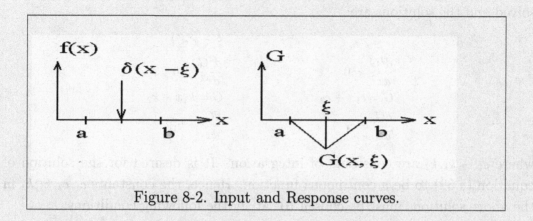

Figure 8-2. Input and Response curves.

The solution $G = G(x;\xi)$ of equation (8.52) can be thought of as the response of the differential equation, over the interval (a,b) to a unit impulse at $x = \xi$. Since $\delta(x - \xi)$ is not continuous, then it is expected the derivatives of G are not continuous. It was shown earlier we could use the interpretation

$$\frac{dH(x-\xi)}{dx} = \delta(x-\xi),\tag{8.53}$$

where H is the Heaviside unit step function. By integrating the differential equation (8.52) we obtain

$$\frac{dG}{dx} = H(x-\xi) + c,\tag{8.54}$$

where c is a constant of integration. In equation (8.54) hold ξ constant and vary x and observe the equation (8.54) implies that there is an abrupt change in the slope of the response curve G at the point $x = \xi$, and as x passes from ξ^- to ξ^+ the derivative changes by unity, or

$$\frac{\partial G(x;\xi)}{\partial x}\bigg|_{x=\xi^-} - \frac{\partial G(x;\xi)}{\partial x}\bigg|_{x=\xi^+} = c - (c+1) = -1.\tag{8.55}$$

The jump discontinuity condition given by equation (8.55) can be expressed as

$$\frac{\partial G(x;\xi)}{\partial x}\bigg|_{x=\xi^-}^{x=\xi^+} = 1.\tag{8.56}$$

For $x \neq \xi$ we have $\delta(x - \xi) = 0$ so the differential equation (8.52) can be considered as two differential equations, one defined over the interval $a \leq x < \xi$, and the other defined over the interval $\xi < x \le b$. These equations are easily solved and the solutions are:

$$a \le x < \xi \qquad\qquad\qquad \xi < x \le b$$
$$\frac{d^2 G}{dx^2} = 0 \qquad\qquad\qquad \frac{d^2 G}{dx^2} = 0$$
$$G = c_1 x + c_2 \qquad\qquad\qquad G = k_1 x + k_2$$
$$\frac{dG}{dx} = c_1 \qquad\qquad\qquad \frac{dG}{dx} = k_1,$$

where c_1, c_2, k_1, k_2 are constants of integration. It is desired for the solution of equation (8.52) to be a continuous function. Hence, the constants c_1, c_2, k_1, k_2 in the above solutions must be chosen to satisfy the following conditions:

(a) The two solutions must meet at $x = \xi$ and hence we have the continuity requirement at $x = \xi$

$$G\big|_{x=\xi^-} = G\big|_{x=\xi^+}.$$

(b) At the point $x = \xi$ we have found the change in the slope must satisfy the jump discontinuity condition

$$\frac{dG}{dx}\Big|_{x=\xi^-} - \frac{dG}{dx}\Big|_{x=\xi^+} = -1.$$

(c) Further we must determine boundary conditions for the Green's functions. It is important to observe the boundary conditions are obtained from an analysis of the Green's identity associated with the operator L_x. For this example the Green's identity is

$$\int_a^b G \frac{d^2 y}{d\xi^2}\, d\xi - \int_a^b y \frac{d^2 G}{d\xi^2}\, d\xi = \left[G \frac{dy}{d\xi} - y \frac{dG}{d\xi} \right]_a^b$$
$$= G(b)y'(b) - y(b)G'(b) - G(a)y'(a) + y(a)G'(a)$$
$$= G(b)y'(b) - G(a)y'(a),$$

where we have used the hypothesis that the boundary conditions on y are $y(a) = y(b) = 0$. In the Green's identity the terms $y'(b)$ and $y'(a)$ are unknown terms and so we must require G to satisfy the conditions $B_2^*[G] = G(b) = 0$ and $B_1^*[G] = G(a) = 0$ as this choice of boundary conditions removes the unknown terms. Then all the terms in the Green's identity can be evaluated.

We now have four conditions, two boundary conditions plus a continuity condition as well as a jump discontinuity condition and these conditions imply the four constants c_1, c_2, k_1, k_2 must satisfy the relations:

$$c_1\xi + c_2 = k_1\xi + k_2 \qquad \text{continuity}$$
$$c_1 - k_1 = -1 \qquad \text{jump discontinuity}$$
$$c_1 a + c_2 = 0 \qquad \text{boundary condition}$$
$$k_1 b + k_2 = 0 \qquad \text{boundary condition.}$$

(8.57)

Solving this system of equations for the constants c_1, c_2, k_1, k_2 there results the solution

$$G = G(x;\xi) = \begin{cases} \frac{(\xi-b)(x-a)}{b-a}, & a \le x < \xi \\ \frac{(\xi-a)(x-b)}{b-a}, & \xi < x \le b \end{cases}$$

(8.58)

which can be expressed in a form involving the unit step function as

$$G(x;\xi) = \frac{(\xi-b)(x-a)}{b-a} + (x-\xi)H(x-\xi).$$

The function G is called the Green's function associated with the operators (L_x, B_1, B_2) where $L_x = \frac{d^2}{dx^2}$ is a differential operator and the boundary operators are B_1, B_2. Knowing the Green's function associated with these operators, then nonhomogeneous differential equations associated with L_x may be solved provided the boundary conditions are compatible. Of course if the boundary conditions B_1, B_2 change, then we must go through another analysis of the Green's identity to find out the new adjoint boundary conditions to assign to the Green's function. This in turn produces another Green's function associated with the operator L_x. Analysis of the boundary conditions is an important step in the process of constructing a Green's function.

Let us return to the two point boundary value problem

$$L_x(y) = \frac{d^2y}{dx^2} = f(x), \quad a \le x \le b$$

(8.59)

subject to the boundary conditions

$$B_1[y] = y(a) = 0, \qquad B_2[y] = y(b) = 0.$$

The solution of equation (8.59) can be related to the solution of the Green's function problem

$$\frac{d^2G}{dx^2} = \delta(x-\xi), \qquad a \le x \le b$$
$$B_1^*[G] = G(a) = 0, \qquad B_2^*[G] = G(b) = 0.$$

(8.60)

442

One can argue as follows: Let $G(x;\xi)$ denote the response due to a unit impulse at $x = \xi$. Using the fact L_x is a linear operator, we then have $f(\xi)G(x;\xi)$ as the response due to the input $f(\xi)\delta(x - \xi)$, because $f(\xi)$ is constant, if ξ is considered constant.

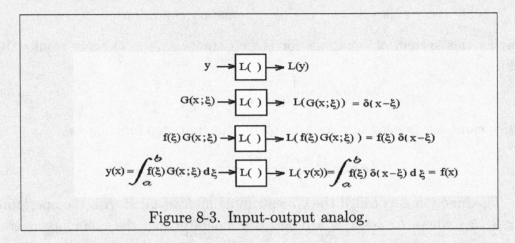

Figure 8-3. Input-output analog.

By summing the responses for all values of ξ between a and b we are using the superposition of the respective solutions and this gives

$$y(x) = \int_a^b f(\xi)G(x;\xi)\,d\xi \qquad (8.61)$$

as the response due to the input

$$\int_a^b f(\xi)\delta(x - \xi)\,d\xi = f(x). \qquad (8.62)$$

This superposition is illustrated in figure 8-3. Thus, we find that the equation (8.61) is the solution of the differential equation (8.59) and satisfies the given boundary conditions. This may be verified by direct differentiation. Writing the equation (8.61) in the expanded form

$$y(x) = \int_a^x f(\xi)\frac{(\xi - a)(x - b)}{b - a}\,d\xi + \int_x^b f(\xi)\frac{(\xi - b)(x - a)}{b - a}\,d\xi \qquad (8.63)$$

there results upon differentiating this function

$$\frac{dy}{dx} = \frac{1}{b - a}\left[\int_a^x f(\xi)(\xi - a)\,d\xi + \int_x^b f(\xi)(\xi - b)\,d\xi\right]$$

and $\qquad \dfrac{d^2y}{dx^2} = f(x).$

Further the boundary conditions $y(a) = y(b) = 0$ are satisfied since G has been constructed to satisfy $G(a; \xi) = G(b; \xi) = 0$.

In terms of our input-output analogy, we can think of $f(x)$ as an input signal which gets operated upon by an integral operator to produce the output. Here $G(x; \xi)$ is called the kernel of the integral operator.

Another way of interpreting the results obtained is in terms of operators. If $L_x(y) = f(x)$, where L_x is a linear differential operator, we would like to obtain an inverse operator L_x^{-1} which has the property

$$L_x^{-1} L_x(y) = y = L_x^{-1}(f(x))$$

so that $y = L_x^{-1}(f(x))$ is the solution of the given differential equation. We can think of the Green's function solution, such as equation (8.61), as being the desired inverse operator associated with the operators (L_x, B_1, B_2) defining the boundary value problem.

Terminology

The function $G(x; \xi)$ is called by different names in different branches of science and engineering. Some of the more popular names are: Green's function; kernel function; point source function; impulse response function; influence function. In the next chapter we discuss Green's functions associated with partial differential equations where the right-hand side input can be a function of more than one variable, for example the input might appear as

$$f(x_1, x_2, \ldots, x_n, t),$$

where there are n space variables and a time variable $t \geq t_0$. The integral operator which describes the output might be represented as

$$y(x_1, \ldots, x_n, t) = \int_{t_0}^{t} \int \cdots \int_{D} G(x_1, \ldots, x_n, t; \xi_1, \ldots, \xi_n, \tau) f(\xi_1, \ldots, \xi_n, \tau) \, d\xi_1 \cdots d\xi_n d\tau,$$

where D is the open set where the solution is desired. In the case of two variables x, t, where the input is $f(x, t)$, the kernel of the integral operator is denoted $G(x, t; \xi, \tau)$. In the case where G does not depend upon the origin of the time variable t, τ but only on the difference $t - \tau$, it is customary to write the Green's function as $G(x, \xi, t - \tau)$. In such cases it is convenient to use Laplace transforms with respect to the time variables and take the time origin as zero. In this case we can write the Laplace transform as

$$\overline{f}(x, s) = \int_0^{\infty} e^{-st} f(x, t) \, dt$$

so that the Laplace transform of G can be written

$$\overline{G}(x,\xi,s) = \int_0^\infty e^{-st} G(x,\xi,t)\, dt.$$

In this case $\overline{G}$ is referred to as a transfer function. If we replace s by $i\omega$ in the above equation, we obtain the Fourier transform with respect to time and in this special case $\overline{G}$ is referred to as a frequency function.

Green's Functions for Other Linear Operators

The previous discussions are now generalized to nth order linear differential equations. A Green's function can be associated with a nth order linear differential equation

$$L_x(y) = a_0(x)\frac{d^n y}{dx^n} + a_1(x)\frac{d^{n-1}y}{dx^{n-1}} + \ldots + a_{n-1}(x)\frac{dy}{dx} + a_n(x)y = 0, \quad a_0(x) \neq 0 \quad (8.64)$$

over an interval $a \leq x \leq b$ with n boundary conditions of the type

$$\begin{aligned}
B_i[y] &= \alpha_{0i}y(a) + \alpha_{1i}y'(a) + \alpha_{2i}y''(a) + \ldots + \alpha_{n-1,i}y^{(n-1)}(a) \\
&+ \beta_{0i}y(b) + \beta_{1i}y'(b) + \beta_{2i}y''(b) + \ldots + \beta_{n-1,i}y^{(n-1)}(b) = 0
\end{aligned} \quad (8.65)$$

for $i = 1, 2, \ldots, n$, where $\alpha_{m,i}$, $\beta_{m,i}$ are constants, for $m = 0, 1, \ldots, n-1$ provided the equation (8.64) possesses no nontrivial continuous solutions with $(n-1)$ continuous derivatives over the interval (a, b). The Green's function together with its first $(n-2)$ derivatives is to be a continuous function. The $(n-1)st$ derivative of the Green's function has a jump discontinuity at $x = \xi$ and the Green's function is to formally satisfy the differential equation (8.64). The Green's function has the properties:

(a) Over the interval (a, b) the Green's function $G(x; \xi)$ must be continuous and its first $(n-2)$ derivatives must be continuous at the point $x = \xi$.

(b) The $(n-1)st$ derivative has a jump discontinuity at the point $x = \xi$ in the interval (a, b) and satisfies:

$$\frac{\partial^{n-1}G(x;\xi)}{\partial x^{n-1}}\bigg|_{x=\xi^-} - \frac{\partial^{n-1}G(x;\xi)}{\partial x^{n-1}}\bigg|_{x=\xi^+} = \frac{-1}{a_0(\xi)} \quad (8.66)$$

where $a_0(x)$ is the coefficient of the highest ordered derivative in the operator $L_x(y)$ given by equation (8.64). The proof of this statement follows from an integration of the differential equation $L_x(G) = \delta(x - \xi)$ and performing an analysis of the results.

(c) The Green's function satisfies the boundary conditions $B_i[G] = 0$, $i = 1, \ldots, n$.

(d) The Green's function, with ξ fixed, must satisfy the differential equation (8.64) over each of the intervals $a \leq x < \xi$ and $\xi < x \leq b$.

(e) If L_x is self-adjoint and $L_x = L_x^*$, then $G(x; \xi) = G(\xi; x)$ and the Green's function is symmetric.

(f) Associated with the differential operator $L_x()$ and boundary operators $B_i[]$ there is an adjoint operator $L_x^*()$ and adjoint boundary conditions $B_i^*[]$ which are determined from an analysis of Green's identity. Sometimes these operators are helpful in constructing the Green's function associated with the operator $L_x()$ and given boundary conditions.

Construction of Green's Function

To construct a Green's function, first obtain a fundamental set of solutions $\{y_1(x), y_2(x), \ldots, y_n(x)\}$ to the differential equation (8.64). The Green's function G is to be a solution of the differential equation (8.64), and so it must be some linear combination of this fundamental set. We can assume a solution over the two intervals $a \leq x < \xi$ and $\xi < x \leq b$ of the form

$$G = G(x; \xi) = \begin{cases} c_1 y_1(x) + c_2 y_2(x) + \cdots + c_n y_n(x), & a \leq x < \xi \\ k_1 y_1(x) + k_2 y_2(x) + \cdots + k_n y_n(x), & \xi < x \leq b \end{cases}, \qquad (8.67)$$

where $c_1, \ldots, c_n, k_1, \ldots, k_n$ are $2n$ constants to be determined.

Consider G and its first $(n-1)$ derivatives over the interval (a, b). The function G and its first $(n-2)$ derivatives must satisfy continuity relations at $x = \xi$ and the $(n-1)st$ derivative must satisfy the jump discontinuity relation. These conditions together with the boundary conditions $B_i[y] = 0$, $i = 1, \ldots, n$, give $2n$ equations from which we can solve for the $2n$ unknowns $c_1, \ldots, c_n, k_1, \ldots, k_n$.

Green's Functions for Second Order Differential Equations

The above generalization is now applied to second order linear differential equations of the form

$$a_0(x)y'' + a_1(x)y' + a_2(x)y = 0, \qquad a_0(x) \neq 0 \qquad (8.68)$$

which can be transformed into the self-adjoint form

$$L(y) = \frac{d}{dx}\left[p(x)\frac{dy}{dx}\right] + q(x)y = 0. \qquad (8.69)$$

For convenience, we assume boundary conditions to be of the type

$$B_1[y] = \alpha_{01}y(a) + \alpha_{11}y'(a) = 0$$
$$B_2[y] = \beta_{02}y(b) + \beta_{12}y'(b) = 0$$

(8.70)

where $\alpha_{01}, \alpha_{11}, \beta_{02}, \beta_{12}$ are constants.

We let $y_1(x)$, $y_2(x)$ denote a fundamental set associated with the differential equation (8.69). It should be observed that the algebra involved in constructing the Green's function solution is much easier if we construct $y_1(x)$ to satisfy the boundary condition at $x = a$ and construct $y_2(x)$ to satisfy the boundary condition at $x = b$ as this reduces the equation (8.67) to the form

$$G = G(x;\xi) = \begin{cases} c_1 y_1(x), & a \le x < \xi \\ k_2 y_2(x), & \xi < x \le b \end{cases},$$

(8.71)

where now there are only two constants c_1 and k_2 to be determined. The continuity condition at $x = \xi$ requires that

$$c_1 y_1(\xi) = k_2 y_2(\xi)$$

(8.72)

and the jump discontinuity condition requires that

$$c_1 y_1'(\xi) - k_2 y_2'(\xi) = \frac{-1}{p(\xi)}.$$

(8.73)

These two simultaneous equations enable us to solve for the constants c_1, k_2 and we find:

$$c_1 = \frac{y_2(\xi)}{p(\xi)W(\xi)} \qquad k_2 = \frac{y_1(\xi)}{p(\xi)W(\xi)},$$

where

$$W(\xi) = \begin{vmatrix} y_1(\xi) & y_2(\xi) \\ y_1'(\xi) & y_2'(\xi) \end{vmatrix}$$

is the Wronskian determinant of the independent solutions in the fundamental set. The equation (8.71) can then be written in the form

$$G = G(x;\xi) = \begin{cases} \frac{y_1(x)y_2(\xi)}{p(\xi)W(\xi)}, & a \le x < \xi \\ \frac{y_1(\xi)y_2(x)}{p(\xi)W(\xi)}, & \xi < x \le b \end{cases}$$

(8.74)

which can also be written in the alternate form

$$G(x;\xi) = \frac{y_1(x)y_2(\xi)}{p(\xi)W(\xi)} + \left[\frac{y_2(x)y_1(\xi) - y_1(x)y_2(\xi)}{p(\xi)W(\xi)} \right] H(x - \xi)$$

(8.75)

involving the unit step function.

Example 8-4. (Greens Function.)

Verify the results given by equation (8.74) for the operator $L_x(y) = \frac{d^2y}{dx^2}$ with boundary conditions $y(a) = y(b) = 0$.

Solution: The general solution of $\frac{d^2y}{dx^2} = 0$ is $y = Ax + B$ with A, B constants. The solution $y_1(x) = x - a$ is constructed to satisfy the boundary condition at $x = a$ and the solution $y_2(x) = x - b$ is constructed to satisfy the boundary condition at $x = b$. Further, the Wronskian of these two solutions is

$$W(\xi) = \begin{vmatrix} \xi - a & \xi - b \\ 1 & 1 \end{vmatrix} = b - a \neq 0.$$

For this problem $p(x) = 1$ and therefore the equation (8.74) can be written in the form

$$G = G(x; \xi) = \begin{cases} \frac{(x-a)(\xi-b)}{b-a}, & a \leq x < \xi \\ \frac{(\xi-a)(x-b)}{b-a}, & \xi < x \leq b. \end{cases}$$

which agrees with our previous result. ∎

Two Point Boundary Value Problem

Consider the problem of determining the solution of the second order, linear, nonhomogeneous differential equation

$$L_x(y) = \frac{d}{dx}\left[p(x)\frac{dy}{dx}\right] + q(x)y = f(x), \quad a \leq x \leq b \tag{8.76}$$

subject to the homogeneous boundary conditions

$$B_1[y] = \alpha y(a) + \beta y'(a) = 0, \qquad \alpha^2 + \beta^2 \neq 0$$
$$B_2[y] = \gamma y(b) + \delta y'(b) = 0, \qquad \gamma^2 + \delta^2 \neq 0 \tag{8.77}$$

where α, β, γ, δ are constants. The solution to this problem can be determined from the Green's identity associated with the operator L_x. This identity can be written

$$\int_a^b [vL_\xi(u) - uL_\xi(v)]\, d\xi = \left[\left(p(\xi)\frac{du}{d\xi} + p'(\xi)u\right)v - \left(p(\xi)\frac{dv}{d\xi} + p'(\xi)v\right)u\right]_a^b.$$

By letting $v = G$ be a solution of $L_\xi(G) = \delta(\xi - x)$ and setting $u = y(\xi)$, with $L_\xi(y) = f(\xi)$, the above Green's identity becomes

$$\int_a^b G(x;\xi)f(\xi)\, d\xi - \int_a^b y(\xi)\delta(x-\xi)\, d\xi = p(b)y'(b)G(b) - p(b)G'(b)y(b)$$
$$- p(a)y'(a)G(a) + p(a)G'(a)y(a).$$

Here the right hand side of this identity can be expressed in terms of the boundary conditions given by equation (8.77) and can be written as:

$$\left[\frac{p(b)}{\delta}(\delta y'(b) + \gamma y(b))\right]G(b) - \left[\frac{p(b)}{\delta}(\delta G'(b) + \gamma G(b))\right]y(b)$$

$$-\left[\frac{p(a)}{\beta}(\beta y'(a) + \alpha y(a))\right]G(a) + \left[\frac{p(a)}{\beta}(\beta G'(a) + \alpha G(a))\right]y(a)$$

$$=\frac{p(b)}{\delta}B_2[y]G(b) - \frac{p(b)}{\delta}B_2[G]y(b) - \frac{p(a)}{\beta}B_1[y]G(a) + \frac{p(a)}{\beta}B_1[G]y(a).$$

By hypothesis $B_1[y] = 0$ and $B_2[y] = 0$ and these terms are zero in the right hand side of the Green's identity. If we choose $B_1[G] = 0$ and $B_2[G] = 0$ we see that the right hand side of the Green's identity is zero. Here $B_1^* = B_1$ and $B_2^* = B_2$. The remaining terms in Green's identity give us the relation

$$y(x) = \int_a^b G(x;\xi)f(\xi)\,d\xi$$

and this solution can be written in the form

$$y(x) = \int_a^b G(x;\xi)f(\xi)\,d\xi = \int_a^x G_2(x;\xi)f(\xi)\,d\xi + \int_x^b G_1(x;\xi)f(\xi)\,d\xi, \qquad (8.78)$$

where

$$G = G(x;\xi) = \begin{cases} G_1, & a \leq x < \xi \\ G_2, & \xi < x \leq b \end{cases} \qquad (8.79)$$

is the Green's function associated with the operator L_x and satisfies

$$L_x(G(x;\xi)) = \delta(x - \xi)$$

which is a shorthand notation for the system of equations

$$L_x(G_1(x;\xi)) = 0, \quad a \leq x < \xi \qquad \text{and} \qquad L_x(G_2(x;\xi)) = 0, \quad \xi < x \leq b.$$

Further these equations are subject to the boundary conditions $B_1[G] = 0$ and $B_2[G] = 0$. To show that equation (8.78) is indeed the solution, we differentiate the relation (8.78) and obtain

$$y' = \int_a^x \frac{\partial G_2}{\partial x}f\,d\xi + \int_x^b \frac{\partial G_1}{\partial x}f\,d\xi + [G_2(x;x^-) - G_1(x;x^+)]f(x)$$

Using the continuity condition this reduces to

$$y' = \int_a^x \frac{\partial G_2}{\partial x}f\,d\xi + \int_x^b \frac{\partial G_1}{\partial x}f\,d\xi = \int_a^b \frac{\partial G}{\partial x}f\,d\xi.$$

Upon differentiating this relation we obtain the second derivative function

$$y'' = \int_a^x \frac{\partial^2 G_2}{\partial x^2} f \, d\xi + \int_x^b \frac{\partial^2 G_1}{\partial x^2} f \, d\xi + \left[\frac{\partial G_2(x; x^-)}{\partial x} - \frac{\partial G_1(x; x^+)}{\partial x} \right] f(x)$$

and using the jump discontinuity condition this reduces to

$$y'' = \int_a^b \frac{\partial^2 G}{\partial x^2} f \, d\xi + \frac{f(x)}{p(x)}.$$

Upon substituting y, y' and y'' into the differential equation (8.76) we obtain

$$L_x(y) = \int_a^x L_x(G_2) f \, d\xi + \int_x^b L_x(G_1) f \, d\xi + f(x) = f(x)$$

since $L_x(G_1) = 0$ and $L_x(G_2) = 0$ by construction. Further, the boundary conditions are satisfied with

$$B_i[y] = \int_a^b B_i[G] f \, d\xi = 0, \quad \text{for} \quad i = 1, 2$$

since $B_i[G] = 0$, for $i = 1, 2$, by construction.

Example 8-5. **(Nonhomogeneous boundary conditions.)**
Let L_x denote the linear differential operator

$$L_x(u) = \frac{d}{dx} \left(p(x) \frac{du}{dx} \right) + q(x)u, \qquad a \le x \le b \tag{8.80}$$

and let B_1, B_2 denote the linear boundary operators

$$B_1[u] = \alpha_0 u(a) + \alpha_1 u'(a)$$

$$B_2[u] = \beta_0 u(b) + \beta_1 u'(b),$$

where $\alpha_0, \alpha_1, \beta_0, \beta_1$ are constants not all zero. We desire to solve the nonhomogeneous boundary value problem with nonhomogeneous boundary conditions:

$$L_x(u) = f(x), \qquad B_1[u] = g_1, \qquad B_2[u] = g_2,$$

where f, g_1, g_2 are given. This can be accomplish by using the principal of superposition. If we let G denote the solution of the boundary value problem

$$L_x(G(x; \xi)) = \delta(x - \xi), \qquad B_1[G] = 0, \qquad B_2[G] = 0,$$

then G can be expressed

$$G(x, \xi) = \begin{cases} \frac{y_1(x) y_2(\xi)}{p(\xi) W(\xi)}, & a \le x < \xi \\ \frac{y_1(\xi) y_2(x)}{p(\xi) W(\xi)}, & \xi < x \le b \end{cases} \tag{8.81}$$

where $\{y_1, y_2\}$ are independent solutions of $L_x(y) = 0$ which satisfy $B_1[y_1] = 0$ and $B_2[y_2] = 0$.

450

We can verify that the boundary value problem

$$L_x(u_1) = f(x), \qquad B_1[u_1] = 0, \qquad B_2[u_1] = 0$$

has the solution

$$u_1(x) = \int_a^b G(x,\xi) f(\xi) \, d\xi.$$

Finally, if we can solve the boundary value problem

$$L_x(u_2) = 0, \qquad B_1[u_2] = g_1, \qquad B_2[u_2] = g_2,$$

then by superposition we construct $u(x) = u_1(x) + u_2(x)$ as the solution of the nonhomogeneous boundary value problem

$$L_x(u) = f(x), \qquad B_1[u] = g_1, \qquad B_2[u] = g_2.$$

Since L_x, B_1, B_2 are linear operators it follows that

$$L_x(u) = L_x(u_1) + L_x(u_2) = f(x)$$
$$B_1[u] = B_1[u_1] + B_2[u_2] = g_1$$
$$B_2[u] = B_2[u_1] + B_2[u_2] = g_2$$

and consequently the principal of superposition is valid. Here we see that the original nonhomogeneous boundary value problem can be broken up into smaller easier to handle boundary value problems. By adding the results from these easier problems we can construct a solution to a more difficult problem.

∎

Series Solutions for Green's Functions

The Green's function can also be represented as a series of orthogonal functions. Consider the boundary value problem to solve

$$L_x(y) = f(x), \quad a \le x \le b, \quad B_1[y] = 0, \quad B_2[y] = 0 \qquad (8.82)$$

where $L_x(y)$ is a linear differential operator and $B_1[y], B_2[y]$ are boundary operators of the type given by the equation (8.65). Each self-adjoint operator $L_x(y)$ has associated with it an eigenvalue problem

$$L_x^*(u_n) = \lambda_n r(x) u_n, \quad a \le x \le b, \quad B_1^*[u_n] = 0, \quad B_2^*[u_n] = 0 \qquad (8.83)$$

with given boundary conditions. Here $L_x^* = L_x$ since L_x is self-adjoint, and λ_n are the eigenvalues, $r(x)$ is a weight function and u_n are the eigenfunctions which are orthogonal over the interval (a, b) with respect to the weight function $r(x)$.

Consider the Green's formula

$$\int_a^b [u_n L_x(y) - y L_x^*(u_n)]\, dx = [P(y, u_n)]_a^b$$

and assume the boundary conditions on $u_n(x)$ are such that the bilinear concomitant P vanishes at the end points a and b. This is the condition which determines the adjoint boundary conditions $B_1^*[u_n] = 0$ and $B_2^*[u_n] = 0$. For these conditions the Green's formula reduces to

$$\int_a^b u_n(x) L_x(y)\, dx = \int_a^b y(x) L_x^*(u_n)\, dx. \tag{8.84}$$

We represent the solution to (8.82) in the form of a Fourier series involving the eigenfunctions u_n, and write

$$y(x) = \sum_{n=1}^{\infty} C_n u_n(x), \tag{8.85}$$

where C_n are the Fourier coefficients to be determined. These Fourier coefficients are given by an inner product divided by a norm squared

$$C_n = \frac{(y, u_n)}{\| u_n \|^2} = \frac{\int_a^b r(x) y(x) u_n(x)\, dx}{\| u_n \|^2}$$

$$C_n = \frac{1}{\lambda_n \| u_n \|^2} \int_a^b y(x) [\lambda_n r(x) u_n]\, dx.$$

Using the relations (8.83) and (8.84) this can be written

$$\lambda_n \| u_n \|^2 C_n = \int_a^b y(x) L_x^*(u_n)\, dx$$

$$\lambda_n \| u_n \|^2 C_n = \int_a^b u_n(x) L_x(y)\, dx \tag{8.86}$$

$$\lambda_n \| u_n \|^2 C_n = \int_a^b u_n(x) f(x)\, dx$$

Hence the solution (8.85) can be expressed

$$y(x) = \int_a^b \left(\sum_{n=1}^{\infty} \frac{u_n(\xi) u_n(x)}{\lambda_n \| u_n \|^2} \right) f(\xi)\, d\xi$$

or

$$y(x) = \int_a^b G(x; \xi) f(\xi)\, d\xi, \tag{8.87}$$

where

$$G = G(x; \xi) = \sum_{n=1}^{\infty} \frac{u_n(\xi) u_n(x)}{\lambda_n \| u_n \|^2} \tag{8.88}$$

is the Green's function associated with the operator L_x. The representation (8.88) is called a bilinear expansion and is valid provided there are no eigenvalues which are zero.

The result given by equation (8.88) is obtainable in another way. Consider the eigenfunctions and eigenvalues determined by (8.83) and consider the Green's function equation

$$L_x^*(G) = \delta(x - \xi), \quad a \le x \le b. \tag{8.89}$$

Assume a solution to (8.89) as a Fourier series of the form

$$G = G(x; \xi) = \sum_{n=1}^{\infty} c_n u_n(x), \tag{8.90}$$

where c_n, $n = 1, 2, \ldots$ are the Fourier coefficients to be determined and $u_n(x)$ are the eigenfunctions associated with the eigenvalue problem given by equation (8.83). We have that

$$L_x^*(G) = \sum_{n=1}^{\infty} c_n L_x^*(u_n) = \sum_{n=1}^{\infty} c_n \lambda_n r(x) u_n(x) = \delta(x - \xi). \tag{8.91}$$

Now make use of the orthogonality properties of the eigenfunctions $u_n(x)$ and multiply both sides of (8.91) by $u_m(x)$ and integrate from a to b. By the integration properties of the delta function we find

$$\sum_{n=1}^{\infty} c_n \lambda_n (u_n, u_m) = \int_a^b u_m(x) \delta(x - \xi)\, dx = u_m(\xi). \tag{8.92}$$

To further simplify this result, we observe the eigenfunctions are orthogonal functions, and consequently the left side of (8.92) has only one nonzero term and reduces to $c_m \lambda_m \| u_m \|^2 = u_m(\xi)$. This enables us to write the Fourier coefficients as $c_m = \dfrac{u_m(\xi)}{\lambda_m \| u_m \|^2}$. The solution (8.90) can then be expressed

$$G = G(x; \xi) = \sum_{n=1}^{\infty} \frac{u_n(\xi) u_n(x)}{\lambda_n \| u_n \|^2}, \quad \lambda_n \ne 0$$

which is the Green's function solution of (8.89) expressed in the form of a generalized Fourier series.

Example 8-6. (Greens function.)

Solve the boundary value problem

$$y'' + \omega^2 y = f(x), \quad 0 \le x \le L \tag{8.93}$$

such that $y(0) = y(L) = 0$.

Solution: Here $L_x(y) = y'' + \omega^2 y$ and $L_x^*(y) = y'' + \omega^2 y$. Associated with equation (8.93) is the eigenvalue problem

$$L_x^*(u_n) = \lambda_n^2 u_n, \quad u_n(0) = u_n(L) = 0$$

or

$$u_n'' + \omega^2 u_n = \lambda_n^2 u_n.$$

Let $\lambda_n^2 = \omega^2 - \alpha_n^2$, then the Sturm-Liouville problem $u_n'' + \alpha_n^2 u_n = 0$ with homogeneous boundary conditions $u_n(0) = u_n(L) = 0$ has the eigenfunctions

$$u_n(x) = \sin(\alpha_n x), \quad n = 1, 2, \ldots$$

with $\alpha_n = \frac{n\pi}{L}$ and norm squared $\| u_n \|^2 = \frac{L}{2}$. The Green's function representation (8.88) becomes

$$G(x; \xi) = \frac{2}{L} \sum_{n=1}^{\infty} \frac{\sin(\frac{n\pi\xi}{L}) \sin(\frac{n\pi x}{L})}{\omega^2 - \frac{n^2\pi^2}{L^2}}.$$

Observe the symmetry of the Green's function and also note the Green's function fails to exist if $\omega = \frac{n\pi}{L}$, $n = 1, 2, \ldots$. These are cases where the homogeneous equation

$$L_x(y) = y'' + \frac{n^2\pi^2}{L^2} y = 0$$

possesses the nontrivial solution

$$y = y_n(x) = \sin(\frac{n\pi x}{L}), \quad n = 1, 2, \ldots$$

which satisfy the boundary conditions. In physical systems, this situation corresponds to the nonexistence of a Green's function and usually implies that there is some sort of resonance phenomena taking place. ∎

Analysis of equation (8.86) gives the following solvability condition for the boundary value problem

$$L(u) = f, \quad a \leq x \leq b, \quad B_1[u] = 0, \quad B_2[u] = 0. \tag{8.94}$$

Associated with the equation (8.94) is the eigenvalue problem given by equation (8.83). If for some value of n we have $\lambda_n = 0$ as an eigenvalue, then the corresponding eigenfunction $u_n(x)$ is a nonzero solution of the homogeneous equation

$$L^*(u_n) = 0, \quad a \leq x \leq b, \quad B_1^*[u_n] = 0, \quad B_2^*[u_n] = 0. \tag{8.95}$$

In this case we are unable to solve for the coefficient C_n in equation (8.86) and so the Green's function does not exist. A solvability condition under these circumstances is to require that the right-hand side of equation (8.86) to satisfy the integral relation

$$\int_a^b u_n(x) f(x)\, dx = 0. \tag{8.96}$$

No solution exists unless the equation (8.86) is satisfied for all zero eigenvalues in which case there are infinitely many solutions. That is, the equation (8.86) will be identically satisfied for arbitrary C_n when equation (8.96) is satisfied. If $\lambda_n = 0$ is not an eigenvalue of equation (8.86), then the trivial solution $u = 0$ is the only solution of the homogeneous equation $L_x^*(u) = 0$ with homogeneous boundary conditions. In this case the equation (8.86) has a unique solution. These solvability conditions are referred to as the Fredholm[†] alternative theorems and can be summarized as follows.

The above solvability conditions or Fredholm alternatives are as follows:

First alternative: For $f = f(x)$ continuous on $a \leq x \leq b$ the boundary value problem

$$L_x(u) = f, \quad a \leq x \leq b, \quad B_1[u] = 0, \quad B_2[u] = 0 \tag{8.97}$$

has a unique solution and $u = 0$ is the only solution of the homogeneous problem

$$L_x(u) = 0, \quad a \leq x \leq b, \quad B_1[u] = 0, \quad B_2[u] = 0. \tag{8.98}$$

Second alternative: The homogeneous boundary value problem

$$L_x^*(u_H) = 0, \quad a \leq x \leq b, \quad B_1^*[u_H] = 0, \quad B_2^*[u_H] = 0 \tag{8.99}$$

[†] See Appendix C

has a nonzero solution u_H in which case one of the following holds true:

(i) The boundary value problem (8.97) does not have a solution, or

(ii) If $\int_a^b u_H(x) f(x)\, dx = 0$ for each function u_H which satisfies the system (8.99), then the boundary value problem (8.97) has an infinite number of solutions.

Example 8-7. (Greens function.)
Solve the boundary value problem

$$L(y) = \frac{d^2 y}{dx^2} + 3\frac{dy}{dx} + 2y = f(x), \quad y(0) = 0, \quad y'(0) = 0 \qquad (8.100)$$

Solution: The general solution to the homogeneous equation $y'' + 3y' + 2y = 0$ is given in terms of linear combinations of functions from the fundamental set of solutions $\{e^{-2x}, e^{-x}\}$. The general solution to the homogeneous problem

$$L(y) = \frac{d^2 y}{dx^2} + 3\frac{dy}{dx} + 2y = 0, \quad y(0) = 0, \quad y'(0) = 0 \qquad (8.101)$$

can be written $y = K_1 e^{-2x} + K_2 e^{-x}$ where K_1, K_2 are constants. The solution satisfying the boundary conditions requires that $K_1 = K_2 = 0$ and so $y = 0$ is the trivial solution and so this is the Fredholm first alternative. The Green's function can be represented in terms of linear combinations of the fundamental solutions. We write

$$G = G(x;\xi) = \begin{cases} G_1(x;\xi) = C_1 e^{-2x} + C_2 e^{-x}, & x < \xi \\ G_2(x;\xi) = C_3 e^{-2x} + C_4 e^{-x}, & \xi < x \end{cases} \qquad (8.102)$$

where C_1, C_2, C_3, C_4 are constants. The constants C_1, C_2 must be selected to satisfy the boundary conditions

$$\begin{aligned} B_1[G] &= G(0;\xi) = 0 \quad \Rightarrow \quad C_1 + C_2 = 0 \\ B_2[G] &= \frac{dG(0;\xi)}{dx} = 0 \quad \Rightarrow \quad -2C_1 - C_2 = 0 \end{aligned} \qquad (8.103)$$

For continuity at $x = \xi$ we require $G(\xi^+;\xi) = G(\xi^-;\xi)$ or

$$C_1 e^{-2\xi} + C_2 e^{-\xi} = C_3 e^{-2\xi} + C_4 e^{\xi}. \qquad (8.104)$$

The jump discontinuity condition requires $\left.\frac{\partial G(x;\xi)}{\partial x}\right|_{x=\xi^+}^{x=\xi^-} = -1$ or

$$-2C_1 e^{-2\xi} - C_2 e^{-\xi} - (-2C_3 e^{-2\xi} - C_4 e^{-\xi}) = -1. \qquad (8.105)$$

Solving the equations (8.103), (8.104), and (8.105) we find

$$C_1 = C_2 = 0, \quad C_3 = -e^{-2\xi}, \quad C_4 = e^{\xi} \qquad (8.106)$$

which produces the Green's function

$$G(x;\xi) = \begin{cases} G_1(x;\xi) = 0, & x < \xi \\ G_2(x;\xi) = e^{\xi-x} - e^{2(\xi-x)}, & \xi < x \end{cases} \qquad (8.107)$$

This result can also be obtained using the method of variation of parameters. Assume a particular solution to the nonhomogeneous equation (8.100) of the form

$$y_p(x) = u(x)e^{-2x} + v(x)e^{-x}$$

where we require that the functions $u(x), v(x)$ satisfy the conditions

$$u'(x)e^{-2x} + v'(x)e^{-x} = 0$$
$$u'(x)e^{-2x}(-2) + v'(x)e^{-x}(-1) = f(x) \qquad (8.108)$$

We solve for $u'(x)$ and $v'(x)$ and find

$$u'(x) = \frac{du}{dx} = -e^{2x}f(x), \qquad v'(x) = \frac{dv}{dx} = e^{x}f(x).$$

An integration of these results gives

$$u = u(x) = -\int_0^x e^{2\xi}f(\xi)\,d\xi, \qquad v = v(x) = \int_0^x e^{\xi}f(\xi)\,d\xi \qquad (8.109)$$

This produces the particular solution

$$y_p = \int_0^x \left[e^{\xi-x} - e^{2(\xi-x)} \right] f(\xi)\,d\xi \qquad (8.110)$$

and the general solution to the boundary value problem (8.100) is given by

$$y = K_1 e^{-2x} + K_2 e^{-x} + \int_0^x \left[e^{\xi-x} - e^{2(\xi-x)} \right] f(\xi)\,d\xi \qquad (8.111)$$

where K_1 and K_2 are constants. The solution satisfying the given boundary conditions requires $K_1 = K_2 = 0$ and so the final solution is given by

$$y = \int_0^x \left[e^{\xi-x} - e^{2(\xi-x)} \right] f(\xi)\,d\xi = \int_0^x G(x;\xi)f(\xi)\,d\xi \qquad (8.112)$$

where $G(x;\xi)$ is given by equation (8.107) is called a one-sided Green's function.

∎

Modified Green's Functions

To solve the boundary value problem

$$L_x(u) = f, \quad a < x < b, \quad B_1[u] = 0, \quad B_2[u] = 0 \tag{8.113}$$

one must solve the one of the Green's function problems

$$L_x(G) = \delta(x - \xi), \quad B_1[G] = 0, \quad B_2[G] = 0$$

or

$$L_\xi^*(G^*) = \delta(\xi - x), \quad a < x < b, \quad B_1^*[G^*] = 0, \quad B_2^*[G^*] = 0 \tag{8.114}$$

where L_ξ^* is the adjoint operator associated with L_ξ and B_1^*, B_2^* are adjoint boundary operators obtained from an analysis of the Green's identity. Note that if the homogeneous form of equation (8.113)

$$L_\xi(u_H) = 0, \quad a < x < b, \quad B_1[u_H] = 0, \quad B_2[u_H] = 0 \tag{8.115}$$

has a nonzero solution, u_H (H for homogeneous solution), then the Green's function associated with the operator L_x might not exist. This is because the Green's identity

$$\int_a^b \left[v L_\xi(u) - u L_\xi^*(v) \right] d\xi = [P(u,v)]_a^b \tag{8.116}$$

would not be satisfied. Observe that with $v = G^*$ and $u = u_H$, the Green's identity given by equation (8.116) reduces to

$$\int_a^b \left[G^* L_\xi(u_H) - u_H L_\xi^*(G^*) \right] d\xi = [P(u_H, G^*)]_a^b. \tag{8.117}$$

The boundary conditions B_1, B_2, B_1^*, B_2^* are such that the right-hand side of equation (8.117) is zero which reduces the Green's identity to the impossible condition

$$\int_a^b \left[G^* \cdot 0 - u_H \delta(\xi - x) \right] d\xi = u_H(x) = 0. \tag{8.118}$$

That is, the homogeneous solution should be identically zero. However, our original assumption was that $u_H \neq 0$ and so a contradiction arises. To overcome this difficulty we consider instead a modified Green's function G_m which satisfies

$$L_\xi^*(G_m^*(\xi; x)) = \delta(\xi - x) + C u_H(\xi), \quad B_1^*[G_m^*] = 0, \quad B_2^*[G_m^*] = 0 \tag{8.119}$$

458

or

$$L_x[G_m(x; \xi)] = \delta(x - \xi) + C u_H(x), \quad B_1[G_m] = 0, \quad B_2[G_m] = 0 \qquad (8.120)$$

where C is selected such that the Green's identity given by equation (8.117) is satisfied. Here we assume that u_H is a nonzero solution to the homogeneous equation $L_\xi(u_H) = 0$, with homogeneous boundary conditions. Now when we substitute $v = G_m^*$ and $u = u_H$ into the Green's identity we obtain

$$\int_a^b [G_m^* \cdot 0 - u_H(\xi)(\delta(\xi - x) + C u_H(\xi))] \, d\xi = 0 \qquad (8.121)$$

which must be satisfied. This requires that C be selected as

$$C = \frac{-u_H(x)}{\int_a^b u_H^2(\xi) \, d\xi}. \qquad (8.122)$$

This gives the modified Green's function equation

$$L_\xi^*(G_m^*) = \delta(\xi - x) - \frac{u_H(x) u_H(\xi)}{\int_a^b u_H^2(\xi) \, d\xi}. \qquad (8.123)$$

If the equation (8.123) is satisfied, then the original boundary value problem given by equation (8.113) can be solved by substituting $v = G_m^*$ into the Green's identity (8.117) to obtain

$$\int_a^b [G_m^*(\xi; x) L_\xi(u) - u(\xi) L_\xi^*(G_m^*)] \, d\xi = [P(u, G_m^*)]_a^b$$

which simplifies to

$$\int_a^b \left[G_m^*(\xi; x) f(\xi) - u(\xi) \left[\delta(\xi - x) - \frac{u_H(x) u_H(\xi)}{\int_a^b u_H(\eta)^2 \, d\eta} \right] \right] \, d\xi = 0$$

which further simplifies to the solution of the boundary value problem (8.113). This solution can be represented

$$u(x) = \int_a^b G_m^*(\xi; x) f(\xi) \, d\xi + \frac{u_H(x) \int_a^b u(\xi) u_H(\xi) \, d\xi}{\int_a^b u_H^2(\eta) \, d\eta} \qquad (8.124)$$

where the ratio

$$K = \frac{\int_a^b u(\xi) u_H(\xi) \, d\xi}{\int_a^b u_H^2(\eta) \, d\eta} \tag{8.125}$$

is just some constant. Observe that the last term in the equation (8.125) is just some constant times the homogeneous solution.

We now turn the problem around and ask the question, "Is the function

$$u = u(x) = \int_a^b G_m^*(\xi; x) f(\xi) \, d\xi + K u_H(x) \tag{8.126}$$

a solution of the boundary value problem given by equation (8.113) ?" We note that $G_m(x; \xi) = G_m^*(\xi; x)$ and let the operator L_x operate on u. We interchange the order of differentiation and integration to obtain

$$L_x(u) = \int_a^b L_x(G_m(x; \xi)) f(\xi) \, d\xi + K L_x(u_H(x)). \tag{8.127}$$

By hypothesis, the homogeneous solution u_H satisfies $L_x(u_H) = 0$ and we know that

$$L_x(G_m(x; \xi)) = L_\xi(G_m^*(\xi; x)) = \delta(\xi - x) - \frac{u_H(x) u_H(\xi)}{\int_a^b u_H^2(\xi) \, d\xi}.$$

Therefore, the equation (8.127) reduces to

$$L_x(u) = \int_a^b \left[\delta(\xi - x) - \frac{u_H(x) u_H(\xi)}{\int_a^b u_H^2(\eta) \, d\eta} \right] f(\xi) \, d\xi$$

$$= f(x) - u_H(x) \frac{\int_a^b u_H(\xi) f(\xi) \, d\xi}{\int_a^b u_H^2(\eta) \, d\eta}. \tag{8.128}$$

In order for the equation (8.128) to simplify to $f(x)$ we require that the nonzero solution $u_H(x)$ satisfy the integral condition

$$\int_a^b u_H(\xi) f(\xi) \, d\xi = 0. \tag{8.129}$$

That is, if the homogeneous equation has a nonzero solution u_H, then in order to construct a modified Green's function the integral condition (8.129) must be satisfied.

Example 8-8. (Modified Greens function.)

Consider the two-point boundary value problem to solve

$$L_x(u) = \frac{d^2u}{dx^2} + u = f(x), \quad 0 < x < \pi, \quad u(0) = 0, \quad u(\pi) = 0 \qquad (8.130)$$

where $f(x)$ is continuous on $0 \le x \le \pi$.

Solution: Associated with the above operator L_x is the Green's first identity

$$\int_0^\pi \left[vL_\xi(u) - uL_\xi^*(v) \right] d\xi = u'(\pi)v(\pi) - v'(\pi)u(\pi) \\ -u'(0)v(0) + v'(0)u(0) \qquad (8.131)$$

In this identity we substitute $v = G^*$, $u(0) = 0$ and $u(\pi) = 0$ to obtain

$$\int_0^\pi \left[G^*L_\xi(u) - uL_\xi^*(G^*) \right] d\xi = u'(\pi)G^*(\pi; x) - u'(0)G^*(0; x) \qquad (8.132)$$

In the equation (8.132) the quantities $u'(\pi)$ and $u'(0)$ are unknown. In order to remove these terms we require that $G^*(0; x) = 0$ and $G^*(\pi; x) = 0$. These are the adjoint boundary conditions. By changing x to ξ and letting

$$L_\xi^*(G^*) = \delta(\xi - x), \qquad G^*(0; x) = 0, \qquad G^*(\pi; x) = 0 \qquad (8.133)$$

the equation (8.132) reduces to our desired solution of the boundary value problem

$$u(x) = \int_0^\pi G^*(\xi; x)f(\xi)\, d\xi \qquad (8.134)$$

provided that we can solve for the Green's function. To solve the equation (8.133) we treat it as two ordinary differential equations. One over the interval $0 < x < \xi$ and the other over the interval $\xi < x < \pi$. This gives the equations

$$\frac{d^2G^*}{d\xi^2} + G^* = 0 \qquad\qquad \frac{d^2G^*}{d\xi^2} + G^* = 0$$
$$0 < \xi < x \qquad\qquad\qquad x < \xi < \pi$$
$$G^*(0; x) = 0 \qquad\qquad\qquad G^*(\pi; x) = 0$$

with solutions

$$G^* = G^*(\xi; x) = \begin{cases} c_1 \sin\xi + c_2 \cos\xi, & 0 < \xi < x \\ c_3 \sin\xi + c_4 \cos\xi, & \xi < x < \pi \end{cases} \qquad (8.135)$$

where c_1, c_2, c_3, c_4 are constants. We require the Green's function to satisfy the continuity condition at $\xi = x$ that

$$c_1 \sin x + c_2 \cos x = c_3 \sin x + c_4 \cos x. \qquad (8.136)$$

Also we require the Green's function satisfy the jump discontinuity condition

$$\frac{dG^*}{d\xi}\bigg|_{\xi=x^-}^{\xi=x^+} = 1 \quad \text{or} \quad c_3 \cos x - c_4 \sin x - c_1 \cos x + c_2 \sin x = 1 \qquad (8.137)$$

together with the boundary conditions

$$G^*(0; x) = c_2 = 0 \quad \text{and} \quad G^*(\pi; x) = c_4 = 0.$$

The boundary conditions of $c_2 = c_4 = 0$ reduces the equations (8.136) and (8.137) to the form

$$c_1 = c_3 \quad \text{and} \quad c_3 \cos x - c_1 \cos x = 1.$$

These equations cannot be solved for c_1 and c_3 since the equations are inconsistent and require the impossible condition that $0 = 1$ and so the Green's function does not exist.

The above result occurs because the homogeneous boundary value problem associated with the equation (8.130) requiring

$$L_x(u_H) = \frac{d^2 u_H}{dx^2} + u_H = 0, \quad 0 < x < \pi, \quad u_H(0) = 0, \quad u_H(\pi) = 0 \qquad (8.138)$$

has the nonzero solution $u_H(x) = \sin x$. The Green's identity, equation (8.131), with $u = u_H$ and $v = G^*$ becomes

$$\int_0^\pi \left[G^*(\xi; x) L_\xi(u_H) - u_H(\xi) L_\xi^*(G^*) \right] d\xi = 0$$

which for $L_\xi(u_H) = 0$ and $L_\xi^*(G^*) = \delta(\xi - x)$ reduces to

$$\int_0^\pi u_H(\xi) \delta(\xi - x) \, d\xi = u_H(x) = 0,$$

which is not satisfied. However, if the right-hand side $f(x)$ of the equation (8.130) satisfies the integral condition $\int_0^\pi f(x) \sin x \, dx = 0$, obtained from the equation (8.129), then we can construct a modified Green's function. Assuming that this integral condition is satisfied, we construct a modified Green's function G_m^* satisfying

$$L_\xi^*(G_m^*) = \delta(\xi - x) - \frac{u_H(x)u_H(\xi)}{\int_0^\pi u_H^2(\eta)\,d\eta}.$$

We know $u_H(x) = \sin x$ and therefore $\int_0^\pi u_H^2(x)\,dx = \int_0^\pi \sin^2 x\,dx = \frac{\pi}{2}$. We therefore solve the modified Green's function equation

$$L_\xi^*(G_m^*) = \frac{d^2 G_m^*}{d\xi^2} + G_m^* = \delta(\xi - x) - \frac{2}{\pi}\sin x \sin \xi \tag{8.139}$$

subject to the continuity and jump discontinuity conditions at $\xi = x$ together with the boundary conditions at $\xi = 0$ and $\xi = \pi$. We precede exactly like we did before and find the Green's function solution

$$G_m^* = G_m^*(\xi; x) = \begin{cases} c_1 \sin\xi + c_2 \cos\xi + \frac{1}{\pi}\xi \sin x \cos\xi, & 0 < \xi < x \\ c_3 \sin\xi + c_4 \cos\xi + \frac{1}{\pi}\xi \sin x \cos\xi, & x < \xi < \pi \end{cases} \tag{8.140}$$

Applying the boundary conditions, the continuity condition and the jump discontinuity condition we find that c_3 is an arbitrary constant, or an arbitrary function of x, and $c_1 = c_3 - \cos x$. This gives the modified Green's function

$$G_m^* = G_m^*(\xi; x) = \begin{cases} c_3 \sin\xi - \sin\xi \cos x + \frac{1}{\pi}\xi \sin x \cos\xi, & 0 < \xi < x \\ c_3 \sin\xi - \cos\xi \sin x + \frac{1}{\pi}\xi \sin x \cos\xi, & x < \xi < \pi \end{cases} \tag{8.141}$$

where c_3 is arbitrary. The solution of the original boundary value problem is obtained from the Green's first identity using $v = G_m^*$, where $L_\xi^*(G_m^*)$ is obtained from the equation (8.139) and u is treated as the solution of $L_\xi(u) = f(\xi)$. One can now construct the solution

$$u(x) = \int_0^\pi G_m^*(\xi; x) f(\xi)\,d\xi - \frac{2}{\pi}\sin x \int_0^\pi u(\xi) \sin\xi\,d\xi$$

which can be written as

$$u(x) = K\sin x + \int_0^\pi G_m^*(\xi; x) f(\xi)\,d\xi \tag{8.142}$$

where K is constant. Note in particular that in order for the equation (8.142) to be a solution of the given boundary value problem it is necessary that the integral condition $\int_0^\pi f(\xi)\sin\xi\,d\xi = 0$ must be satisfied.

■

Exercises 8

▶ **1.** Show using the method of variation of parameters that the general solution of the one point initial value problem

$$L_x(y) = a_0(x)y'' + a_1(x)y' + a_2(x)y = r(x), \quad a \le x \le b$$

satisfying $y(a) = A$ and $y'(a) = B$ is given by :

$$y(x) = Ay_1(x) + By_2(x) + \int_a^x r(\xi)G_2(x,\xi)\,d\xi,$$

where

$$G_2(x,\xi) = \frac{y_2(x)y_1(\xi) - y_1(x)y_2(\xi)}{a_0(\xi)W(\xi)}, \quad a_0(x) \ne 0$$

and where $W(\xi)$ is the Wronskian, and y_1, y_2 is a normalized fundamental set which satisfies the conditions:

$$y_1(a) = 1, \qquad y_1'(a) = 0$$
$$y_2(a) = 0, \qquad y_2'(a) = 1$$

▶ **2.** Solve the one point initial value problem

$$L_x(y) = y'' + \omega^2 y = f(x), \quad y(2) = 1, \quad y'(2) = 3$$

▶ **3.** Solve the one point initial value problem

$$L_x(y) = y'' = f(x), \quad y(0) = y_0, \quad y'(0) = y_1$$

▶ **4.** Solve the one point initial value problem

$$L_x(u) = u'' + 3u' + 2u = e^x, \quad u(0) = 1, \quad u'(0) = 0$$

▶ **5.** Solve the two point boundary value problem

$$L_x(y) = \frac{d^2y}{dx^2} = f(x), \quad a < x < b$$

subject to the boundary conditions $y(a) = g_a$ and $y(b) = g_b$ where g_a and g_b are nonzero constants.

464

▶ **6.** Consider the Green's function associated with the differential equation

$$L_x(u) = \frac{d^2u}{dx^2} - k^2 u = f(x), \quad 0 < x < L$$

with k constant, where u is subject to the boundary conditions $u(0) = 0$ and $u(L) = 0$.

(a) Use Laplace transforms to find the Green's function associated with the above operator.

(b) Solve the given boundary value problem using Green's function methods.

▶ **7.** Solve the two point boundary value problem

$$L_x(y) = y'' = f(x), \quad y(a) = 0, \quad y'(b) = 0, \quad a \le x \le b$$

▶ **8.** Solve the two point boundary value problem

$$L_x(y) = y'' = f(x), \quad B_1[y] = y(0) + y(1) = 0, \quad B_2[y] = y'(0) + y'(1) = 0, \quad 0 \le x \le 1$$

▶ **9.** Solve the two point boundary value problem

$$L_x(y) = y'' + \omega^2 y = f(x), \quad y(a) = 0, \quad y(b) = 0, \quad a \le x \le b$$

▶ **10.** Solve the two point boundary value problem

$$L_x(u) = u'' + u = f(x), \quad u'(0) = 0, \quad u(\pi) = 0, \quad 0 \le x \le \pi$$

▶ **11.** Solve the two point boundary value problem

$$L_x(v) = v'' - \omega^2 v = f(x), \quad v(0) = 0, \quad v'(L) = 0, \quad 0 \le x \le L$$

▶ **12.** Solve the two point boundary value problem

$$L_x(y) = xy'' + y' = f(x), \quad y(1) = 0, \quad y(x) \text{ bounded as } x \to 0, \quad 0 \le x \le 1$$

▶ **13.** Solve the two point boundary value problem

$$L_x(y) = \frac{d^3y}{dx^3} = f(x), \quad 0 \le x \le 1$$
$$B_1[y] = y(0) + y(1) = 0, \quad B_2[y] = y(0) - y(1) = 0, \quad B_3[y] = y'(0) = 0$$

▶ **14.** Solve the two point boundary value problem

$$L_x(y) = \frac{d^4y}{dx^4} = f(x), \quad 0 \le x \le 1$$
$$y(0) = y'(0) = 0, \quad y(1) = y'(1) = 0$$

▶ **15.**

(a) Find the modified Green's function for the boundary value problem $L_x(u) = \dfrac{d^2u}{dx^2} = f(x)$ on the interval $(0, 2\pi)$ and subject to the boundary conditions $u'(0) = 0$ and $u'(2\pi) = 0$. What condition must $f(x)$ satisfy in order for the modified Green's function to exist?

(b) Interpret the equation $\frac{d^2u}{dx^2} = f(x)$ as a steady state heat equation with source term over the interval $(0, 2\pi)$.

(i) Interpret the boundary conditions $u'(0) = 0$ and $u'(2\pi) = 0$.

(ii) In terms of energy input, what condition must be satisfied in order for an equilibrium temperature to exist?

▶ **16.** Show the solution of the two point boundary value problem

$$L_x(y) = y'' + \beta(x)y' + \alpha(x)y = f(x), \quad a \le x \le b$$

satisfying $y(a) = A$ and $y(b) = B$ is given by

$$y(x) = AY_1(x) + BY_2(x) + \int_a^b \mu(\xi)f(\xi)G(x;\xi)\,d\xi,$$

where $\mu(x) = \exp\left[\int_a^x \beta(x)\,dx\right]$ and where Y_1, Y_2 is a fundamental set of the homogeneous equation satisfying

$$Y_1(a) = 1, \quad Y_2(a) = 0, \quad Y_1(b) = 0, \quad Y_2(b) = 1.$$

Find $G(x;\xi)$ in terms of Y_1 and Y_2.

▶ **17.**   Use the results from problem 16 and solve the two point boundary value problem

$$y'' - 4y = f(x), \quad 0 \le x \le 1, \quad y(0) = 1, \quad y(1) = 3$$

▶ **18.**   Use the results from problem 16 and solve the two point boundary value problem

$$x^2 y'' - 3xy' + 3y = x^5, \quad -1 \le x \le 2, \quad y(-1) = 0, \quad y(2) = 4$$

▶ **19.**   If the Green's function is symmetric with

$$G(x;\xi) = \begin{cases} G_1(x,\xi), & a \le x < \xi \\ G_2(x,\xi), & \xi < x \le b \end{cases} \quad \text{show} \quad G(\xi;x) = \begin{cases} G_2(x,\xi), & a \le \xi < x \\ G_1(x,\xi), & x < \xi \le b \end{cases}$$

▶ **20.**   For the equation

$$L_x(G) = a_0(x)\frac{d^n G}{dx^n} + a_1(x)\frac{d^{n-1}G}{dx^{n-1}} + \cdots + a_n(x)G = \delta(x - \xi)$$

(a) Show using integration by parts that

$$a_0(x)\frac{d^{n-1}G}{dx^{n-1}} + \left\{ \begin{array}{c} \text{continuous functions of} \\ G,\ G',\ldots,G^{(n-2)} \end{array} \right\} = H(x - \xi) + c,$$

where H is the unit step function.

(b) From the results in (a) verify the limit (8.66)

▶ **21.**   Solve for the deflection $y(x)$ of a beam of length L which has a load distribution $\omega = \omega(x)$ lbs/ft by solving the load equation

$$L_x(y) = EI\frac{d^4 y}{dx^4} = \omega(x), \quad E,\ I \text{ constants}$$

by use of Green's functions for the boundary conditions given.

(a)   Fixed ends: $y(0) = y'(0) = 0,\ y(L) = y'(L) = 0$

(b)   Cantilever beam: $y(0) = y'(0) = 0,\ y''(L) = y'''(L) = 0$

(c)   Simple supported ends: $y(0) = y''(0) = 0,\ y(L) = y''(L) = 0$

▶ **22.**

    (a) Find the Green's function by the Fourier series method

$$L_x(G) = G'' + \omega^2 G = \delta(x - \xi), \quad 0 \le x \le L, \quad \omega \ne \frac{n\pi}{L}$$

$$G(0) = G(L) = 0$$

    (b) Solve by Laplace transforms

    (c) Show the results from parts (a) and (b) are equivalent.

▶ **23.**   Consider the two point boundary value problem

$$L_x(y) = y'' = f(x), \quad 0 \le x \le 1, \quad y(0) = y(1) = 0$$

Assume a solution of the form

$$y(x) = \int_a^b G(x;\xi) f(\xi)\, d\xi = \int_0^{x^-} G_2(x,\xi) f(\xi)\, d\xi + \int_{x^+}^1 G_1(x,\xi) f(\xi)\, d\xi,$$

where $G(x;\xi) = \begin{cases} G_1(x,\xi), & 0 \le x < \xi \\ G_2(x,\xi), & \xi < x \le 1 \end{cases}$

    (a) Find $\frac{dy}{dx}$ using Leibnitz'[†] rule for differentiation.

    (b) Find $\frac{d^2 y}{dx^2}$.

    (c) In (a) and (b) show where the continuity condition and jump discontinuity condition of the derivative occurs.

    (d) Show from (b) that if $y'' = f(x)$ it is necessary for $\frac{\partial^2 G_2}{\partial x^2} = 0$ and $\frac{\partial^2 G_1}{\partial x^2} = 0$ and consequently $G_1 = A_1 x + B_1$ and $G_2 = A_2 x + B_2$.

    (e) By applying the continuity and jump discontinuity conditions determine the constants A_1, A_2, B_1, B_2 in (d) and construct the Green's function.

▶ **24.**

    (a) Find a Green's function for $L_x(y) = \frac{d^2 y}{dx^2}, \quad y(0) = y(1) = 0$ by the Fourier series method.

    (b) Expand the function $G(x;\xi) = \begin{cases} (\xi - 1)x, & 0 \le x < \xi \\ (x - 1)\xi, & \xi < x \le 1 \end{cases}$ in a Fourier sine series and show this is the same result as in part (a).

▶ **25.**   Solve by Green's function methods

$$L_x(y) = x^2 y'' + x y' - n^2 y = f(x), \quad 0 \le x \le 1, \quad n > 0$$

satisfying the conditions $y(1) = 0$,   and $\lim\limits_{x \to 0} y(x)$ exists.

[†]   See Appendix C

▶ **26.** Given

$$y'' + \beta(x)y' + \alpha(x)y = g(x), \quad a \le x$$

Find $y(x)$ satisfying $y(a) = A$, $y'(a) = B$ as follows:

(a) Integrate the given equation from a to x and use integration by parts to show

$$y' = \beta y - \int_a^x (\alpha - \beta')y \, dx + \int_a^x g \, dx + \beta(a)A + B.$$

(b) Integrate the result in part(a) from a to x and show $y(x)$ must satisfy the integral equation

$$y(x) = f(x) + \int_a^x G(x,t)y(t) \, dt$$

where

$$f(x) = \int_a^x (x - t)g(t) \, dt + [\beta(a)A + B](x - a) + A$$

and

$$G(x,t) = (t - x)[\alpha - \beta'] - \beta$$

▶ **27.** Solve the boundary value problem

$$x^2 y'' + xy' + (x^2 - \nu^2)y = f(x), \quad 0 < a \le x \le b$$

subject to the boundary conditions $y(a) = y(b) = 0$, $\nu \ne$ an integer.

▶ **28.** Consider

$$L_x(y) = y'' + \beta(x)y' + \alpha(x)y = r(x), \qquad 28(a)$$

where β, α are continuous and $r(x)$ is sectionally continuous. Let $Y_1(x;\xi), Y_2(x;\xi)$ be a fundamental set of $L(y) = 0$ satisfying

$$Y_1(x;\xi)\big|_{x=\xi} = 1, \quad Y_2(x;\xi)\big|_{x=\xi} = 0. \quad Y_1'(x;\xi)\big|_{x=\xi} = 0, \quad Y_2'(x;\xi)\big|_{x=\xi} = 1.$$

A solution of equation 28(a) can then be represented

$$y(x) = AY_1(x;\xi) + BY_2(x;\xi) + \int_a^x Y_2(x;u)r(u) \, du \qquad 28(b)$$

(a) Show that the Wronskian satisfies

$$W = \begin{vmatrix} Y_1(x;a) & Y_2(x;a) \\ Y_1'(x;a) & Y_2'(x;a) \end{vmatrix} = \exp\left(-\int_a^x \beta(v) \, dv\right)$$

(b) When $r(x)$ is replaced by the function $\delta_\epsilon(x - \xi)$ illustrated in figure 8-4, let $y(x;\epsilon)$ denote the solution to equation 28(a) satisfying $y(a) = 0$ and $y'(a) = 0$

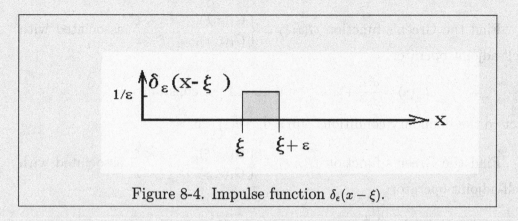

Figure 8-4. Impulse function $\delta_\epsilon(x - \xi)$.

Use variation of parameters and show a solution of the form

$$y(x; \epsilon) = u(x)Y_1(x; a) + v(x)Y_2(x; a)$$

requires

$$u'(x) = -\frac{Y_2(x; a)\delta_\epsilon(x - \xi)}{W} \qquad \text{and} \qquad v'(x) = \frac{Y_1(x; a)\delta_\epsilon(x - \xi)}{W}$$

(c) Show the solution $y(x; \epsilon)$ satisfying $y(a) = 0, y'(a) = 0$ can be written

$$y(x; \epsilon) = \left\{ \frac{1}{\epsilon} \int_\xi^x Q(x, t)\, dt \right\} H(x - \xi), \qquad 28(c)$$

where

$$Q(x, t) = \begin{vmatrix} Y_1(t; a) & Y_2(t; a) \\ Y_1(x; a) & Y_2(x; a) \end{vmatrix} \exp\left(\int_a^t \beta(v)\, dv \right)$$

(d) Show for $x > \xi + \epsilon$

$$y(x; \epsilon) = \frac{1}{\epsilon} \int_\xi^{\xi+\epsilon} Q(x, t)\, dt \qquad \text{and} \qquad \lim_{\epsilon \to 0} y(x; \epsilon) = Q(x, \xi)$$

is the Green's function response due to a unit impulse at $x = \xi$.

(e) Show $Q(x, \xi)$ as a function of x is a solution of the homogeneous differential equation $L(y) = 0$. Show $Q(\xi, \xi) = 0$ and $Q'(\xi, \xi) = 1$.

(f) Why do the results of (e) imply $Q(x, \xi) = Y_2(x; \xi)$?

(g) Use superposition and show the sum of the responses of the system due to an impulse $r(\xi)\delta(x - \xi)$ is

$$y(x) = \int_a^x Y_2(x; \xi)r(\xi)\, d\xi$$

Here $Y_2(x; \xi)$ is called the Green's function response due to a unit impulse.

▶ **29.** Find the Green's function $G(x; \xi) = \begin{cases} G_1(x; \xi), & x < \xi \\ G_1(\xi; x), & \xi < x \end{cases}$ associated with the self-adjoint operator

$$L_x(u) = \frac{d^2u}{dx^2} + \omega^2 u, \qquad 0 < x < 1 \quad \omega \text{ constant}$$

subject to the boundary conditions $u(0) = 0, \quad u(1) = 0$.

▶ **30.** Find the Green's function $G(x; \xi) = \begin{cases} G_1(x; \xi), & x < \xi \\ G_1(\xi; x), & \xi < x \end{cases}$ associated with the self-adjoint operator

$$L_x(u) = \frac{d^2u}{dx^2} + \omega^2 u, \qquad -1 < x < 1 \quad \omega \text{ constant}$$

subject to the boundary conditions $u(-1) = u(1), \quad u'(-1) = u'(1)$.

▶ **31.** Find the Green's function $G(x; \xi) = \begin{cases} G_1(x; \xi), & x < \xi \\ G_1(\xi; x), & \xi < x \end{cases}$ associated with the self-adjoint operator $L_x(u) = \dfrac{d^2u}{dx^2}$ on the interval $(0, 1)$ and subject to the boundary conditions $u(0) = 0, \quad u'(1) = 0$.

▶ **32.** Find the Green's function $G(x; \xi) = \begin{cases} G_1(x; \xi), & x < \xi \\ G_1(\xi; x), & \xi < x \end{cases}$ associated with the self-adjoint operator

$$L_x(u) = \frac{d^2u}{dx^2} + \omega^2 u, \qquad 0 < x < 1, \quad \omega \text{ constant}$$

subject to the boundary conditions $u(0) = 0, \quad u'(1) = 0$.

▶ **33.** Find the Green's function $G(x; \xi) = \begin{cases} G_1(x; \xi), & x < \xi \\ G_1(\xi; x), & \xi < x \end{cases}$ associated with the self-adjoint operator

$$L_x(u) = \frac{d^2u}{dx^2}, \qquad -1 < x < 1$$

subject to the boundary conditions $u(-1) = 0, \quad u(1) = 0$.

▶ **34.** Find the Green's function $G(x; \xi) = \begin{cases} G_1(x; \xi), & x < \xi \\ G_1(\xi; x), & \xi < x \end{cases}$ associated with the self-adjoint operator

$$L_x(u) = \frac{d^2u}{dx^2} - \omega^2 u, \qquad 0 < x < 1, \quad \omega \text{ constant}$$

subject to the boundary conditions $u(0) = 0, \quad u(1) = 0$.

Chapter 9
Green's Functions for PDE's

Recall that for certain linear homogeneous partial differential equations with homogeneous boundary conditions it is possible to use the method of separation of variables to obtain a series solution. The method of separation of variables together with appropriate homogeneous boundary conditions produces Sturm-Liouville problems, eigenvalues, eigenfunctions and required generalized Fourier series expansions to represent initial conditions. When the given partial differential equation is nonhomogeneous or there exists nonhomogeneous boundary conditions we must develop alternative methods for constructing the solution. One such method which is applicable for both nonhomogeneous forcing terms and for nonhomogeneous boundary conditions is the Green's function method. This method is similar to the Green's functions that have been developed for the study of nonhomogeneous ordinary differential equations. Associated with each partial differential operator and boundary conditions there is associated a Green's function. The Green's function associated with a partial differential equation represents a response at a point P due to a unit impulse applied at a point Q. Various notations are used to represent the Green's function associated with partial differential equations. The following table 9-1 list some of the more standard notations for denoting a Green's function associated with ordinary and partial differential equations.

Table 9-1 Notation for Green's function representation			
Notation	Response point	Impulse point	Dimension
$G(x;\xi)$	x	ξ	1
$G(x,y;\xi,\eta)$	(x,y)	(ξ,η)	2
$G(x,y,z;\xi,\eta,\zeta)$	(x,y,z)	(ξ,η,ζ)	3
$G(\vec{x};\vec{\xi})$	$\vec{x}=(x_1,x_2,\ldots,x_n)$	$\vec{\xi}=(\xi_1,\xi_2,\ldots,\xi_n)$	n
$G(P;Q)$	P	Q	general

In those instances where a time variable is one of the variables in the partial differential equation, then x_n is replaced by t and ξ_n is replaced by τ in the n-dimensional notation. For example in one space dimension and a time dimension the Green's function notation would be $G = G(x,t;\xi,\tau)$. The impulse function representation uses a similar notation. For example, in one space dimension and a time dimension we use $\delta(x-\xi)\delta(t-\tau)$ to denote the unit impulse at the point (ξ,τ). Similarly, in n-dimensions we use the notation

$$\delta(\vec{x}-\vec{\xi}) = \delta(x_1-\xi_1)\delta(x_2-\xi_2)\cdots\delta(x_n-\xi_n) \tag{9.1}$$

to denote a unit impulse at the point $\vec{x} = \vec{\xi}$. In some problems the unit impulse is applied as a forcing function, whereas in other problems the unit impulse is applied to the initial conditions. The sifting property for the Dirac delta function also extends to more than one dimension. For example, in two dimensions we use the sifting property that for $(x,y) \in R$ and $(\xi,\eta) \in R$ one can write

$$\iint_R f(x,y)\delta(x-\xi)\delta(y-\eta)\,dxdy = f(\xi,\eta). \tag{9.2}$$

The following is a brief introduction to Green's functions for partial differential equations over finite and infinite regions. We begin with second order partial differential equations. Let $u = u(x,y)$ denote a function of x,y and let $L_{xy}(u)$ denote the linear partial differential operator

$$L_{xy}(u) = A\frac{\partial^2 u}{\partial x^2} + 2B\frac{\partial^2 u}{\partial x\partial y} + C\frac{\partial^2 u}{\partial y^2} + D\frac{\partial u}{\partial x} + E\frac{\partial u}{\partial y} + Fu, \tag{9.3}$$

where A, B, C, D, E, F are functions of x and y which are continuous functions over some simply connected region R of the x,y plane which is bounded by a simple closed curve $C = \partial R$. Often it is convenient to replace y by t in equation (9.3), then x can be considered as a space variable and t can be considered as a time variable. Also there are times when we desire to replace the variables x,y in equation (9.3) by new variables ξ,η and represent the solution in the form $u = u(\xi,\eta)$.

There is associated with the operator L_{xy} of equation (9.3) a Green's formula in two dimensions which can be expressed as

$$\int\int_R \left[vL_{xy}(u) - uL_{xy}^*(v)\right]\,dxdy = \oint_C M(x,y)\,dx + N(x,y)\,dy$$
$$= \oint_C (N\hat{\imath} - M\hat{\jmath})\cdot\hat{n}\,ds \tag{9.4}$$

with

$$L_{xy}^*(v) = \frac{\partial^2}{\partial x^2}(Av) + 2\frac{\partial^2}{\partial x\partial y}(Bv) + \frac{\partial^2}{\partial y^2}(Cv) - \frac{\partial}{\partial x}(Dv) - \frac{\partial}{\partial y}(Ev) + Fv \tag{9.5}$$

the adjoint operator associated with equation (9.3) and

$$M(x,y) = \left[\frac{\partial B}{\partial x} + \frac{\partial C}{\partial y} - E\right] uv + B\left[u\frac{\partial v}{\partial x} - v\frac{\partial u}{\partial x}\right] + C\left[u\frac{\partial v}{\partial y} - v\frac{\partial u}{\partial y}\right]$$

$$N(x,y) = \left[D - \frac{\partial A}{\partial x} - \frac{\partial B}{\partial y}\right] uv + A\left[v\frac{\partial u}{\partial x} - u\frac{\partial v}{\partial x}\right] + B\left[v\frac{\partial u}{\partial y} - u\frac{\partial v}{\partial y}\right].$$

(9.6)

In the equation (9.4), the integral on the left is an area integral over the region R and $dxdy$ denotes an element of area inside the region R. The integral on the right hand side of equation (9.4) is a line integral around the boundary of the region R taken in the positive sense, where $C = \partial R$ denotes the boundary curve. The region R must be a simply connected region and the boundary curve $C = \partial R$ is some simple closed curve. An example of such a region is illustrated in the figure 9-1. The functions u and v in the Green's formula given by equation (9.4) are assumed to be continuous and differentiable both within the region R and on the boundary curve $C = \partial R$. If either of the functions u or v has a singularity at a point (x_0, y_0) within R, then one must place a small disk $D_\epsilon = \{x,y \mid (x - x_0)^2 + (y - y_0)^2 \leq \epsilon^2\}$ of radius ϵ about the singular point and replace the region R in equation (9.4) with the region $R - D_\epsilon$ which represents the region R with the circular disk removed. The new boundary curve then consists of an inner boundary around the small circle plus the original outer boundary. Any results obtained from the Green's formula, with this modified region, then becomes a function of the radius ϵ of the small disk. One can then investigate the Green's formula in the limit as ϵ tends to zero. Note the right-hand side of equation (9.4) can be expressed in terms of unit vectors $\hat{i}$ and $\hat{j}$ in the direction of the x and y axes. In this line integral $\hat{n}$ is a unit exterior normal vector at a point on the boundary curve $C = \partial R$ and ds is an element of arc length along the boundary curve. If the parametric equations defining the boundary curve C are expressed in terms of arc length and have the form $x = x(s)$, $y = y(s)$, then the vector $\vec{r} = x(s)\hat{i} + y(s)\hat{j}$, for $0 \leq s \leq s_0$, is the position vector to a point on the boundary curve C. The vector $\frac{d\vec{r}}{ds} = \frac{dx}{ds}\hat{i} + \frac{dy}{ds}\hat{j}$ is a unit tangent vector to the boundary curve C and the exterior unit normal to the boundary curve C is given by $\hat{n} = \frac{dy}{ds}\hat{i} - \frac{dx}{ds}\hat{j} = \frac{d\vec{r}}{ds} \times \hat{k}$ where $\hat{i}, \hat{j}$, and $\hat{k}$ are orthogonal unit vectors in the x, y and z directions.

In the use of the Green's formula it is often times convenient to change the variables x, y to some new variables like ξ, η and so one should be careful to observe the notation indicating the dummy variables of integration that are being used. For example, one can write the Green's identity given by the equation (9.4) in the form

$$\int\int_R \left[vL_{\xi\eta}(u) - uL_{\xi\eta}^*(v)\right] d\xi d\eta = \oint_C M(\xi,\eta)\, d\xi + N(\xi,\eta)\, d\eta$$

$$= \oint_C (N\hat{\imath} - M\hat{\jmath}) \cdot \hat{n}\, ds$$

(9.7)

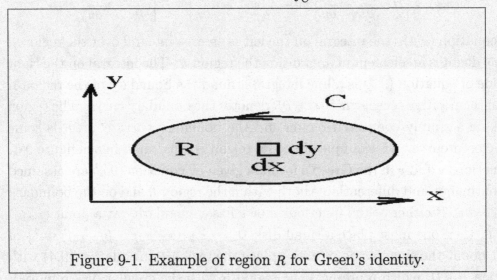

Figure 9-1. Example of region R for Green's identity.

In two-dimensions the Green's function is represented $G = G(x,y;\xi,\eta)$ and satisfies

$$L_{xy}G = \delta(x-\xi)\delta(y-\eta)$$

(9.8)

and is interpreted as the response at the point (x,y) due to an impulse applied at the point (ξ,η). The adjoint Green's function $G^*(x,y;\xi,\eta)$ satisfies the equation

$$L_{xy}^* G^* = \delta(x-\xi)\delta(y-\eta)$$

(9.9)

which has the similar interpretation as a response due to an impulse. Analogous to what has been done for the one-dimensional Green's function, we can show that the Green's function from equation (9.8) and the adjoint Green's function from the equation (9.9), both of which satisfy homogeneous boundary conditions, are symmetric and satisfy the symmetry condition $G^*(x,y;x_0,y_0) = G(x_0,y_0;x,y)$. To establish this result let $u = G$ and $v = G^*$ in the Green's formula. Assume that $G = G(\xi,\eta;x,y)$ satisfies

$$L_{\xi\eta}G(\xi,\eta;x,y) = \delta(\xi-x)\delta(\eta-y) \quad x,y \in R$$

(9.10)

with homogeneous boundary conditions. Let $G^* = G^*(\xi,\eta;x_0,y_0)$ satisfy the adjoint equation

$$L_{\xi\eta}^* G^*(\xi,\eta;x_0,y_0) = \delta(\xi-x_0)\delta(\eta-y_0) \quad x_0,y_0 \in R$$

(9.11)

also with homogeneous boundary conditions. The homogeneous boundary conditions are that both G and G^* are zero on the boundary of the region R. These conditions make the right-hand side of the Green's formula zero so that the equation (9.7) becomes

$$\iint_R [G^*(\xi,\eta;x_0,y_0)\delta(\xi-x)\delta(\eta-y) - G(\xi,\eta;x,y)\delta(\xi-x_0)\delta(\eta-y_0)]\, d\xi d\eta = 0 \qquad (9.12)$$

which reduces by the sifting property to the symmetry condition

$$G^*(x,y;x_0,y_0) = G(x_0,y_0;x,y). \qquad (9.13)$$

In Green's formula, given by equation (9.7), we replace v by a Green's function $G^* = G^*(\xi,\eta;x,y)$ which is the solution of the adjoint equation

$$L^*_{\xi\eta}(G^*) = \delta(\xi-x)\delta(\eta-y), \qquad x,y \in R \qquad (9.14)$$

where the variables x,y in the operator L^*_{xy} have been replaced by ξ,η. Observe that when v is replaced by the Green's function G^*, the second term in the Green's identity (9.7) gets simplified by using the sifting property of the delta function. One finds

$$\iint_R u L^*_{\xi\eta}(G^*)\, d\xi d\eta = \iint_R u(\xi,\eta)\delta(\xi-x)\delta(\eta-y)\, d\xi d\eta = u(x,y). \qquad (9.15)$$

The remaining terms in equation (9.7) involve the Green's function $G^*(\xi,\eta;x,y)$ and its derivatives. These remaining terms must be analyzed to determine the adjoint boundary conditions that are to be imposed upon the partial differential equation (9.14) defining the Green's function . If all the terms in the Green's identity are properly defined, then we can make use of the Green's identity in two dimensions to construct solutions to boundary value problems in partial differential equations. The boundary conditions on the Green's function or any of its derivatives are obtained from an analysis of the Green's identity. This is similar to what has been done for the Green's identity in ordinary differential equations. The only difference is that we are now dealing with line integrals and area integrals.

It should be observed that there are Green's formulas for differential operators in dimensions higher than two which involve integrations over surfaces and volumes . However, most of the examples given in this section are for two-dimensional operators of the form given by equation (9.3).

476

Example 9-1. (Laplacian operator.)

Show how one would employ the Green's identity (9.7) to solve the Dirichlet boundary value problem associated with the Poisson equation

$$L_{xy}(u) = \nabla^2 u = \frac{\partial^2 u}{\partial x^2} + \frac{\partial^2 u}{\partial y^2} = f(x,y) \qquad x,y \in R$$

such that the solution satisfies $u(x,y) = 0$ for x,y on the boundary $C = \partial R$ of a simply connected region R. We will treat the region R as being arbitrary and set up all necessary equations needed to obtain the desired solution.

Solution: Here the operator $L_{xy}(u) = \nabla^2 u$ is the self-adjoint Laplace operator, and the Green's identity given by equation (9.7) for this elliptic equation can be written as

$$\iint_R [vL_{\xi\eta}(u) - uL_{\xi\eta}(v)]\,d\xi d\eta = \oint_C \left[u\frac{\partial v}{\partial \eta} - v\frac{\partial u}{\partial \eta} \right] d\xi + \left[v\frac{\partial u}{\partial \xi} - u\frac{\partial v}{\partial \xi} \right] d\eta, \qquad (9.16)$$

where $C = \partial R$ is a simple closed curve enclosing the region R. With reference to the figure 9-2, the position vector to a general point on the boundary can be represented in terms of the unit vectors $\hat{\imath}$ and $\hat{\jmath}$. Let $\vec{r} = \xi\hat{\imath} + \eta\hat{\jmath}$ denote the position vector to a point (ξ, η) on the boundary C of R.

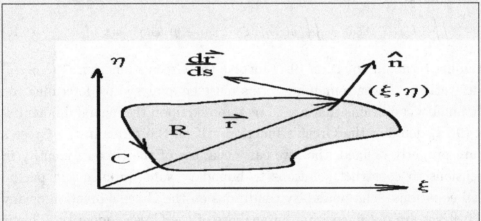

Figure 9-2. Tangent and normal to region R for Green's identity.

The vector

$$\hat{t} = \frac{d\vec{r}}{ds} = \frac{d\xi}{ds}\hat{\imath} + \frac{d\eta}{ds}\hat{\jmath}$$

is a unit tangent vector to the boundary, where s is arc length along the boundary. The unit normal vector to a point (ξ, η) on the boundary can be expressed as

$$\vec{n} = \vec{t} \times \vec{k} = \frac{d\eta}{ds}\hat{\imath} - \frac{d\xi}{ds}\hat{\jmath}$$

and this enables us to express the right-hand side of equation (9.16) as

$$\oint_C v \operatorname{grad} u \cdot \hat{n} \, ds - u \operatorname{grad} v \cdot \hat{n} \, ds.$$

The Green's formula can therefore be written in the alternative form

$$\iint_R \left[v \nabla^2 u - u \nabla^2 v \right] d\xi d\eta = \oint_C \left[v \frac{\partial u}{\partial n} - u \frac{\partial v}{\partial n} \right] ds, \tag{9.17}$$

where $\frac{\partial u}{\partial n} = \operatorname{grad} u \cdot \hat{n}$ and $\frac{\partial v}{\partial n} = \operatorname{grad} v \cdot \hat{n}$ are normal derivatives. The normal derivatives are just directional derivatives evaluated on the boundary curve C, in the direction of the outward normal to the region R.

The equation (9.17) can now be used to solve the given boundary value problem. If we make the substitution $v = G^*(\xi, \eta; x, y)$, where G^* is a Green's function which satisfies

$$L_{\xi\eta}^*(G^*) = \nabla^2 G^* = \frac{\partial^2 G^*}{\partial \xi^2} + \frac{\partial^2 G^*}{\partial \eta^2} = \delta(\xi - x)\delta(\eta - y) \tag{9.18}$$

and we let $u = u(\xi, \eta)$ denote the solution of the given boundary value problem, $L_{\xi\eta}(u) = f(\xi, \eta)$, then the Green's identity given by equation (9.17) can be written as

$$\iint_R G^* f(\xi, \eta) \, d\xi d\eta = \iint_R u(\xi, \eta) \delta(\xi - x) \delta(\eta - y) \, d\xi d\eta + \oint_C \left(G^* \frac{\partial u}{\partial n} - u \frac{\partial G^*}{\partial n} \right) ds. \tag{9.19}$$

The line integral term in equation (9.19) vanishes if we require that $G^* = 0$ on the boundary of R in addition to the given condition that $u = 0$ on the boundary curve C. This reduces equation (9.19) to

$$u(x, y) = \iint_R G^*(\xi, \eta; x, y) f(\xi, \eta) \, d\xi d\eta$$

as the solution of the given boundary value problem, where G^* satisfies the equation (9.18) with $G^* = 0$ for (ξ, η) on the boundary of R. Here L_{xy} is a self-adjoint operator and consequently the Green's function is symmetric and satisfies $G(x, y; \xi, \eta) = G(\xi, \eta; x, y)$. This result allows us to represent the solution in the alternate form

$$u(x, y) = \iint_R G(x, y; \xi, \eta) f(\xi, \eta) \, d\xi d\eta. \tag{9.20}$$

The solution of the equation (9.18) for some arbitrary region R is extremely difficult. However, for certain special regions, the solution to equation (9.18) can be constructed using transform methods. In other special regions the Green's function can be constructed in terms of functions which are orthogonal over the region R.

Note for nonhomogeneous boundary conditions, say $u(x,y) = g(x,y)$ for $x, y \in C = \partial R$, we will have in general that one or more of the line integral terms in equation (9.19) do not vanish and so under these circumstances additional terms must be included as part of the solution. In such cases the solution becomes dependent upon the boundary conditions along the boundary curve C. This example illustrates that the line integral terms must be analyzed for each boundary value problem involving the Laplace operator.

Note also the general form of the equation (9.17) can be altered by adding and subtracting the term $\dfrac{\beta}{\alpha} \dfrac{\partial u}{\partial n} \dfrac{\partial v}{\partial n}$ to the line integral on the right-hand side. This produces the more general result

$$\iint_R \left[v\nabla^2 u - u\nabla^2 v \right] d\xi d\eta = \oint_C \frac{1}{\alpha} \left[\left(\beta \frac{\partial v}{\partial n} + \alpha v \right) \frac{\partial u}{\partial n} - \left(\beta \frac{\partial u}{\partial n} + \alpha u \right) \frac{\partial v}{\partial n} \right] ds, \qquad (9.21)$$

whereby Robin type boundary conditions can be considered.

For example, to solve the boundary value problem

$$\text{PDE} \qquad \nabla^2 u = f(x,y), \qquad x, y \in R$$

$$\text{BC} \qquad \beta \frac{\partial u}{\partial n} + \alpha u = g(x,y), \qquad x, y \in \partial R$$

one can employ the equation (9.21) as follows. In equation (9.21) let $u = u(\xi, \eta)$ be a solution of $\nabla^2 u = f(\xi, \eta)$ for $\xi, \eta \in R$ and subject to the boundary condition $\beta \frac{\partial u}{\partial n} + \alpha u = g(\xi, \eta)$ for $\xi, \eta \in \partial R$ and let $v = G^*(\xi, \eta; x, y)$ denote a Green's function which satisfies $\nabla^2 G^* = \delta(\xi - x)\delta(\eta - y)$ for $\xi, \eta \in R$ and subject to the boundary condition $\beta \frac{\partial G^*}{\partial n} + \alpha G^* = 0$ for $\xi, \eta \in \partial R$. The equation (9.21) then produces the solution function

$$u = u(x,y) = \int_R G(x,y;\xi,\eta) f(\xi,\eta) \, d\xi d\eta + \int_{\partial R} \frac{1}{\alpha} g(\xi,\eta) \frac{\partial G(x,y;\xi,\eta)}{\partial n} \, ds \qquad (9.22)$$

where we have made use of the symmetry property and replaced G^* by G.

∎

Example 9-2.   (Heat equation.)
Show how to use the Green's identity (9.7) to solve the following Dirichlet problem associated with the nonhomogeneous heat equation representing heat flow

in a thin rod of length ℓ. We desire to solve

$$L_{xt}(u) = \frac{\partial u}{\partial t} - \alpha^2 \frac{\partial^2 u}{\partial x^2} = F(x,t), \quad 0 < x < \ell, \quad t > 0 \tag{9.23}$$

subject to the end point boundary conditions $u(0,t) = 0$, $u(\ell,t) = 0$ and the initial condition $u(x,0) = f(x)$.

Solution: Here L_{xt} denotes the differential operator for the heat equation. For this parabolic operator the Green's formula given by equation (9.7) can be written

$$\iint_R \left[v L_{\xi\tau}(u) - u L^*_{\xi\tau}(v) \right] d\xi d\tau = \oint_C \left[-\alpha^2 (v \frac{\partial u}{\partial \xi} - u \frac{\partial v}{\partial \xi}) \hat{i} + uv\hat{j} \right] \cdot \hat{n} \, ds \tag{9.24}$$

The Green's function $G(x,t;\xi,\tau) = G^*(\xi,\tau;x,t)$ is to be zero for all $t < \tau$. If we select $v = G^*$ to be a Green's function $G^* = G^*(\xi,\tau;x,t)$ as a function of ξ,τ and a solution of the adjoint equation

$$L^*_{\xi\tau}(G^*) = -\frac{\partial G^*}{\partial \tau} - \alpha^2 \frac{\partial^2 G^*}{\partial \xi^2} = \delta(\xi - x)\delta(\tau - t),$$

with the requirement $G^* = 0$ for $\tau > t$, then the equation (9.24) reduces to

$$u(x,t) = \int\int_R G^*(\xi,\tau;x,t) L_{\xi\tau}(u) \, d\xi d\tau + \oint_C \left\{ \alpha^2 \left[G^* \frac{\partial u}{\partial \xi} - u \frac{\partial G^*}{\partial \xi} \right] \hat{i} - uG^*\hat{j} \right\} \cdot \hat{n} \, ds \tag{9.25}$$

and the line integral around the boundary can be analyzed once a region R and boundary C have been selected.

For purposes of illustration we select the rectangular region R illustrated in figure 9-3 to be associated with the given boundary value problem.

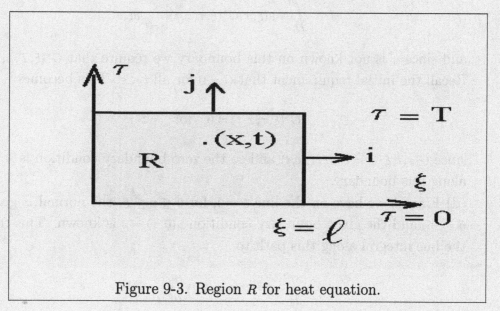

Figure 9-3. Region R for heat equation.

In equation (9.25) we let $u = u(\xi, \eta)$ denote the solution of

$$L_{\xi\tau}(u) = F(\xi, \tau), \qquad 0 \leq \xi \leq \ell,$$

and we analyze the line integral term in equation (9.25) as we move around the boundary of the rectangle in figure 9-3.

(a) On the line $\tau = 0$, for $0 \leq \xi \leq \ell$, we have $u(\xi, 0) = f(\xi)$, and $\hat{n} = -\hat{\jmath}$ so that this portion of the line integral reduces to

$$\int_0^\ell f(\xi) G^*(\xi, 0; x, t) \, d\xi$$

(b) On the line $\xi = \ell$, for $0 \leq \tau \leq T$, we have $\hat{n} = \hat{\imath}$, and $u(\ell, \tau) = 0$ so that this portion of the line integral reduces to

$$\int_0^T \alpha^2 G^*(\ell, \tau; x, t) \frac{\partial u}{\partial \xi}\bigg|_{\xi=\ell} d\tau$$

and since $\frac{\partial u}{\partial \xi}\big|_{\xi=\ell}$ is not known, we enforce the condition that $G^*(\ell, \tau; x, t) = 0$ along this boundary.

(c) On the line $\tau = T$ we have $\hat{n} = \hat{\jmath}$ so that the line integral term on the right-hand side of equation (9.25) becomes

$$\int_\ell^0 -u(\xi, \tau) G^*(\xi, \tau; x, t)\bigg|_{\tau=T} d\xi,$$

and since u is not known on this boundary we require that $G^*(\xi, T; x, t) = 0$. Recall the initial requirement that $G = 0$ for all $t < \tau$. This becomes

$$G^*(\xi, \tau; x, t) = 0 \quad \text{for} \quad \tau > t$$

since $G(x, t; \xi, \tau) = G^*(\xi, \tau; x, t)$ and so the zero boundary condition is satisfied along this boundary.

(d) Finally, we have on the line $\xi = 0$, for $0 \leq \tau \leq T$, the normal is given by $\hat{n} = -\hat{\imath}$, and the given boundary condition $u(0, \tau) = 0$ is known. This reduces the line integral along this path to

$$\int_T^0 -\alpha^2 G^*(0, \tau; x, t) \frac{\partial u}{\partial \xi}\bigg|_{\xi=0} d\tau$$

and since $\frac{\partial u}{\partial \xi}\big|_{\xi=0}$ is not known on this boundary, we enforce the condition that $G^*(0,\tau;x,t) = 0$. Combining the above line integrals produces the solution to the boundary value problem in the form

$$u(x,t) = \int_0^t \int_0^\ell G^*(\xi,\tau;x,t)F(\xi,\tau)\,d\xi d\tau + \int_0^\ell f(\xi)G^*(\xi,0;x,t)\,d\xi, \qquad (9.26)$$

where the Green's function is to satisfy

$$L_{\xi\tau}^*(G^*) = -\frac{\partial G^*}{\partial \tau} - \alpha^2 G^* = \delta(\xi - x)\delta(\tau - t)$$

$$G^* = 0 \quad \text{on} \quad \xi = 0 \qquad \text{and} \qquad G^* = 0 \quad \text{on} \quad \xi = \ell \qquad (9.27)$$

$$G^* = 0 \quad \text{for} \quad \tau > t, \quad 0 < \xi < \ell.$$

The Green's function is symmetric $G(x,t;\xi,\tau) = G^*(\xi,\tau;x,t)$ and when treated as a function of x,t it is a solution of the equation

$$L_{xt}(G) = \frac{\partial G}{\partial t} - \alpha^2 \frac{\partial^2 G}{\partial x^2} = \delta(x - \xi)\delta(t - \tau), \qquad t > \tau$$

with $G = 0$ for $t < \tau$. The boundary conditions for this problem are given by $G(0,t;\xi,\tau) = 0$, and $G(\ell,t;\xi,\tau) = 0$. We also require that G be zero for $t < \tau$. That is G is zero before the source term acts. The solution given by equation (9.26) can therefore be written in the alternative form

$$u(x,t) = \int_0^t \int_0^\ell G(x,t;\xi,\tau)F(\xi,\tau)\,d\xi d\tau + \int_0^\ell f(\xi)G(x,t;\xi,0)\,d\xi. \qquad (9.28)$$

Thus, if we can solve the Green's function problem (9.27), then it becomes possible to construct the solution to the given boundary value problem. The solution is then represented in the form given by either equation (9.26) or (9.28). The main point of this example is that some type of region R must be constructed and analyzed for each boundary value problem associated with the heat equation operator.

If we take the limit as $\ell \to \infty$, the solution given by equation (9.26) then becomes

$$u(x,t) = \int_0^t \int_0^\infty G^*(\xi,\tau;x,t)F(\xi,\tau)\,d\xi d\tau + \int_0^\infty f(\xi)G^*(\xi,0;x,t)\,d\xi \qquad (9.29)$$

where now boundedness conditions on G^* are assumed to exist as $\xi \to \infty$.

Example 9-3. (**The wave equation.**)

Show how to use the Green's identity (9.7) to solve the following boundary and initial value problem associated with the wave equation. Let $u = u(x, t)$ satisfy

$$L_{xt}(u) = \frac{\partial^2 u}{\partial t^2} - \alpha^2 \frac{\partial^2 u}{\partial x^2} = F(x, t), \qquad -\ell < x < \ell, \quad t > 0$$

subject to the boundary conditions $u(-\ell, t) = 0$ and $u(\ell, t) = 0$ and initial conditions

$$u(x, 0) = f(x), \qquad \frac{\partial u(x, 0)}{\partial t} = g(x).$$

Solution: Here L_{xt} is a hyperbolic operator and the Green's formula given by equation (9.7) can be written

$$\int\!\!\int_R \left[v L_{\xi\tau}(u) - u L_{\xi\tau}^*(v) \right] d\xi d\tau = \oint_c \left\{ -\alpha^2 \left[v \frac{\partial u}{\partial \xi} - u \frac{\partial v}{\partial \xi} \right] \hat{\imath} - \left[u \frac{\partial v}{\partial \tau} - v \frac{\partial u}{\partial \tau} \right] \hat{\jmath} \right\} \cdot \hat{n} \, ds. \quad (9.30)$$

By choosing $v = G^*$ to be a Green's function satisfying

$$L_{\xi\tau}^*(G^*) = \frac{\partial^2 G^*}{\partial \tau^2} - \alpha^2 \frac{\partial^2 G^*}{\partial \xi^2} = \delta(\xi - x)\delta(\tau - t)$$

we can make use of the sifting property of the Dirac delta function to simplify the equation (9.30) to the form

$$u(x, t) = \int\!\!\int_R G^*(\xi, \tau; x, t) L_{\xi\tau}(u) \, d\xi d\tau$$
$$+ \oint_C \left\{ \alpha^2 \left[G^* \frac{\partial u}{\partial \xi} - u \frac{\partial G^*}{\partial \xi} \right] \hat{\imath} + \left[u \frac{\partial G^*}{\partial \tau} - G^* \frac{\partial u}{\partial \tau} \right] \hat{\jmath} \right\} \cdot \hat{n} \, ds. \quad (9.31)$$

In order for the Green's formula (9.31) to be useful for the construction of solutions to the wave equation we must be able to choose a region R with boundary curve C that can be analyzed. For the above problem, we use the region R illustrated in the figure 9-4. Note that in this figure we could let $\ell \to \infty$ to extend this region over an infinite domain.

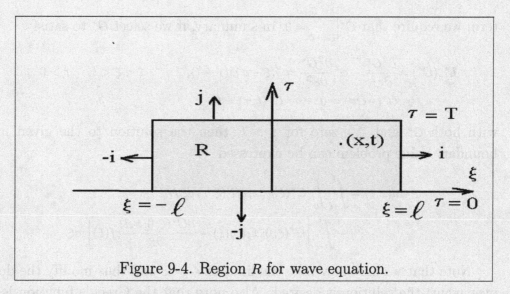

Figure 9-4. Region R for wave equation.

We let $u = u(\xi, \tau)$ satisfy the equation

$$L_{\xi\tau}(u) = \frac{\partial^2 u}{\partial \tau^2} - \alpha^2 \frac{\partial^2 u}{\partial \xi^2} = F(\xi, \tau), \qquad u(-\ell, \tau) = 0, \qquad u(\xi, 0) = f(\xi)$$
$$-\ell < \xi < \ell, \quad t > 0 \qquad u(\ell, \tau) = 0, \qquad \frac{\partial u(\xi, 0)}{\partial \tau} = g(\xi)$$

and then expand the line integral term in equation (9.31). The Green's identity given by equation (9.31) is applied to the region R selected and one finds

$$u(x,t) = \int_{-\ell}^{\ell} \int_0^T G^*(\xi, \tau; x, t) F(\xi, \tau) \, d\tau d\xi + \int_{-\ell}^{\ell} \left[G^* \frac{\partial u}{\partial \tau} - u \frac{\partial G^*}{\partial \tau} \right]_{\tau=0} d\xi$$
$$+ \int_0^T \alpha^2 \left[G^* \frac{\partial u}{\partial \xi} - u \frac{\partial G^*}{\partial \xi} \right]_{\xi=\ell} d\tau + \int_{\ell}^{-\ell} \left[u \frac{\partial G^*}{\partial \tau} - G^* \frac{\partial u}{\partial \tau} \right]_{\tau=T} d\xi$$
$$+ \int_T^0 \alpha^2 \left[u \frac{\partial G^*}{\partial \xi} - G^* \frac{\partial u}{\partial \xi} \right]_{\xi=-\ell} d\tau$$

where each line integral term must now be analyzed. On the line $\tau = 0$, we have the given boundary conditions $u(\xi, 0) = f(\xi)$ and $\frac{\partial u(\xi,0)}{\partial \tau} = g(\xi)$. On the line $\xi = \ell$, we have $u(\ell, \tau) = 0$ as a known term and $\frac{\partial u(\ell,\tau)}{\partial \xi}$ as an unknown term. To remove the unknown term we require that $G^*\big|_{\xi=\ell} = 0$. On the line $\tau = T$ the terms $u(\xi, \tau)$ and $\frac{\partial u(\xi,\tau)}{\partial \tau}$ are unknown and so to remove these terms we require that both $\frac{\partial G^*}{\partial \tau}$ and G^* be zero along this line. This can be accomplished by requiring that both $\frac{\partial G^*}{\partial \tau}$ and G^* be zero for all $\tau > t$. On the line $\xi = -\ell$ we have the known condition $u(-\ell, \tau) = 0$, but we also have $\frac{\partial u(-\ell,\tau)}{\partial \xi}$ which is unknown. To remove the unknown term we require that $G^*\big|_{\xi=-\ell} = 0$. In summary, if we select G^* to satisfy

$$L^*_{\xi\tau}(G^*) = \frac{\partial^2 G^*}{\partial \tau^2} - \alpha^2 \frac{\partial^2 G^*}{\partial \xi^2} = \delta(\xi - x)\delta(\tau - t), \qquad -\ell < \xi < \ell, \quad \tau > 0$$

$$G^*(-\ell, \tau) = 0, \qquad G^*(\ell, \tau) = 0$$

(9.32)

with both G^* and $\frac{\partial G^*}{\partial \tau}$ zero for $\tau > t$, then the solution to the given initial-boundary value problem can be expressed

$$u(x,t) = \int_{-\ell}^{\ell} \int_0^t G^*(\xi, \tau; x, t) F(\xi, \tau)\, d\xi d\tau$$

$$+ \int_{-\ell}^{\ell} \left[G^*(\xi, 0; x, t) g(\xi) - \frac{\partial G^*(\xi, 0; x, t)}{\partial \tau} f(\xi) \right] d\xi.$$

(9.33)

Note that we may now take the limit as $\ell \to \infty$ and thus modify the domain over which the solution is desired. Also note that the Green's function is symmetric and satisfies $G(x, t; \xi, \tau) = G^*(\xi, \tau; x, t)$ and so one can solve the alternative Green's function problem

$$L_{xt}(G) = \frac{\partial^2 G}{\partial t^2} - \alpha^2 \frac{\partial^2 G}{\partial x^2} = \delta(x - \xi)\delta(t - \tau), \quad -\ell < x < \ell, \quad t > 0$$

$$G(-\ell, t; \xi, \tau) = 0, \quad G(\ell, t; \xi, \tau) = 0$$

(9.34)

with both G and $\frac{\partial G}{\partial t}$ zero for $t < \tau$.

The solution given by equation (9.33) is dependent upon whether one can solve either of the Green's function problems given by the equations (9.32) or (9.34).

■

Fourier Series Representation

Consider the nonhomogeneous partial differential equation

$$L_{xy}(u) = f(x, y), \quad x, y \in R$$

(9.35)

defined over a simply connected region R which is bounded by a simple closed curve $C = \partial R$. Associated with the operator L_{xy} is the eigenvalue problem

$$L^*_{xy}(\phi_{nm}) = \lambda_{nm} r(x, y)\phi_{nm}, \qquad \phi_{nm} = \phi_{nm}(x, y), \quad x, y \in R$$

(9.36)

with weight function $r(x, y)$. We shall assume that the eigenvalue problem given by equation (9.36) has homogeneous boundary conditions. We wish to solve the equation (9.36) and obtain a set of eigenfunctions $\phi_{nm}(x,y)$ which are orthogonal with respect to the weight function $r(x,y)$ over the region R. That is, the inner product of two eigenfunctions can be expressed

$$(\phi_{nm}, \phi_{ij}) = \int\int_R r(x,y)\phi_{nm}(x,y)\phi_{ij}(x,y)\,dxdy = \begin{cases} \| \phi_{nm} \|^2, & (i,j) = (n,m) \\ 0, & \text{otherwise} \end{cases} \quad (9.37)$$

Let us assume that the boundary conditions associated with equations (9.35) and (9.36) are such that the line integral terms in the Green's identity (9.7) vanish when v is replaced by ϕ_{nm}. In this special case the equation (9.7) reduces to

$$\int\int_R vL_{\xi\eta}(u)d\xi d\eta = \int\int_R uL_{\xi\tau}^*(v)d\xi d\eta \quad (9.38)$$

which is a convenient form to be used shortly. Assume a solution to equation (9.35) in the form of a Fourier series

$$u = \sum_{m=1}^{\infty}\sum_{n=1}^{\infty} B_{nm}\phi_{nm}(x,y),$$

where B_{nm} are Fourier coefficients. We use the orthogonality property of the set ϕ_{nm} over the region R and express the Fourier coefficients by an inner product divided by a norm squared

$$B_{nm} = \frac{(u, \phi_{nm})}{\|\phi_{nm}\|^2} = \frac{1}{\| \phi_{nm} \|^2} \int\int_R r(\xi,\eta)u(\xi,\eta)\phi_{nm}(\xi,\eta)\,d\xi d\eta. \quad (9.39)$$

We use the relation given by equation (9.36), to simplify the equation (9.39) to the form

$$B_{nm} = \frac{1}{\lambda_{nm}\| \phi_{nm} \|^2} \int\int_R u(\xi,\eta)\lambda_{nm}r(\xi,\eta)\phi_{nm}(\xi,\eta)\,d\xi d\eta$$

$$= \frac{1}{\lambda_{nm}\| \phi_{nm} \|^2} \int\int_R u(\xi,\eta)L_{\xi\eta}^*(\phi_{nm})\,d\xi d\eta$$

which by equation (9.38) further simplifies to

$$B_{nm} = \frac{1}{\lambda_{nm}\| \phi_{nm} \|^2} \int\int_R \phi_{nm}(\xi,\eta)\,L_{\xi\eta}(u)\,d\xi d\eta$$

$$= \frac{1}{\lambda_{nm}\| \phi_{nm} \|^2} \int\int_R \phi_{nm}(\xi,\eta)f(\xi,\eta)\,d\xi d\eta \quad (9.40)$$

We substitute the results from equation (9.40) into the assumed series solution. Then interchange the order of summation and integration to obtain

$$u(x,y) = \int\int_R G(x,y;\xi,\eta)f(\xi,\eta)\,d\xi d\eta, \quad (9.41)$$

where G is the Green's function given by the series

$$G(x,y;\xi,\eta) = \sum_{m=1}^{\infty}\sum_{n=1}^{\infty} \frac{\phi_{nm}(x,y)\phi_{nm}(\xi,\eta)}{\lambda_{nm}\parallel \phi_{nm}\parallel^2}. \qquad (9.42)$$

where $\lambda_{nm} \neq 0$. This result is analogous to the series representation we found earlier for the one-dimensional Green's function.

Example 9-4.   (Poisson's equation.)
 Solve the boundary value problem

$$L_{xy}(u) = \frac{\partial^2 u}{\partial x^2} + \frac{\partial^2 u}{\partial y^2} = f(x,y),$$

where $x,y \in R = \{x,y\,|\,0 \leq x \leq a,\ 0 \leq y \leq b\}$ subject to the homogeneous boundary conditions

$$u(0,y) = u(a,y) = u(x,0) = u(x,b) = 0$$

Solution: The eigenvalue problem associated with the operator L_{xy} and the region R is

$$L_{xy}(\phi) = \frac{\partial^2 \phi}{\partial x^2} + \frac{\partial^2 \phi}{\partial y^2} = \lambda\phi, \qquad x,y \in R$$

where $\phi = 0$ on the boundary of R. By using the method of separation of variables this problem has the eigenvalues

$$\lambda = \lambda_{mn} = -\pi^2\left(\frac{n^2}{a^2} + \frac{m^2}{b^2}\right)$$

and eigenfunctions

$$\phi = \phi_{mn} = \sin(\frac{n\pi x}{a})\sin(\frac{m\pi y}{b}), \qquad m,n = 1,2,\dots$$

with the norm squared given by

$$\parallel \phi_{mn}\parallel^2 = \int_0^a \int_0^b \phi_{mn}^2\,dxdy = \frac{ab}{4}.$$

The solution corresponding to equation (9.42) is given by

$$G(x,y;\xi,\eta) = \frac{4}{ab}\sum_{m=1}^{\infty}\sum_{n=1}^{\infty} \frac{\phi_{mn}(\xi,\eta)\phi_{mn}(x,y)}{\lambda_{mn}}.$$

Substituting this solution into equation (9.41) and interchanging the order of summation and integration, we find the solution can be represented

$$u(x,y) = -\frac{4}{ab} \sum_{m=1}^{\infty} \sum_{n=1}^{\infty} B_{nm} \frac{\sin(\frac{n\pi x}{a})\sin(\frac{m\pi y}{b})}{\pi^2 \left(\frac{n^2}{a^2} + \frac{m^2}{b^2}\right)},$$

where

$$B_{nm} = \int_0^a \int_0^b f(\xi,\eta) \sin(\frac{n\pi\xi}{a}) \sin(\frac{m\pi\eta}{b}) \, d\xi d\eta.$$

Here $u(x,y)$ has been represented as a double Fourier series and B_{nm} are the Fourier coefficients.   ∎

Green's function formulas for the heat and wave equation.

The use of Green's functions to solve partial differential equations can be described in a general way. Let $L_{x_1 x_2 \cdots x_n t}(u)$ denote a nth order linear partial differential operator to be applied to a region R of n-dimensional space. Using integration by parts one can obtain the general integral relationship

$$\int_0^T \int \cdots \int_{V_R} G L_{\xi_1 \cdots \xi_n \tau}(u) dV d\tau = \int_0^T \int \cdots \int_{V_R} u L_{\xi_1 \cdots \xi_n \tau}^*(G) dV d\tau + \begin{cases} \text{boundary} \\ \text{terms} \end{cases} \quad (9.43)$$

where L^* is an adjoint operator associated with L, V_R is the volume of the spatial region R, $d\tau$ is an element of time. When the boundary terms in equation (9.43) are found, then an analysis will determine the adjoint boundary conditions that one should apply to the Green's function. The boundary terms are obtained by integrating the left-hand side of equation (9.43) using integration by parts. The following examples develop the Green's function representations for the two and three dimensional heat and wave equation operators for selected regions R.

Example 9-5. (Heat equation infinite region.)

Consider the heat equation operator $L_{xyt}(u) = \frac{\partial u}{\partial t} - \kappa \left(\frac{\partial^2 u}{\partial x^2} + \frac{\partial^2 u}{\partial y^2}\right)$ for the region $R = \{x,y \mid -\infty < x < \infty, -\infty < y < \infty\}$. We replace (x,y,t) by (ξ,η,τ) and evaluate the integral

$$I = \int_0^T \int_{-\infty}^{\infty} \int_{-\infty}^{\infty} G L_{\xi\eta\tau}(u) d\xi d\eta d\tau \quad (9.44)$$

using integration by parts. We find that by assigning suitable zero boundary conditions at infinity on G and $\frac{\partial G}{\partial \xi}$, the relation

$$\int_0^T \int_{-\infty}^\infty \int_{-\infty}^\infty GL_{\xi\eta\tau}(u)\,d\xi\,d\eta\,d\tau = \int_0^T \int_{-\infty}^\infty \int_{-\infty}^\infty uL_{\xi\eta\tau}^*(G)\,d\xi\,d\eta\,d\tau$$
$$+ \int_{-\infty}^\infty \int_{-\infty}^\infty [Gu]_{\tau=0}^{\tau=T}\,d\xi\,d\eta \tag{9.45}$$

Observe that if we select $G = 0$ and $\frac{\partial G}{\partial \tau} = 0$ for all $\tau > t$ and require that the Green's function satisfy

$$L_{\xi\eta\tau}^*(G) = \delta(\xi - x)\delta(\eta - y)\delta(\tau - t) \tag{9.46}$$

with $G = G(x, y, t; \xi, \eta, 0) = 0$, for all $\tau > t$, then the equation (9.45) reduces to the useful form

$$u = u(x, y, t) = \int_0^t \int_{-\infty}^\infty \int_{-\infty}^\infty GL_{\xi\eta\tau}(u)\,d\xi\,d\eta\,d\tau + \int_{-\infty}^\infty \int_{-\infty}^\infty [Gu]_{\tau=0}\,d\xi\,d\eta \tag{9.47}$$

where now the solution depends upon the given initial conditions. For example, we can use the equation (9.47) to solve the nonhomogeneous initial value problem

$$\frac{\partial u}{\partial t} - k\left(\frac{\partial^2 u}{\partial x^2} + \frac{\partial^2 u}{\partial y^2}\right) = F(x, y, t), \quad x, y \in R$$

where $R = \{x, y \mid -\infty < x < \infty, \ -\infty < y < \infty\}$ and $u(x, y, 0) = f(x, y)$.

Boundary conditions at infinity are implied and so the equation (9.47) gives the solution

$$u = u(x, y, t) = \int_0^t \int_{-\infty}^\infty \int_{-\infty}^\infty G(x, y, t; \xi, \eta, \tau)F(\xi, \eta, \tau)\,d\xi\,d\eta\,d\tau + \int_{-\infty}^\infty \int_{-\infty}^\infty G\Big|_{\tau=0} f(\xi, \eta)\,d\xi\,d\eta$$

where G is the Green's function associated with the operator L over the two-dimensional infinite region R.

∎

Example 9-6. (Heat equation finite region.)

Consider the heat equation operator $L_{xyzt}(u) = \frac{\partial u}{\partial t} - \kappa\left(\frac{\partial^2 u}{\partial x^2} + \frac{\partial^2 u}{\partial y^2} + \frac{\partial^2 u}{\partial x^2}\right)$ for $x, y, z \in R$ where R is some finite connected region of space. Note that if we let

$$L_{\xi\eta\zeta\tau}^*(G) = \delta(\xi - x)\delta(\eta - y)\delta(\zeta - z)\delta(\tau - t) \tag{9.48}$$

and assume that

$$L_{\xi\eta\zeta\tau}(u) = F(\xi, \eta, \zeta, \tau) \tag{9.49}$$

then instead of integrating $\int_0^T \iiint_{V_R} GL_{\xi\eta\zeta\tau}(u)dVd\tau$ using integration by parts, we can instead multiply equation (9.48) by u and multiply equation (9.49) by G and then subtract the results. An integration over the spatial region V_R and integrating with respect to τ from 0 to T produces the integral

$$\int_0^T \iiint_{V_R} \left[GL_{\xi\eta\zeta\tau}(u) - uL_{\xi\eta\zeta\tau}^*(G) \right] dVd\tau = \int_0^T \iiint_{V_R} GF\, dVd\tau$$
$$- \int_0^T \iiint_{V_R} u\delta(\xi - x)\delta(\eta - y)\delta(\zeta - z)\delta(\tau - t)\, dVd\tau$$

(9.50)

The integral on the left-hand side of equation (9.50) can be written

$$\int_0^T \iiint_{V_R} \left[G\left(\frac{\partial u}{\partial \tau} - \kappa\nabla^2 u \right) - u\left(-\frac{\partial G}{\partial \tau} - \kappa\nabla^2 G \right) \right] dVd\tau$$
$$= \int_0^T \iiint_{V_R} \left(G\frac{\partial u}{\partial \tau} + u\frac{\partial G}{\partial \tau} \right) dVd\tau - \kappa \int_0^T \iiint_{V_R} \left(G\nabla^2 u - u\nabla^2 G \right) dVd\tau$$

(9.51)

$$= \iiint_{V_R} \int_0^T \frac{\partial}{\partial \tau}(Gu)\, d\tau dV - \kappa \int_0^T \iint_S \left(G\frac{\partial u}{\partial n} - u\frac{\partial G}{\partial n} \right) d\sigma d\tau$$

where $d\sigma$ is an element of surface area and S denotes the surface enclosing the volume V_R. One finds the equation (9.50) can be written in the form

$$u = u(x, y, z, t) = \int_0^T \iiint_{V_R} GF\, dVd\tau - \iiint_{V_R} [Gu]_{\tau=0}^{\tau=T}\, dV$$
$$+ \kappa \int_0^T \iint_S \left(G\frac{\partial u}{\partial n} - u\frac{\partial G}{\partial n} \right) d\sigma d\tau$$

(9.52)

The condition that $G = 0$ for $\tau > t$ reduces the equation (9.52) to the form

$$u = u(x, y, z, t) = \int_0^t \iiint_{V_R} GF\, dVd\tau + \iiint_{V_R} (Gu)\Big|_{\tau=0}\, dV$$
$$+ \kappa \int_0^t \iint_S \left(G\frac{\partial u}{\partial n} - u\frac{\partial G}{\partial n} \right) d\sigma d\tau$$

(9.53)

which can be used to analyze Dirichlet, Neumann or Robin type boundary conditions associated with the boundary surface S. For appropriate conditions on G and $\frac{\partial G}{\partial n}$ going to zero at infinity, then the surface integral term will vanish over unbounded regions and then equation (9.53) will also be valid for all of unbounded space.

Example 9-7. **(Wave equation infinite region.)**

Consider the wave equation operator $L_{xyt}(u) = \dfrac{\partial^2 u}{\partial t^2} - c^2 \left(\dfrac{\partial^2 u}{\partial x^2} + \dfrac{\partial^2 u}{\partial y^2} \right)$ for the region $R = \{x, y \mid -\infty < x < \infty, -\infty < y < \infty\}$. We replace (x, y, t) by (ξ, η, τ) and evaluate the integral

$$I = \int_0^T \int_{-\infty}^\infty \int_{-\infty}^\infty G L_{\xi\eta\tau}(u) \, d\xi \, d\eta \, d\tau \tag{9.54}$$

using integration by parts. We find, with suitable zero boundary conditions at infinity, the relation

$$\int_0^T \int_{-\infty}^\infty \int_{-\infty}^\infty G L_{\xi\eta\tau}(u) \, d\xi \, d\eta \, d\tau = \int_0^T \int_{-\infty}^\infty \int_{-\infty}^\infty u L_{\xi\eta\tau}^*(G) \, d\xi \, d\eta \, d\tau$$
$$- \int_{-\infty}^\infty \int_{-\infty}^\infty \left[u \frac{\partial G}{\partial \tau} - G \frac{\partial u}{\partial \tau} \right]_{\tau=0}^{\tau=T} d\xi \, d\eta. \tag{9.55}$$

Observe that if we require $G = 0$ and $\frac{\partial G}{\partial \tau} = 0$ for all $\tau > t$ and the Green's function is selected to satisfy

$$L_{\xi\eta\tau}^*(G) = \frac{\partial^2 G}{\partial \tau^2} - c^2 \left(\frac{\partial^2 G}{\partial \xi^2} + \frac{\partial^2 G}{\partial \eta^2} \right) = \delta(\xi - x)\delta(\eta - y)\delta(\tau - t) \tag{9.56}$$

then the equation (9.55) reduces to the useful form

$$u = u(x, y, t) = \int_0^t \int_{-\infty}^\infty \int_{-\infty}^\infty G L_{\xi\eta\tau}(u) \, d\xi \, d\eta \, d\tau - \int_{-\infty}^\infty \int_{-\infty}^\infty \left[u \frac{\partial G}{\partial \tau} - \frac{\partial u}{\partial \tau} G \right]_{\tau=0} d\xi \, d\eta \tag{9.57}$$

involving the initial conditions.

∎

Example 9-8. **(Wave equation finite region.)**

Consider the three-dimensional wave equation operator

$$L_{xyzt}(u) = \frac{\partial^2 u}{\partial t^2} - c^2 \left(\frac{\partial^2 u}{\partial x^2} + \frac{\partial^2 u}{\partial y^2} + \frac{\partial^2 u}{\partial z^2} \right)$$

for $x, y, z \in R$ where R is some finite connected region of space. One can readily verify with the help of the Gauss divergence theorem that

$$\int_0^T \iiint_{V_R} G L_{\xi\eta\zeta\tau}(u) \, dV \, d\tau = \int_0^T \iiint_{V_R} u L_{\xi\eta\zeta\tau}^*(G) \, dV \, d\tau$$
$$+ \iiint_{V_R} \left[u \frac{\partial G}{\partial \tau} - G \frac{\partial u}{\partial \tau} \right]_{\tau=0}^{\tau=T} dV + c^2 \int_0^T \iint_S \left(G \frac{\partial u}{\partial n} - u \frac{\partial G}{\partial n} \right) d\sigma \, d\tau \tag{9.58}$$

where $d\sigma$ is an element of surface area, and S again denotes the surface enclosing the volume V_R. Observe that by selecting G to be a solution of

$$L^*_{\xi\eta\zeta\tau}(G) = \delta(\xi - x)\delta(\eta - y)\delta(\zeta - z)\delta(\tau - t) \tag{9.59}$$

where $G = 0$ for $\tau > t$, then the equation (9.58) reduces to the useful form

$$u = u(x, y, z, t) = \int_0^t \iiint_{V_R} G L_{\xi\eta\zeta\tau}(u)\, dV d\tau$$
$$+ \iiint_{V_R} \left(u\frac{\partial G}{\partial \tau} - G\frac{\partial u}{\partial \tau}\right)\Big|_{\tau=0} dV - c^2 \int_0^t \iint_S \left(G\frac{\partial u}{\partial n} - u\frac{\partial G}{\partial n}\right) d\sigma d\tau \tag{9.60}$$

which can be analyzed for Dirichlet, Neumann and Robin type boundary conditions associated with the boundary surface S.

∎

Green's Functions for an Infinite Domain

Many times the Green's function is broken up into two parts and written as

$$G(x, y; \xi, \eta) = G_f(x, y; \xi, \eta) + G_r(x, y; \xi, \eta) \tag{9.61}$$

where $G_f(x, y; \xi, \eta) = G^*_f(\xi, \eta; x, y)$ is called a fundamental solution or free-space Green's function and satisfies

$$L_{xy}G_f = \delta(x - \xi)\delta(y - \eta), \qquad -\infty < x, y < \infty. \tag{9.62}$$

where the equation defining the fundamental solution is without boundary conditions. The function G_r is called the regular part of the Green's function and is added for those problems which have a finite boundary $C = \partial R$. The regular part G_r is chosen in such a way that the Green's function $G = G_f + G_r$ satisfies the assigned boundary conditions on the finite boundary $C = \partial R$. For example, if one wanted to solve the Dirichlet problem $L_{xy}(G) = \delta(x - \xi)\delta(y - \eta)$ for $x, y \in R$ subject to the boundary condition that $G = g(x, y)$ for $x, y \in C = \partial R$, we could use the free space Green's function associated with the operator L_{xy} as a first guess to the Green's function associated with the operator L_{xy} with a finite boundary. One would require that

$$L_{xy}(G) = L_{xy}(G_f + G_r) = L_{xy}(G_f) + L_{xy}(G_r)$$
$$= \delta(x - \xi)\delta(y - \eta) + 0.$$

Note that when the operator L_{xy} operates on the free-space Green's function, then the impulse function is obtained and so it is customary to require that G_r satisfy the homogeneous equation $L_{xy}(G_r) = 0$ with the boundary condition that $G_r = g(x,y) - G_f$ for $x, y \in C = \partial R$. We would then have the desired result that

$$G\Big|_{x,y \in C} = (G_f + G_r)\Big|_{x,y \in C} = (G_f + g(x,y) - G_f)\Big|_{x,y \in C} = g(x,y) \quad \text{for } x, y \in C = \partial R.$$

The free-space Green's function is usually singular whereas the regular part is nonsingular and satisfies a homogeneous equation. In general the Green's function associated with a partial differential operator is dependent not only upon the boundary conditions but also upon the region over which the solution is desired.

Free-space Green's function for the Laplacian operator

In certain special cases we can translate axes and make changes of variables to reduce a partial differential equation to an ordinary differential equation. This idea is illustrated in the next three examples where we develop the free-space Green's functions in one, two and three dimensions for the Laplacian operator.

Example 9-9.   (Laplace operator in one-dimension.)
Find the free-space Green's function associated with the ordinary differential operator $L_x(u) = \dfrac{d^2u}{dx^2}$. We wish to solve the equation

$$L_x(G) = \frac{d^2G}{dx^2} = \delta(x - \xi), \qquad -\infty < x < \infty$$

which defines the Green's function. We make the change of variable $t = x - \xi$ and solve $\dfrac{d^2G}{dt^2} = \delta(t)$ subject to the condition that $G(0) = 0$. We obtain

$$G = \begin{cases} c_1 t + c_2, & t < 0 \\ k_1 t + k_2, & t > 0 \end{cases}.$$

The condition that $G = 0$ at $t = 0$ requires that $c_2 = k_2 = 0$. The jump discontinuity condition requires that

$$\frac{dG(t^+)}{dt} - \frac{dG(t^-)}{dt} = 1 \quad \text{or} \quad k_1 - c_1 = 1.$$

Replacing t by $x - \xi$ we obtain the Green's function

$$G = G(x; \xi) = \begin{cases} c_1 (x - \xi), & x < \xi \\ (1 + c_1)(x - \xi), & x > \xi \end{cases}.$$

We require the free-space Green's function to be symmetric in the variables x and ξ and satisfy $G(x;\xi) = G(\xi;x)$. This requires that

$$c_1(x - \xi) = (1 + c_1)(\xi - x) \qquad \text{or} \qquad (2c_1 + 1)(x - \xi) = 0$$

which implies that $c_1 = -1/2$. The free-space Green's function can therefore be written as

$$G = G(x;\xi) = \begin{cases} -\frac{1}{2}(x - \xi), & x < \xi \\ \frac{1}{2}(x - \xi), & x > \xi \end{cases}. \tag{9.63}$$

This can also be expressed in the form

$$G = G(x;\xi) = \frac{1}{2}|x - \xi|. \tag{9.64}$$

This result is consistent with the equations (8.51) with the boundary conditions replaced by $G = 0$ at $x = \xi$ and requiring that G be symmetric in the variables x and ξ. Note also this Green's function is not singular at $x = \xi$ and so technically a free-space Green's function for the one-dimensional Laplacian operator does not exist. However, one can write the solution to the differential equation

$$\frac{d^2u}{dx^2} = f(x) \quad \text{in the form} \quad u = u(x) = \frac{1}{2}\int_{-\infty}^{\infty} |x - \xi| f(\xi)\, d\xi = \int_{-\infty}^{\infty} G(x;\xi) f(\xi)\, d\xi. \tag{9.65}$$

The solution given by equation (9.65) is equivalent to the solution written in the form $u = u(x) = \int_0^x \left(\int_0^t f(\xi)\, d\xi \right) dt$ obtained using two integrations of the differential equation. Note that if $\xi < x$, then $|x - \xi| = x - \xi$ and if $\xi > x$, then $|x - \xi| = -(x - \xi)$. Therefore, if $|f(\xi)| \to 0$ as $|\xi| \to \infty$, then one can write

$$u = u(x) = \frac{1}{2}\int_{-\infty}^{\infty} |x - \xi| f(\xi)\, d\xi = \frac{1}{2}\int_{-\infty}^{x} (x - \xi) f(\xi)\, d\xi + \frac{1}{2}\int_{x}^{\infty} -(x - \xi) f(\xi)\, d\xi$$

then $\quad \dfrac{du}{dx} = \dfrac{1}{2}\int_{-\infty}^{x} f(\xi)\, d\xi - \dfrac{1}{2}\int_{x}^{\infty} f(\xi)\, d\xi$

and $\quad \dfrac{d^2u}{dx^2} = \dfrac{1}{2}f(x) - \dfrac{1}{2}f(x)(-1) = f(x).$

If we drop the requirement that G be singular, then we can refer to G as a free-space Green's function for the one-dimensional Laplacian.

Example 9-10. (Laplace operator in two-dimensions.)

Find the free-space Green's function associated with the two-dimensional Laplacian operator

$$L_{xy}(u) = \nabla^2 u = \frac{\partial^2 u}{\partial x^2} + \frac{\partial^2 u}{\partial y^2}.$$

Solution: In order to solve for the Green's function which satisfies

$$L_{xy}(G) = \nabla^2 G = \frac{\partial^2 G}{\partial x^2} + \frac{\partial^2 G}{\partial y^2} = \delta(x - \xi)\delta(y - \eta), \qquad -\infty < x, y < \infty \qquad (9.66)$$

we make the change of variables $x - \xi = r\cos\theta$ and $y - \eta = r\sin\theta$ and assume symmetry with respect to theta. We use the equations (6.5) and (7.9) to write the equation (9.66) in the form

$$\nabla^2 G = \frac{d^2 G}{dr^2} + \frac{1}{r}\frac{dG}{dr} = \frac{1}{2\pi r}\delta(r) \qquad (9.67)$$

where $r = \sqrt{(x - \xi)^2 + (y - \eta)^2}$. Recall the two-dimensional form of the Gauss divergence theorem given by equation (6.31)

$$\iint_R \text{div}\,\vec{F}\,d\sigma = \oint_C \vec{F}\cdot\hat{n}\,ds \qquad (9.68)$$

and let $\vec{F} = \nabla G$ to obtain

$$\iint_R \nabla^2 G\,d\sigma = \oint_C \nabla G\cdot\hat{n}\,ds \qquad (9.69)$$

Let R denote a small circle of radius r with C its boundary having the outward unit normal $\hat{n} = \hat{e}_r$ and arc length element $ds = rd\theta$. In the special case $G = G(r)$ we have $\nabla G = \frac{\partial G}{\partial r}\hat{e}_r$ so that the equation (9.69) reduces to

$$\iint_R \nabla^2 G\,d\sigma = \int_0^r \int_0^{2\pi} \frac{\delta(r)}{2\pi r}\,rdrd\theta = 1 = \int_0^{2\pi} \frac{\partial G}{\partial r}\,rd\theta = 2\pi r\frac{\partial G}{\partial r}$$

which simplifies to

$$1 = 2\pi r\frac{\partial G}{\partial r}. \qquad (9.70)$$

This gives the singularity condition

$$\lim_{r \to 0} r\frac{\partial G}{\partial r} = \frac{1}{2\pi}. \qquad (9.71)$$

The equation (9.70) can be integrated to obtain

$$G = \frac{1}{2\pi}\log r$$

where we have selected the constant of integration to be zero. This gives the free-space Green's function

$$G = G(x, y; \xi, \eta) = \frac{1}{2\pi} \log \sqrt{(x-\xi)^2 + (y-\eta)^2} \tag{9.72}$$

which is singular at the point (ξ, η). This Green's function can also be written in the form

$$G = \frac{1}{2\pi} \log |\vec{r} - \vec{\xi}| \tag{9.73}$$

where $\vec{r} = (x, y)$ and $\vec{\xi} = (\xi, \eta)$ are position vectors.

■

Example 9-11.   (Laplace operator in three-dimensions.)
Find the free-space Green's function associated with the three-dimensional Laplacian operator

$$L_{xyz}(u) = \frac{\partial^2 u}{\partial x^2} + \frac{\partial^2 u}{\partial y^2} + \frac{\partial^2 u}{\partial z^2} = \nabla^2 u, \qquad -\infty < x, y, z < \infty$$

Solution: We wish to solve

$$\nabla^2 G = \frac{\partial^2 G}{\partial x^2} + \frac{\partial^2 G}{\partial y^2} + \frac{\partial^2 G}{\partial z^2} = \delta(x-\xi)\delta(y-\eta)\delta(z-\zeta), \qquad -\infty < x, y, z < \infty \tag{9.74}$$

We make the change of variable

$$x - \xi = \rho \sin\theta \cos\phi, \qquad y - \eta = \rho \sin\theta \sin\phi, \qquad z - \zeta = \rho \cos\theta$$

and assume G is spherically symmetric about the origin. The equation (9.74) can then be written in the form

$$\nabla^2 G = \frac{d^2 G}{d\rho^2} + \frac{2}{\rho}\frac{dG}{d\rho} = \frac{\delta(\rho)}{4\pi\rho^2} \tag{9.75}$$

The Gauss divergence theorem in three-dimensions, from equation (6.30), is represented

$$\iiint_V \operatorname{div} \vec{F} \, d\tau = \iint_S \vec{F} \cdot \hat{n} \, d\sigma \tag{9.76}$$

In this equation we let $\vec{F} = \nabla G$ to obtain

$$\iiint_V \nabla^2 G \, d\tau = \iint_S \nabla G \cdot \hat{n} \, d\sigma \tag{9.77}$$

496

Let V denote a small sphere of radius ρ and write $\nabla G = \frac{\partial G}{\partial \rho}\hat{e}_\rho$ so that the equation (9.77) becomes

$$\iiint_V \nabla^2 G \, d\tau = \int_0^{2\pi}\int_0^\pi \int_0^\rho \frac{\delta(\rho)}{4\pi\rho^2}\rho^2 \sin\theta \, d\rho \, d\theta d\phi = \int_0^{2\pi}\int_0^\pi \frac{\partial G}{\partial \rho}\rho^2 \sin\theta d\theta d\phi \qquad (9.78)$$

which simplifies to

$$1 = 4\pi\rho^2 \frac{\partial G}{\partial \rho}. \qquad (9.79)$$

This gives the singularity condition

$$\lim_{\rho \to 0}\rho^2 \frac{\partial G}{\partial \rho} = \frac{1}{4\pi}. \qquad (9.80)$$

Integrating the equation (9.79) gives the free-space Green's function

$$G = -\frac{1}{4\pi\rho} = \frac{-1}{4\pi\sqrt{(x-\xi)^2+(y-\eta)^2+(z-\zeta)^2}},$$

where the constant of integration has been chosen to be zero. This Green's function has a singularity at the point (ξ,η,ζ) and can be written in the alternative form

$$G = \frac{-1}{4\pi|\vec{r}-\vec{\xi}|} = \frac{-1}{4\pi}|\vec{r}-\vec{\xi}|^{-1} \qquad (9.81)$$

where $\vec{r} = (x,y,z)$ and $\vec{\xi} = (\xi,\eta,\zeta)$ are position vectors.

∎

The previous three examples are summarized with the following table 9-2.

Table 9-2
Free-space Green's function for Laplace's equation

Dimension	Green's function
$n=1$	$G(x;\xi) = \frac{1}{2}\,|x-\xi|$
$n=2$	$G(x,y;\xi,\eta) = \frac{1}{2\pi}\log\sqrt{(x-\xi)^2+(y-\eta)^2}$ $\qquad = \frac{1}{2\pi}\log|\vec{r}-\vec{\xi}|$ $\vec{r}=(x,y)\quad \vec{\xi}=(\xi,\eta)$
$n=3$	$G(x,y,z;\xi,\eta,\zeta) = \frac{-1}{4\pi}\frac{1}{\sqrt{(x-\xi)^2+(y-\eta)^2+(z-\zeta)^2}}$ $\qquad = \frac{-1}{4\pi}|\vec{r}-\vec{\xi}|^{-1}$ $\vec{r}=(x,y,z)\quad \vec{\xi}=(\xi,\eta,\zeta)$

Example 9-12. (Poisson equation.)

Solve the Poisson equation

$$\nabla^2 u = \frac{\partial^2 u}{\partial x^2} + \frac{\partial^2 u}{\partial y^2} + \frac{\partial^2 u}{\partial z^2} = F(x, y, z), \quad -\infty < x, y, z < \infty$$

subject to the boundary conditions $u \to 0$ as x, y or z increases without bound.

Solution: The free-space Green's function is obtained from the table 9-2. Observe that the representation theorem (6.41) in the limit as $\rho \to \infty$ shows that the solution can be written in the form

$$u = u(x, y, z) = \frac{-1}{4\pi} \int_{-\infty}^{\infty} \int_{-\infty}^{\infty} \int_{-\infty}^{\infty} \frac{F(\xi, \eta, \zeta)\, d\xi d\eta d\zeta}{\sqrt{(x - \xi)^2 + (y - \eta)^2 + (z - \zeta)^2}}$$

$$u = u(x, y, z) = \int_{-\infty}^{\infty} \int_{-\infty}^{\infty} \int_{-\infty}^{\infty} G(x, y, z; \xi, \eta, \zeta) F(\xi, \eta, \zeta)\, d\xi d\eta d\zeta \tag{9.82}$$

■

Example 9-13. (Poisson equation.)

Solve the Poisson equation

$$\nabla^2 u = \frac{\partial^2 u}{\partial x^2} + \frac{\partial^2 u}{\partial y^2} = f(x, y), \quad x, y \in R = \{x, y \mid -\infty < x < \infty, \quad y > 0\}$$

subject to the boundary condition $u(x, 0) = h(x)$.

Solution: We use Green's second identity and write

$$\iint_R \left(u \nabla^2 G - G \nabla^2 u \right) dx dy = \oint_C (u \nabla G - G \nabla u) \cdot \hat{n}\, ds \tag{9.83}$$

where R is the region illustrated in the figure 9-5.

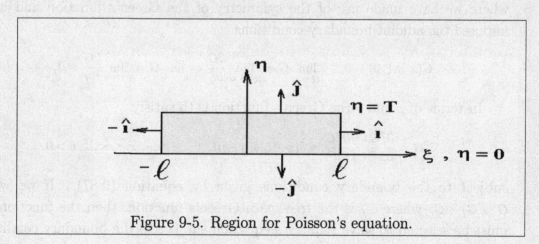

Figure 9-5. Region for Poisson's equation.

The equation (9.83) can be expanded and written in terms of dummy variables ξ and η to obtain

$$\int_{-\ell}^{\ell}\int_{0}^{T}\left(u\nabla_{\xi}^{2}G-G\nabla_{\xi}^{2}u\right)d\eta d\xi = \int_{-\ell}^{\ell}\left(u\nabla_{\xi}G-G\nabla_{\xi}u\right)\cdot(-\hat{j})\Big|_{\eta=0}d\xi$$
$$+\int_{0}^{T}\left(u\nabla_{\xi}G-G\nabla_{\xi}u\right)\cdot(\hat{i})\Big|_{\xi=\ell}d\eta$$
$$+\int_{\ell}^{-\ell}\left(u\nabla_{\xi}G-G\nabla_{\xi}u\right)\cdot(\hat{j})\Big|_{\eta=T}d\xi \qquad (9.84)$$
$$+\int_{T}^{0}\left(u\nabla_{\xi}G-G\nabla_{\xi}u\right)\cdot(-\hat{i})\Big|_{\xi=-\ell}d\eta$$

Letting

$$\nabla_{\xi}^{2}G = \delta(\xi-x)\delta(\eta-y) \quad\text{and}\quad \nabla_{\xi}^{2}u = f(\xi,\eta)$$

the equation (9.84) simplifies to the form

$$u = u(x,y) = \int_{-\ell}^{\ell}\int_{0}^{T}G(\xi,\eta;x,y)f(\xi,\eta)\,d\eta d\xi + \int_{-\ell}^{\ell}\left(-u\frac{\partial G}{\partial\eta}+G\frac{\partial u}{\partial\eta}\right)\Big|_{\eta=0}d\xi$$
$$+\int_{0}^{T}\left(u\frac{\partial G}{\partial\xi}-G\frac{\partial u}{\partial\xi}\right)\Big|_{\xi=\ell}d\eta + \int_{-\ell}^{\ell}\left(-u\frac{\partial G}{\partial\eta}+G\frac{\partial u}{\partial\eta}\right)\Big|_{\eta=T}d\xi \qquad (9.85)$$
$$+\int_{0}^{T}\left(u\frac{\partial G}{\partial\xi}-G\frac{\partial u}{\partial\xi}\right)\Big|_{\xi=-\ell}d\eta$$

In the limit as $\ell\to\infty$ and $T\to\infty$ the equation (9.85) becomes

$$u = u(x,y) = \int_{-\infty}^{\infty}\int_{0}^{\infty}G(x,y;\xi,\eta)f(\xi,\eta)\,d\eta d\xi - \int_{-\infty}^{\infty}h(\xi)\frac{\partial G}{\partial\eta}\Big|_{\eta=0}d\xi \qquad (9.86)$$

where we have made use of the symmetry of the Green's function and have imposed the adjoint boundary conditions

$$G(x,y;\xi,0)=0, \quad \lim_{|\xi|\to\infty}G=\lim_{|\xi|\to\infty}\frac{\partial G}{\partial\xi}=\lim_{\eta\to\infty}G=\lim_{\eta\to\infty}\frac{\partial G}{\partial\eta}=0. \qquad (9.87)$$

In terms of x and y the Green's function is to satisfy

$$\nabla^{2}G = \frac{\partial^{2}G}{\partial x^{2}}+\frac{\partial^{2}G}{\partial y^{2}} = \delta(x-\xi)\delta(y-\eta), \qquad -\infty<x<\infty, \quad y>0$$

subject to the boundary conditions given by equation (9.87). If we write $G=G_{f}+G_{r}$ where G_{f} is the free-space Green's function, then the function G_{r} must be selected to solve Laplace's equation subject to the boundary condition

$G_r(x,y;\xi,0) = -G_f(x,y;\xi,0)$ together with limiting conditions for large values of the arguments. We obtain from the table 9-2 the free-space Green's function $G_f = \frac{1}{2\pi} \log |\vec{r} - \vec{\xi}|$ which has a singularity at the point (ξ,η) somewhere inside the region R. We want to construct a function G_r to have the following properties; (i) G_r is harmonic in R and, (ii) G_r evaluated at $\eta = 0$ is the negative of G_f evaluated at $\eta = 0$. This suggests that G_r should have approximately the same form as G_f and so we select G_r to have the form

$$G_r = -\frac{1}{2\pi} \log |\vec{r} - \vec{\xi}^*|$$

where $\vec{\xi}^*$ is a singular point outside the region R (because we want G_r to be harmonic inside the region R.) and the negative sign takes care of the initial condition.

The method of images is a technique borrowed from the theory of electrostatic potentials. If one views the Green's function as an electrostatic potential at point (x,y) which is created by a unit test charge placed at the point (ξ,η), then $G_f(x,y;\xi,0)$ is the charge on the bounding surface produced by the unit test charge. The method of electrostatic images views G_r as the potential created by an imaginary test charge, located at some symmetric point, which when added to G_f along the boundary gives a zero potential. For example if $\vec{\xi} = (\xi,\eta)$ is the point of the original test charge, then one can try placing additional test charges at one of the symmetric points $\vec{\xi}^* = (\xi,-\eta)$, $\vec{\xi}^* = (-\xi,\eta)$ or $\vec{\xi}^* = (-\xi,-\eta)$. If we select the point $(\xi,-\eta)$ this gives the Green's function

$$\begin{aligned} G &= \frac{1}{2\pi} \log \sqrt{(x-\xi)^2 + (y-\eta)^2} - \frac{1}{2\pi} \log \sqrt{(x-\xi)^2 + (y+\eta)^2} \\ G &= \frac{1}{4\pi} \log \left[\frac{(x-\xi)^2 + (y-\eta)^2}{(x-\xi)^2 + (y+\eta)^2} \right] \end{aligned} \tag{9.88}$$

which satisfies the boundary conditions given by equation (9.87). The equation (9.88) produces the derivative

$$\frac{\partial G}{\partial \eta} = \frac{1}{4\pi} \left[\frac{-2(y-\eta)}{(x-\xi)^2 + (y-\eta)^2} - \frac{2(y+\eta)}{(x-\xi)^2 + (y+\eta)^2} \right] \tag{9.89}$$

so that

$$\left. \frac{\partial G}{\partial \eta} \right|_{\eta=0} = \frac{-1}{4\pi} \frac{4y}{(x-\xi)^2 + y^2} \tag{9.90}$$

The solution for the Poisson problem for the upper half plane is given by equation (9.86) where G is defined by equation (9.88) and $\frac{\partial G}{\partial \eta}\big|_{\eta=0}$ is given by equation (9.90). One can now construct the solution

$$u(x,y) = \int_{-\infty}^{\infty} \int_{0}^{\infty} G(x,y;\xi,\eta) f(\xi,\eta)\, d\eta d\xi - \frac{y}{\pi} \int_{-\infty}^{\infty} \frac{h(\xi)\, d\xi}{(x-\xi)^2 + (y-\eta)^2} \qquad (9.91)$$

■

Example 9-14. (Poisson equation.)
Solve the Poisson equation

$$\nabla^2 u = \frac{\partial^2 u}{\partial r^2} + \frac{1}{r}\frac{\partial u}{\partial r} + \frac{1}{r^2}\frac{\partial^2 u}{\partial \theta^2} = f(r,\theta), \quad r,\theta \in R = \{r,\theta \mid 0 < r < a,\, 0 < \theta < 2\pi\}$$

subject to the boundary condition $u(a,\theta) = g(\theta)$.
Solution: The Green's second identity for the region R is written

$$\iint_R \left(u\nabla^2 G - G\nabla^2 u \right) r dr d\theta = \oint_C (u\nabla G - G\nabla u) \cdot \hat{n}\, ds \qquad (9.92)$$

Here we represent the Green's second identity in polar coordinates (r,θ) where

$$\hat{n} = \hat{e}_r, \quad \nabla u \cdot \hat{n} = \frac{\partial u}{\partial r}, \quad \nabla G \cdot \hat{n} = \frac{\partial G}{\partial r}, \quad ds = a d\theta$$

and we let

$$\nabla^2 G = \delta(r - r_0)\delta(\theta - \theta_0) \qquad \text{with} \qquad \nabla^2 u = f(r,\theta).$$

These substitutions allow us to represent the Green's second identity in the form

$$u(r_0,\theta_0) = \int_0^a \int_0^{2\pi} G(r_0,\theta_0;r,\theta) f(r,\theta)\, r dr d\theta + \int_0^{2\pi} \left(u\frac{\partial G}{\partial r} - G\frac{\partial u}{\partial r} \right)\Big|_{r=a} a d\theta. \qquad (9.93)$$

We impose the adjoint boundary condition $G = 0$ on the boundary $r = a$ and interchange (r_0,θ_0) and (r,θ) and find the solution can be represented in the form

$$u = u(r,\theta) = \int_0^a \int_0^{2\pi} G(r,\theta;r_0,\theta_0) f(r_0,\theta_0)\, r_0 dr_0 d\theta_0 + \int_0^{2\pi} g(\theta_0)\frac{\partial G}{\partial r_0}\Big|_{r_0=a} a d\theta_0 \qquad (9.94)$$

where G is the Green's function satisfying

$$\nabla^2 G = \delta(r - r_0)\delta(\theta - \theta_0), \qquad r,\theta \in R$$

and $G = 0$ on the boundary $r = a$. We let $G = G_f + G_r$ where G_f is the free-space Green's function obtained from the table 9-2 and try to construct a function G_r

which is harmonic in R and modifies the free-space Green's function in order to achieve the desired boundary condition of $G = 0$ on $r = a$. We find

$$G_f = \frac{1}{2\pi} \log |\vec{r} - \vec{\xi}| = \frac{1}{4\pi} \log \left[(x - \xi)^2 + (y - \eta)^2 \right] \tag{9.95}$$

which we write in polar form. We substitute

$$x = r \cos\theta, \quad y = r \sin\theta, \quad \xi = r_0 \cos\theta_0, \quad \eta = r_0 \sin\theta_0 \tag{9.96}$$

into the equation (9.95) and find

$$G_f = \frac{1}{4\pi} \log \left[r^2 + r_0^2 - 2rr_0 \cos(\theta - \theta_0) \right] \tag{9.97}$$

Note that the free-space Green's function does not satisfy the condition $G = 0$ on the boundary $r = a$. We try to construct the function G_r to be harmonic in R and satisfy the boundary condition $G_r \big|_{r=a} = -G_f \big|_{r=a}$. One method of accomplishing this is the use the method of images. The method of images utilizes the principle of superposition, symmetry and sometimes scaling in order to achieve the desired boundary condition. For example, if (r_0, θ_0) is the singular point in R we make use of the symmetry properties given by equations (6.63) and (6.64) as this inversion property considers an image source placed at (r_0^*, θ_0^*) which by proper scaling produces a harmonic function. Constants are also solutions of the Laplace equation and so we can assume that the harmonic function G_r has the form

$$G_r = -G_f^* + c \tag{9.98}$$

where c is a constant and

$$G_f^* = \frac{1}{4\pi} \log |\vec{r} - \vec{\xi}^*| = \frac{1}{4\pi} \log \left[r^2 + r_0^{*2} - 2rr_0^* \cos(\theta - \theta_0^*) \right] \tag{9.99}$$

where $r_0^* = a^2/r_0$ and $\theta_0^* = \theta_0$ define the image point of inversion with respect to the circle $r = a$. The constant c in equation (9.98) can be selected as a scaled logarithm function and so equation (9.98) can be written in the form

$$G_r = \frac{-1}{4\pi} \log \left[C \frac{a^2}{r_0^2} \left(\frac{r^2 r_0^2}{a^2} + a^2 - 2rr_0 \cos(\theta - \theta_0) \right) \right] \tag{9.100}$$

where C is some new constant. Therefore,

$$G = G_f + G_r = \frac{1}{4\pi} \log \left[\frac{r^2 + r_0^2 - 2rr_0 \cos(\theta - \theta_0)}{C \frac{a^2}{r_0^2} \left(\frac{r^2 r_0^2}{a^2} + a^2 - 2rr_0 \cos(\theta - \theta_0) \right)} \right] \tag{9.101}$$

and the condition $G = 0$ when $r = a$ requires that

$$\frac{1}{4\pi} \log \left[\frac{a^2 + r_0^2 - 2ar_0 \cos(\theta - \theta_0)}{C\frac{a^2}{r_0^2}(r_0^2 + a^2 - 2ar_0 \cos(\theta - \theta_0))} \right] = 0. \qquad (9.102)$$

This requires that the scaling factor C be selected as $C = r_0^2/a^2$. This produces the Green's function

$$G = G(r, \theta; r_0, \theta_0) = G_f + G_r = \frac{1}{4\pi} \log \left[a^2 \left(\frac{r^2 + r_0^2 - 2rr_0 \cos(\theta - \theta_0)}{r^2 r_0^2 + a^4 - 2rr_0 a^2 \cos(\theta - \theta_0)} \right) \right]. \qquad (9.103)$$

This function has the derivative

$$\frac{\partial G}{\partial r_0} = \frac{1}{4\pi} \left[\frac{2r_0 - 2r \cos(\theta - \theta_0)}{r^2 + r_0^2 - 2rr_0 \cos(\theta - \theta_0)} - \frac{2r^2 r_0 - 2ra^2 \cos(\theta - \theta_0)}{r^2 r_0^2 + a^4 - 2rr_0 a^2 \cos(\theta - \theta_0)} \right]$$

and consequently

$$\frac{\partial G}{\partial r_0}\Big|_{r_0=a} a d\theta_0 = \frac{1}{2\pi} \left[\frac{a^2 - r^2}{r^2 + a^2 - 2ra \cos(\theta - \theta_0)} \right] d\theta_0. \qquad (9.104)$$

The solution given by equation (9.94) can be written as

$$u = u(r, \theta) = \int_0^a \int_{-\pi}^\pi G(r, \theta; r_0, \theta_0) f(r_0, \theta_0) r_0 dr_0 d\theta_0$$
$$+ \frac{a^2 - r^2}{2\pi} \int_0^{2\pi} \frac{g(\theta_0) d\theta_0}{r^2 + a^2 - 2ra \cos(\theta - \theta_0)} \qquad (9.105)$$

Note in the special case where $f(r_0, \theta_0) = 0$ there results the Poisson integral formula

$$u = u(r, \theta) = \int_0^{2\pi} g(\theta_0) \frac{\partial G}{\partial r_0}\Big|_{r_0=a} a d\theta_0$$
$$u = u(r, \theta) = \frac{a^2 - r^2}{2\pi} \int_0^{2\pi} \frac{g(\theta_0) d\theta_0}{r^2 + a^2 - 2ra \cos(\theta - \theta_0)} \qquad (9.106)$$

developed in an earlier example.

■

Free-space Green's Function for the Heat Equation

Consider the heat equation in one dimension and let $g_1(x, t; \xi, \tau)$ denote the Green's function for the initial value problem

$$L(g_1) = \frac{\partial g_1}{\partial t} - \alpha^2 \frac{\partial^2 g_1}{\partial x^2} = 0, \quad -\infty < x < \infty, t > \tau$$
$$g_1(x, \tau; \xi, \tau) = \delta(x - \xi) \qquad (9.107)$$
$$g_1(x, t; \xi, \tau) = 0, \quad t < \tau,$$

where $g_1(x, t; \xi, \tau)$ denotes the temperature in an infinitely long bar which has been given an initial 'squirt' of heat at the point $x = \xi$ at the initial time $t = \tau$. Assume a solution to equation (9.107) of the form $g_1 = F(x)h(t)$. This produces two ordinary differential equations $F'' + \omega^2 F = 0$ and $h' + \omega^2 \alpha^2 h = 0$, where $-\omega^2$ is the separation constant. These equations are easily solved and we obtain the solution

$$g_1 = [A\cos(\omega x) + B\sin(\omega x)]\, e^{-\omega^2 \alpha^2 t}$$

for all values of $\omega > 0$. The constants A and B are considered to be functions of ω so that by superposition we have

$$g_1 = \int_0^\infty [A(\omega)\cos\omega x + B(\omega)\sin\omega x]\, e^{-\omega^2 \alpha^2 t}\, d\omega \qquad (9.108)$$

is also a solution. Applying the initial condition requires

$$g_1(x, \tau; \xi, \tau) = \delta(x - \xi) = \int_0^\infty [A(\omega)\cos\omega x + B(\omega)\sin\omega x]\, e^{-\omega^2 \alpha^2 \tau}\, d\omega.$$

We recognize this as a Fourier integral problem, where A and B are the Fourier integral coefficients to be determined. We have previously shown these coefficients are given by

$$A = A(\omega) = \frac{\cos(\omega \xi)}{\pi} e^{\omega^2 \alpha^2 \tau}$$

$$B = B(\omega) = \frac{\sin(\omega \xi)}{\pi} e^{\omega^2 \alpha^2 \tau}.$$

This enables us to express the solution as

$$g_1(x, t; \xi, \tau) = \int_0^\infty \cos\omega(x - \xi)\, e^{-\omega^2 \alpha^2 (t - \tau)}\, d\omega.$$

Using the result

$$\int_0^\infty e^{-A^2 r^2} \cos(Br)\, dr = \frac{\sqrt{\pi}}{2A} e^{-\frac{B^2}{4A^2}}$$

the solution to the boundary value problem can be written in the form

$$g_1(x, t; \xi, \tau) = \begin{cases} \dfrac{1}{2\alpha\sqrt{\pi(t-\tau)}} \exp\left[\dfrac{-(x-\xi)^2}{4\alpha^2(t-\tau)}\right], & t > \tau \\ 0, & t < \tau \end{cases}$$

This result can also be expressed in the form

$$g_1(x, t; \xi, \tau) = \frac{1}{2\alpha\sqrt{\pi(t-\tau)}} \exp\left[\frac{-(x-\xi)^2}{4\alpha^2(t-\tau)}\right] H(t-\tau), \qquad (9.109)$$

where H is the unit step function. The function g_1 is called the Green's function for the initial value problem and has the properties

$$\int_{-\infty}^{\infty} g_1(x,t;\xi,\tau)\, dx = 1$$

and

$$\lim_{\substack{t \to \tau \\ t > \tau}} g_1 = 0, \quad \text{for} \quad x \neq \xi.$$

■

Example 9-15. (Heat equation.)

From the problem 49, in the exercises for chapter 4, we obtained the solution to the nonhomogeneous heat equation

$$L(u) = \frac{\partial u}{\partial t} - \alpha^2 \frac{\partial^2 u}{\partial x^2} = f(x), \quad -\infty < x < \infty, t > 0$$

with initial condition $u(x,0) = 0$. This solution was

$$u(x,t) = \int_0^t \int_{-\infty}^{\infty} \frac{1}{2\alpha\sqrt{\pi(t-\tau)}} \exp\left[-\frac{(x-\xi)^2}{4\alpha^2(t-\tau)}\right] f(\xi)\, d\xi d\tau.$$

From this solution we can infer the Green's function associated with the heat operator $L(u)$ is given by

$$g_1(x,t;\xi,\tau) = \frac{1}{2\alpha\sqrt{\pi(t-\tau)}} \exp\left[-\frac{(x-\xi)^2}{4\alpha^2(t-\tau)}\right], \quad -\infty < x < \infty, \quad t > \tau > 0.$$

This is the same Green's function calculated in the previous example. This Green's function can be viewed as a solution of the boundary value problem

$$L^*(u) = -\frac{\partial u}{\partial \tau} - \alpha^2 \frac{\partial^2 u}{\partial \xi^2} = \delta(\xi - x)\delta(\tau - t), \quad \tau < t, -\infty < x < \infty \tag{9.110}$$
$$u = 0, \quad \text{for } \tau > t.$$

It can also be viewed as a solution of the boundary value problem

$$L(u) = \frac{\partial u}{\partial t} - \alpha^2 \frac{\partial^2 u}{\partial x^2} = \delta(x - \xi)\delta(t - \tau), \quad t > \tau, -\infty < x < \infty \tag{9.111}$$
$$u = 0, \quad \text{for } t < \tau.$$

Physically the Green's function represents the temperature distribution in an infinite rod at a position x and time t due to a unit heat pulse released at time $t = \tau$ at the position $x = \xi$.

■

Nonhomogeneous Heat Equation

Consider the nonhomogeneous heat equation

$$L_{xt}(u) = \frac{\partial u}{\partial t} - \alpha^2 \frac{\partial^2 u}{\partial x^2} = F(x,t), \quad -\infty < x < \infty, \quad t > 0$$

with the initial condition $u(x,0) = f(x)$. We write the Green's formula for the rectangular region R illustrated in figure 9-5 and then take the limit as ℓ increases without bound. This produces the Greens formulas for the infinite region associated with the heat equation operator.

The Green's formula for the finite rectangle illustrated in figure 9-5 is represented in terms of the dummy variables ξ, τ and can be expressed in the form

$$\int_0^T \int_{-\ell}^{\ell} g_1 L_{\xi\tau}(u)\,d\xi d\tau = -\int_{-\ell}^{\ell} g_1 u \big|_{\tau=0}\,d\xi - \int_0^T \alpha^2 \left(g_1 \frac{\partial u}{\partial \xi} - u \frac{\partial g_1}{\partial \xi} \right)\Big|_{\xi=\ell}\,d\tau$$

$$+ \int_{-\ell}^{\ell} g_1 u \big|_{\tau=T}\,d\xi + \int_0^T \alpha^2 \left(g_1 \frac{\partial u}{\partial \xi} - u \frac{\partial g_1}{\partial \xi} \right)\Big|_{\xi=-\ell}\,d\tau \quad (9.112)$$

$$+ \int_0^T \int_{-\ell}^{\ell} u L_{\xi\tau}^*(g_1)\,d\xi d\tau.$$

In equation (9.112) we let

$$L_{\xi\tau}^*(g_1) = \delta(\xi - x)\delta(\tau - t),$$

then the last term of equation (9.112) becomes

$$\lim_{\ell \to \infty} \int_0^T \int_{-\ell}^{\ell} u L_{\xi\tau}^*(g_1)\,d\xi d\tau = \int_0^T \int_{-\infty}^{\infty} u(\xi,\tau)\delta(\xi-x)\delta(\tau-t)d\xi d\tau = u(x,t).$$

We now consider the line integral where τ is zero. We have on the boundary $\tau = 0$ the condition $u(\xi,0) = f(\xi)$ and hence we can write

$$\lim_{\ell \to \infty} \int_{-\ell}^{\ell} g_1 u \big|_{\tau=0}\,d\xi = \int_{-\infty}^{\infty} g_1(x,t;\xi,0)f(\xi)\,d\xi.$$

By hypothesis $L_{\xi\tau}(u) = F(\xi,\tau)$ and hence the left hand side of equation (9.112) becomes

$$\lim_{\ell \to \infty} \int_0^T \int_{-\ell}^{\ell} g_1 L_{\xi\tau}(u)\,d\xi d\tau = \int_0^T \int_{-\infty}^{\infty} g_1(x,t;\xi,\tau)F(\xi,\tau)\,d\xi d\tau.$$

To analyze the remaining terms in the Green's formula we desire that

$$g_1(x,t;\xi,\tau) = 0 \quad \text{for } \tau > t$$

and also

$$\lim_{\xi \to \infty} g_1(x, t; \xi, \tau) = 0.$$

These results enable us to show the remaining line integrals in Green's formula approach zero as L increases without bound. The Green's formula reduces to

$$u(x, t) = \int_{-\infty}^{\infty} g_1(x, t; \xi, 0) f(\xi) \, d\xi + \int_{0}^{t} \int_{-\infty}^{\infty} g_1(x, t; \xi, \tau) F(\xi, \tau) \, d\xi d\tau \qquad (9.113)$$

which is the solution of the given boundary value problem.

For the special case where $F(x, t) = 0$ the equation (9.113) reduces to

$$u(x, t) = \int_{-\infty}^{\infty} g_1(x, t; \xi, 0) f(\xi) \, d\xi \qquad (9.114)$$

which represents the solution of the initial value problem

$$L_{xt}(u) = u_t - \alpha^2 u_{xx} = 0, \quad -\infty < x < \infty, \, t > 0, \qquad (9.115)$$

where $u(x, 0) = f(x)$. Observe that in order for the equation (9.114) to satisfy the initial condition $u(x, 0) = f(x)$, it is required that

$$u(x, 0) = \int_{-\infty}^{\infty} g_1(x, 0; \xi, 0) f(\xi) \, d\xi = f(x). \qquad (9.116)$$

For this to happen the Green's function $g_1(x, 0; \xi, 0)$ in equation (9.114) must act like a Dirac delta function. We find that this is indeed the case as can be seen by an examination of the illustration in the figure 9-6.

In the special case $f(x) = 0$ and $F \neq 0$ the solution (9.113) reduces to

$$u(x, t) = \int_{0}^{t} \int_{-\infty}^{\infty} g_1(x, t; \xi, \tau) F(\xi, \tau) \, d\xi d\tau.$$

This represents the solution of the nonhomogeneous heat equation

$$L_{xt}(u) = u_t - \alpha^2 u_{xx} = F(x, t), \quad -\infty < x < \infty, \, t > 0 \qquad (9.117)$$

with the initial condition $u(x, 0) = 0$.

The equation (9.113) is an example of the superposition of the solutions from two different but related problems. This superposition is observed by making the substitution $u = u_1 + u_2$ into the given boundary value problem. Then let u_1 satisfy the equation (9.115) and let u_2 satisfy the equation (9.117).

The above solutions require that we know the Green's function for the region R under investigation. The free-space Green's function is derived in the following example.

Example 9-16. (Heat equation.)
Solve for the Green's function $G = G(x, t; \xi, \tau)$ which satisfies

$$\frac{\partial G}{\partial t} - \alpha^2 \frac{\partial^2 G}{\partial x^2} = \delta(x - \xi)\delta(t - \tau), \quad -\infty < x < \infty, \quad t > \tau \qquad (9.118)$$

where $G = 0$ for $t < \tau$.

Solution: The geometry suggests we use a Fourier exponential transform with respect to the x-variable. Define

$$\widetilde{G} = \mathcal{F}_e\{G; x \to \omega\} = \frac{1}{2\pi} \int_{-\infty}^{\infty} G(x; t)e^{i\omega x}\, dx$$

and take the Fourier exponential transform of equation (9.118) to obtain

$$\frac{d\widetilde{G}}{dt} - \alpha^2(-i\omega)^2 \widetilde{G} = \frac{1}{2\pi} e^{i\omega \xi}\delta(t - \tau) \qquad (9.119)$$

with $\widetilde{G} = 0$ for $t < \tau$. The equation (9.119) represents the two equations

$$\frac{d\widetilde{G}}{dt} + \alpha^2 \omega^2 \widetilde{G} = 0, \quad t < \tau$$

$$\frac{d\widetilde{G}}{dt} + \alpha^2 \omega^2 \widetilde{G} = 0, \quad t > \tau$$

with solutions

$$\widetilde{G} = \begin{cases} Ae^{-\alpha^2 \omega^2 t}, & t < \tau \\ Be^{-\alpha^2 \omega^2 t}, & t > \tau \end{cases} \qquad (9.120)$$

where A, B are constants to be determined. We integrate equation (9.119) from τ^- to τ^+ to find that

$$\widetilde{G}\Big|_{t=\tau^-}^{t=\tau^+} + \alpha^2 \omega^2 \int_{\tau^-}^{\tau^+} G\, dt = \frac{1}{2\pi} e^{i\omega \xi} H(t - \tau)\Big|_{t=\tau^-}^{t=\tau^+} \qquad (9.121)$$

We substitute the results from equation (9.120) into the equation (9.121) and find that G must satisfy the relation

$$Be^{-\alpha^2\omega^2\tau} - Ae^{-\alpha^2\omega^2\tau} = \frac{1}{2\pi}e^{i\omega\xi} \tag{9.122}$$

The condition that $\widetilde{G} = 0$ for $t < \tau$ requires that $A = 0$ and so the equation (9.122) gives us that

$$B = \frac{1}{2\pi}e^{i\omega\xi}e^{\alpha^2\omega^2\tau} \tag{9.123}$$

This gives the transform of the Green's function as

$$\widetilde{G} = \begin{cases} 0, & t < \tau \\ \frac{1}{2\pi}e^{i\omega\xi}e^{-\alpha^2\omega^2(t-\tau)}, & t > \tau \end{cases} \tag{9.124}$$

The Fourier exponential transform table 7-4 aids us in obtaining the inverse Fourier exponential transform. We find

$$\mathcal{F}_e^{-1}\{e^{-\alpha^2\omega^2(t-\tau)}\} = \frac{\sqrt{\pi}}{\alpha\sqrt{(t-\tau)}}e^{-x^2/4\alpha^2(t-\tau)}. \tag{9.125}$$

The shifting property of the Fourier exponential transform gives us

$$\mathcal{F}_e^{-1}\{\frac{1}{2\pi}e^{i\omega\xi}e^{-\alpha^2\omega^2(t-\tau)}\} = \frac{1}{2\alpha\sqrt{\pi(t-\tau)}}e^{-(x-\xi)^2/4\alpha^2(t-\tau)}. \tag{9.126}$$

This gives us the Green's function

$$G = G(x,t;\xi,\tau) = \begin{cases} 0, & t < \tau \\ \frac{1}{2\alpha\sqrt{\pi(t-\tau)}}e^{-(x-\xi)^2/4\alpha^2(t-\tau)}, & t > \tau \end{cases} \tag{9.127}$$

which can also be written in the form

$$G = G(x,t;\xi,\tau) = \frac{1}{2\alpha\sqrt{\pi(t-\tau)}}\exp\left[-(x-\xi)^2/4\alpha^2(t-\tau)\right]H(t-\tau) \tag{9.128}$$

where $H(t-\tau)$ is the Heaviside unit step function. This is the same function which we found to be a solution of the heat equation having an initial impulse at time $t = \tau$. We can then write $g_1(x,t;\xi,\tau) = G(x,t;\xi,\tau)$ as this notation will prove to be more convenient to analyze the two and three-dimensional heat equation. Note also that as t approaches τ the Green's function approaches zero everywhere except at the point where $x = \xi$ which is called a singular point for the Green's function. A graph of this function in the neighborhood of the point $t = \tau$ and $x = \xi$ is illustrated in the figure 9-6.

Note that as a function of τ the Green's function given by equation (9.128) is zero for all $\tau > t$. This is an important piece of information which we will need in order to solve the adjoint equation which defines G as a function of ξ and τ. Here the adjoint equation which defines G as a function of ξ and τ is given by

$$-\frac{\partial G}{\partial \tau} - \alpha^2 \frac{\partial^2 G}{\partial \xi^2} = \delta(\xi - x)\delta(\tau - t) \tag{9.129}$$

This equation is to be solved for $\tau < t$ and satisfy the condition $G = 0$ for all $\tau > t$. To solve this equation we will again employ the Fourier exponential transform. Define $\mathcal{F}_e\{G; \xi \to \omega\} = \widetilde{G}$. Taking the Fourier exponential transform of the equation (9.129) we obtain

$$-\frac{d\widetilde{G}}{d\tau} - \alpha^2(-i\omega)^2\widetilde{G} = \frac{1}{2\pi}e^{i\omega x}\delta(\tau - t) \tag{9.130}$$

which can also be written in the form

$$\frac{d\widetilde{G}}{d\tau} - \alpha^2\omega^2\widetilde{G} = -\frac{1}{2\pi}e^{i\omega x}\delta(\tau - t) \tag{9.131}$$

$$G = G(x, t; \xi, \tau) = \frac{1}{2\alpha\sqrt{\pi(t - \tau)}}\exp\left[-(x - \xi)^2/4\alpha^2(t - \tau)\right]H(t - \tau)$$

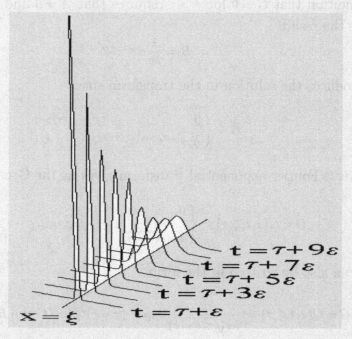

Figure 9-6. Delta function response in neighborhood of $t = \tau$ and $x = \xi$.

The equation (9.131) represents the two equations

$$\frac{d\widetilde{G}}{d\tau} - \alpha^2\omega^2\widetilde{G} = 0, \quad \tau > t$$

$$\frac{d\widetilde{G}}{d\tau} - \alpha^2\omega^2\widetilde{G} = 0, \quad \tau < t \tag{9.132}$$

which has the solution

$$\widetilde{G} = \begin{cases} Ae^{\alpha^2\omega^2\tau}, & \tau > t \\ Be^{\alpha^2\omega^2\tau}, & \tau < t \end{cases} \tag{9.133}$$

where A, B are constants to be determined. We integrate the equation (9.131) from τ^- to τ^+ to obtain

$$\widetilde{G}\Big|_{\tau=t^-}^{\tau=t^+} - \alpha^2\omega^2 \int_{t^-}^{t^+} \widetilde{G}\, d\tau = -\frac{1}{2\pi} e^{i\omega x} H(\tau - t)\Big|_{\tau=t^-}^{\tau=t^+} \tag{9.134}$$

This implies that the solution given by equation (9.133) must satisfy the condition

$$Ae^{\alpha^2\omega^2 t} - Be^{\alpha^2\omega^2 t} = -\frac{1}{2\pi} e^{i\omega x} \tag{9.135}$$

The condition that $\widetilde{G} = 0$ for $\tau > t$ requires that $A = 0$ and so equation (9.135) gives us the result

$$B = \frac{1}{2\pi} e^{i\omega x} e^{-\alpha^2\omega^2 t} \tag{9.136}$$

This produces the solution in the transform space

$$\widetilde{G} = \begin{cases} 0, & \tau > t \\ \frac{1}{2\pi} e^{i\omega x} e^{-\alpha^2\omega^2(\tau-t)}, & \tau < t \end{cases} \tag{9.137}$$

The inverse Fourier exponential transform gives us the Green's function

$$G = G(x, t; \xi, \tau) = \begin{cases} 0, & \tau > t \\ \frac{1}{2\alpha\sqrt{\pi(t-\tau)}} e^{-(x-\xi)^2/4\alpha^2(t-\tau)}, & \tau < t \end{cases} \tag{9.138}$$

which can also be written in the form

$$G = G(x, t; \xi, \tau) = \frac{1}{2\alpha\sqrt{\pi(t-\tau)}} \exp\left[-(x-\xi)^2/4\alpha^2(t-\tau)\right] H(t - \tau) \tag{9.139}$$

Two-dimensional Heat Equation

We now examine the two-dimensional homogeneous heat equation

$$L_{xyt}(u) = \frac{\partial u}{\partial t} - \alpha^2 \left(\frac{\partial^2 u}{\partial x^2} + \frac{\partial^2 u}{\partial y^2} \right) = 0 \tag{9.140}$$

$$-\infty < x, y < \infty, \quad t > \tau$$

with the initial conditions

$$u(x, y, \tau) = \delta(x - \xi)\delta(y - \eta)$$

$$u(x, y, t) = 0 \quad \text{for } t < \tau.$$

The solution of this initial value problem is denoted

$$u = g_2(x, y, t; \xi, \eta, \tau) = \frac{1}{4\pi\alpha^2(t - \tau)} \exp\left[-\frac{(x - \xi)^2 + (y - \eta)^2}{4\alpha^2(t - \tau)} \right] H(t - \tau). \tag{9.141}$$

It is now demonstrated that

$$g_2(x, y, t; \xi, \eta, \tau) = g_1(x, t; \xi, \tau)g_1(y, t; \eta, \tau). \tag{9.142}$$

That is, the Green's function for the two dimensional initial value problem is a product of one dimensional Green's functions. This is verified by differentiating the equation (9.142) and showing that it satisfies the two dimensional heat equation (9.140). The various partial derivatives of g_2 are

$$\frac{\partial g_2}{\partial t} = g_1(x, t; \xi, \tau)\frac{\partial g_1(y, t, ; \eta, \tau)}{\partial t} + \frac{\partial g_1(x, t; \xi, \tau)}{\partial t}g_1(y, t; \eta, \tau)$$

$$\frac{\partial^2 g_2}{\partial x^2} = \frac{\partial^2 g_1(x, t; \xi, \tau)}{\partial x^2}g_1(y, t; \eta, \tau)$$

$$\frac{\partial^2 g_2}{\partial y^2} = g_1(x, t; \xi, \tau)\frac{\partial^2 g_1(y, t; \eta, \tau)}{\partial y^2}$$

When these derivatives are substituted into the equation (9.140) we obtain

$$g_1(x, t; \xi, \tau)\left[\frac{\partial g_1(y, t; \eta, \tau)}{\partial t} - \alpha^2\frac{\partial^2 g_1(y, t; \eta, \tau)}{\partial y^2} \right]$$

$$+ g_1(y, t; \eta, \tau)\left[\frac{\partial g_1(x, t; \xi, \tau)}{\partial t} - \alpha^2\frac{\partial^2 g_1(x, t; \xi, \tau)}{\partial x^2} \right] = 0,$$

where the terms inside the brackets are zero since we have shown that g_1 is a solution of the one dimensional heat equation.

The solution of the nonhomogeneous two-dimensional heat equation

$$L_{xyt}(u) = \frac{\partial u}{\partial t} - \alpha^2 \left(\frac{\partial^2 u}{\partial x^2} + \frac{\partial^2 u}{\partial y^2} \right) = F(x, y, t) \tag{9.143}$$

$$-\infty < x, y < \infty, \quad t > 0$$

with initial conditions

$$u(x, y, 0) = f(x, y)$$

is given by

$$u(x, y, t) = \int_{-\infty}^{\infty} \int_{-\infty}^{\infty} g_2(x, y, t; \xi, \eta, 0) f(\xi, \eta) \, d\xi d\eta$$

$$+ \int_0^t \int_{-\infty}^{\infty} \int_{-\infty}^{\infty} g_2(x, y, t; \xi, \eta, \tau) F(\xi, \eta, \tau) \, d\xi d\eta d\tau \tag{9.144}$$

and can be derived in a manner similar to how equation (9.113) was derived. In three-dimensions, the corresponding Green's function for the heat equation operator is a product of one-dimensional Green's function and the two-dimensional Green' function. We write

$$g_3(x, y, z, t; \xi, \eta, \zeta) = g_2(x, y, t; \xi, \eta, \tau) g_1(z, t; \zeta, \tau)$$

which expands to

$$g_3 = \frac{1}{8\alpha^3 \left[\pi(t - \tau) \right]^{3/2}} \exp \left\{ -\frac{(x - \xi)^2 + (y - \eta)^2 + (z - \zeta)^2}{4\alpha^2(t - \tau)} \right\} H(t - \tau). \tag{9.145}$$

The solution of the three-dimensional nonhomogeneous heat equation

$$L_{xyzt}(u) = \frac{\partial u}{\partial t} - \alpha^2 \left(\frac{\partial^2 u}{\partial x^2} + \frac{\partial^2 u}{\partial y^2} + \frac{\partial^2 u}{\partial z^2} \right) = F(x, y, z, t) \tag{9.146}$$

$$-\infty < x, y, z < \infty, \quad t > 0$$

with initial condition $u(x, y, z, 0) = f(x, y, z)$ is given by

$$u(x, y, z, t) = \int_{-\infty}^{\infty} \int_{-\infty}^{\infty} \int_{-\infty}^{\infty} g_3(x, y, z, t; \xi, \eta, \zeta, 0) f(\xi, \eta, \zeta) \, d\xi d\eta d\zeta$$

$$+ \int_0^t \int_{-\infty}^{\infty} \int_{-\infty}^{\infty} \int_{-\infty}^{\infty} g_3(x, y, z, t; \xi, \eta, \zeta, \tau) F(\xi, \eta, \zeta, \tau) \, d\xi d\eta d\zeta d\tau. \tag{9.147}$$

Notice the pattern associated with the one, two and three dimensional solutions to the heat equation. This pattern can be extended to n dimensions.

The previous results are summarized by the following table 9-3.

Table 9-3		
Free-space Green's function for the heat equation		
Dimension	Green's function	
$n = 1$	$G(x, t; \xi, \tau) = \frac{1}{2\alpha\sqrt{\pi(t-\tau)}} \exp\left[-\frac{(x-\xi)^2}{4\alpha^2(t-\tau)}\right] H(t-\tau)$	
$n = 2$	$G(x, y, t; \xi, \eta, \tau) = \frac{1}{4\alpha^2\pi(t-\tau)} \exp\left[-\frac{(x-\xi)^2+(y-\eta)^2}{4\alpha^2(t-\tau)}\right] H(t-\tau)$	
$n = 3$	$G(x, y, z, t; \xi, \eta, \zeta, \tau) = \frac{1}{8\alpha^3[\pi(t-\tau)]^{3/2}} \exp\left[-\frac{(x-\xi)^2+(y-\eta)^2+(z-\zeta)^2}{4\alpha^2(t-\tau)}\right] H(t-\tau)$	
$n = m > 3$	$G(\vec{r}, t; \vec{\xi}, \tau) = \left[4\alpha^2\pi(t-\tau)\right]^{-m/2} \exp\left[\frac{-\|\vec{x}-\vec{\xi}\|^2}{4\alpha^2(t-\tau)}\right] H(t-\tau)$ $\vec{x} = (x_1, x_2, \dots, x_m), \quad \vec{\xi} = (\xi_1, \xi_2, \dots, \xi_m)$	

One-dimensional wave equation

The one-dimensional wave operator

$$L_{xt}(u) = \frac{\partial^2 u}{\partial t^2} - c^2\frac{\partial^2 u}{\partial x^2}, \qquad -\infty < x < \infty, \quad t > 0 \tag{9.148}$$

has associated with it the Green's function G which satisfies

$$L_{xt}(G) = \frac{\partial^2 G}{\partial t^2} - c^2\frac{\partial^2 G}{\partial x^2} = \delta(x-\xi)\delta(t-\tau)$$
$$G = G(x, t; \xi, \tau) = 0 \quad \text{for} \quad t < \tau. \tag{9.149}$$

We use Laplace transforms to solve the equation (9.149). Let

$$\widetilde{G} = \widetilde{G}(x, s; \xi, \tau) = \mathcal{L}\{G(x, t; \xi, \tau); t \to s\}$$

denote the Laplace transform of the Green's function. We then take the Laplace transform of equation (9.149)to obtain

$$s^2\widetilde{G} - c^2\frac{d^2\widetilde{G}}{dx^2} = \delta(x-\xi)e^{-s\tau} \tag{9.150}$$

where all initial conditions have been set to zero. The equation (9.150) represents two ordinary differential equations

$$\frac{d^2\widetilde{G}}{dx^2} - \frac{s^2}{c^2}\widetilde{G} = 0, \qquad x < \xi$$
$$\frac{d^2\widetilde{G}}{dx^2} - \frac{s^2}{c^2}\widetilde{G} = 0, \qquad x > \xi \tag{9.151}$$

These equations have the solutions

$$\widetilde{G} = \begin{cases} c_1 e^{sx/c} + c_2 e^{-sx/c}, & x < \xi \\ k_1 e^{sx/c} + k_2 e^{-sx/c}, & x > \xi \end{cases} \tag{9.152}$$

where c_1, c_2, k_1, k_2 are constants. We require that $c_2 = 0$ and $k_1 = 0$ for bounded solutions as $x \to \infty$ and $x \to -\infty$. This reduces the solutions to the form

$$\widetilde{G} = \begin{cases} c_1 e^{sx/c}, & x < \xi \\ k_2 e^{-sx/c}, & x > \xi \end{cases} \tag{9.153}$$

The continuity condition requires

$$c_1 e^{s\xi/c} = k_2 e^{-s\xi/c} \tag{9.154}$$

and the jump discontinuity condition requires

$$-k_2 \frac{s}{c} e^{-s\xi/c} - c_1 \frac{s}{c} e^{s\xi/c} = -\frac{1}{c^2} e^{-s\tau}. \tag{9.155}$$

The equations (9.154) and (9.155) enable us to solve for the constants c_1 and k_2 to obtain

$$c_1 = \frac{1}{2cs} e^{-s\tau} e^{-s\xi/c} \qquad k_2 = \frac{1}{2cs} e^{-s\tau} e^{s\xi/c} \tag{9.156}$$

This gives the transform solution

$$\widetilde{G} = \begin{cases} \frac{1}{2cs} e^{-s\tau} e^{-s\xi/c} e^{sx/c}, & x < \xi \\ \frac{1}{2cs} e^{-s\tau} e^{s\xi/c} e^{-sx/c}, & x > \xi \end{cases} \tag{9.157}$$

From table 7-2 we use the inverse Laplace transforms

$$\mathcal{L}^{-1}\{\frac{1}{s} e^{-s(\tau - (x-\xi))/c}\} = H(t - \tau + (x - \xi)/c)$$
$$\mathcal{L}^{-1}\{\frac{1}{s} e^{-s(\tau + (x-\xi))/c}\} = H(t - \tau - (x - \xi)/c) \tag{9.158}$$

This gives the Green's function

$$G = G(x, t; \xi, \tau) = \begin{cases} \frac{1}{2c} H(t - \tau + (x - \xi)/c), & x < \xi \\ \frac{1}{2c} H(t - \tau - (x - \xi)/c), & x > \xi \end{cases} \tag{9.159}$$

The Green's function given by the equation (9.158) can also be expressed in the alternative forms

$$G(x, t; \xi, \tau) = \frac{1}{2c} H(c(t - \tau) - |x - \xi|) \tag{9.160}$$

or

$$G(x,t;\xi,\tau) = \begin{cases} \frac{1}{2c}, & |x-\xi| < c(t-\tau) \\ 0, & |x-\xi| > c(t-\tau) \end{cases} \qquad (9.161)$$

or

$$G(x,t;\xi,\tau) = \frac{1}{2c}\left[H(x-\xi+c(t-\tau)) - H(x-\xi-c(t-\tau))\right] \qquad (9.162)$$

A graphic illustration of the Green's function for the one-dimensional wave equation is given in the figure 9-7.

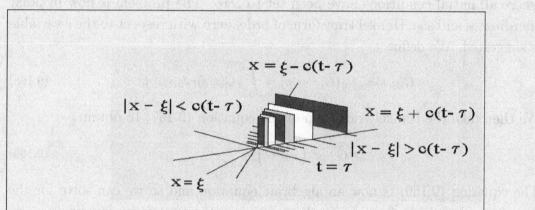

Figure 9-7. Green's function for the one-dimensional wave operator.

Two-dimensional wave equation

The wave operator in two-dimensions is given by

$$L_{xyt}(u) = \frac{\partial^2 u}{\partial t^2} - c^2\left(\frac{\partial^2 G}{\partial x^2} + \frac{\partial^2 G}{\partial y^2}\right), \qquad -\infty < x,y < \infty, \quad t > 0 \qquad (9.163)$$

and has associated with it a Green's function G which satisfies

$$L_{xyt}(G) = \frac{\partial^2 G}{\partial t^2} - c^2\left(\frac{\partial^2 G}{\partial x^2} + \frac{\partial^2 G}{\partial y^2}\right) = \delta(x-\xi)\delta(y-\eta)\delta(t-\tau) \qquad (9.164)$$

for $-\infty < x,y < \infty$, $t > \tau$ and we require that $G = 0$ for all $t < \tau$. We make the change of variables

$$x - \xi = r\cos\theta, \qquad y - \eta = r\sin\theta$$

and assume symmetry with respect to the variable θ and write the equation (9.164) in the form

$$\frac{\partial^2 G}{\partial t^2} - c^2\left(\frac{\partial^2 G}{\partial r^2} + \frac{1}{r}\frac{\partial G}{\partial r}\right) = \frac{1}{2\pi r}\delta(r)\delta(t-\tau). \qquad (9.165)$$

We use Laplace transforms and define

$$\widetilde{G}(r,s) = \mathcal{L}\{G(r,t); t \to s\} = \int_0^\infty G(r,t)e^{-st}\,dt. \tag{9.166}$$

We then take the Laplace transform of the equation (9.165) and find

$$s^2\widetilde{G} - c^2\left(\frac{d^2\widetilde{G}}{dr^2} + \frac{1}{r}\frac{d\widetilde{G}}{dr}\right) = \frac{1}{2\pi r}\delta(r)e^{-s\tau} \tag{9.167}$$

where all initial conditions have been set to zero. The problem is now in polar coordinates and so a Hankel transform of order zero with respect to the r variable is suggested. We define

$$\widetilde{\widetilde{G}}(p,s) = \mathcal{H}_0\{\widetilde{G}; r \to p\} = \int_0^\infty \widetilde{G}(r,s)rJ_0(rp)\,dr. \tag{9.168}$$

We then take the Hankel transform of the equation (9.167) to obtain

$$s^2\widetilde{\widetilde{G}} - c^2\left(-p^2\widetilde{\widetilde{G}}\right) = \frac{1}{2\pi}e^{-s\tau}. \tag{9.169}$$

The equation (9.169) is now an algebraic equation and so we can solve for the transformed variable $\widetilde{\widetilde{G}}$ to obtain the solution in the transform domain

$$\widetilde{\widetilde{G}} = \frac{e^{-s\tau}}{2\pi(s^2 + c^2p^2)}. \tag{9.170}$$

We use the Laplace transform table 7-2 and find the inverse Laplace transform

$$\mathcal{L}^{-1}\{\frac{1}{s^2 + c^2p^2}\} = \frac{1}{cp}\sin cpt \tag{9.171}$$

and by the shift theorem we obtain

$$\mathcal{L}^{-1}\{\frac{e^{-s\tau}}{s^2 + c^2p^2}\} = \frac{1}{cp}\sin cp(t-\tau)\,H(t-\tau). \tag{9.172}$$

We can now use the Hankel transform table 7-7 and find the inverse Hankel transform by table look up or by actually evaluating the inversion integral. For example, one can find by using a good table of integrals that

$$\mathcal{H}_0^{-1}\left\{\frac{1}{cp}\sin cp(t-\tau)\right\} = \int_0^\infty \frac{1}{cp}\sin cp(t-\tau)pJ_0(rp)\,dp$$
$$= \frac{1}{c}\int_0^\infty \sin cp(t-\tau)J_0(rp)\,dp \tag{9.173}$$
$$= \begin{cases} 0, & c(t-\tau) < r \\ \frac{1}{c}\frac{1}{\sqrt{c^2(t-\tau)^2 - r^2}}, & c(t-\tau) > r \end{cases}$$

The Green's function solution is therefore given by

$$G = G(r,t) = \begin{cases} 0, & c(t-\tau) < r \\ \frac{1}{2\pi c}\frac{H(t-\tau)}{\sqrt{c^2(t-\tau)^2-r^2}}, & c(t-\tau) > r \end{cases} \tag{9.174}$$

which can also be written in the form

$$G = G(r,t) = \begin{cases} 0, & t < \tau \\ \frac{1}{2\pi c}\frac{1}{\sqrt{c^2(t-\tau)^2-r^2}}, & t > \tau \end{cases} \tag{9.175}$$

In terms of the original (x,y,t) variables the solution of the equation (9.164) can be written in either of the forms

$$G = G(x,y,t;\xi,\eta,\tau) = \begin{cases} 0, & t < \tau \\ \frac{1}{2\pi c}\frac{1}{\sqrt{c^2(t-\tau)^2-(x-\xi)^2-(y-\eta)^2}}, & t > \tau \end{cases} \tag{9.176}$$

or $\quad G = G(x,y,t;\xi,\eta,\tau) = \frac{1}{2\pi c}\frac{1}{\sqrt{c^2(t-\tau)^2-|\vec{r}-\vec{\xi}|}}H(c(t-\tau)-|\vec{r}-\vec{\xi}|)$

where $\vec{r} = (x,y)$ and $\vec{\xi} = (\xi,\eta)$.

Three-dimensional wave equation

The three-dimensional wave operator is given by

$$L_{xyzt}(u) = \frac{\partial^2 G}{\partial t^2} - c^2\left(\frac{\partial^2 G}{\partial x^2} + \frac{\partial^2 G}{\partial y^2} + \frac{\partial^2 G}{\partial z^2}\right) \tag{9.177}$$

and the Green's function G associated with this operator is a solution of the equation

$$L_{xyzt}(G) = \frac{\partial^2 G}{\partial t^2} - c^2\left(\frac{\partial^2 G}{\partial x^2} + \frac{\partial^2 G}{\partial y^2} + \frac{\partial^2 G}{\partial z^2}\right) = \delta(x-\xi)\delta(y-\eta)\delta(z-\zeta)\delta(t-\tau) \tag{9.178}$$

for $-\infty < x,y,z < \infty$, $t > \tau$ and satisfying $G(x,y,z,t;\xi,\eta,\zeta,\tau) = 0$ for $t < \tau$. To solve this equation we use the Fourier exponential transform and define

$$\widetilde{G}(\omega_1,y,z,t) = \mathcal{F}_e\{G(x,y,z,t); x \to \omega_1\}$$
$$\widetilde{\widetilde{G}}(\omega_1,\omega_2,z,t) = \mathcal{F}_e\{\widetilde{G}(\omega_1,y,z,t); y \to \omega_2\} \tag{9.179}$$
$$\widetilde{\widetilde{\widetilde{G}}}(\omega_1,\omega_2,\omega_3,t) = \mathcal{F}_e\{\widetilde{\widetilde{G}}(\omega_1,\omega_2,z,t); z \to \omega_2\}$$

as transforms with respect to the x,y and z variables. We then take, one at a time, the Fourier transform of equation (9.178) with respect to the x,y and z

variables and interchange the orders of differentiation and integration to obtain the ordinary differential equation

$$\frac{d^2\widetilde{\widetilde{\widetilde{G}}}}{dt^2} + c^2 \left(\omega_1^2 + \omega_2^2 + \omega_3^2\right)\widetilde{\widetilde{\widetilde{G}}} = \frac{e^{i\omega_1\xi}}{2\pi}\frac{e^{i\omega_2\eta}}{2\pi}\frac{e^{i\omega_3\zeta}}{2\pi}\delta(t-\tau) \qquad (9.180)$$

where $\widetilde{\widetilde{\widetilde{G}}} = 0$ for $t < \tau$. The equation (9.180) has the solution

$$\widetilde{\widetilde{\widetilde{G}}} = \begin{cases} 0, & t < \tau \\ k_1 \sin c\omega(t-\tau) + k_2 \cos c\omega(t-\tau), & t > \tau \end{cases} \qquad (9.181)$$

where $\omega = \sqrt{\omega_1^2 + \omega_2^2 + \omega_3^2}$ and k_1, k_2 are constants. The equation (9.181) must satisfy the condition $\widetilde{\widetilde{\widetilde{G}}} = 0$ for $t = \tau$. This requires that $k_2 = 0$. The equation (9.181) must also satisfy the jump discontinuity condition

$$\frac{d\widetilde{\widetilde{\widetilde{G}}}}{dt}\Big|_{t=\tau^-}^{t=\tau^+} = \frac{e^{i\omega_1\xi}}{2\pi}\frac{e^{i\omega_2\eta}}{2\pi}\frac{e^{i\omega_3\zeta}}{2\pi}$$

which implies that

$$k_1 = \frac{1}{c\omega}\frac{e^{i\omega_1\xi}}{2\pi}\frac{e^{i\omega_2\eta}}{2\pi}\frac{e^{i\omega_3\zeta}}{2\pi} \qquad (9.182)$$

This produces the solution in the transform domain as

$$\widetilde{\widetilde{\widetilde{G}}} = \begin{cases} 0, & t < \tau \\ \frac{1}{c\omega}\sin c\omega(t-\tau)\frac{e^{i\omega_1\xi}}{2\pi}\frac{e^{i\omega_2\eta}}{2\pi}\frac{e^{i\omega_3\zeta}}{2\pi}, & t > \tau \end{cases} \qquad (9.183)$$

The inverse Fourier exponential transform produces the results

$$\widetilde{\widetilde{G}} = \widetilde{\widetilde{G}}(\omega_1, \omega_2, z, t) = \mathcal{F}_e^{-1}\{\widetilde{\widetilde{\widetilde{G}}}(\omega_1, \omega_2, \omega_3, t); \omega_3 \to z\}$$

$$\widetilde{G} = \widetilde{G}(\omega_1, y, z, t) = \mathcal{F}_e^{-1}\{\widetilde{\widetilde{G}}(\omega_1, \omega_2, z, t); \omega_2 \to y\}$$

$$G = G(x, y, z, t) = \mathcal{F}_e^{-1}\{\widetilde{G}(\omega_1, y, z, t); \omega_1 \to x\}.$$

This gives the solution

$$G = \int_{-\infty}^{\infty}\int_{-\infty}^{\infty}\int_{-\infty}^{\infty}\widetilde{\widetilde{\widetilde{G}}}e^{-i(\omega_1 x + \omega_2 y + \omega_3 z)}\,d\omega_3 d\omega_2 d\omega_1 \qquad (9.184)$$

The solution given by equation (9.184) requires integration over all of the $\omega_1, \omega_2, \omega_3$ space. One finds that by changing this integral to spherical coordinates that the integral can be evaluated. Define the quantities

$$\begin{array}{ll} \vec{r} = (x, y, z) & \vec{\omega} = (\omega_1, \omega_2, \omega_3) \\ \vec{\xi} = (\xi, \eta, \zeta) & \rho = |\vec{r} - \vec{\xi}| \end{array} \qquad \omega = |\vec{\omega}| \qquad (9.185)$$

and introduce the spherical coordinates (ω, θ, ϕ) where $0 < \theta < \pi$ and $0 < \phi < 2\pi$.

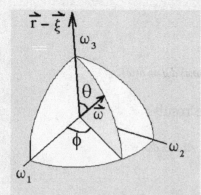

One can introduce the spherical coordinates such that the direction ω_3 is in the direction $\vec{r} - \vec{\xi}$ as illustrated. We then have

$$\vec{\omega} \cdot (\vec{r} - \vec{\xi}) = \omega \rho \cos \theta$$

$$d\omega_1 d\omega_2 d\omega_3 = \omega^2 \sin \theta \, d\theta d\phi d\omega$$

The integral given by equation (9.184) can now be written in the form

$$G = \frac{1}{(2\pi)^3} \int_{-\infty}^{\infty} \int_{-\infty}^{\infty} \int_{-\infty}^{\infty} \frac{1}{c\omega} e^{-i\vec{\omega} \cdot (\vec{r} - \vec{\xi})} \sin c\omega(t - \tau) \, d\omega_3 d\omega_2 d\omega_3 \qquad (9.186)$$

which in spherical coordinates becomes

$$G = \frac{1}{c(2\pi)^3} \int_0^{\infty} \int_0^{2\pi} \int_0^{\pi} \frac{1}{\omega} e^{-i\omega\rho \cos \theta} \sin c\omega(t - \tau)\omega^2 \sin \theta \, d\theta d\phi d\omega. \qquad (9.187)$$

The inner integral gives

$$\int_0^{\pi} e^{-i\omega\rho \cos \theta} \omega \sin \theta \, d\theta = \frac{1}{i\rho} e^{-i\omega\rho \cos \theta} \Big|_0^{\pi} = \frac{2}{\rho} \sin \omega\rho. \qquad (9.188)$$

The middle integral gives $\int_0^{2\pi} d\phi = 2\pi$ and so the integral given by equation (9.187) can be reduced to the form

$$G = \frac{2}{c\rho(2\pi)^2} \int_0^{\infty} \sin c\omega(t - \tau) \sin \omega\rho \, d\omega \qquad (9.189)$$

The product of sine functions can be written as a difference of cosine functions and so the integral in equation (9.189) can also be written in the form

$$G = \frac{1}{c\rho(2\pi)^2} \int_0^{\infty} [\cos \omega(\rho - c(t - \tau)) - \cos \omega(\rho + c(t - \tau))] \, d\omega. \qquad (9.190)$$

Observe that the cosine function is an even function and so the integral (9.190) has the alternate form

$$G = \frac{1}{2c\rho(2\pi)^2} \int_{-\infty}^{\infty} [\cos \omega(\rho - c(t - \tau)) - \cos \omega(\rho + c(t - \tau))] \, d\omega. \qquad (9.191)$$

The Fourier exponential transform table 7-4 gives the result

$$\mathcal{F}_e^{-1}\left\{\frac{1}{2\pi} e^{-i\omega x_0}\right\} = \int_{-\infty}^{\infty} \frac{1}{2\pi} e^{i\omega x_0} e^{-i\omega x} \, d\omega = \delta(x - x_0)$$

and in the special case $x_0 = 0$ we have

$$\int_{-\infty}^{\infty} \frac{1}{2\pi} e^{-i\omega x} \, d\omega = \frac{1}{2\pi} \int_{-\infty}^{\infty} (\cos \omega x - i \sin \omega x) \, d\omega = \delta(x).$$

We equate the real and imaginary parts and use the results

$$\frac{1}{2\pi} \int_{-\infty}^{\infty} \cos \omega x \, d\omega = \delta(x), \qquad \frac{1}{2\pi} \int_{-\infty}^{\infty} \sin \omega x \, d\omega = 0$$

to simplify the equation (9.191) to the final form

$$G = G(x,y,z,t;\xi,\eta,\zeta,\tau) = \frac{1}{4\pi c \rho} \left[\delta(\rho - c(t-\tau)) - \delta(\rho + c(t-\tau)) \right].$$

The second delta function is zero for $\rho > 0$ and $t > \tau$ and so the free-space Green's function for the three-dimensional wave operator can be represented

$$G = G(x,y,z,t;\xi,\eta,\zeta,\tau) = \frac{1}{4\pi c \rho} \delta(\rho - c(t-\tau)) \qquad (9.192)$$

where

$$\rho = |\vec{r} - \vec{\xi}| = \sqrt{(x-\xi)^2 + (y-\eta)^2 + (z-\zeta)^2}. \qquad (9.193)$$

This result can also be represented in the form

$$G = \frac{1}{4\pi c |\vec{r} - \vec{\xi}|} \delta(|\vec{r} - \vec{\xi}| - c(t-\tau)).$$

The previous results are summarized by the following table 9-4.

Table 9-4.
Free-space Green's function for the wave equation

Dimension	Green's function
$n = 1$	$G(x,t;\xi,\tau) = \frac{1}{2c} H(c(t-\tau) - |x - \xi|)$
$n = 2$	$G(x,y,t;\xi,\eta,\tau) = \frac{1}{2\pi c \sqrt{c^2(t-\tau)^2 - |\vec{r} - \vec{\xi}|}} H(c(t-\tau) - |\vec{x} - \vec{\xi}|)$ $\vec{x} = (x,y), \qquad \vec{\xi} = (\xi,\eta)$
$n = 3$	$G(x,y,z,t;\xi,\eta,\zeta,\tau) = \frac{1}{4\pi c |\vec{x} - \vec{\xi}|} \delta(|\vec{x} - \vec{\xi}| - c(t-\tau))$ $\vec{x} = (x,y,z), \qquad \vec{\xi} = (\xi,\eta,\zeta)$

Example 9-17. (Method of images.)

Solve the one-dimensional heat equation over the semi-infinite region $0 < x < \infty$ given by

$$\frac{\partial u}{\partial t} - \alpha^2 \frac{\partial^2 u}{\partial x^2} = F(x,t), \quad 0 < x < \infty, \quad t > 0$$

subject to the boundary condition $u(0,t) = f(t)$, with $|u(x,t)| < M$ as $x \to \infty$, and initial condition $u(x,0) = g(x)$.

Solution: The Green's function for this problem $G(x,t;\xi,\tau) = G^*(\xi,\tau;x,t)$ can be associated with the region R given in the figure 9-3 in the limit as $\ell \to \infty$. The Green's function is to satisfy the equation

$$L_{\xi\tau}^* = -\frac{\partial G^*}{\partial \tau} - \alpha^2 \frac{\partial^2 G^*}{\partial x^2} = \delta(\xi - x)\delta(\tau - t) \tag{9.194}$$

and the Green's identity for the region R from figure 9-3 is obtain from equation (9.25) in the limit as $\ell \to \infty$. We write the modified equation (9.25) in the form

$$u(x,t) = \lim_{\ell \to \infty} \int_0^t \int_0^\ell G^*(\xi,\tau;x,t) L_{\xi\tau}(u) \, d\xi d\tau$$
$$+ \lim_{\ell \to \infty} \oint_C \left\{ \alpha^2 \left[G^* \frac{\partial u}{\partial \xi} - u \frac{\partial G^*}{\partial \xi} \right] \hat{\imath} - uG^*\hat{\jmath} \right\} \cdot \hat{n} \, ds. \tag{9.195}$$

Expand the line integral and write the equation (9.195) in the form

$$u(x,t) = \int_0^T \int_0^\infty G^*(\xi,\tau;x,t) F(\xi,\tau) \, d\xi d\tau + \int_0^\infty uG^* \Big|_{\tau=0} d\xi$$
$$+ \lim_{\ell \to \infty} \int_0^T \alpha^2 \left[G^* \frac{\partial u}{\partial \xi} - u \frac{\partial G^*}{\partial \xi} \right]_{\xi=\ell} d\tau + \int_\infty^0 uG^* \Big|_{\tau=T} (-d\xi)$$
$$+ \int_T^0 -\alpha^2 \left[G^* \frac{\partial u}{\partial \xi} - u \frac{\partial G^*}{\partial \xi} \right]_{\xi=0} (-d\tau).$$

We can now analyze each term to determine the adjoint boundary conditions.

(a) On the line $\tau = 0$ we have $u(\xi,0) = g(\xi)$ as the boundary condition and so the line integral along this portion of the boundary becomes $\int_0^\infty g(\xi)G^*(\xi,0;x,t) \, d\xi$.

(b) On the line $\xi = \ell$ we want $G^* \to 0$ and $\frac{\partial G^*}{\partial \xi} \to 0$ as $\ell \to \infty$ and so this portion of the line integral vanishes.

(c) On the line $\tau = T$ we want G^* to be zero since $u(\xi,T)$ is not known on this boundary. This condition is satisfied since we have required that $G = 0$ for all $t < \tau$.

(d) On the line $\xi = 0$, we have the boundary condition $u(0,\tau) = f(\tau)$. The term $\frac{\partial u}{\partial \xi}\big|_{\xi=0}$ is not known and so we require that $G^*\big|_{\xi=0} = 0$. Note that $G^* = 0$ for all $\tau > t$ and so this portion of the line integral can be written $\int_0^t \alpha^2 f(\tau) \frac{\partial G^*}{\partial \xi}\big|_{\xi=0} d\tau$. One can now express the solution in the form

$$u(x,t) = \int_0^t \int_0^\infty G^*(\xi,\tau;x,t)F(\xi,\tau)\, d\xi d\tau + \int_0^\infty g(\xi)G^*(\xi,0;x,t)\, d\xi$$
$$+\alpha^2 \int_0^t f(\tau)\frac{\partial G^*(0,\tau;x,t)}{\partial \xi}\, d\tau \tag{9.196}$$

This solution can also be written in the form

$$u(x,t) = \int_0^t \int_0^\infty G(x,t;\xi,\tau)F(\xi,\tau)\, d\xi d\tau + \int_0^\infty g(\xi)G(x,t;\xi,0)\, d\xi$$
$$+\alpha^2 \int_0^t f(\tau)\frac{\partial G(x,t;0,\tau)}{\partial \xi}\, d\tau \tag{9.197}$$

The equation (9.197) is the representation of the solution to the given Dirichlet problem provided one can construct the appropriate Green's function subject to the boundary condition that $G = 0$ at $x = 0$. Recall that the one-dimensional free-space Green's function for the heat equation on $-\infty < x < \infty$ is given by

$$G_f(x,t;\xi,\tau) = \frac{1}{2\alpha\sqrt{\pi(t-\tau)}} \exp\left[-\frac{(x-\xi)^2}{4\alpha^2(t-\tau)}\right] H(t-\tau) \tag{9.198}$$

and is illustrated in the figure 9-8 for various values of $t > \tau$.

Recall that one can construct Green's functions in the form $G = G_f + G_r$ where G_f is the free-space Green's function and G_r is called the regular part and selected to satisfy appropriate boundary conditions. Here we can use G_f as a first guess to the Green's function for our problem. However, G_f is not zero at $x = 0$ for $t > \tau$ and so we can select a regular part to achieve this result. This can be accomplished by using a negative image source at $x = -\xi$ as illustrated in the figure 9-9.

By selecting $G_r = -G_f(x,t;-\xi,\tau)$ as the negative image of the free-space Green's function, one can achieve the desired value of $G = 0$ at $x = 0$. This gives the desired Green's function for the original Dirichlet problem as

$$G(x,t;\xi,\tau) = \frac{1}{2\alpha\sqrt{\pi(t-\tau)}} \left\{ \exp\left[-\frac{(x-\xi)^2}{4\alpha^2(t-\tau)}\right] - \exp\left[-\frac{(x+\xi)^2}{4\alpha^2(t-\tau)}\right] \right\} H(t-\tau) \tag{9.199}$$

Note that over the region $0 < x < \infty$ we have $L_{xt}(G_f) = \delta(x - \xi)\delta(t - \tau)$ and $L_{xt}(G_r) = 0$. That is, L_{xt} operating on the free-space Green's function produces the Dirac delta function and L_{xt} operating on the regular part of the Green's function gives zero since the impulse is outside the domain of interest. It is the regular part which modifies the free-space Green's function to achieve the proper boundary condition.

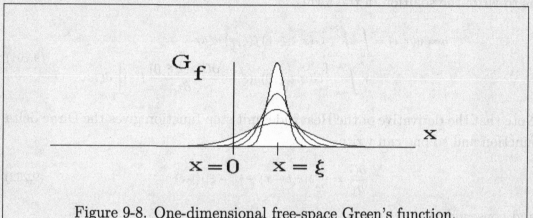

Figure 9-8. One-dimensional free-space Green's function.

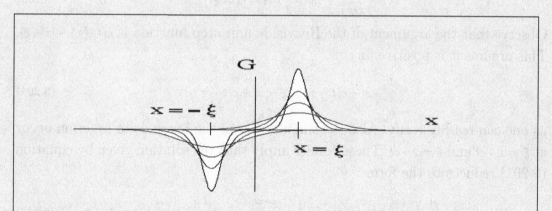

Figure 9-9. Negative image source added to free-space Green's function.

Example 9-18.   (Wave equation.)

Use Green's functions to solve the one-dimensional wave equation

$$\frac{\partial^2 u}{\partial t^2} - c^2 \frac{\partial^2 u}{\partial x^2} = F(x,t), \quad -\infty < x < \infty, \quad t > 0$$

subject to the initial conditions $u(x,0) = f(x)$ and $\dfrac{\partial u(x,0)}{\partial t} = g(x)$ over the interval $-\infty < x < \infty$.

Solution: The table 9-4 gives the free-space one-dimensional Green's function

$$G = G(x,t;\xi,\tau) = \frac{1}{2c}H(c(t-\tau) - |x-\xi|) \tag{9.200}$$

where H is the Heaviside unit step function. We use the region R illustrated in the figure 9-4 and take the limit as $\ell \to \infty$ and $T \to \infty$. The equation (9.33) allows us to write the solution in the form

$$u = u(x,t) = \int_0^t \int_{-\infty}^{\infty} G(x,t;\xi,\tau)f(\xi,\tau)\,d\xi d\tau$$
$$+ \int_{-\infty}^{\infty} \left[G(x,t;\xi,0)g(\xi) - \frac{\partial G(x,t;\xi,0)}{\partial \tau}f(\xi) \right]\,d\xi. \tag{9.201}$$

Note that the derivative of the Heaviside unit step function gives the Dirac delta function and so one can write

$$\frac{\partial G}{\partial \tau} = \frac{1}{2c}\delta(c(t-\tau) - |x-\xi|)(-c) \tag{9.202}$$

and consequently

$$\left.\frac{\partial G}{\partial \tau}\right|_{\tau=0} = -\frac{1}{2}\delta(ct - |x-\xi|). \tag{9.203}$$

Observe that the argument of the Heaviside unit step function is $c(t-\tau) - |x-\xi|$. This argument is positive for

$$x - c(t-\tau) < \xi < x + c(t-\tau) \tag{9.204}$$

as one can readily verify. Also the singularities of the Dirac delta function occur at $\xi = x - ct$ and $\xi = x + ct$. These results imply that the solution given by equation (9.201) reduces to the form

$$u = u(x,t) = \frac{1}{2c}\int_0^t \int_{x-ct}^{x+ct} F(\xi,\tau)\,d\xi d\tau + \frac{1}{2c}\int_{x-ct}^{x+ct} g(\xi)\,d\xi + \frac{1}{2}\left[f(x-ct) + f(x+ct) \right] \tag{9.205}$$

which is recognized as the D'Alembert solution previously considered.

∎

Example 9-19.   (Wave equation.)
The solution of the two-dimensional wave equation

$$\frac{\partial^2 u}{\partial t^2} = c^2 \left(\frac{\partial^2 u}{\partial x^2} + \frac{\partial^2 u}{\partial y^2} \right), \quad -\infty < x, y < \infty$$

subject to the initial conditions $u(x,y,0) = f(x,y)$ and $\dfrac{\partial u(x,y,0)}{\partial t} = g(x,y)$ is obtained from equation (9.57) and can be written

$$u(x,y,t) = \int_{-\infty}^{\infty}\int_{-\infty}^{\infty}\left[g(\xi,\eta)G - f(\xi,\eta)\frac{\partial G}{\partial \tau}\right]_{\tau=0} d\xi d\eta$$

where G is found from table 9-4 for the case $n = 2$. Note that $\dfrac{\partial G}{\partial \tau} = -\dfrac{\partial G}{\partial t}$ so that the solution can be represented in the form

$$u(x,y,t) = \frac{1}{2\pi c}\int_{-\infty}^{\infty}\int_{-\infty}^{\infty}\frac{g(\xi,\eta)H(ct - |\vec{r}-\vec{\xi}|)\,d\xi d\eta}{[c^2t^2 - (x-\xi)^2 - (y-\eta)^2]^{1/2}}$$
$$+ \frac{1}{2\pi c}\frac{\partial}{\partial t}\int_{-\infty}^{\infty}\int_{-\infty}^{\infty}\frac{f(\xi,\eta)H(ct - |\vec{r}-\vec{\xi}|)\,d\xi d\eta}{[c^2t^2 - (x-\xi)^2 - (y-\eta)^2]^{1/2}}$$

which reduces to the solution given by equation (5.105) previously considered. ∎

Example 9-20. (Wave equation.)
Solve the three-dimensional wave equation for $u = u(x,y,z,t)$ which satisfies

$$\frac{\partial^2 u}{\partial t^2} - c^2\nabla^2 u = \frac{\partial^2 u}{\partial t^2} - c^2\left(\frac{\partial^2 u}{\partial x^2} + \frac{\partial^2 u}{\partial y^2} + \frac{\partial^2 u}{\partial z^2}\right) = F(x,y,z,t)$$

for $-\infty < x,y,z < \infty$, and subject to the initial conditions

$$u(x,y,z,0) = 0 \quad \text{and} \quad \frac{\partial u(x,y,z,0)}{\partial t} = 0.$$

Solution: Using the free-space Green's function from the table 9-4, the solution is obtain from the equation (9.60) using a limiting process and can be written

$$u = u(x,y,z,t) = \frac{1}{4\pi c}\int_0^t\int_{-\infty}^{\infty}\int_{-\infty}^{\infty}\int_{-\infty}^{\infty}\frac{F(\xi,\eta,\zeta,\tau)}{|\vec{r}-\vec{\xi}|}\delta(|\vec{r}-\vec{\xi}| - c(t-\tau))\,d\xi d\eta d\zeta d\tau \quad (9.206)$$

The argument of the Dirac delta function is zero when

$$|\vec{r}-\vec{\xi}| - ct + c\tau = 0 \quad \text{or} \quad \tau = t - \frac{1}{c}|\vec{r}-\vec{\xi}|.$$

This is the time at which the effect of the disturbance at time $\tau = t$ is first experienced. By the sifting property of the Dirac delta function the solution given by equation (9.206) simplifies to

$$u = u(x,y,z,t) = \frac{1}{4\pi c}\int_{-\infty}^{\infty}\int_{-\infty}^{\infty}\int_{-\infty}^{\infty}\frac{F(\xi,\eta,\zeta,t - \frac{1}{c}|\vec{r}-\vec{\xi}|)}{|\vec{r}-\vec{\xi}|}\,d\xi d\eta d\zeta \quad (9.207)$$

This type of solution is called a retarded potential since the integrand is evaluated at the delay time $\tau = t - \frac{1}{c}|\vec{r}-\vec{\xi}|$ rather than at the time $\tau = t$. Here the quantity $\frac{1}{c}|\vec{r}-\vec{\xi}|$ represents the time the disturbance takes to travel, at speed c, the distance $|\vec{r}-\vec{\xi}|$. ∎

Exercises 9

▶ **1.**

Let $L_{xt}(u) = \dfrac{\partial u}{\partial t} - \alpha^2 \dfrac{\partial^2 u}{\partial x^2}$, $\quad x, t \in R$, where $R = \{x, t \mid -\ell \le x \le \ell, t > 0\}$ is the region given in figure 9-4.

(a) Find $L_{xt}^*(v)$

(b) Show the Green's formula for the region R becomes

$$\int\int_R \left[v L_{\xi\tau}(u) - u L_{\xi\tau}^*(v) \right] d\xi d\tau = \oint_C -uv\, d\xi - \alpha^2 \left[v\frac{\partial u}{\partial \xi} - u\frac{\partial v}{\partial \xi} \right] d\tau \qquad (9.208)$$

▶ **2.**

Let $L_{xt}(u) = \dfrac{\partial^2 u}{\partial t^2} - \alpha^2 \dfrac{\partial^2 u}{\partial x^2}$, $\quad x, t \in R$, where R is the region in figure 9-4.

(a) Find $L_{xt}^*(v)$

(b) Show the Green's formula for the region R of figure 9-4 can be expressed

$$\int\int_R \left[v L_{\xi\tau}(u) - u L_{\xi\tau}^*(v) \right] d\xi d\tau = \oint_C \left[(u\frac{\partial v}{\partial \tau} - v\frac{\partial u}{\partial \tau})\, d\xi + \alpha^2 (u\frac{\partial v}{\partial \xi} - v\frac{\partial u}{\partial \xi})\, d\tau \right] \qquad (9.209)$$

▶ **3.**

Solve the eigenvalue problem $L_{xy}(\phi) = \dfrac{\partial^2 \phi}{\partial x^2} + \dfrac{\partial^2 \phi}{\partial y^2} = \lambda\phi$ for $x, y \in R$ where the region R is the rectangle $R = \{x, y \mid 0 \le x \le a, 0 \le y \le b\}$ subject to the boundary conditions that ϕ is zero on the boundary of the region R. Show there exists eigenvalues

$$\lambda = \lambda_{mn} = -\pi^2 \left(\frac{n^2}{a^2} + \frac{m^2}{b^2} \right)$$

and eigenfunctions

$$\phi = \phi_{mn} = \sin(\frac{m\pi x}{a}) \sin(\frac{n\pi y}{b}), \quad m, n = 1, 2, \dots$$

which satisfy the orthogonality conditions

$$(\phi_{mn}, \phi_{jk}) = \int_0^a \int_0^b \phi_{nm} \phi_{jk}\, dx dy = \begin{cases} 0, & (m, n) \ne (i, j) \\ \parallel \phi_{mn} \parallel^2 = \frac{ab}{4}, & (m, n) = (i, j) \end{cases}$$

▶ **4.**

Solve the Green's function problem

$$L_{xy}(G) = \frac{\partial^2 G}{\partial x^2} + \frac{\partial^2 G}{\partial y^2} = \delta(x - \xi)\delta(y - \eta)$$

for $x, y \in R = \{x, y \,|\, 0 \le x \le a, 0 \le y \le b\}$ satisfying $G = 0$ on the boundary of the region R by assuming a solution

$$G = G(x, y; \xi, \eta) = \sum_{m=1}^{\infty} \sum_{n=1}^{\infty} B_{nm}\phi_{nm}(x, y),$$

where the functions ϕ_{nm} are given in the previous problem.

▶ **5.**

(a) Consider the one dimensional Sturm–Liouville problem

$$\frac{d^2 v_n}{dx^2} = \lambda_n v_n, \quad x_0 < x < x_1$$

with boundary conditions

$$A_0 \frac{dv_n}{dx} + B_0 v_n = 0 \quad \text{at } x = x_0, \quad \text{and} \quad E_0 \frac{dv_n}{dx} + F_0 v_n = 0 \quad \text{at } x = x_1$$

for $n = 0, 1, 2, \ldots$, and where λ_n are the eigenvalues and v_n are the corresponding eigenfunctions. Also consider the one dimensional Sturm–Liouville problem

$$\frac{d^2 w_m}{dy^2} = \mu_m w_m, \quad y_0 < y < y_1$$

with boundary conditions

$$A_1 \frac{dw_m}{dy} + B_1 w_m = 0 \quad \text{at } y = y_0 \quad \text{and} \quad E_1 \frac{dw_m}{dy} + F_1 w_m = 0 \quad \text{at } y = y_1,$$

where μ_m are the eigenvalues and w_m are the eigenfunctions. Show the functions

$$u_{nm}(x, y) = v_n(x)w_m(y)$$

satisfy the partial differential equation

$$L_{xy}(u_{nm}) = \frac{\partial^2 u_{nm}}{\partial x^2} + \frac{\partial^2 u_{nm}}{\partial y^2} = (\lambda_n + \mu_m)u_{nm}.$$

(b) To determine the Green's function which satisfies

$$L_{xy}(G) = \frac{\partial^2 G}{\partial x^2} + \frac{\partial^2 G}{\partial y^2} = \delta(x-\xi)\delta(y-\eta), \quad x, y \in R = \{x, y \,|\, x_0 \le x \le x_1, \ y_0 \le y \le y_1\}$$

assume a solution

$$G(x, y) = \sum_{n=0}^{\infty} \sum_{m=0}^{\infty} A_{nm} v_n(x) w_m(y)$$

and show

$$G = G(x, y; \xi, \eta) = \sum_{m=0}^{\infty} \sum_{n=0}^{\infty} \frac{v_n(\xi) v_n(x) w_m(\eta) w_m(y)}{(\lambda_n + \mu_m) \,\|\, v_n \,\|^2 \|\, w_m \,\|^2}$$

provided that $(\lambda_n + \mu_m) \ne 0$. What boundary conditions does G satisfy?

▶ **6.**

(a) Solve the Sturm–Liouville problem $\dfrac{d^2 v_n}{dx^2} = \lambda_n v_n$, $\quad 0 \le x \le a$ with boundary conditions $v_n(0) = v_n(a) = 0$.

(b) Solve the Sturm–Liouville problem $\dfrac{d^2 w_m}{dy^2} = \mu_m w_m$, $\quad 0 \le y \le b$ with boundary conditions $w_m(0) = w_m(b) = 0$

(c) Solve the Green's function problem

$$L_{xy}(G) = \frac{\partial^2 G}{\partial x^2} + \frac{\partial^2 G}{\partial y^2} = \delta(x - \xi)\delta(y - \eta), \quad x, y \in R,$$

where R is the rectangular region $0 \le x \le a$ and $0 \le y \le b$ with $G = 0$ on the boundary of R.

▶ **7.**

Consider the Green's function problem

$$L_{xy}(G) = \nabla^2 G = \frac{\partial^2 G}{\partial x^2} + \frac{\partial^2 G}{\partial y^2} = \delta(x - \xi)\delta(y - \eta)$$

for x, y in the rectangular region $0 \le x \le a, 0 \le y \le b$ and subject to the condition $G = 0$ on the boundary of R.

(a) Expand $\delta(x-\xi)$, $\quad 0 < x < a$ in a Fourier sine series of the form $\sum b_n \sin(\frac{n\pi x}{a})$ and then assume a solution to the Green's function problem of the form

$$G = \sum_{n=1}^{\infty} w_n(y) \sin(\frac{n\pi x}{a})$$

and show w_n must satisfy $\qquad \dfrac{d^2 w_n}{dy^2} - \dfrac{n^2 \pi^2}{a^2} w_n = b_n \delta(y - \eta), \quad 0 < y < b$

with $w_n(0) = w_n(b) = 0$. Solve this equation using Laplace transforms and find the Green's function.

(b) Expand $\delta(y - \eta)$ in a Fourier sine series of the form $\sum c_m \sin(\frac{m \pi y}{b})$ and assume a solution to the Green's function problem of the form

$$G = \sum_{m=1}^{\infty} v_m(x) \sin(\frac{m \pi y}{b})$$

and show v_m must satisfy $\qquad \dfrac{d^2 v_m}{dx^2} - \dfrac{m^2 \pi^2}{b^2} v_m = c_m \delta(x - \xi), \quad 0 < x < a$

with $v_m(0) = v_m(a) = 0$. Solve this equation using Laplace transforms and find the Green's function.

▶ 8.

Solve the boundary value problem

$$L_{xy}(u) = \nabla^2 u = \frac{\partial^2 u}{\partial x^2} + \frac{\partial^2 u}{\partial y^2} = f(x, y), \quad x, y \in R$$

with R the rectangular region $0 \le x \le a$, $0 \le y \le b$, where u is subject to the condition $u = 0$ on the boundary of R.

▶ 9.

(a) Solve the Dirichlet problem

$$L_{xy}(u) = \nabla^2 u = \frac{\partial^2 u}{\partial x^2} + \frac{\partial^2 u}{\partial y^2} = f(x, y), \quad x, y \in R,$$

where R is the rectangular region $0 \le x \le a$, $0 \le y \le b$ and where the solution is subject to the condition $u = g(x, y)$ for $x, y \in \partial R$.

(b) Illustrate the results of part (a) by solving

$$L_{xy}(u) = \nabla^2 u = \frac{\partial^2 u}{\partial x^2} + \frac{\partial^2 u}{\partial y^2} = f(x, y), \quad x, y \in R$$

with the following boundary conditions:

$$u(0, y) = T_4, \qquad u(x, 0) = T_1, \qquad u(a, y) = T_2, \qquad u(x, b) = T_3,$$

where T_1, T_2, T_3, T_4 are constants. Give a physical interpretation of this problem. Hint: Consider the line integral around the region illustrated in figure 9-10.

530

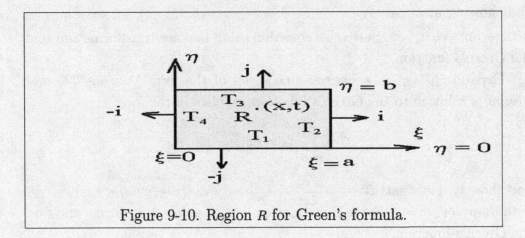

Figure 9-10. Region R for Green's formula.

▶ 10.

Show the eigenvalue problem

$$L_{r\theta}(u) = \nabla^2 u = \frac{\partial^2 u}{\partial r^2} + \frac{1}{r}\frac{\partial u}{\partial r} + \frac{1}{r^2}\frac{\partial^2 u}{\partial \theta^2} = -\lambda^2 u$$

for $r, \theta \in R = \{r, \theta \,|\, 0 < r < a, \, 0 \le \theta \le 2\pi\}$ with the boundary conditions

$$u(a, \theta) = 0, \quad u(r, \theta + 2\pi) = u(r, \theta)$$

possesses the eigenvalues $\lambda = \lambda_{nm} = \dfrac{\alpha_{nm}}{a}$, with $J_n(\alpha_{nm}) = 0$, where α_{nm} are the zeros of the nth order Bessel function $J_n(x)$ and the eigenfunctions are

$$c_{nm} = J_n(\frac{\alpha_{nm}r}{a})\cos(n\theta) \qquad s_{nm} = J_n(\frac{\alpha_{nm}r}{a})\sin(n\theta)$$

for $n = 0, 1, 2, \dots$ and $m = 1, 2, \dots$. Show these eigenfunctions have the inner products

$$(c_{nm}, c_{jk}) = \int_0^a \int_0^{2\pi} c_{nm}c_{jk}\, r\,dr\,d\theta = 0, \quad (n, m) \ne (j, k)$$

$$(s_{nm}, s_{jk}) = \int_0^a \int_0^{2\pi} s_{nm}s_{jk}\, r\,dr\,d\theta = 0, \quad (n, m) \ne (j, k)$$

$$(c_{nm}, s_{nm}) = \int_0^a \int_0^{2\pi} c_{nm}s_{jk}\, r\,dr\,d\theta = 0, \quad \text{for all } n, m, j, k$$

and the square norms

$$\| c_{nm} \|^2 = (c_{nm}, c_{nm}) = \frac{\pi a^2}{2}J_{n+1}^2(\alpha_{nm})$$

$$\| s_{nm} \|^2 = (s_{nm}, s_{nm}) = \frac{\pi a^2}{2}J_{n+1}^2(\alpha_{nm})$$

$$\| c_{0m} \|^2 = (c_{0m}, c_{0m}) = \pi a^2 J_1^2(\alpha_{0m})$$

for $n, m = 1, 2, 3, \dots$

► **11.**

Solve the Green's function problem

$$L_{r\theta}(G) = \nabla^2 G = \frac{\partial^2 G}{\partial r^2} + \frac{1}{r}\frac{\partial G}{\partial r} + \frac{1}{r^2}\frac{\partial^2 G}{\partial \theta^2} = \frac{\delta(r - r_0)\delta(\theta - \theta_0)}{r}$$

for $0 \le r \le a$, $0 \le \theta \le 2\pi$ and for G satisfying $G(a, \theta; r_0, \theta_0) = 0$. Assume a solution

$$G(r, \theta; r_0, \theta_0) = \sum_{n=0}^{\infty}\sum_{m=1}^{\infty} A_{nm}c_{nm} + B_{nm}s_{nm},$$

where c_{nm} and s_{nm} are given in problem 10. Show the coefficients are given by

$$A_{0m} = -\frac{J_0\left(\frac{\alpha_{0m}r_0}{a}\right)}{\pi a^2 \lambda_{0m}^2 J_1^2(\alpha_{0m})}, \quad m = 1, 2, 3, \ldots$$

$$A_{mn} = -\frac{2J_n\left(\frac{\alpha_{nm}r_0}{a}\right)\cos(n\theta_0)}{\pi a^2 \lambda_{nm}^2 J_{n+1}^2(\alpha_{nm})}, \quad n, m = 1, 2, 3, \ldots$$

$$B_{mn} = -\frac{2J_n\left(\frac{\alpha_{nm}r_0}{a}\right)\sin(n\theta_0)}{\pi a^2 \lambda_{nm}^2 J_{n+1}^2(\alpha_{nm})},$$

where $\lambda_{nm} = \frac{\alpha_{nm}}{a}$ and α_{nm} is the mth zero of the nth order Bessel function of the first kind.

► **12.**

Use the results of problem 11 and the appropriate Green's formula to solve the boundary value problem

$$L_{r\theta}(u) = \nabla^2 u = \frac{\partial^2 u}{\partial r^2} + \frac{1}{r}\frac{\partial u}{\partial r} + \frac{1}{r^2}\frac{\partial^2 u}{\partial \theta^2} = F(r, \theta)$$

for $r, \theta \in R = \{r, \theta \mid 0 \le r \le a, \ 0 \le \theta \le 2\pi\}$ with the boundary condition $u(a, \theta) = f(\theta)$.

► **13.**

Consider the Green's function problem

$$L_{xt}(G) = \frac{\partial G}{\partial t} - \alpha^2 \frac{\partial^2 G}{\partial x^2} = \delta(x - \xi)\delta(t - \tau)$$

for $0 < x < a$, $t > \tau$ and satisfying

$$G = 0 \quad \text{at} \quad x = 0, \ t > \tau \qquad (13a)$$

$$G = 0 \quad \text{at} \quad x = a$$

$$G = 0 \quad \text{for} \quad 0 \le x \le a, \ t < \tau$$

which represents a solution due to an instantaneous heat source of strength unity applied at $x = \xi$ at the time $t = \tau$.

(a) Assume a solution to equation (13a) of the form

$$G = \sum_{n=1}^{\infty} z_n(t) \sin\left(\frac{n\pi x}{a}\right)$$

and expand the delta function as a Fourier sine series

$$\delta(x - \xi) = \sum_{n=1}^{\infty} b_n \sin\left(\frac{n\pi x}{a}\right)$$

and show

$$\frac{dz_n}{dt} + \alpha^2 \omega_n^2 z_n = \frac{2}{a} \sin(\omega_n \xi)\delta(t - \tau),$$

where $\omega_n = \frac{n\pi}{a}$, $n = 1, 2, 3, \ldots$ Solve this equation using Laplace transform methods and show

$$G(x, t; \xi, \tau) = \frac{2}{a} \sum_{n=1}^{\infty} e^{-\omega_n^2 \alpha^2 (t - \tau)} \sin(\omega_n x) \sin(\omega_n \xi) H(t - \tau)$$

(b) Solve the boundary value problem

$$L_{xt}(u) = \frac{\partial u}{\partial t} - \alpha^2 \frac{\partial^2 u}{\partial x^2} = 0, \quad 0 \le x \le a, \ t > 0,$$

where $u(0, t) = 0$ and $u(a, t) = 0$ on the boundaries and the initial condition is $u(x, 0) = f(x)$

(c) Show the solution of the boundary value problem

$$L_{xt}(u) = \frac{\partial u}{\partial t} - \alpha^2 \frac{\partial^2 u}{\partial x^2} = F(x, t), \quad 0 \le x \le a, \ t > 0$$

$$u(0, t) = \phi_1(t), \quad u(a, t) = \phi_2(t), \quad u(x, 0) = f(x)$$

is given by

$$u(x, t) = \int_0^t \int_0^a G(x, t; \xi, \tau) F(\xi, \tau) \, d\xi d\tau + \int_0^a G(x, t; \xi, 0) f(\xi) \, d\xi$$

$$+ \alpha^2 \int_0^t \left[\frac{\partial G(x, t; \xi, \tau)}{\partial \xi} \bigg|_{\xi=0} \phi_1(\tau) - \frac{\partial G(x, t; \xi, \tau)}{\partial \xi} \bigg|_{\xi=a} \phi_2(\tau) \right] d\tau,$$

where G is given in part (a).

▶ **14.**

Proceed as in problem 13 and solve:

(a) The Green's function problem

$$L_{xt}(G) = \frac{\partial G}{\partial t} - \alpha^2 \frac{\partial^2 G}{\partial x^2} = \delta(x - \xi)\delta(t - \tau), \quad 0 < x < a, \ t > \tau$$

$$\frac{\partial G}{\partial x} = 0 \quad \text{at} \quad x = 0, \quad t > \tau$$

$$\frac{\partial G}{\partial x} = 0 \quad \text{at} \quad x = a, \quad t > \tau$$

$$G = 0 \quad \text{for} \quad 0 < x < a, \ t < \tau$$

and show

$$G(x, t; \xi, \tau) = \left[\frac{1}{a} + \frac{2}{a} \sum_{n=1}^{\infty} e^{-\omega^2 \alpha^2 (t - \tau)} \cos(\omega_n x) \cos(\omega_n \xi) \right] H(t - \tau),$$

where $\omega_n = \frac{n\pi}{a}$, $n = 1, 2, \ldots$

(b) Find the solution of the boundary value problem

$$\frac{\partial u}{\partial t} - \alpha^2 \frac{\partial^2 u}{\partial x^2} = F(x, t), \quad 0 < x < a, \ t > 0$$

satisfying $\quad \dfrac{\partial u(0, t)}{\partial x} = \phi_1(t), \qquad \dfrac{\partial u(a, t)}{\partial x} = \phi_2(t), \qquad u(x, 0) = f(x)$

▶ **15.**

(a) Solve the Green's function problem

$$\frac{\partial G}{\partial t} - \alpha^2 \frac{\partial^2 G}{\partial x^2} = \delta(x - \xi)\delta(t - \tau), \quad 0 < x < a, \ t > \tau$$

$$\frac{\partial G}{\partial x} = 0 \quad \text{at} \quad x = a$$

$$G = 0 \quad \text{at} \quad x = 0$$

$$G = 0 \quad 0 < x < a, \ t < \tau$$

(b) Solve the boundary value problem

$$\frac{\partial u}{\partial t} - \alpha^2 \frac{\partial^2 u}{\partial x^2} = F(x, t), \quad 0 < x < a, \ t > 0$$

satisfying $\quad \dfrac{\partial u(a, t)}{\partial x} = \phi_2(t), \qquad u(0, t) = \phi_1(t), \qquad u(x, 0) = f(x)$

▶ **16.**

Find the Green's function for the operator

$$L_{xyzt}(u) = \frac{\partial u}{\partial t} - \alpha^2 \left(\frac{\partial^2 u}{\partial x^2} + \frac{\partial^2 u}{\partial y^2} + \frac{\partial^2 u}{\partial z^2} \right)$$

for $\quad x, y, z \in R = \{x, y, z \mid 0 < x < a, \ 0 < y < b, \ 0 < z < c\}$,

where G is zero on the boundary of R.

▶ **17.** (Method of images)

(a) Consider a heat source $\delta(x - \xi)$ applied at $x = \xi$ at time $t = 0$ and a heat sink $-\delta(x + \xi)$ applied at $x = -\xi$ at time $t = 0$. Solve the partial differential equation

$$\frac{\partial u}{\partial t} - \alpha^2 \frac{\partial^2 u}{\partial x^2} = 0, \quad -\infty < x < \infty, \; t > 0$$
$$u(x, 0) = \delta(x - \xi) - \delta(x + \xi) \quad \text{and} \quad u(x, t) = 0, \quad t < 0$$

Hint: use superposition.

(b) Show the solution in (a) satisfies $u(0, t) = 0$ for all $t > 0$ and interpret this result physically.

(c) Solve

$$\frac{\partial u}{\partial t} - \alpha^2 \frac{\partial^2 u}{\partial x^2} = 0, \quad 0 < x < \infty, \; t > 0 \tag{17a}$$

such that $u(0, t) = 0$ and $u(x, 0) = f(x)$, $x > 0$ using the concept that a source at $x = \xi$ of strength $f(\xi)$ can be offset by a sink at $x = -\xi$ of strength $-f(-\xi)$ in order to maintain a zero boundary condition. Show the solution can be written

$$u(x, t) = \frac{1}{2\alpha\sqrt{\pi t}} \left[\int_{-\infty}^{0} -f(-\xi) \exp\left\{ -\frac{(x - \xi)^2}{4\alpha^2 t} \right\} d\xi + \int_{0}^{\infty} f(\xi) \exp\left\{ -\frac{(x - \xi)^2}{4\alpha^2 t} \right\} d\xi \right]$$

(This is equivalent to extending $f(x)$, $x > 0$ as an odd function.)

(d) In the first integral in part (c), replace ξ by $-\xi$ and show

$$u(x, t) = \int_{0}^{\infty} \left[g_1(x, t; \xi, 0) - g_1(x, t; -\xi, 0) \right] f(\xi) \, d\xi \tag{17b}$$

Verify that equation (17b) satisfies the boundary and initial conditions.

▶ **18.** (Method of images)

(a) Consider the heat source $\delta(x - \xi)$ applied at $x = \xi$ at time $t = 0$ and a heat sink $\delta(x + \xi)$ applied at $x = -\xi$ at time $t = 0$. Solve

$$\frac{\partial u}{\partial t} - \alpha^2 \frac{\partial^2 u}{\partial x^2} = 0, \quad -\infty < x < \infty, \; t > 0$$
$$u(x, 0) = \delta(x - \xi) + \delta(x + \xi) \quad \text{and} \quad u(x, 0) = 0, \; t < 0$$

(b) Show the solution in (a) satisfies $\frac{\partial u(0, t)}{\partial x} = 0$ for all $t > 0$

(c) Solve

$$\frac{\partial u}{\partial t} - \alpha^2 \frac{\partial^2 u}{\partial x^2} = 0, \quad -\infty < x < \infty, \; t > 0 \tag{18a}$$

such that $\frac{\partial u(0,t)}{\partial x} = 0$ and $u(x,0) = f(x)$, $x > 0$ using the concept that a source at $x = \xi$ of strength $f(\xi)$ is to be offset by a source of strength $f(-\xi)$ at $x = -\xi$. Thus, the flux at the boundary $x = 0$ due to the source at $x = \xi$ is added to the flux from the image source and the resulting flux in maintained to be zero at all times. Show the solution can be written

$$u(x,t) = \int_0^\infty [g_1(x,t;\xi,0) + g_1(x,t;-\xi,0)] f(\xi)\, d\xi \qquad (18b)$$

This is equivalent to extending $f(x), x < 0$ as an even function. Verify equation (18b) satisfies the boundary and initial conditions.

▶ **19.**

Modify the Green's function solution given by equation (9.113) to solve the boundary value problem

$$\frac{\partial u}{\partial t} - \alpha^2 \frac{\partial^2 u}{\partial x^2} = F(x,t), \quad 0 < x < \infty,\ t > 0$$

such that $u(x,0) = 0$ and $u(0,t) = 0$.

▶ **20.**

Modify the Green's function solution given by equation (9.113) to solve

$$\frac{\partial u}{\partial t} - \alpha^2 \frac{\partial^2 u}{\partial x^2} = F(x,t), \quad 0 < x < \infty,\ t > 0$$

such that $u(x,0) = 0$ and $\frac{\partial u(0,t)}{\partial x} = 0$.

▶ **21.**

A semi infinite rod $x > 0$ is initially at temperature T_0 throughout the rod. For all $t > 0$, the face $x = 0$ is maintained at a zero temperature. Solve the heat equation

$$\frac{\partial u}{\partial t} - \alpha^2 \frac{\partial^2 u}{\partial x^2} = 0, \quad 0 < x < \infty,\ t > 0$$

such that $u(x,0) = T_0$ and $u(0,t) = 0$.
(a) Find $u(x,t)$ by the method of images.
(b) Let $w(x,t) = T_0 - u(x,t)$ and find the boundary value problem satisfied by w and interpret the problem physically.
(c) Find the solution for $w(x,t)$.

▶ **22.**

It was previously shown that $u(x,t) = \frac{1}{2}[f(x+\alpha t) + f(x-\alpha t)]$ is a general solution to the wave equation

$$L_{xt}(u) = \frac{\partial^2 u}{\partial t^2} - \alpha^2 \frac{\partial^2 u}{\partial x^2} = 0$$

satisfying $u(x,0) = f(x)$ and $\dfrac{\partial u(x,0)}{\partial t} = 0, -\infty < x < \infty,\ t > 0$. The lines defined by $x + \alpha t = c_1,\ x - \alpha t = c_2$ with $c_1,\ c_2$ constants, are called characteristic lines associated with the wave equation. To solve the nonhomogeneous equation

$$L_{xt}(u) = \frac{\partial^2 u}{\partial t^2} - \alpha^2 \frac{\partial^2 u}{\partial x^2} = F(x,t), \quad -\infty < x < \infty,\ t > 0$$

satisfying $u(x,0) = f(x),\ \dfrac{\partial u(x,0)}{\partial t} = g(x)$ proceed as follows:

(a) With reference to figure 9-11 show the characteristics through the point (x,t) are given by $\xi + \alpha \tau = x + \alpha t$ and $\xi - \alpha t = x - \alpha t$.

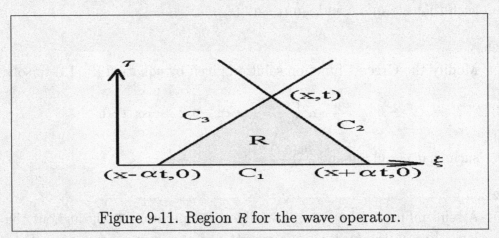

Figure 9-11. Region R for the wave operator.

(b) For the region R in figure 9-11 which is bounded by the lines $c_1,\ c_2$ and c_3 show the Green's formula with $v = 1$ reduces to

$$\iint_R L_{\xi\tau}(u)\,d\xi d\tau = \oint_c \left[-\frac{\partial u}{\partial \tau}\,d\xi - \alpha^2 \frac{\partial u}{\partial \xi}\,d\tau \right] \tag{9.210}$$

(c) on the line c_1 we have $\tau = 0,\ d\tau = 0$ and hence show

$$\int_{c_1} -\frac{\partial u}{\partial \tau}\,d\xi - \alpha^2 \frac{\partial u}{\partial \xi}\,d\tau = \int_{c_1} -\frac{\partial u(\xi,0)}{\partial \tau}\,d\xi = \int_{x-\alpha t}^{x+\alpha t} -g(\xi)\,d\xi$$

Observe on the line c_2 we have $d\xi = -\alpha d\tau$ and hence show

$$\int_{c_1} -\frac{\partial u}{\partial \tau} d\xi - \alpha^2 \frac{\partial u}{\partial \xi} d\tau = \alpha \int_{(x+\alpha t,0)}^{(x,t)} du = \alpha \left[u(x,t) - f(x+\alpha t) \right]$$

Observe on the line c_3 we have $d\xi = \alpha d\tau$ and hence show

$$\int_{c_1} -\frac{\partial u}{\partial \tau} d\xi - \alpha^2 \frac{\partial u}{\partial \xi} d\tau = -\alpha \int_{(x,t)}^{(x-\alpha t,0)} du = -\alpha \left[f(x-\alpha t) - u(x,t) \right]$$

(d) Show with $L(u) = F(\xi, \tau)$, the Green's formula (9.210) implies the solution to the nonhomogeneous wave equation can be represented

$$u(x,t) = \frac{1}{2} \left[f(x+\alpha t) + f(x-\alpha t) \right] + \frac{1}{2} \int_{x-\alpha t}^{x+\alpha t} g(\xi)\, d\xi + \frac{1}{2\alpha} \int_{\eta=0}^{\eta=t} \int_{\alpha\eta+x-\alpha t}^{\alpha\eta+x+\alpha t} F(\xi,\eta)\, d\xi d\eta$$

▶ **23.**

Show the Green's function satisfying

$$L_{xt}(G) = \frac{\partial^2 G}{\partial t^2} - \alpha^2 \frac{\partial^2 G}{\partial x^2} = \delta(x-\xi)\delta(t-\tau), \quad 0 \le x \le \ell$$

with boundary conditions $G = 0$ at $x = 0$ and $x = \ell$ and initial conditions $G = \frac{\partial G}{\partial t} = 0$ at $t = \tau$ is given by

$$G(x,t;\xi,\tau) = \frac{2}{\pi\alpha} \sum_{n=1}^{\infty} \frac{1}{n} \sin(\omega_n x) \sin(\omega_n \xi) \sin\left[\omega_n \alpha(t-\tau)\right] H(t-\tau),$$

where $\omega_n = \frac{n\pi}{L}$, $n = 1, 2, 3, \dots$.

Hint: Use Laplace transforms.

▶ **24.**

(a) For the operator $L_{xt}(u) = \frac{\partial^2 u}{\partial t^2} - c^2 \frac{\partial^2 u}{\partial x^2}$ find the Green's formula for the region R illustrated in figure 9-4.

(b) In Green's formula let $L_{\xi\tau}(u) = F(\xi,\tau)$ and $L_{\xi\tau}^*(G) = \delta(\xi-x)\delta(\tau-t)$ and show

$$u(x,t) = \int_0^t \int_{-\ell}^\ell GF(\xi,\tau)\, d\xi d\tau - \int_{-\ell}^\ell \left(u \frac{\partial G}{\partial \tau} - G \frac{\partial u}{\partial \tau} \right)\Big|_{\tau=0} d\xi$$

$$+ c^2 \int_0^t \left(G \frac{\partial u}{\partial \xi} - u \frac{\partial G}{\partial \xi} \right)\Big|_{\xi=\ell} d\tau + c^2 \int_0^t \left(G \frac{\partial u}{\partial \xi} - u \frac{\partial G}{\partial \xi} \right)\Big|_{\xi=-\ell} d\tau$$

Hint: $G, \frac{\partial G}{\partial \tau} = 0$ on $\tau = T$.

(c) What is the solution to the boundary value problem:

$$L_{xt}(u) = \frac{\partial^2 u}{\partial t^2} - c^2 \frac{\partial^2 u}{\partial x^2} = F(x,t), \quad -\ell \leq x \leq \ell, \quad t > 0$$
$$u(-\ell, t) = \phi_1(t), \qquad u(\ell, t) = \phi_2(t)$$
$$u(x, 0) = f(x), \qquad u_t(x, 0) = g(x)$$

(d) Determine the differential equation and boundary conditions which defines the Green's function?

▶ **25.** Let $L_{xt}(u) = \frac{\partial u}{\partial t} - K\frac{\partial^2 u}{\partial x^2} = F(x,t), \quad -\infty < x < \infty,\ t > 0$ with initial condition given by $u(x, 0) = f(x)$, where $x, t \in R$.

(a) Write the Green's formula for the region R of figure 9-4.

(b) Show the solution can be written in the form

$$u(x,t) = \int_{-\infty}^{\infty} G(x,t;\xi,0) f(\xi)\, d\xi + \int_0^t \int_{-\infty}^{\infty} G(x,t;\xi,\tau) F(\xi,\tau)\, d\xi d\tau$$

where $\quad G(x,t;\xi,\tau) = \exp\left[-(x-\xi)^2/\sqrt{4\pi K(t-\tau)}\right] H(t-\tau)$

(c) Show that $u = u_1 + u_2$ where u_1 satisfies $\dfrac{\partial u}{\partial t} = K\dfrac{\partial^2 u}{\partial x^2},\quad u(x,0) = f(x)$ and u_2 satisfies $\dfrac{\partial u}{\partial t} = K\dfrac{\partial^2 u}{\partial x^2} + F(x,t),\quad u(x,0) = 0$.

▶ **26.**

(a) Show the solution for the vibrating string

$$L_{xt}(u) = \frac{\partial^2 u}{\partial t^2} - c^2 \frac{\partial^2 u}{\partial x^2} = F(x,t), \quad 0 \leq x \leq \ell,\ t > 0$$
$$u(0, t) = \phi_1(t), \qquad u(\ell, t) = \phi_2(t)$$
$$u(x, 0) = f(x), \qquad u_t(x, 0) = g(x)$$

can be expressed

$$u(x,t) = \int_0^t \int_0^\ell G(x,t;\xi,\tau) F(\xi,\tau)\, d\xi d\tau + c^2 \int_0^t \phi_1(\tau) G(x,t;0,\tau)\, d\tau$$
$$- c^2 \int_0^t \phi_2(\tau) \frac{\partial G(x,t;\ell,\tau)}{\partial \xi}\, d\tau + \int_0^\ell G(x,t;\xi,0) g(\xi)\, d\xi$$
$$+ \int_0^\ell \frac{\partial G(x,t;\xi,0)}{\partial \tau} f(\xi)\, d\xi$$

(b) Determine the differential equation and boundary conditions which defines the Green's function?

▶ **27.**

Let $u = u(x_1, x_2, \ldots, x_n)$ denote a function of n-independent variables and define the linear operator

$$L(u) = \sum_{i=1}^{n} \sum_{j=1}^{n} a_{ij} \frac{\partial^2 u}{\partial x_i \partial x_j} + \sum_{i=1}^{n} b_i \frac{\partial u}{\partial x_i} + cu$$

with adjoint operator

$$L^*(u) = \sum_{i=1}^{n} \sum_{j=1}^{n} \frac{\partial^2 (a_{ij} u)}{\partial x_i \partial x_j} - \sum_{i=1}^{n} \frac{\partial (b_i u)}{\partial x_i} + cu$$

where c, a_{ij} and b_i for $i, j = 1, 2, \ldots n$, are also functions of the same n-independent variables. Here the functions a_{ij} and b_i are assumed to be continuous and differentiable so that all of the above derivatives exist.

(a) Show that for continuous and differentiable functions u and v there exists the Lagrange identity

$$vL(u) - uL^*(v) = \sum_{i=1}^{n} \frac{\partial}{\partial x_i} \left[\sum_{j=1}^{n} a_{ij} \left(v \frac{\partial u}{\partial x_j} - u \frac{\partial v}{\partial x_j} \right) + uv \left(b_i - \sum_{j=1}^{n} \frac{\partial a_{ij}}{\partial x_j} \right) \right].$$

(b) Show that by defining the n-dimensional vector $\vec{P} = P_1 \hat{e}_1 + P_2 \hat{e}_2 + \ldots + P_n \hat{e}_n$ where

$$P_i = \sum_{j=1}^{n} a_{ij} \left(v \frac{\partial u}{\partial x_j} - u \frac{\partial v}{\partial x_j} \right) + uv \left(b_i - \sum_{j=1}^{n} \frac{\partial a_{ij}}{\partial x_j} \right),$$

the Lagrange identity can be written in the form

$$vL(u) - uL^*(v) = \nabla \cdot \vec{P}$$

where ∇ is the n-dimensional gradient operator

$$\nabla = \frac{\partial}{\partial x_1} \hat{e}_1 + \frac{\partial}{\partial x_2} \hat{e}_2 + \cdots + \frac{\partial}{\partial x_n} \hat{e}_n.$$

(c) Show that an integration of the Lagrange identity produces the generalized Green's identity

$$\int_R (vL(u) - uL^*(v)) \, d\tau = \int_S \vec{P} \cdot \hat{n} \, d\sigma$$

where $d\tau$ is an element of volume in the n-dimensional region R, the vector $\hat{n}$ is a unit outward normal vector to the surface $S = \partial R$ which represents the boundary of the region R, and $d\sigma$ is an element of surface area on S. Hint: Consider the n-dimensional form of the Gauss divergence theorem.

▶ **28.** Verify the following:

(a) The free-space Green's function associated with the Helmholtz equation is formed by solving the equation

$$\nabla^2_\xi G + k^2 G = \delta(\xi - x)\delta(\eta - y)\delta(\zeta - z) \quad k \text{ constant} \tag{28a}$$

We shift the origin to the point (ξ, η, ζ) and introduce the spherical coordinates (ρ, θ, ϕ) and assume symmetry with respect to the θ and ϕ variables. This reduces the equation (28a) to the form

$$\frac{1}{\rho^2} \frac{\partial}{\partial \rho} \left(\rho^2 \frac{\partial G}{\partial \rho} \right) + k^2 G = \frac{\delta(\rho)}{4\pi \rho^2} \tag{28b}$$

which can also be written in the form

$$\frac{1}{\rho} \left[\frac{\partial^2 (\rho G)}{\partial \rho^2} + k^2 (\rho G) \right] = \frac{\delta(\rho)}{4\pi \rho^2} \tag{28c}$$

(b) In the region $\rho > 0$, the Dirac delta function $\delta(\rho)$ is zero and so the equation (28c) reduces to

$$\frac{\partial^2 (\rho G)}{\partial \rho^2} + k^2 (\rho G) = 0 \tag{28d}$$

which has the solution

$$\rho G = C_1 e^{ik\rho} + C_2 e^{-ik\rho}$$

where C_1 and C_2 are constants. For convenience we select the constants C_1 and C_2 to have the form $C_1 = c_1/4\pi$ and $C_2 = c_2/4\pi$ and write the general solution in the form

$$G = \frac{c_1}{4\pi \rho} e^{ik\rho} + \frac{c_2}{4\pi \rho} e^{-ik\rho} \tag{28e}$$

Note that by assuming a solution to the wave equation $\dfrac{\partial^2 W}{\partial t^2} = c^2 \dfrac{\partial^2 W}{\partial x^2}$ having the form $W = He^{-i\omega t}$, with $\omega = kc$, then H must be a solution of the Helmholtz equation. Hence one can interpret the function

$$W = He^{-i\omega t} = \rho G e^{-i\omega t} = \frac{c_1}{4\pi} e^{ik(\rho - ct)} + \frac{c_2}{4\pi} e^{-ik(\rho - ct)}$$

as a wave motion. Recall the problem 26 in the exercises from chapter 5. The function $e^{ik(\rho - ct)}$ represents an expanding or outgoing wave which moves away from the origin and $e^{-ik(\rho + ct)}$ represents a contracting wave moving toward the origin. The equation (28b) has the impulse at the origin and so we want only outgoing waves. This requires that $c_2 = 0$. This gives the free-space Green's function associated with the three-dimensional Helmholtz equation as having the form

$$G = G(x, y, z; \xi, \eta, \zeta) = \frac{c_1}{4\pi\rho} e^{ik\rho} = \frac{c_1}{4\pi|\vec{r} - \vec{\xi}|} e^{ik|\vec{r} - \vec{\xi}|} \tag{28f}$$

We select $c_1 = -1$ so that the equation (28f) reduces to the free-space Green's function associated with the Laplace equation. This gives the free-space Green's function

$$G = \frac{-\exp\left(ik\sqrt{(x - \xi)^2 + (y - \eta)^2 + (z - \zeta)^2}\right)}{4\pi\sqrt{(x - \xi)^2 + (y - \eta)^2 + (z - \zeta)^2}} \tag{28g}$$

(c) Write the solution to the nonhomogeneous Helmholtz equation

$$\nabla^2 u + k^2 u = F(x, y, z) \quad -\infty < x, y, z < \infty.$$

(d) Use the method of descent to derive the two-dimensional and one-dimensional Green's function associated with the Helmholtz equation. These free-space Green's functions are given in the table 9-5. Note that there is no continuity with respect to parameter variation associated with the two-dimensional and one-dimensional Helmholtz equation. (i.e. The free space Green's function results do not reduce down to the Laplace equation form as $k \to 0$.) Hint: Use the Appendix D.

Table 9-5.	
Free-space Green's function for the Helmholtz equation	
Dimension	Green's function
$n = 1$	$G(x, \xi) = \dfrac{-i}{2k} \exp(ik\,\vert\, x - \xi\,\vert\,)$
$n = 2$	$G(x, y; \xi, \eta) = \dfrac{-i}{4} H_0^{(1)}(k\sqrt{(x - \xi)^2 + (y - \eta)^2})$
$n = 3$	$G(x, y, z; \xi, \eta, \zeta) = \dfrac{-\exp\left(ik\sqrt{(x - \xi)^2 + (y - \eta)^2 + (z - \zeta)^2}\right)}{4\pi\sqrt{(x - \xi)^2 + (y - \eta)^2 + (z - \zeta)^2}}$

Bibliography

- Abramowitz, M. , Stegun, A. **Handbook of Mathematical Functions**, 10th edition, New York: Dover Publications, 1972.

- Andrews, L.C., **Elementary Partial Differential Equations with Boundary Value Problems**, Orlando, Fl.: Academic Press, 1986.

- Brown, J.W., Churchill R.V., **Fourier Series and Boundary Value Problems**, Sixth Edition, New York: McGraw-Hill Book Company, 2001.

- Campbell, G.A., Foster, R.M., **Fourier Integrals for Practical Application** , John Wiley & Sons, N.Y., 1948

- Chester, C.R., **Techniques in Partial Differential Equations**, New York: McGraw-Hill Book Company, 1971.

- Churchill, R.V., **Operational Mathematics**, New York: McGraw-Hill Book Company, 1958.

- Dennemeyer, R., **Introduction to Partial Differential Equations and Boundary Value Problems**, New York: McGraw-Hill Book Company, 1968.

- Duchateau, P., Zachmann, D.W., **Partial Differential Equations**, Schaum's Outline, McGraw-Hill Book Company, 1986.

- Epstein, B., **Partial Differential Equations**, New York: McGraw-Hill Book Company, 1962.

- Erdelyi,A.(Ed.), **Tables of Integral Transforms**, 2 volumes, Bateman Manuscript Project, McGraw Hill Book Co.,N.Y.,1954.

- Farlow, S.J., **Partial Differential Equations for Scientists & Engineers**, New York: John Wiley and Sons, 1982.

- Gradshteyn, I.S., Ryzhik,, I.M., **Table of Integrals, Series, and Products**, New York: Academic Press, 1980.

- Greenberg, M.D., **Applications of Green's Functions in Science and Engineering**, Englewood Cliffs, N.J.: Prentice Hall, 1971.

- Guenther, R.B., Lee, J.W., **Partial Differential Equations of Mathematical Physics and Integral Equations**, Englewood Cliffs, N.J.: Prentice Hall, 1988.

- Hanna, J.R., **Fourier Series and Integrals of Boundary Value Problems**, New York: John Wiley and Sons, 1982.

- Haberman, R., **Elementary Applied Partial Differential Equations with Fourier Series and Boundary Value Problems**, Third Edition, Englewood Cliffs, N.J.: Prentice Hall, 1998.

Bibliography

- John, F., **Partial Differential Equations**, 4th Edition, New York: Springer Verlang, 1981.

- Johnson, D.E, Johnson, J.R., **Mathematical Methods in Engineering and Physics**, New York: The Ronald Press Company, 1965.

- Kovach, L.D., **Boundary-Value Problems**, Reading, Mass.: Addison-Wesley Publishing Company, 1984.

- Koshlyakov, N.S., Smirnov, M.M., Gliner, E.B., **Differential Equations of Mathematical Physics**, Amsterdam: North-Holland Publishing Company, 1964.

- Lebedev, N.N., Skalskaya, I.P., Uflyand, Y.S., **Worked Problems in Applied Mathematics**, New York: Dover Publications, 1965.

- Magnus, W, Oberhettinger, F., **Formulas and Theorems for the Functions of Mathematical Physics**, Chelsea Publishing Co., N.Y., 1954.

- Miller, K.S., **Partial Differential Equations in Engineering Problems**, Englewood Cliffs, N.J.: Prentice Hall, 1959.

- Roach, G.F., **Green's Functions**, Second Edition, London: Cambridge University Press, 1982.

- Rogosinski, W. **Fourier Series**, Chelsea Publishing Company, N.Y., 1959.

- Sneddon, I.N., **Elements of Partial Differential Equations**, New York: McGraw-Hill Book Company, 1957.

- Sneddon, I.N., **The Use of Integral Transforms**, New York: McGraw-Hill Book Company, 1972.

- Sneddon, I.N., **Fourier Transforms**, New York: McGraw-Hill Book Company, 1951.

- Snider, A.D., **Partial Differential Equations**, Upper Saddle River, N.J.: Prentice Hall, 1999.

- Sobolev, S.L., **Partial Differential Equations of Mathematical Physics**, New York: Dover Publications, 1964.

- Spiegel, M.R., **Vector Analysis**, Schaum's Outline Series, McGraw Hill Book Co., N.Y., 1959.

- Stakgold, I. **Green's Functions and Boundary Value Problems**, New York: John Wiley & Sons, 1979.

- Strauss, W.A., **Partial Differential Equations An Introduction**, New York: John Wiley and Sons, 1992.

- Zachmanoglou, E.C., Thoe, D.W., **Introduction to Partial Differential Equations with Applications**, Baltimore: The Williams & Wilkins Company, 1976.

APPENDIX A

Units of Measurement

The following units, abbreviations and prefixes are from the
Système International d'Unitès   (designated SI in all Languages.)

Prefixes.

Abbreviations		
Prefix	Multiplication factor	Symbol
tera	10^{12}	T
giga	10^9	G
mega	10^6	M
kilo	10^3	K
hecto	10^2	h
deka	10	da
deci	10^{-1}	d
centi	10^{-2}	c
milli	10^{-3}	m
micro	10^{-6}	μ
nano	10^{-9}	n
pico	10^{-12}	p

Basic Units.

Basic units of measurement		
Unit	Name	Symbol
Length	meter	m
Mass	kilogram	kg
Time	second	s
Electric current	ampere	A
Temperature	degree Kelvin	$^\circ$K
Luminous intensity	candela	cd

Supplementary units		
Unit	Name	Symbol
Plane angle	radian	rad
Solid angle	steradian	sr

Appendix A

DERIVED UNITS		
Name	Units	Symbol
Area	square meter	m^2
Volume	cubic meter	m^3
Frequency	hertz	Hz (s^{-1})
Density	kilogram per cubic meter	kg/m^3
Velocity	meter per second	m/s
Angular velocity	radian per second	rad/s
Acceleration	meter per second squared	m/s^2
Angular acceleration	radian per second squared	rad/s^2
Force	newton	N ($kg \cdot m/s^2$)
Pressure	newton per square meter	N/m^2
Kinematic viscosity	square meter per second	m^2/s
Dynamic viscosity	newton second per square meter	$N \cdot s/m^2$
Work, energy, quantity of heat	joule	J ($N \cdot m$)
Power	watt	W (J/s)
Electric charge	coulomb	C ($A \cdot s$)
Voltage, Potential difference	volt	V (W/A)
Electromotive force	volt	V (W/A)
Electric force field	volt per meter	V/m
Electric resistance	ohm	Ω (V/A)
Electric capacitance	farad	F ($A \cdot s/V$)
Magnetic flux	weber	Wb ($V \cdot s$)
Inductance	henry	H ($V \cdot s/A$)
Magnetic flux density	tesla	T (Wb/m^2)
Magnetic field strength	ampere per meter	A/m
Magnetomotive force	ampere	A

Physical Constants:

- $4 \arctan 1 = \pi = 3.14159\,26535\,89793\,23846\,2643\ldots$
- $\lim_{n \to \infty} \left(1 + \frac{1}{n}\right)^n = e = 2.71828\,18284\,59045\,23536\,0287\ldots$
- Euler's constant $\quad \gamma = 0.57721\,56649\,01532\,86060\,6512\ldots$
- $\gamma = \lim_{n \to \infty} \left(1 + \frac{1}{2} + \frac{1}{3} + \cdots + \frac{1}{n} - \log n\right)$
- Speed of light in vacuum $= 2.997925(10)^8\ m\ s^{-1}$
- Electron charge $= 1.60210(10)^{-19}\ C$
- Avogadro's constant $= 6.02252(10)^{23}\ mol^{-1}$
- Plank's constant $= 6.6256(10)^{-34}\ J\,s$
- Universal gas constant $= 8.3143\ J\,K^{-1}\,mol^{-1} = 8314.3\ J\,Kg^{-1}\,K^{-1}$
- Boltzmann constant $= 1.38054(10)^{-23}\ J\,K^{-1}$
- Stefan–Boltzmann constant $= 5.6697(10)^{-8}\ W\,m^{-2}\,K^{-4}$
- Gravitational constant $= 6.67(10)^{-11}\ N\,m^2 kg^{-2}$

Appendix A

APPENDIX B
Solutions to Selected Problems
Solutions Chapter 1

▶ 1.   (a) $u = c_1 x + c_2$ (b) $u = \dfrac{T_b - T_a}{b - a}(x - a) + T_a$

▶ 3.   (a) $y_p = \dfrac{F_0(k - m\omega^2)\sin\omega t}{(k - m\omega^2)^2 + c^2\omega^2} - \dfrac{F_0 c\omega\cos\omega t}{(k - m\omega^2)^2 + c^2\omega^2},$ $y = y_c + y_p.$

        (i) $c^2 > 4mk, \quad m_{1,2} = (-c \pm \sqrt{c^2 - 4mk})/2m < 0$

             $y_c = c_1 e^{m_1 t} + c_2 e^{m_2 t}$

        (ii) $c^2 = 4mk, \quad m_1 = m_2 = -c/2m$

             $y_c = c_1 e^{m_1 t} + c_2 t e^{m_1 t}$

        (iii) $c^2 < 4mk, \quad m_{1,2} = (-c \pm i\sqrt{4mk - c^2})/2m$

             $y_c = e^{-ct/2m}(c_1 \cos\lambda t + c_2 \sin\lambda t), \quad \lambda = \sqrt{4mk - c^2}/2m, \quad i^2 = -1.$

▶ 5. $F = c_1 r^n + c_2 r^{-(n+1)}$

▶ 7.   (b) $y = J_0(nr)/J_0(nb)$

▶ 9. $y = A e^{-Kt} + \displaystyle\int_0^t f(\xi) e^{-K(t-\xi)}\, d\xi$

▶ 12.   (a) $\theta = \sqrt{x}\left[c_1 J_{1/3}\left(\frac{2}{3}\sigma x^{3/2}\right) + c_2 J_{-1/3}\left(\frac{2}{3}\sigma x^{3/2}\right) \right]$

       $\theta'(0) = 0 \implies c_1 = 0$

       $\theta(\ell) = c_2 \ell^{1/2} J_{-1/3}\left(\frac{2}{3}\sigma \ell^{3/2}\right) = 0 \implies \frac{2}{3}\sigma \ell^{3/2}$ is first zero of $J_{-1/3}$.

▶ 15.   (b) $u = \dfrac{T_b - T_a}{\ln b/a}\ln r/a + T_a$

▶ 18. $F = \dfrac{1}{\omega}\displaystyle\int_0^t h(\tau)\sin\omega(t - \tau)\, d\tau$

▶ 21.   (a) $z = xy + f(x)$ (d) $z = \frac{1}{2}x^2 + xf(y) + g(y)$

▶ 23.   (a) $z = xf(x/y)$ (d) $(x + y)e^{-z} = f(x^2 - y^2)$

▶ 25.

                 (a) $u = f(x + y) + g(x - y) + \dfrac{x^3}{6}$

                 (c) $u = f(x + y) + xg(x + y)$

                 (e) $u = e^{-y}f(x - y) + e^x f(x - y)$

▶ 29.   (a) $u = (x - 1)\cos y + y$ (b) $u = \frac{1}{2}x^2 y^2 + x(\cos y - y - \frac{1}{2}y^2) + y$

▶ 30. $u = f(3x - y) + g(x - y) + 2e^{3x+y}$

Solutions Chapter 2

▶ 1.
$$(f_n, f_m) = \int_a^b r(x) \frac{g_n}{\| g_n \|} \frac{g_m}{\| g_m \|} \, dx = \begin{cases} 0, & m \neq n \\ \frac{\|g_n\|^2}{\|g_n\|^2} = 1, & m = n \end{cases}$$

▶ 3. Orthonormal set $f_0 = \frac{1}{\sqrt{2\pi}}$, $f_n = \frac{\cos nx}{\sqrt{\pi}}$ $n = 1, 2, 3, \dots$

▶ 6.
$$\begin{aligned} \phi_0(x) &= 1, & \| \phi_0 \|^2 &= 1 \\ \phi_1(x) &= 1 - 2x, & \| \phi_1 \|^2 &= 1/3 \\ \phi_2(x) &= x^2 - x + 1/6, & \| \phi_2 \|^2 &= 1/180 \\ \phi_3(x) &= x^3 - \frac{3}{2}x^2 + \frac{1}{2}x, & \| \phi_3 \|^2 &= 1/840 \end{aligned}$$

▶ 8. $\lambda_n = \omega_n^2$ where $\omega_n = (2n-1)\pi/2L$ and $y_n(x) = \sin \omega_n(L - x)$, $n = 1, 2, 3, \dots$
$$(y_n, y_m) = \int_0^L \sin \omega_n(L - x) \sin \omega_m(L - x) \, dx = \begin{cases} 0, & m \neq n \\ \frac{L}{2}, & m = n \end{cases}$$

▶ 10. $\lambda_n = \omega_n^2$, $\omega_n = n\pi$ and $y_n(x) = \cos \omega_n \pi x$
$$(y_n, y_m) = \int_0^1 \cos n\pi x \cos m\pi x \, dx = \begin{cases} 0, & m \neq n \\ \frac{1}{2}, & m = n \end{cases}$$

▶ 12. $\lambda_n = \omega_n^2$ where $\omega_n = \frac{(2n+1)\pi}{2 \ln b}$ and $y_n(x) = \cos \omega_n \ln x$
$$(y_n, y_m) = \int_1^b \frac{1}{x} \cos \omega_n \ln x \cos \omega_m \ln x \, dx = \begin{cases} 0, & m \neq n \\ \frac{1}{2} \ln b, & m = n \end{cases}$$

▶ 14. For $1 - 4\lambda = -\omega^2$ we find $\omega = \omega_n = \frac{2n\pi}{\ln b}$ and $y_n(x) = \frac{1}{\sqrt{x}} \sin \frac{n\pi}{\ln b} \ln x$
$$(y_n, y_m) = \int_1^b \frac{1}{x} \sin \frac{n\pi}{\ln b} \ln x \sin \frac{m\pi}{\ln b} \ln x \, dx = \begin{cases} 0, & m \neq n \\ \frac{1}{2} \ln b, & m = n \end{cases}$$

▶ 17. $r^2 \frac{d^2 y}{dr^2} + 2r \frac{dy}{dr} - \ell(\ell + 1)y = -\lambda r^2 y$ Consider the cases $\lambda = -K^2$, $\lambda = 0$, $\lambda = +K^2$, $j_\ell(Kb) = 0$ implies $K = K_n = \xi_{\ell n}/b$
$$(j_\ell(K_n r), j_\ell(K_m r)) = \int_0^b r^2 j_\ell(K_n r) j_\ell(K_m r) \, dr = \begin{cases} 0, & m \neq n \\ \frac{b^3}{2} j_{\ell+1}^2(K_n b), & m = n \end{cases}.$$

▶ 19. $y_n = \sin \omega_n x$ where $\tan \omega_n L = -\omega_n/h$, for $n = 1, 2, 3, \ldots$

$$(y_n, y_m) = \int_0^L \sin \omega_n x \sin \omega_m x \, dx = \begin{cases} 0, & m \neq n \\ \| y_n \|^2, & m = n \end{cases}$$

where $\| y_n \|^2 = \dfrac{L}{2} + \dfrac{h}{2\omega_n^2} \sin^2 \omega_n L$.

▶ 21. $y_n(x) = e^{-x/2} \sin \dfrac{\omega_n}{2} x$, $\omega_n = \dfrac{2n\pi}{L}$, $r(x) = e^x$, $\lambda_n = \dfrac{1}{4}(1 + \omega_n^2)$

$$(y_n, y_m) = \int_0^L r(x) y_n(x) y_m(x) \, dx = \int_0^L \sin \frac{n\pi x}{L} \sin \frac{m\pi x}{L} \, dx = \begin{cases} 0, & m \neq n \\ \frac{L}{2}, & m = n \end{cases}$$

▶ 23. $y_m = P_{2m}(x)$, $\quad P_{2m}'(0) = 0$

$$(y_m, y_n) = \int_0^1 P_{2m}(x) P_{2n}(x) \, dx = \begin{cases} 0, & m \neq n \\ \frac{1}{4n+1}, & m = n \end{cases}$$

▶ 25. $\lambda_n = \omega_n^2$ where $\tan \omega_n \pi = -\omega_n$ for $n = 1, 2, 3, \ldots$ We find $\omega_1 \approx .78764$
$y_n(x) = \sin \omega_n x$ where

$$(y_n, y_m) = \int_0^\pi \sin \omega_n x \sin \omega_m x \, dx = \begin{cases} 0, & m \neq n \\ \frac{\pi}{2} + \frac{1}{2} \cos^2 \pi \omega_n, & m = n \end{cases}$$

▶ 29. $y_n(x) = \sin \omega_n(x - a)$ where $\tan \omega(b - a) = -\omega/h$ $\lambda_n = \omega_n^2$ For $a = 0, b = 1, h = 8$ we find the first ten values satisfying $\tan \omega = -\omega/8$ are given by the approximate values

$\omega_1 = 2.80443$	$\omega_3 = 8.60307$	$\omega_5 = 14.6374$	$\omega_7 = 20.7877$	$\omega_9 = 26.9917$
$\omega_2 = 5.66687$	$\omega_4 = 11.5993$	$\omega_6 = 17.7032$	$\omega_8 = 23.8851$	$\omega_{10} = 30.1049$

▶ 32. $F_n(r) = J_0(\omega_n r)$ where $J_0(\omega a) = 0$ gives $\omega = \omega_n = \xi_{0n}/a$ for $n = 1, 2, 3, \ldots$.

$$(F_n, F_m) = \int_0^a r J_0(\omega_n r) J_0(\omega_m r) \, dr = \begin{cases} 0, & m \neq n \\ \frac{a^2}{2} J_1^2(\omega_n a), & m = n \end{cases}$$

Solutions Chapter 2

Solutions Chapter 3

► 1.

(a) $\tilde{f}(x) = 1$

(b) $\tilde{f}(x) = \sin x$

(c) $\tilde{f}(x) = \sin 5x$

► 3.

(a)

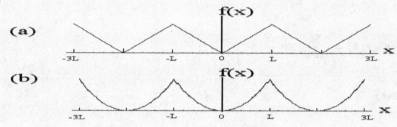

(b)

► 4.

(a) $F(x) = \frac{2}{\pi}\left(\sin \pi x + \frac{1}{2}\sin 2\pi x + \frac{1}{3}\sin 3\pi x + \cdots\right)$

(c) $F(x) = \pi - \sum_{n=1}^{\infty}\frac{2}{n}[1-(-1)^n]\sin\frac{nx}{2}$

(e) $F(x) = \frac{2}{\pi} - \frac{2}{\pi}\sum_{n=1}^{\infty}\frac{[1-(-1)^n]}{n^2-1}\cos nx$

► 5. (b) $h(-x) = F_e(-x)F_o(-x) = -F_e(x)F_o(x) = -h(x)$, hence $h(x)$ is odd.

► 6. (b) $H(-x) = \frac{F(-x)-F(x)}{2} = -\frac{F(x)-F(-x)}{2} = -H(x)$, hence $H(x)$ is odd.

► 7. (a)

$$\int_{-L}^{L} F(x)\,dx = \int_{-L}^{0} F(x)\,dx + \int_{0}^{L} F(x)\,dx$$

$$= \int_{L}^{0} F(-\xi)\,(-d\xi) + \int_{0}^{L} F(\xi)\,d\xi$$

$$= -\int_{0}^{L} F(\xi)\,d\xi + \int_{0}^{L} F(\xi)\,d\xi = 0$$

► 8. (b)

$$I_1 = \int_{-L}^{L} F(x)\,dx = \int_{-L}^{0} F(x)\,dx + \int_{0}^{2L} F(x)\,dx + \int_{2L}^{L} F(x)\,dx$$

$$= \int_{-L}^{0} F(x)\,dx + \int_{0}^{2L} F(x)\,dx + \int_{0}^{-L} F(\xi+2L)\,d\xi$$

also $I_1 = \int_{-L}^{L} F(x)\,dx = \int_{-L}^{\alpha} F(x)\,dx + \int_{\alpha}^{\alpha+2L} F(x)\,dx + \int_{\alpha+2L}^{L} F(x)\,dx$

$$= \int_{-L}^{\alpha} F(x)\,dx + \int_{\alpha}^{\alpha+2L} F(x)\,dx + \int_{\alpha}^{-L} F(\xi+2L)\,d\xi$$

► 9.

$$y = y_c + y_p = c_1\cos\omega t + c_2\sin\omega t + \sum_{n=1}^{\infty}\frac{2}{\pi}\frac{[1-(-1)^n]}{\omega^2-n^2\pi^2}\sin n\pi x \quad \omega \neq n\pi$$

550

▶ 11.

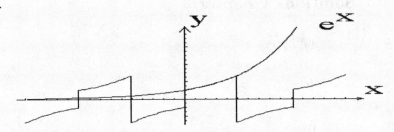

▶ 13. (b) $f(x) = \sum_{n=1}^{\infty} b_n \sin \dfrac{n\pi x}{L}$ where $b_n = \dfrac{8h}{Ln^2\pi^2} \sin \dfrac{n\pi}{2}$.

▶ 15. $f(x) = 5/2 - \sum_{n=1}^{\infty} \dfrac{1}{n\pi} \sin 2n\pi x$

▶ 16. $\sin x = \dfrac{2}{\pi} - \dfrac{2}{\pi} \sum_{n=2}^{\infty} \dfrac{[1+(-1)^n]}{n^2-1} \cos nx$

▶ 18. $f(x) = 3\sin \dfrac{5\pi x}{L} + 4\cos \dfrac{5\pi x}{L}$

▶ 19.

(b) $H(x) = \begin{cases} 1, & 0 < x < 1 \\ -1, & -1 < x < 0 \end{cases}$

(c) $G(x) = \begin{cases} x, & 0 < x < 1 \\ -x, & -1 < x < 0 \end{cases}$

▶ 20. $C_n = \dfrac{(f, P_n)}{\| P_n \|^2}$ gives $C_0 = 1/2$, $C_1 = 3/4$, $C_2 = 0$, $C_3 = -7/16$

▶ 23. (b) $f_e(x) = 4\pi^2 + x^2$ (c) $f_o(x) = 4\pi x$.

▶ 24. $1 = \sum_{n=1}^{\infty} C_n J_0(\lambda_n x)$, $C_n = \dfrac{2h}{(h^2+\lambda_m^2)J_0(\lambda_m)}$

Solutions Chapter 3

▶ 27.

$$f(x) = \frac{7}{3} + \sum_{n=1}^{\infty} \left(\frac{2[1+(-1)^n]}{n^2\pi^2} \cos n\pi x - \frac{3[1+(-1)^n]}{n\pi} \sin n\pi x \right) \quad -1 < x < 1$$

$$f(x) = \begin{cases} (x+1)^2, & 0 < x < 1 \\ (x+2)^2, & -1 < x < 0 \end{cases}$$

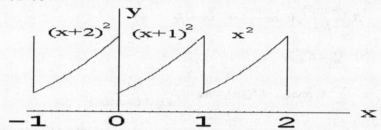

Another form for the solution is

$$f(x) = \frac{7}{3} + \frac{1}{\pi^2} \sum_{n=1}^{\infty} \frac{1}{n^2} \cos 2n\pi x - \frac{3}{\pi} \sum_{n=1}^{\infty} \frac{1}{n} \sin 2n\pi x, \quad -1/2 < x < 1/2$$

▶ 28. $I_2 = \frac{1}{2}\sqrt{\frac{\pi}{\beta}}$

▶ 29. $I = \frac{1}{2}\sqrt{\frac{\pi}{\alpha}} e^{-\beta^2/4\alpha}$

▶ 30. (a) $f(x) = \frac{1}{x^2+4} = \frac{1}{2} \int_0^{\infty} e^{-2\omega} \cos \omega x \, d\omega$

▶ 31.

(a) $f(x) = \int_0^{\infty} A(\omega) \cos \omega x \, d\omega, \qquad A(\omega) = \frac{2}{\pi(1-\omega^2)} \cos \frac{\omega\pi}{2}$

(b) $f(x) = \int_0^{\infty} A(\omega) \cos \omega x \, d\omega, \qquad A(\omega) = \frac{2}{\pi(1-\omega^2)}(1 + \cos \pi\omega)$

▶ 34. $a_i = \frac{2}{\xi_{0i} J_1(\xi_{0i})}$

▶ 36.

(b) $x^2 = \frac{1}{3} P_0(x) + \frac{2}{3} P_2(x)$

(c) $x^4 = \frac{1}{5} P_0(x) + \frac{4}{7} P_2(x) + \frac{8}{35} P_4(x)$

▶ 39.

(a) $f(x) = \int_0^{\infty} (A(\omega) \cos \omega x + B(\omega) \sin \omega x) \, d\omega$

$A(\omega) = \frac{1}{\pi} \left[\frac{-1}{\omega^2} + \frac{\cos \omega + \omega \sin \omega}{\omega^2} \right], \qquad B(\omega) = \frac{1}{\pi} \left[\frac{\sin \omega - \omega \cos \omega}{\omega^2} \right]$

(b) $f(x) = \int_0^{\infty} (A(\omega) \cos \omega x + B(\omega) \sin \omega x) \, d\omega$

$A(\omega) = \frac{\sin 2\omega}{\pi\omega}, \qquad B(\omega) = \frac{1}{\pi} \left(\frac{1}{\omega} - \frac{\cos 2\omega}{\omega} \right)$

(c) $f(x) = \int_0^{\infty} (A(\omega) \cos \omega x + B(\omega) \sin \omega x) \, d\omega$

$A(\omega) = \frac{1}{\pi} \left(\frac{1}{1+\omega^2} \right), \qquad B(\omega) = \frac{1}{\pi} \left(\frac{\omega}{1+\omega^2} \right)$

Solutions Chapter 3

Solutions Chapter 4

▶ 4.

(a) $\dfrac{\partial v}{\partial t} = K\dfrac{\partial^2 v}{\partial x^2}$ (b) $u = e^{-\beta t}v, \quad v = v(x,t) = \displaystyle\sum_{n=1}^{\infty} B_n e^{-\omega_n^2 Kt}\sin\dfrac{n\pi x}{L}$

$B_n = \dfrac{2}{L}\displaystyle\int_0^L f(\xi)\sin\dfrac{n\pi\xi}{L}\,d\xi, \quad \omega_n = \dfrac{n\pi}{L}.$

▶ 5. (a) $\dfrac{\partial v}{\partial t} = K\dfrac{\partial^2 v}{\partial x^2}.$

▶ 9. $u = \displaystyle\sum_{n=1}^{\infty} A_n\cosh\xi_{0n}z\,J_0(\xi_{0n}r), \quad A_n = \dfrac{2T_0}{\xi_{0n}\cosh\xi_{0n}J_1(\xi_{0n})}.$

▶ 11. $b\omega_i J_0'(\omega_i b) + hJ_0(\omega_i b) = 0, \; i = 1,2,3,\ldots \quad u = u(r,t) = \sum_{n=1}^{\infty} A_n J_0(\omega_n r)e^{-\omega_n^2 Kt}$

$A_n = \dfrac{2\omega_n^2}{(\omega_n^2 b^2 + h^2)J_0^2(\omega_n b)}\displaystyle\int_0^b rf(r)J_0(\omega_n r)\,dr$

▶ 13.

(a) $u = u(x,t) = \dfrac{1}{2\alpha\sqrt{\pi t}}\displaystyle\int_{-\infty}^{\infty} f(\xi)e^{-(x-\xi)^2/4\alpha^2 t}\,d\xi$

(b) $u = u(x,t) = \dfrac{T_0}{2}\left[\operatorname{erf}\left(\dfrac{1-x}{2\alpha\sqrt{t}}\right) + \operatorname{erf}\left(\dfrac{1+x}{2\alpha\sqrt{t}}\right)\right]$

▶ 15. $u = u(x,t) = \displaystyle\sum_{m=1}^{\infty}\left[\left(\dfrac{4}{\pi m^3}(1-(-1)^m) - \dfrac{4}{K\pi m^4}\sin\dfrac{m\pi}{2}\right)e^{-Km^2 t} + \dfrac{4}{K\pi m^4}\sin\dfrac{m\pi}{2}\right]\sin mx$

▶ 17. $u = u(r,t) = \displaystyle\sum_{i=1}^{\infty} A_i J_0(\omega_i r)e^{-\omega_i^2\alpha^2 t}$ where $A_i = \dfrac{1}{2J_1^2(\xi_{0i})}\displaystyle\int_0^2 rf(r)J_0(\omega_i r)\,dr$

and $\omega_i = \xi_{0i}/2$ with $J_0(\xi_{0i}) = 0.$

▶ 19. $u = u(\rho) = \dfrac{(T_1-T_0)}{\left(\frac{1}{\rho_1} - \frac{1}{\rho_0}\right)}\left(\dfrac{1}{\rho} - \dfrac{1}{\rho_0}\right) + T_0$

▶ 21. $u = u(x,t) = \operatorname{erf}\left(\dfrac{x}{2\alpha\sqrt{t}}\right)$

▶ 23. $u = w + v, \quad v = \alpha(t)(1-x/L), \quad w = w(x,t) = \displaystyle\sum_{n=1}^{\infty} B_n\sin\dfrac{n\pi x}{L}$ where

$B_n(t) = B_n(0)e^{-K(m\pi/L)^2 t} - \dfrac{2}{m\pi}e^{-K(m\pi/L)^2 t}\displaystyle\int_0^t \alpha'(\tau)e^{K(m\pi/L)^2\tau}\,d\tau$

$B_n(0) = \dfrac{2}{L}\displaystyle\int_0^L [f(x) - \alpha(0)(1-x/L)]\sin\dfrac{n\pi x}{L}\,dx$

► 25. $u = u(r,z) = \sum_{n=1}^{\infty} B_n I_0(\omega_n r) \sin \omega_n z$ where $\omega_n = \frac{n\pi}{h}$ for $n = 1, 2, 3, \ldots$ and

$B_n = \dfrac{2}{h I_0(\omega_n h)} \displaystyle\int_0^h f(z) \sin \omega_n z \, dz.$

► 27. $u = u(x,t) = A_0 + \sum_{n=1}^{\infty} A_n e^{-Kn^2 t} \cos nx$ where $A_0 = \dfrac{1}{\pi} \displaystyle\int_0^{\pi} f(x) \, dx$

and $A_n = \dfrac{2}{\pi} \displaystyle\int_0^{\pi} f(x) \cos nx \, dx$

► 29. $u = u(x,t) = \sum_{n=1}^{\infty} A_n(t) \cos \omega_n x$ where $\tan \omega_n L = \dfrac{h}{\omega_n}$ for $n = 1, 2, 3, \ldots$

$A_n(t) = A_n(0) e^{-K\omega_n^2 t} + \dfrac{2he^{-K\omega_n^2 t} \sin \omega_n L}{\omega_n (hL + \sin^2 \omega_n L)} \displaystyle\int_0^t q(\tau) e^{K\omega_n^2 \tau} \, d\tau$

$A_n(0) = \dfrac{2h}{hL + \sin^2 \omega_n L} \displaystyle\int_0^L f(x) \cos \omega_n x \, dx$

► 31. $u = u(x,t) = e^{-\beta t} e^{-K(\pi/L)^2 t} \sin \dfrac{\pi x}{L}$

► 33. $u = u(r,t) = c_0 + \sum_{n=1}^{\infty} c_n e^{-\lambda_n^2 t} J_0(\lambda_n r)$ where $J_1(\lambda_n) = 0$ for $n = 1, 2, 3, \ldots$

$c_0 = 2 \displaystyle\int_0^1 r f(r) \, dr, \quad c_n = \dfrac{1}{J_0^2(\lambda_n)} \int_0^1 r f(r) J_0(\lambda_n r) \, dr$

► 35. $u = u(x,t) = T_0 + \sum_{n=1}^{\infty} C_n G_n(t) F_n(x)$ where $G_n(t) = e^{-(\lambda_n/L)^2 Kt}$

and $F_n(x) = \cos \dfrac{\lambda_n x}{L} + \dfrac{hL}{\lambda_n} \sin \dfrac{\lambda_n x}{L}$

$C_n = \dfrac{1}{\| F_n \|^2} \displaystyle\int_0^L (f(x) - T_0) F_n(x) \, dx$ where $\| F_n \|^2 = \dfrac{L}{2} \left[1 + \dfrac{c^2}{\lambda_n^2} + \dfrac{2c}{\lambda_n^2} \right]$ and $c = hL$.

► 37. $u = \dfrac{1}{\rho} v, \quad v = v(\rho,t) = \sum_{n=1}^{\infty} C_n G_n(t) F_n(\rho)$ where $G_n(t) = e^{-\lambda_n^2 Kt/b^2}$ and

$F_n(\rho) = \sin \dfrac{\lambda_n \rho}{b}$ The eigenvalues satisfy $\tan \lambda_n = \dfrac{-\lambda_n}{hb - 1}$ for $n = 1, 2, 3, \ldots$ and the coefficients are given by

$C_n = \dfrac{1}{\frac{b}{2}\left[1 + \frac{c}{\lambda_n^2 + c^2}\right]} \displaystyle\int_0^b \rho f(\rho) \sin \dfrac{\lambda_n \rho}{b} \, d\rho$ with $c = hb - 1$.

Solutions Chapter 4

▶ 39. $\quad u = u(r,t) = \sum_{n=1}^{\infty} B_n(t) J_0(\xi_{0n} \frac{r}{b})$ where $J_0(\xi_{0n}) = 0$ for $n = 1, 2, 3, \ldots$

$B_n(t) = e^{-K\lambda_n^2 t} \left[B_n(0) + \int_0^t q_n(\tau) e^{K\lambda_n^2 \tau} \, d\tau \right]$ where $q_n(t) = \frac{2}{b^2 J_1^2(\xi_{0n})} \int_0^b r J_0(\xi_{0n} \frac{r}{b}) q(r,t) \, dr$ and $B_n(0) =$

$\frac{2}{b^2 J_1^2(\xi_{0n})} \int_0^b r f(r) J_0(\xi_{0n} \frac{r}{b}) \, dr$

▶ 41. $\quad u = u(r,t) = \sum_{n=1}^{\infty} B_n(t) U_0(\lambda_n r)$ where $B_n(t) = e^{-K\lambda_n^2 t} \left[B_n(0) + \int_0^t q_n(\tau) e^{K\lambda_n^2 \tau} \, d\tau \right]$

$q_n(t) = \frac{1}{\| U_0(\lambda_n r) \|^2} \int_0^b r q(r,t) U_0(\lambda_n r) \, dr$ and $B_n(0) = \frac{1}{\| U_0(\lambda_n r) \|^2} \int_0^b r f(r) U_0(\lambda_n r) \, dr$ $\quad$ (See (2.51))

▶ 43. $\quad$ For $g(r) = T_0 + \frac{(T_1 - T_0)}{\ln(b/a)} \ln(r/a)$ the solution is given by

$u = u(r,t) = g(r) + U(r,t)$ where $U(r,t)$ is a solution of problem 41 with $f(r)$ replaced by $F(r) = f(r) - g(r)$.

▶ 45.

(a) $\quad$ Problem 1. $u_1 = u_1(x,y) = \sum_{n=1}^{\infty} B_n \sinh \omega_n x \sin \omega_n y$ where $\omega_n = \frac{n\pi}{b}$ for $n = 1, 2, 3, \ldots$

$B_n = \frac{2}{b \sinh \omega_n a} \int_0^b f_1(y) \sin \omega_n y \, dy$

(b) $\quad$ Problem 3. $u_3 = u_3(x,y) = \sum_{n=1}^{\infty} B_n \sinh \omega_n y \sin \omega_n x$ where $\omega_n = \frac{n\pi}{a}$

and $B_n = \frac{2}{b \sinh \omega_n b} \int_0^a g_3(x) \sin \omega_n x \, dx$

▶ 47.

$u = u(x,t) = \sum_{n=1}^{\infty} B_n(t) \sin nx$ where $B_n(t) = B_n(0) e^{-Kn^2 t} + \frac{4}{\pi K n^4} \sin \frac{n\pi}{2} (1 - e^{-Kn^2 t})$ with $B_n(0) =$

$\frac{4}{\pi} \frac{(1 - (-1)^n)}{n^3}$.

▶ 51.

(a) $\quad u = u(x,t) = \frac{1}{2} \mathrm{erf} \left(\frac{x}{2\sqrt{Kt}} \right) + \frac{1}{2}$

(b) $\quad u = u(x,t) = \frac{1}{2} \left[\mathrm{erf} \left(\frac{L+x}{2\sqrt{Kt}} \right) + \mathrm{erf} \left(\frac{L-x}{2\sqrt{Kt}} \right) \right]$

Index

Index

Printed in the United States
By Bookmasters